STRUCTURE OF SPACE AND THE SUBMICROSCOPIC DETERMINISTIC CONCEPT OF PHYSICS

STRUCTURE OF SPACE AND THE SUBMICROSCOPIC DETERMINISTIC CONCEPT OF PHYSICS

Written By

Volodymyr Krasnoholovets, PhD

Senior Research Scientist, Department of Theoretical Physics, Institute of Physics, National Academy of Sciences of Ukraine, Kyiv, Ukraine

AAP | APPLE ACADEMIC PRESS

Apple Academic Press Inc.
3333 Mistwell Crescent
Oakville, ON L6L 0A2 Canada

Apple Academic Press Inc.
9 Spinnaker Way
Waretown, NJ 08758 USA

First issued in paperback 2021

Exclusive worldwide distribution by CRC Press, a member of Taylor & Francis Group

ISBN 13: 978-1-77-463671-8 (pbk)
ISBN 13: 978-1-77-188530-0 (hbk)

Library and Archives Canada Cataloguing in Publication

Krasnoholovets, Volodymyr, author

Structure of space and the submicroscopic deterministic concept of physics / written by Volodymyr Krasnoholovets, PhD (Senior Research Scientist, Department of Theoretical Physics, Institute of Physics, National Academy of Sciences of Ukraine, Kyiv, Ukraine).

Includes bibliographical references and index.

Issued in print and electronic formats.

ISBN 978-1-77188-530-0 (hardcover).--ISBN 978-1-315-36552-7 (PDF)

1. Particles (Nuclear physics). 2. Quantum theory. 3. Space and time. 4. Gravitation. 5. Cosmology. I. Title.

QC721.K73 2017 539.7'21 C2017-901587-7 C2017-901588-5

CIP data on file with US Library of Congress

Apple Academic Press also publishes its books in a variety of electronic formats. Some content that appears in print may not be available in electronic format. For information about Apple Academic Press products, visit our website at **www.appleacademicpress.com** and the CRC Press website at **www.crcpress.com**

CONTENTS

CONTENTS

DEDICATION

This book is dedicated to the memory of three prominent French scientists—Henri Poincaré, Louis de Broglie, and Michel Bounias—who made the greatest contribution to understanding the structure and properties of real physical space.

Henri Poincarè

"... the space revealed to us by our senses is absolutely different from the space of geometry. Is geometry derived from experience? Careful discussion will give the answer—no! We therefore conclude that the principles of geometry are only conventions; but these conventions are not arbitrary... as soon as measurement is introduced into the continuum we have just defined, the continuum becomes space, and geometry is born."

Louis de Broglie

"...The current theory, which uses only a ψ-function, says that some property *g* of a particle is distributed throughout a density wave $|\psi|^2 d\tau$ and the wave is deprived of singularity. in the double solution theory, however, the quantity *g* is certainly concentrated in a very small region occupied by the particle. ... in reality the particle's quantity *g* is totally concentrated in the singularity."

Michel Bounias

"Space is made of {objects + distances} and all comes from the same origin: manifold of sets and nothing else!.. The association of discrete sets (macro → cosmic) whose interior is continuous although covered by discrete subparts (micro), as derived from the empty set Ø provides a wonderfully organized fundamental 'sub-strate,' i.e., a mathematical lattice."

LIST OF ABBREVIATIONS

BCS	Bardeen-Cooper-Schrieffer theory
CMB	cosmic microwave background radiation
FLRW	Friedmann–Lemaître–Robertson–Walker
LHC	Large Hadron Collider
M	mapping
MJ	moments of junction
QCD	quantum chromodynamics
QD	quark dynamics theory
SETI	search for extraterrestrial intelligence
SI	International System of Units
SUSY	supersymmetry
TC	Teslar chip

PREFACE

This book presents a very new approach to the foundation of physics. The book shows that fundamental physics did not start three minutes after the enigmatic Big Bang, which allegedly happened 13.8 billion years ago, but arose from pure mathematical constructions stemming from set theory, topology and fractal geometry. These disciplines make it possible to formulate a mathematical space in deeper terms than expressed by B. Riemann and H. Helmholtz whose ideas came to dominate physics and mathematical physics since the end of the 19th century. Their approach suggests that the motion of a 'rigid' object through space does not change the shape of the object, but rather that the space around the object becomes curved. The suggestion of a 'rigidity' of objects is crucial for physical mathematics, because it allows the introduction of symmetries in the study of the equations of motion of these objects. Today such an approach is in progress in all aspects of theoretical physics, from the microcosm of elementary particles to cosmology.

Nevertheless, a more profound mathematical study of space reveals a founding lattice within it. This mathematical lattice highlights the creation of matter (i.e., particles in the form of both leptons and quarks) and the generation of the laws of physics from that fundamental mathematical lattice. This means that space not only wraps around a rigid object, but that the object itself is the product of space. Therefore, *the object interacts with the space*, which must influence the shape and properties of the moving object.

Such an interesting mathematical approach permits one to look behind the barrier erected by the quantum mechanical formalism. Beyond the formalism we can see that a particle appears as a local distortion of the mathematical lattice, namely, a cell of the lattice may decrease or increase in size under a relevant rule. This in turn further deforms the surrounding cells. The moving particle creates a cloud of excitations, since it moves in the lattice and not in an absolute emptiness. Therefore, an appropriate theory of the motion of a particle is a submicroscopic theory; it is characterized by short-range action, which automatically means the introduction of a new kind of carriers, i.e., carriers of the quantum mechanical force.

This book recounts how the submicroscopic mechanics of particles is developed and how it coexists with and supplements the formalism of conventional quantum mechanics. The theory of space unambiguously answers such challenging issues as: what is mass?, what is charge?, what is a lepton?, what is a quark?, what is a neutrino?, what are the nuclear forces?, and so on. The submicroscopic concept uncovers new peculiar properties of quantum systems, especially the dynamics of particles within a section equal to the particle's de Broglie wavelength, which is fundamentally impossible for quantum mechanics. The concept allows one to study complex problems in quantum optics and quantum electrodynamics in detail. Besides, for the first time particle physics and nuclear physics open out their inner world exposing the structure of quarks and nucleons in real space. We can investigate the interaction of quarks both with space and between themselves in new terms that naturally explain such phenomena, as the confinement of quarks, asymptotic freedom of quarks inside a hadron, transformation of particles and the nuclear forces that act between nucleons.

The existence of carriers of the quantum mechanical interaction, which were named "inertons" (because they carry inert properties of matter), was indeed verified in a number of experiments. These experiments are described in this book.

The final chapters are devoted to the phenomenon of gravity and cosmology. The reader will see that both the lattice structure of space and the carriers of local deformations (i.e., inertons, carriers of mass) cause the reasons for the phenomenon of gravity. The phenomenon of gravity is derived from the first principles: gravitation appears due to the transfer of local deformation of space by inertons. It is these inertons that play the role of universal carriers of quantum mechanical, inertial, nuclear and gravitational properties of matter. Inertons are responsible also for such a cryptic substance as dark matter. Solar inertons can be measured in a terrestrial laboratory and the reader will learn how we can do that.

Some prospects for further possible fundamental and applied studies are shortly described in the last chapter. It seems that the recent advances at the frontier of physics will already be able to lead us to a qualitative change in several directions within a few years time.

I am very thankful to many researchers for their remarks and comments to my research papers, unselfish assistance in preparation of talks and the useful correspondence from which I have benefitted; the list is long and I cannot name them all here. Nevertheless, I would like to thank Georges Lochak who considerably helped me to enlarge upon Louis de Broglie's views on the nature of things, my nearest colleagues Yuri Zabulonov and Ivan Gandzha with who we carry out R&D work in both fundamental and applied science, for many discussions and helpful suggestions, Daniel M. Dubois, Michael C. Duffy, Stephen Crothers, Jean-Pierre Robitaille and Victor Christianto who have a global perspective and broad interests, which contributed to improvements of the material presented in this book. Special thanks to Timothy Hooker and August Keder, who read the whole manuscript providing me with many valuable advices, which significantly improved the content. Finally, I am very thankful to our Mother-Father Space for setting me on subtle sensations of the laws of their essence. Glory to Space!

—*Volodymyr Krasnoholovets, PhD*
November 1, 2016

ABOUT THE AUTHOR

Dr. V. Krasnoholovets was born in Kyiv, Ukraine. He graduated from a mathematical school and then became a student of the Kyiv's Taras Shevchenko National University, Department of Physics, Faculty of Theoretical Physics; he received a master's degree in 1979. For the next several years he worked as an experimentalist in the area of superconductivity at the Institute for Metal Physics, National Academy of Sciences of Ukraine, Kyiv. Since the end of 1981 and to now, he has been working at the Department of Theoretical Physics, Institute of Physics, National Academy of Sciences, Kyiv. His PhD thesis was defended in 1987; it was devoted to the study of a proton polaron model in compounds with hydrogen bonds including biological systems.

At the Department, he focuses on condensed matter physics. Since 1993 he has been a Senior Research Scientist. In the mid-1980s he also began to take an interest in the foundations of physics. The first paper in this field was published in 1993. In 1998–2003 Dr. Krasnoholovets actively worked with one of the classical French mathematicians, Prof. Michel Bounias (1943–2003). Together with Prof. Bounias, a theory of real physical space was developed, which started from pure mathematical principles, namely, set theory, topology, and fractal geometry. Another professional interest of Dr. Krasnoholovets is applied physics. In 2006 he co-founded a company in Belgium devoted to the development of technologies proposed by Ukrainian scientists. The company was named Indra Scientific, and it has been gradually developing new areas of applications (the production of biodiesel, recycling of industrial waste, organic waste to energy by using a new design of a gasifier, cleaning of wastewater, infrared heating thin films, measuring devices, ecological chemistry, etc.).

He was an editor of several books and collections of works dealing with quantum physics and gravity. Dr. Krasnoholovets has published over 80 research papers.

INTRODUCTION

Modern science has become so differentiated that its unification into a coherent integral form seems practically impossible. In any area of physics including the foundations of physics, researchers demonstrate a form of tunnel vision. And even when one talks about the Standard Model of Particle Physics or the Standard Model of Cosmology we can see that these unified models are very imperfect. For example, the Grand Unified Theory is a model in which at high energy levels the three gauge interactions of the Standard Model that defines the electromagnetic, weak, and strong interactions are merged into one single force. The researchers try to unify these interactions into one unified constant, which has several force carriers, in the framework of one large gauge symmetry. However, the nature of particles, their origin, mutation of one particle to another, etc., are still outside the general consideration of physicists.

Of course one can say that the Standard Model is ideally accounting for the microscopic world, especially after the discovery of a new particle associated with a Higgs boson in 2012, which is now treated as the cornerstone of the Standard Model [1]: it is assumed that space devoid of particles and radiation is filled with a material medium that endows the *W* and *Z* bosons with mass and this space-filling material is associated with a Higgs boson. Particles have different masses because they interact with the Higgs field to different degrees. Higgs bosons must be accompanied by supersymmetric particles (so-called 'sparticles'); but not one of such 'sparticles' has been found.

In modern physics leptons and quarks are treated as point particles [2]. Such a view in physics resulted in the foundation of supersymmetry (SUSY), a theory of particle physics in which spacetime symmetry plays an important role and bosons (which have an integer-valued spin) are related to appropriate fermions (which have a half-integer spin). Each particle from one group is associated with a particle from the other group, known as its superpartner and its spin differs by a half-integer. However, no superpartners have been observed yet, even at the Large Hadron Collider, which is paradoxical for the Higgs like particle, because the Higgs mechanism only works when/if supersymmetry is included in the Standard Model. So, did the researchers really record the decay of the Higgs boson?

Since the fundamental theory of elementary particles failed experimental tests on SUSY, Shifman [3] called on his colleagues to stop modifying supersymmetry. He recommends starting to think about and to develop new ideas, while he notes that major difficulties would be connected with the reeducation of high-energy theoretical physicists who have never worked on any issues beyond supersymmetry-based phenomenology or string theory. So the crises (or, at least, huge question marks) in these two areas could pose a serious problem in the community of physicists.

In relation with this, we have to mention a known fact, which so far has been completely ignored by high-energy theoretical physicists. In 1923, Compton [4] demonstrated that a canonical elementary particle has a radius of hardness, which was later named in his honor—the Compton wavelength. Nevertheless, particle physicists still continue to consider particles as points or point-like multi-dimensional strings, which, according to Shifman [3], already resulted in a crisis of particle theories.

Quantum electrodynamics is considered as the best-developed doctrine of modern theoretical physics. Nevertheless, in his lectures on quantum electrodynamics, Feynman [5] noted that we don't yet have a good model to explain partial reflection of light by two surfaces; we can only calculate the probability that a photomultiplier will be hit by a photon reflected from a sheet of glass. Then he continued: "…the more you see how strangely Nature behaves, the harder it is to make a model that explains how even the simplest phenomena actually work. So theoretical physics has given up on that."

Some researchers would agree on these last mentioned words of Richard Feynman, but others wouldn't. If one agrees, one automatically closes the possibility for the development of science and will continue to walk in a vicious circle, being constantly in a state of crisis. And doing so one will never answer questions about the origin of the electron, photon, quark, their shapes, their inner dynamics and inner properties.

A detailed study of the mobility of electrons in graphene's honeycomb lattice allowed the researchers [6] to state that half-integer spin like that carried by the quarks and leptons can derive from a hidden substructure, not from the particles themselves, but rather from the space in which these particles exist. In other words, it seems the particle's spin is not a pure particle property but rather an attribute induced by the space through which the particle is moving.

Does this mean that a vacuum is able to influence moving particles?

Some physicists say that physics will not be complete until it can explain how space and time emerge from something more fundamental [7]. Nevertheless, the directions of studies listed in Ref. [7] are also narrowly focused, because the quoted authors do not apply their ideas widely to the whole of physics, but only wish to use their ideas for describing this or that discipline of physics, such as general relativity, radiation from a black hole, visualization of what happened before the Big Bang, the emergence of space-time based on the holographic principle, space emerging from points in a similar way to temperature emerging from the kinetic energy of atoms, and so on. However, at the same time the researchers emphasize that adding causality changed everything; quantum entanglement and space-time are the same thing; quantum is the most fundamental, and space-time emerges from it; space-time looks rather continuous all the way down to the Planck scale $\ell_P = \sqrt{\hbar G / c^3} \approx 1.616 \times 10^{-35}$ m and space-time should therefore be granulated at this size.

Wilczek [8] says the following about space: "…this is the effervescent Grid… Matter is not what it used to be. It consists of small, more-or-less stable patterns of disturbance in the Grid… Usually the metric field is taken to be fundamental, but in many ways it resembles a condensate, and that view of it may become important…. What we ordinarily call matter consists of more-or-less stable patterns of excitation in the Grid, which is more fundamental. At least, that's how things look today."

In the well-known work on the dynamics of the electron by Poincaré [9] we find an interesting remark, which so far has been overlooked or underestimated by the majority of researchers: an electron is moving, surrounded by excitations of the ether.

If these excitations are real, can they be organized in a wave? De Broglie [10] hypothesized that a real wave should guide a particle, which resulted in the famous relationship between the wave and the particle, $\lambda = h/(m\text{v})$. Other leading physicists of that time reinterpreted de Broglie's result introducing the concept of a wave-particle, which resulted in the creation of quantum mechanics with its statistical description of quantum systems. However, de Broglie [11] returned to his initial views in 1952 (after two papers by Bohm [12]) and he continued to develop the double solution theory until his death. His major goal was the search for a physical interpretation of the wave ψ-function instead of its pure statistical explication.

Can we revise and put in order all the difficulties associated with quantum physics, particles, space and gravity? For a positive solution, obviously, we need new ideas. But how to elaborate them?

There are a few sources from which they could emerge.

First, there are studies by dissidents in quantum physics (see, e.g., Ref. [13]) and the gravitation phenomenon resting on alternative viewpoints (see, e.g., Ref. [14]). in particular, in quantum physics, most intensive exhaustive studies are associated with so-called "hidden variables" and the de Broglie's double solution theory.

Second, modern theories of particle physics combine all fundamental interactions in a unified theory (*the theory of everything*). However, in doing so the theory of everything rests on complete undetermined basic notions, such as particle, mass, charge, lepton, quark, Compton wavelength, de Broglie wavelength, wave-particle, matter waves, wave ψ-function, spin, uncertainty relations, Pauli's exclusion principle, etc. So, we would initially have to develop *a theorem of something*, which would clarify the fundamental notions of quantum physics mentioned above.

Third, the notion of space in which all physical processes occur is also beyond understanding, though some of the researchers lean toward the idea of a fine-scale morphology, which incorporates particles, as we have mentioned above.

Fourth, pure classical mathematics can also be explored as a possible source for the knowledge base. It is known that in fundamental physics such mathematical disciplines as group theory, topology and geometry (in particular, fractal geometry) play an important role.

Group theory is the study of symmetry. If an object possesses symmetry, group theory helps to analyze the behavior of the object under some transformations. Group theory, known as the theory of representations, is most often used in physics. In this theory matrices act on the members of a vector space and certain members of the space emerge as this or that kind of symmetry. Group theory is widely used in applications of quantum mechanics and quantum field theories (see, e.g., Refs. [15, 16]). In particular, Lie groups, Poincaré groups and others, which are invariant in respect to symmetric transformations, are widely applicable to investigations of multidimensional systems of nonlinear differential equations in partial derivatives.

Topology is the mathematical analysis of shapes and spaces; in particular, topology studies how properties of space change under continuous deformations of given types. Topology is a field of study emerging from geometry and set theory. Topology investigates such concepts as space, dimension, and transformation. A topology allows one to tell how elements of a set are related spatially to each other. The same set can have different topologies. A topological invariant is a property of a topological space that is invariant under homeomorphisms. Topology is broadly used in physics. A topological quantum field theory is a quantum field theory that computes topological invariants (for knots, links, n-manifolds arising from operator algebras, and so on). In quantum gravity, topology plays the role of a mathematical tool for understanding networks (e.g., spin networks) that arise from the study of the loop states of quantum gravity. So basically, topology is used as a discipline that further formalizes the formalism of physical mathematical theories (see, e.g., Refs. [17–19]).

A geometric fractal is a mathematical set that displays self-similar patterns, which means that the shape is made of smaller copies of itself. The copies are similar to the whole: same shape but different size. Fractals can have an "irregular" and "fractured" appearance. Fractal geometry has been applied for the description of many physical systems, from the quantum world to general relativity [20–23].

Thus the common feature of all these fundamental mathematical theories is to provide service to physical mathematics and different fields of physics.

However, mathematics can play a principally different role. Namely, mathematics itself is able to create fundamental physics and we will see in the present book how it is occurring. The reader will learn how the ordinary physical space appears from pure mathematical constructions and how space generates matter and laws of physics.

This book begins with Chapter 1 describing a purely mathematical theory of the real physical space. The reader will see how physical space is organized and in which way it is possible to formalize all the fundamentals of physics. Subsequently in Chapter 2, the reader will see how Poincaré's excitations (which were collected in a wave by de Broglie) form a submicroscopic deterministic mechanics in the real space. The excitations of space were named *inertons* because they represent the field of inertia of the particle. Chapter 3 discloses inner peculiarities of inerton dynamics that penetrates much deeper into the submicroscopic world than the conventional quantum mechanical formalism. Chapter 4 derives electrodynamics from the real space. Chapter 5 discusses issues associated with the presence of inertons in condensed media (Bose-Einstein condensation, diffraction, double-slit experiments, anomalous photoelectric effect, electron droplet, sonoluminescence, and a manifestation in biophysics). In Chapter 6, the emergence and inner dynamics of quarks and hadrons is studied in the framework of the real space. Chapter 7 is devoted to nuclei, the nuclear forces and new nuclear reactions in condensed media. The submicroscopic theory is applied to gravity in Chapter 8. Submicroscopic interpretations of four phenomena predicted by the formalism of general relativity are given in Chapter 9. Modern problems of cosmology, as seen from the submicroscopic viewpoint, are investigated in Chapter 10. Finally, Chapter 11 lists prospects for further studies and applications.

CHAPTER 1

SPACE: FOUNDING PRINCIPLES

CONTENTS

1.1 THE NOTION OF SPACE IN GENERAL PHYSICS

In physics, space is defined via measurement and the standard space interval, called a standard meter or simply meter, which is the distance traveled by light in a vacuum per a specific period of time and in this determination the velocity of light c is treated as constant.

In astronomy, space refers collectively to the relatively empty parts of the universe; any area outside of a celestial object can be considered as space.

In classical physics, space is a three-dimensional Euclidean space where any position can be described using three coordinates. In relativistic physics researchers operate with the notion of space-time in which matter is able to influence space. Here, the old idea of Riemann is exploited. Riemann [24] expressed the idea that the geometry of space depends on bodies available in the space and therefore the measure of the curvature can have any random value. Riemann determined the general topological properties of space showing that the topological properties of figures are invariant under any transformation, in particular motion. He introduced the concept of manifolds in which a manifold is treated as a topological space that resembles Euclidean space near each point, i.e., a point of an n-dimensional manifold has a neighborhood that is homeomorphic to the nD Euclidean space. Or in other words, a manifold is represented as a geometric object obtained by sticking together some open balls. Those made it possible to link different mathematical branches like topology, differential geometry and algebraic geometry covering not only pure mathematical problems but also the knowledge base of physical space and the nature of the constitution and interaction of matter [24–26].

Helmholtz [26–28] further developed Riemann's ideas regarding physical space. He wondered which are the most common geometrical axioms that explain the empirical measurement of objects. He distinguished physical space from geometric space, as possessing its own properties. Physical space is distinguished from all other *n*-fold extended manifolds for two reasons: it is not only 3D, but also metrically infinitesimally Euclidean. Objects "occupy" space and hence the space must satisfy a condition of "free mobility" for these objects. Space allows accidental continuous motions of rigid bodies. Helmholtz presented the idea of "rigid motions" to account for invariant properties of figures/bodies. Namely, these "rigid motions" preserve a set of properties of the objects, which allowed Lie [29] to describe the set of rigid motions in the mathematical terms of group theory; Lie group theory is today embedded in the context of all quantum and particle physics.

Poincaré [30] emphasized that geometry could not have been developed without solids; namely, he wrote: "if there were no solid bodies in nature, there would be no geometry." He also noted an important role in geometry of non-Euclidean displacements and deformations. Thus space has no geometrical structure until we take into account the physical properties of matter within it, and that this structure can be determined only by measurement. According to Poincaré physical matter establishes the geometrical structure of space. The quintessence of Poincaré's views on the theory of space are perfectly expressed by Kuyk [31]: "...without external bodies in motion, we would not be able to form an empirical notion of space and geometry. In forming these empirical notions, our bodies function as the axes of reference, the source of coordination. We go from this representational space to a *mathematical space* by *reasoning* about the objects in representational space as if they were given in a mathematical space (or continuum). The external bodies become rigid figures and the motion of these bodies, i.e., the succession of sensations, is idealized into the mathematical concept of a *group* of motions. But, many different laws of succession of sensations are possible; accordingly, there is no one geometry that 'fits' the world."

Later on researchers introduced some new details and generalizations to the mathematical concept elaborated by Riemann, Helmholtz, Lie and Poincaré. The process of generalization, refining and concretely defining the concept followed various directions; for example, such concepts as Riemannian space, topological space, vector space, Hilbert space, metric space, function space, and multidimensional space (in which topological spaces acquire a new notion of dimension when seen as measurable spaces, namely fractals) were developed. But interestingly, the study went further by examining the differential equations that describe the movement of "solids" in space, the symmetry of these equations, exact solutions, etc.

Bunge and Mayenez [32] carried out a consideration of physical space in terms of things and a sort of network of relations among things. The set of things together with the topology of separation functions was called the physical space. In such a theory the family of balls lying between any two things was postulated as satisfying Huntington's axioms for solid geometry.

Thus, researchers consider physical space as something that consisted of objects and intervals. However, mathematicians have overlooked that a "solid" itself is an element of space and because of that a moving "solid" has also to be considered as part of the space (cf. with the situation when chunks of ice are floating in water: whether the pieces of ice are not water?). In fairness, we should note that only Clifford [33] surmised that matter is the result of a local curvature of space. But so far his guess has not been further investigated in terms applicable for consideration in real physical systems (though some abstract geometrizations in establishing particle properties by using special exotic solutions of nonlinear equations of general relativity have been made [34, 35]).

In contemporary mathematics a space is defined as a set of objects, which are called the points of the space [36]: These objects may be geometric figures, functions, or the states of a physical system. When we consider a set of objects as a space, we deal not with the individual properties of the objects but with only those properties of the set that are determined by relations that we wish to take into account or that we introduce by definition. These relations between points and various configurations, or sets of points, determine the geometry of the space. When the geometry is constructed axiomatically, the basic properties of these relations are expressed in the corresponding axioms.

Here are examples of three spaces [36]: metric spaces, spaces of events, and phase spaces.

In a metric space, the distance between points is defined. Thus, the continuous functions $f(x)$ on an interval $[a, b]$ form a metric space—whose points are the functions $f(x)$ – when the distance between $f_1(x)$ and $f_2(x)$ is defined as the maximum of the absolute value of the difference between the two functions: $r = \max |f_1(x) - f_2(x)|$.

The concept of space of events plays an important role in the geometric interpretation of the theory of relativity. Every event is characterized by its position—the coordinates x, y, and z—and the time t of its occurrence. The set of all possible events is thus a four-dimensional space, in which an event or point is defined by the four coordinates x, y, z, and t.

Phase spaces are studied in theoretical physics and mechanics. The phase space of a physical system is the set of all the possible states of the system. The states are the points of the space. In particular, quantum mechanics was constructed in the phase space. The question of which mathematical space reflects most accurately the general properties of physical space is answered experimentally.

It is interesting to read Vernadsky's [37] work who, back in the 1920–1930s widely developed the notion of *noosphere* (from Greek *nous* — mind and *sphaira* — ball): a spherical arena of interaction between people and nature. In particular, Vernadsky noted: "In discussing the state of space, I will be dealing with the state of empirical or physical space, which has only in part been assimilated by geometry. Grasping it geometrically is a task for the future." Vernadsky introduced such notion as *the state of space*, which in his opinion has to be closely connected with the concept of a physical field that plays such an important role in contemporary theoretical physics (the noosphere collects subtle vibrations of all the people and is able to influence individuals, which is in some aspects similar to the phenomenon of homeopathy). Planck [38] also noted: "There is no matter as such! All matter originates and exists only by virtue of a force that brings the particles of an atom to vibration and holds this most minute solar system of the atom together. We must assume behind this force the existence of a conscious and intelligent Mind. This Mind is the matrix of all matter."

Now let us come to the physics of the microscopic world. In microscopic physics, or quantum physics, the notion of space is associated with an "arena of actions" in which all physical processes and phenomena take place. This arena of actions we feel subjectively as a "receptacle for subjects." The measurement of physical space has long been important. The International System of Units (SI) is today the most common system of units used in the measuring of space, and is almost universally used within physics.

However, let us critically look at the determination of physical space as an "arena of actions." In such a determination there exists, first, subjectivity and, second, objects themselves that play in processes can not be examined at all (for instance size, shape and the inner dynamics of the electron; what is a photon?; what is the particle's Compton wavelength $\lambda_{\rm Com}$?; what is the particle's de Broglie wavelength λ and how does the particle behave along this section of λ?; how to understand the notion, or phenomenon of "wave-particle duality?"; what is spin?; what is the mechanism that forms Newton's gravitational potential $-Gm/r$ around an object with mass m?; what does the notion of 'mass' mean exactly?, etc.).

Especially interesting are the next three examples of motion "in the arena of action, as a reservoir for objects":

1. When a vehicle suddenly jams on the brakes, an experienced physicist sitting in the vehicle will feel that something pushes him forward;
2. Our experienced physicist comes to a playground and decides to go on the merry-go-round. The physicist raises the marry-go-round to a high speed and jumps onto it. However, holding hands tightly over the merry-go-round, he suddenly feels that something unseen grabs his legs pulling them out of the merry-go-round;
3. The experienced physicist wishes to have fun with a gyroscope and taking it in his hands, when the gyroscope's rotor reaches the speed of about 20,000 revolutions per minute, he feels that this quietly functioning plaything for some reasons goes out of hand, and injures the muscles and tendons of his arms.

These three examples clearly give evidence of the existence of otherworldly forces at the scene of action among macroscopic subjects.

The source of these forces is probably hidden in a sub microscopic structure of space. Because it looks as though a body at its "rigid motion" is able to disturb the ordinary space locally and the space in its turn responds to its disturbance.

The "arena of actions" can be entirely formalized, such that those mystical forces (veiled under the force of inertia and the centrifugal force) will unravel explicitly, because fundamental physical notions and interactions will be derived from pure mathematical constructions.

All this means that physical space is a peculiar substrate that is subject to certain subtle laws, which as will been seen below, are purely mathematical (though hitherto the laws have been out of sight of examiners). Such a view allows us to completely remove any subjectivity and all the figurants of fundamental physical processes will be one hundred percent defined. In this chapter we elucidate those *somethings* that form a primordial physical substrate and determine its mathematical properties. Three new features are introduced to the physical space, namely:

- structure of space from subquantum to cosmic scales;
- matter as stable local irregularities of space;
- space and matter interact and influence each other reciprocally.

1.2 THE CONCEPT OF MEASURE AND DISTANCES

In a series of works Bounias [39–51] (partly in collaboration with Bonaly) developed a mathematical theory of space from a canonical particle to the universe. In particular, he demonstrated that any property of an object must be consistent with the characteristics of the corresponding embedding space [48].

A similar problem about a sub microscopic constitution of real space was also raised in my first papers [52, 53] devoted to the submicroscopic mechanics of a particle.

Later in 1998–2003, working together, we [54–57] examined in depth founding mathematical conditions for a scientific scanning of a physical world. In particular, we considered the main determining properties of ordinary space, namely, measure, distances and dimensionality in a broad topological sense.

1.2.1 MEASURE AND DISTANCES

The concept of measure takes into account the existence of mappings and the indexation of collections of subsets on natural integers. By classical determination, a measure is a comparison of the measured object with some unit taken as a standard [58]. However, sets (or spaces) and functions are measurable under various conditions, which are cross-connected. A mapping f of a set E into a topological space $\mathcal{T}$ is measurable if the reciprocal image of an open $\mathcal{T}$ by f is measurable in E. A set measure on E is a mapping m of a tribe B of sets of E in the interval $[0, \infty]$ exhibits denumerable additivity for any sequence of disjoint subsets (b_n) of B, and denumerable finiteness

$$m\left(\bigcup_{n=0}^{\infty} b\right) = \sum_{n=0}^{\infty} m(b_n) \tag{1.1}$$

where $\exists An, An \in B, E = \cup An, \forall n \in N, m(An)$ is fine.

Unit, or gauge J, which usually has non-zero real values and used as a standard, is a function defined on all bounded sets of the space studied [59]:

- a singleton has measure naught: $\forall x, J(\{x\}) = 0$;
- J is continued with respect to Hausdorff distance;
- J is growing: $E \subset F \Rightarrow J(E) \subset J(F)$;
- J is linear: $F(r{\cdot}E)$: $r{\cdot}J(E)$.

This rule set represents the concept of distance. Terms for the distance are a diameter, a size, or a deviation; such a distance is applied on totally ordered sets. The Caratheodory measure (μ^*) poses some conditions that also involve the necessity of a common gauge:

- $A \subset B \Rightarrow \mu^*(A)\ {''}\ \mu^*(B)$;
- For a sequence of subsets (Ei): $\mu^* \cup (Ri)\ {''}\sum \mu^*(Ri)$;
- $\angle(A, B)$, $A \cap B = \varnothing$: $\mu^* A \cup B = = \mu^*(A) + \mu^*(B)$ in consistency with Eq. (1.1);
- $\mu^*(E) = \mu^*(A \cap E) + \mu^*(\complement_E A \cap E)$.

The Jordan and Lebesgue measures involve respective mappings (I) and (m^*) on spaces that must be provided with operations of union $\cup$, intersection $\cap$, and complement $\complement$. Spaces of the $\mathbb{R}^n$ type include tessellation of balls [44], which demands the need of a distance for the measure of the diameters of intervals.

A set E of measure naught defined by Borel in 1912, is a linear set, such that given a number (e) as small as needed, all points of E can be contained in intervals whose sum is lower than (e).

Following Borel, the length of an interval $F = [a, b]$ is

$$L(F) = (b - a) = \sum_n L(Cn) \tag{1.2}$$

where Cn is the adjoined (or open) interval inserted in the fundamental segment. The distance (1.2) is required in the Hausdorff distances of sets E and F: let $E(\varepsilon)$ and $F(\varepsilon)$ be the covers of E and F by balls $B(x, \varepsilon)$, respectively, for $x \in E$ or $x \in F$,

$$\text{dist}_{\text{H}}(E, F) = \inf\{\text{e} : E \subset F(\text{e}) \wedge F \subset E(\text{e})\} \tag{1.3}$$

$$\text{dist}_{\text{H}}(E, F) = (x \in E,\ y \in F(\varepsilon) : \inf \text{dist}(x, y)) \tag{1.4}$$

Such a distance is not necessarily compatible with the topological properties of the space under consideration. The set (E) should be covered with a sequence of intervals (Un), but the intervals can be replaced by topological balls. But doing so we should evaluate the diameter of these balls, which in turn requires an appropriate general definition of a distance (the other approach [60] involves a path $\varphi(x, y)$, such that $\varphi(0) = x$ and $\varphi(1) = y$).

For the case of sets A and B in a partly ordered space, the symmetric difference $\Delta(A, B) = \complement_{A \cup B}(A \cap B)$ has been proved to be a true distance also holding for more than two sets [44, 45, 48, 49]. However, if $A \cap B = \varnothing$, the situation requires a clarification, which is done below in terms of a separating distance versus an intrinsic distance.

How can one assess the space dimension? In a space E a fundamental interval (AB) and intervals $Li = [Ai, A(i + 1)]$ permit one to compose a generator of such intervals: $G = \cup_{(i \in [1, n])}(Li)$. The similarity coefficients can be defined for each interval by a ratio $\rho i = \text{dist}(Ai,\ A(i+1)) / \text{dist}(AB)$.

The similarity exponent of Bouligand is e, such that for a generator with n parts

$$\sum_{(i \in [1,\, n])} (\rho i)^{\mathrm{e}} = 1 \tag{1.5}$$

When all intervals have practically the same size, then in accord with Bouligand, Minkowski, Hausdorff and Besicovitch, various dimension approaches come to the resulting relation

$$n \cdot (\rho)^{\mathrm{e}} \approx 1 \tag{1.6}$$

and hence $\mathrm{e} \approx \log n / \log \rho$. When e is an integer, it reflects a topological dimension, because this signifies that the fundamental space E is tesselated with an entire number of identical balls B that exhibit a similarity with the space E upon the coefficient ρ.

1.2.2 DISTANCES AND DIMENSIONS REVISED

In our approach distances are compatible with both the involved topologies and the scanning of objects whose characteristics are not known in the space studied. Indeed we [54] demonstrated that a generalized distance between spaces A and B within their common embedding space E is provided by the intersection of a path-set $\varphi(A, B)$ joining each member of A to each member of B with the complement of A and B in E. In this case $\mathrm{j}\ (A, B) = \bigcup_{a \in A,\, b \in B} \mathrm{j}\ (a, b)$ in which all elements are defined on a sequence of interval $[0, f^n(x)]$, $x \in E$, becomes a continuous sequence of a function f of a gauge J that belongs to the ultrafilter of topologies on $\{E, A, B, \ldots\}$. Thus the relative distance of A and B in E is contained in $\varphi(A, B)$: $\Lambda_E(A, B) \subseteq \varphi(A, B)$. If we denote the interior of E by E°, then:

$$\min\{\varphi(A, B) \cap E^{\circ}\} \text{ is a geodesic of space } E \text{ connecting } A \text{ to } B \tag{1.7}$$

$$\max\{\varphi(A, B) \cap \complement_{E^{\circ}}(A \cup B)\} \text{ is a tessellation of } E \text{ out of } A \text{ and } B \tag{1.8}$$

Here, the relation (1.7) refers to $\dim \Lambda = \dim \varphi$, while in relation (1.8) the dimension of the probe is that of the scanned sets.

If a closed set D contains a closed subset $D_1 \subset D$, then there is a member of $\varphi(A, B)$ that intersects D_1. $\varphi(A, B) \cap E^\circ$ is the growing function defined for any Jordan point, which is a characteristic of a gauge.

Let one path $\varphi(a, b)$ meet an empty space $\varnothing$. Then a discontinuity takes place and there exists some i, such that $\varphi(f^i(b)) = \varnothing$. If all $\varphi(A, B)$ meet $\{\varnothing\}$, then no distance is measurable. So for any singleton $\{x\}$ one has $\varphi\{f(x)\} = \varnothing$.

These properties are two other characteristics of a gauge.

In the case of a totally ordered space the distance between A and B is represented by the relation

$$d(A, B) \subseteq \text{dist}(\inf A, \inf B) \cap \text{dist}(\sup A, \sup B) \tag{1.9}$$

with the distance evaluated through either classical forms or even the set-distance $\Delta(A, B)$.

1.2.3 THE CASE OF TOPOLOGICAL SPACES

In the case of topological spaces, a space can be subdivided into two main classes – objects and distances. The set-distance is the symmetric difference between sets and it owns all the properties of a true distance [44], which can be extended to manifolds of sets [45]. In a topologically closed space, these distances are the open complement of closed intersections called "instances." The intersection of closed sets is closed and the intersection of sets with non-equal dimensions is always closed, which permits identifying the instances with closed objects. Distances, as being their complements, form the alternative class. Therefore a physical-like space may be globally subdivided into objects and distances as full components.

The properties of the set-distance allow the claim that any topological space is metrizable as provided with the set-distance (Δ) as a natural metric. Indeed, union and intersection allow the introduction of the symmetric difference between two sets A_i and A_j.

$$\Delta(A_i, A_j) = \underset{\cup\{A_i\}}{\complement} \cup_{i \neq j} (A_i \cap A_j) \tag{1.10}$$

i.e., the complement of the intersection of these sets in their union. Symmetric difference satisfies the following properties: $\Delta(A_i, A_j) = 0$ if $A_i = A_j$, $\Delta(A_i, A_j) = \Delta(A_j, A_i)$ and $\Delta(A_i, A_j)$ is contained in union of $\Delta(A_i, A_j)$ and $\Delta(A_j, Ak)$. This means, it is a true distance and it can also be extended to the distance of three, four, etc. sets in one, namely, $\Delta(A_i, A_j, A_k, A_l, \ldots)$. Since the definition of a topology implies the definition of such a set distance, every topological space is endowed with this set metric.

Thus all topological spaces are kinds of metric spaces called "delta-metric spaces." The norm of symmetric distance is provided by the relation $\|\Delta(A)\| = \Delta(A, \varnothing)$. Hence any topology provides the set-distance that can be named a topological distance. A topological space always assures a self-mapping of any of its parts into any one metric. Hence any topological space is metrisable.

Reciprocally, given the set-distance, since it is constructed on the complement of the intersection of sets in their union, it is compatible with the existence of a topology. That is why a topological space is always a "delta-metric" space.

The set-distance $\Delta(A, B)$ is a kind of an intrinsic case $[\Lambda_{(A, B)}(A, B)]$ of $\Lambda_E(A, B)$, while $\Lambda_E(A, B)$ is called "separating distance." This separating distance also stands for a topological metric. Thus if a physical space is a topological space, it will always be measurable.

An important case is measuring open sets. Even a continuous path cannot scan an open component of the separating distance between the two sets, since a path has, in general, no closed intersections with an open set with the same dimension. This is consistent with the exclusion of open adjoined intervals in the Borel measure. Hence, a primary topology is a topology of open sets, because a primary topological space cannot be a physically measurable space.

An intersection of a closed set C with a path φ having a non-equal dimension assures a non-empty intersection, which implies that the general conditions of filter membership are fulfilled.

1.2.4 "SCANNING" MEASURE OF AN ABSTRACT SET

The information from the explored space stands like parts of an apparatus being spread on the desk before assembly of the original object. Bounias proposed to call this situation an informational display, likely composed of elements with dimensions lower than or equal to the dimensions of the real object.

A tuple is a finite ordered list of elements. An n-tuple is a sequence of elements where n is a non-negative integer. Bounias gave the following robust definition for an ordered n-tuple: an expression noted $(abc...z)$ is an ordered n-tuple if

$$(abc...z) = (x1, x2, x3, \ldots xn) \quad a = x1, b = x2, c = x3, \ldots z = xn \tag{1.11}$$

As known, Von Neumann provided an equipotent form using replicates at the construction of the set N of natural integers:

$$0 = \varnothing, 1 = \{\varnothing\}, 2 = \{\varnothing, \{\varnothing\}\}, 3 = \{\varnothing, \{\varnothing\}, \{\varnothing, \{\varnothing\}\}\},$$
$$4 = \{\varnothing, \{\varnothing\}, \{\varnothing, \{\varnothing\}\}, \{\varnothing, \{\varnothing\}, \{\varnothing, \{\varnothing\}\}\}\}, \ldots \tag{1.12}$$

A Von Neumann set is isomorphic with none of its parts, which is thus a Mirimanoff first-kind set. This construction associated with the application of Morgan's laws to $(\varnothing)$ allows the empty set to be attributed to an infinite descent of infinite descents. That is why the infinite descent should be clarified as a member of the hypersets family [47].

Two kinds of orders can be imbued in an abstract set: one with respect to the identification of a maximum or minimum, and one with respect to ordered N-tuples. These two conditions become equivalent upon additional conditions on the nature of the involved singleton. This solves the problem of the identification of orders in the set of parts of a set, as compared with components of a simplex set. Besides, a set can be ordered through the rearrangement of its exact members and singletons in a way permitted by the structure of the set of its parts. This permits one to justify that for a set like (En), the set of its (2^n) parts has dimension $D \leq D(E^n)$ [54].

The size of sets with ill-defined structure and order allows an evaluation. If E is a non-ordered set, Id is the identity set map of E, and f is the difference self map of E, then the following relation gives a kind of diameter:

$$\text{diam}_f(E) \oplus \{(x, y) \in E\text{: } \max f^i \,(\text{Id}\,(x) \cap \max f^i \,(\text{Id}\,(y)\} \tag{1.13}$$

which gives a $(N-2)$ members parameter.

If E is well-ordered (the lower boundary can be identified among members of E, such as a singleton $\{m\}$), then an alternative record can be offered. In this case the set E is presented by two members: $E = \{m, Z\}$ with $Z = \complement_E(m)$ – the complement of m in E. Then the measure M_Δ:

$$M_\Delta(E) = \{m = \min(E),\ \Delta(m, Z)\} \tag{1.14}$$

where Δ is the symmetric difference. This gives an $(N-1)$ member parameter that in turn provides a derived kind of diameter diam Δ by repeating the measure for two members m' and m''. Indeed let $E' = \{m',\ \complement_E(m')\}$ and E'': $\{\complement_E(m'')\}$, then

$$\mathrm{diam}(E) \subseteq \{m',\ m'' \in E:\ \max \Delta(E',\ E'')\} \tag{1.15}$$

If both upper and lower boundaries can be identified, i.e., the set E is totally ordered, then the distance separating two segments A and B of E is

$$(A, B) \subset E,\ \mathrm{dist}_E(A,\ B) = \{\mathrm{dist}\,(\inf \mathrm{A}, \inf \mathrm{B}) \cap \mathrm{dist}\,(\sup \mathrm{A}, \sup \mathrm{B})\} \tag{1.16}$$

where a set distance Δ is again provided by the symmetric difference.

A separating distance is an extrinsic form of the set-distance as an intrinsic form. A diameter is evaluated on E as the following limit:

$$\mathrm{diam}\,(E) = \{\inf A \to \inf E,\ \sup A \to \min E,\ \inf B \to \max E,\ \sup \mathrm{B} \to \sup \mathrm{E} |\ \lim \mathrm{dist}_E(A,\ B)\} \tag{1.17}$$

These approaches allow the measure of the size of tessellating balls as well as that of the tessellated spaces, with reference to the calculation of their dimension. A diagonal-like part of an abstract space can be identified with and logically derived as a diameter [54].

Remark [54]: If a measure is obtained each time from a system, this means that no absolutely empty part is present as an adjoined segment on the trajectory of the exploring path; thus, no space accessible to some sort of measure is strictly empty in both mathematical and physical sense, which supports the validity of the quest of quantum mechanics for a structure of the void.

In physical exploration it is important to discern between the detected objects, which are equivalent with abstract ordered N-tuples within their embedding space. In a space composed of members identified with abstract components tessellating balls could have different diameters. A ball with two different members would not have the diameter defined in Eqs. (1.13) or (1.16). Then a measure should be used as a probe for evaluation of the coefficient of size ratio (ρ) needed for the calculation of a diameter.

A three-object has the dimension 2 if the measure of the longer side A^1_{max}

$$M(A^1_{max}) < M(A^1_2) + M(A^1_3) \tag{1.18}$$

where M denotes an appropriate measure. For a space X being an N object:

$$M(A^k_{max}) < \bigcup_{i=1}^{N-1} \{M(A^k_i)\} \tag{1.19}$$

where $k = N - 2$ and A^k_{max} is the k-face with maximum size in X.

According to inequality (1.19), for N = 2 (a 2-object), the space $X = \{x1, x2\}$ has dimension 1 if $x1 < (x1+x2)$, i.e., if $x1 \neq x2$. This qualifies the lower state of an existing space X^1.

Let the space X be decomposed into the union of balls represented by D-faces A^D proved to have dimension $\mathrm{Dim}(A^D) = D$ by inequality (1.19) and size $M(A^1)$ for a 1-face, such that a D-face is a D-simplex S_j whose size, as a ball, is evaluated by $M(A^1_{max})^D = S^D_j$. Let ℵ be the number of such balls that can be filled in a space H, so that

$$\bigcup_{i=1}^{\aleph} \{S^D_j\} \subseteq (H \approx L^d_{max}) \tag{1.20}$$

where H is identified as a ball whose size is estimated by L^d and L is the size of a 1-face of H, and d is the dimension of H. Then if for $\forall S_j$, $S_j \oplus S_o$ and the dimension of H is

$$d(H) \approx (D \cdot \log S_o + \log \aleph) / \log L^1_{max} \tag{1.21}$$

The relationship (1.21) stands for a kind of interior measure in the Jordan's sense. In contrast, if one poses that the reunion of balls covers the space H, then d(H) represents the capacity dimension, which remains an evaluation of a fractal property [50].

For example, Feder [61] demonstrated an application of fractal geometry for the case of the measurement of the coastline of Norway. He used the following expression for the calculation of the length of the coastline, which is similar to relation (1.21): $L(\delta) = a \cdot \delta^{1-D}$ where L is the measurable distance, a is the ruler distance, δ is the radius of each of the N balls that cover the whole line, and D is the dimensionality of the curve. The dimensionality was estimated as D = 1.52, which shows that the curve dimension is strongly beyond the dimension of a straight line D = 1.

The results obtained allow an extension to an ill-defined space E. The problem consists of identifying first a 1-face component of E from k-faces ($k > 1$), which implies to identify ordered N-tuples. Components of E should be decomposed into appropriate simplexes. The number of these simplexes can be evaluated in E, such that relations (1.20) and (1.21) will be applied.

Depending on $\aleph$, non-integer d(H) can be obtained from a fractal or a fractal-like H.

1.3 THE FOUNDING ELEMENT AND THE FOUNDING LATTICE

In the absence of a preliminary postulate about the existence of the so-called "matter" and related concepts, the existence of the empty set is a necessary and sufficient condition for the existence of abstract mathematical spaces (W^n) endowed with topological dimensions (n) as great as needed [46]. Therefore, the empty set appears as a set without members though containing empty parts. The empty set owns the property of a non-well-founded set. Non-well-foundedness means suspending the axiom of foundation. The axiom of foundation says that no set is a member of itself. A non-well-founded set, i.e., hyperset, is infinitely recursive; it has no bottom. Availability of the non-well-foundedness and the expression of self-similarity at all scales and nowhere is derivable are two peculiarities that characterize fractal structures.

These findings result in additional features of physical interest. First of all the empty set is able to ensure an intellectual support in existence of some sort of space.

1.3.1 THE FOUNDING ELEMENT

In abstract algebra, a magma is a basic kind of algebraic structure. Specifically, a magma consists of a set E equipped with a single binary operation $E \times E \rightarrow E$. The binary operation must be closed by definition but no other properties are imposed. The set E above is treated as a nonempty set.

However, the notion of a magma may also be introduced for the empty set ($\varnothing$) in a very specific case.

The strong postulate that there is some set is widely accepted among mathematicians (see, e.g., Ref. [62]). The postulate was reduced to the axiom of existence of the empty set by Bounias and Bonaly [50, 47]: providing the empty set ($\varnothing$) with additional properties – the availability of an element of set membership ($\in$) and the proper subset ($\subset$), as the combination rules, which also possesses the property of complement ($\complement$), results in the definition of a magma.

This permits a consistent application of Morgan's first law without violating the axiom of foundation if the empty set is seen as a hyperset, i.e., a non-well-founded set [63, 64]. (Note that Morgan's first law states that the complement of the union of two sets is equal to the intersection of complements of the sets.) The axiom of foundation is stated as a set that contains no infinitely descending (membership) sequence, or a set that contains a (membership) minimal element, i.e., there is an element of the set that shares no member with the set [65, 66].)

Thus several inconsistencies about the empty set ($\varnothing$) properties were solved, which allowed us [54, 55] to consider the empty set as a founding element in the mathematical construction of physical space.

1.3.2 THE FOUNDING LATTICE

We may designate a magma by the notion $\varnothing^{\varnothing}$ (instead of the usual $E_{\varnothing}$) [54], because in Bourbaki notation E^F means the space of functions of set F in set E. Therefore, writing ($\varnothing^{\varnothing}$) denotes that the magma reflects the set of all self-mappings of ($\varnothing$), which emphasizes the forthcoming results. Among these results is first of all a theorem stated and proved in Ref. [54]:

The magma $\varnothing^{\varnothing}=\{\varnothing, \complement\}$ constructed with the empty hyperset ($\varnothing$) and the axiom of availability is a fractal lattice.

This theorem was proved by Bounias [54] in the following way.

Lemma 1. The space constructed with the empty set cells of $E_{\varnothing}$ is a Boolean lattice.

Proof: (i) Let $\cup(\varnothing) = \mathrm{L}$ denote a simple partition of $(\varnothing)$. Suppose that there exists an object (ε) included in a part of L, then necessarily $(\varepsilon) = \varnothing$ and it belongs to the partition; (ii) Let $\mathrm{P} = \{\varnothing, \varnothing\}$ denote a part bounded by $\sup \mathrm{P} = \mathrm{L}$ and $\inf \mathrm{P} = \{\varnothing\}$. The combination rules $\cup$ and $\cap$ provided with commutativity, associativity and absorption are holding. In effect: $\varnothing \cup \varnothing = \varnothing$, $\varnothing \cap \varnothing = \varnothing$ and thus necessarily $\varnothing \cup \varnothing(\varnothing \cap \varnothing) = \varnothing$, $\varnothing \cup \varnothing(\varnothing \cup \varnothing) = \varnothing$. Therefore, it appears that space $\{\mathrm{P}(\varnothing), (\cup, \cap)\}$ is a lattice.

The null member is $\varnothing$ and the universal member is $2^{\varnothing}$ that should be denoted by $\aleph_{\varnothing}$. Since in addition, by the founding property $\complement_{\varnothing}(\varnothing) = \varnothing$, and that the space of $(\varnothing)$ is distributive, then $\mathrm{L}(\varnothing)$ is a Boolean lattice.

Lemma 2. $\mathrm{L}(\varnothing)$ is provided with a topology of discrete space.

Proof: (i) The lattice $\mathrm{L}(\varnothing)$ owns a topology. In effect, it is stable upon union and with a intersection, and it contains $(\varnothing)$; (ii) Let $\mathrm{L}(\varnothing)$ denote a set of closed units. Two units $\varnothing_1$, $\varnothing_2$ separated by a unit $\varnothing_3$ compose a part $\{\varnothing_1, \varnothing_2, \varnothing_3\}$. Owing to the fact that the complement of a closed set is an open set: $\complement_{\{\varnothing_1, \varnothing_2, \varnothing_3\}}\{\varnothing_1, \varnothing_3\} = \varnothing_2$, then $\varnothing_2$ is an open set. Thus, by recurrence, $\{\varnothing_1, \varnothing_3\}$ are surrounded by an open $[\varnothing]$ and in parts of these open, there exist distinct neighborhoods for $(\varnothing_1)$ and $(\varnothing_3)$. The space $\mathrm{L}(\varnothing)$ is therefore Hausdorff separated. Units $(\varnothing)$ formed with these parts thus constitute a topology $(\mathcal{T}_{\varnothing})$ of discrete space.

Lemma 3. The magma of empty hyperset $\varnothing^{\varnothing}$ is endowed with self-similar ratios.

The Von Neumann notation, which is associated with the axiom of availability, when applied to $(\varnothing)$, provides the existence of sets $(N^{\varnothing})$ and $(Q^{\varnothing})$ equipotent to the natural and the rational numbers [46, 47]. Sets Q and N can thus be used for the purpose of a proof. Consider a Cartesian product $En \times En$ of a section of $(Q^{\varnothing})$ of n integers. The amplitude of the available intervals range from 0 to n, with two particular cases: interval [0, 1] and any of the minimal intervals $[1/(n-1), 1/n]$. Consider now the open section [0, 1]; it is an empty interval, denoted by $\varnothing_1$. Similarly, note

$\varnothing_{\min} = [0, 1/(n(n-1))]$. Since interval $[0, 1/(n(n-1))]$ is contained in [0, 1], then $\varnothing_{\min} \subset \varnothing_1$ is valid. Empty sets constitute the founding cells of the lattice L(∅), which is proved in Lemma 1. That is why the lattice is tessellated with cells (or balls) with homothetic-like ratios of at least $r = n(n-1)$. The absence of unfilled areas is further supported by introduction of the 'set with no parts' (see Subsection 1.6.1).

Definition: A lattice of tessellation balls is named a ***tessellattice***.

Lemma 4. The magma of an empty hyperset is a fractal tessellattice.

Proof: (i) From relations on set-distance, as a natural metric above, it follows that $(\varnothing) \cup (\varnothing) = (\varnothing, \varnothing) = (\varnothing)$; (ii) it is straightforward that $(\varnothing) \cap (\varnothing) = (\varnothing)$; (iii) the magma $(\varnothing^{\varnothing}) = \{\varnothing, \complement\}$ represents the generator of the final structure, since (∅) acts as the "initiator polygon," and complement ∁ as the rule of construction.

These three properties stand for the major features that characterize a fractal object [58].

Finally, the axiom of the existence of the empty set, added with the axiom of availability in turn provides existence to a lattice L(∅) that constitutes a discrete fractal Hausdorff space, and the proof is complete.

Thus Subsections 1.3.1 and 1.3.2 show how the notion of an empty hyperset is used in Michel Bounias reasoning about physical phenomena.

1.4 DEFINING A PROBATIONARY SPACE

The inconsistency of Lorentz covariance used by Maxwell, Einstein, Minkowski, Mach, Poincaré and others, together with Lorentz invariance used by Dirac, Wigner, Feynman, Yang and others remains unsolved [67]. Quantum mechanics is still failing to account for macroscopic phenomena including hydrodynamics. Quantum mechanics and Newton's and Einstein's theories of gravity suffer from the problem of action-at-a-distance. There are still enormous gaps in the knowledge of what the universe could really be, because current cosmological theories remain contradictive with astronomical observations (see, e.g., Refs. [68–72]). In cosmology matter is spread into an undefined vacuum and distances are postulated without reference to objects. Is space independent from matter or is matter a peculiar deformation of space [73, 74, 54–57]?

A conventional theoretical approach postulates the existence of some "corpuscles" though without describing the structure and properties of any recognized embedding medium because that is treated as forbidden [74]. The relativistic theory postulates the existence of frames of reference and the validity of some particular cases of measure; frames of reference are used as classical metric again without the existence of any embedding medium. Physics also postulates the primary existence of the parameter of time and the consistency of the possibility of motion in an undefined space. Although such a space is identified with "void" or some undetermined "vacuum," which, nevertheless, possesses some properties.

Bounias [45] introduced the notion of a probationary space, as a space fulfilling exactly the conditions required for a property to hold, in terms of: (i) identification of set components; (ii) identification of combinations of rules; (iii) identification of the reasoning system. All these components are necessary to provide the whole system with options.

Thus, pure physics cannot cope with the problem of physical space. However, mathematics can do this. It has been demonstrated above how founding mathematical principles make it possible to assess accurately the space of magmas, i.e., the sets, combination rules and structures. Further development of this formalism opens a gateway from the existence of abstract (i.e., pure mathematical) spaces to the justification of a distinction between parts of a physical space that can be said to be empty and parts that can be considered to be filled with particles.

1.5 FOUNDATIONS OF SPACE-TIME

An observed object is able to interact with the observer. To distinguish these two terms from those typically used in relativity theory, we may refer to the "perceiver" and the "perceived object." The conjecture implies that perceived objects should be topologically closed because, if the perceived objects are topologically open, a probe will not be able to reflect their shapes.

It was proved [55] that the intersection of two connected spaces with non-equal dimension is topologically closed. The closed 3D intersections of parts of an n-space with $n \geq 3$ own the properties of Poincaré sections [42]. Given a manifold of such sections, the mappings of one into another section ensures an ordered sequence of corresponding spaces in which closed topological structures are to be found and this accounts for a time-like arrow. As the Jordan curve theorem [76] states, any path connecting the interior of a closed system to an outside exterior point has a non-empty intersection with the frontier of the closed system, hence interactions between closed objects are allowed and this accounts for physical interactions. If such a path is connected to a converging sequence of mappings, the fixed points will stand for perceptions of the outside exterior. The Brouwer's fixed-point theorem (one of the most fundamental theorems of topology), states that in a closed system all continuous mappings have a fixed point; therefore the brain interprets a compact complete space in which mappings from a topological space to a discrete space are continuous and an associate set of fixed points exist representing the self [46, 47]. Accordingly, spaces of topologically closed parts account for interaction and perception and hence they meet the properties of physical spaces.

1.5.1 SPACE-TIME AS FULFILLING A NONLINEAR CONVOLUTION RELATION

The fundamental metric of the ordinary space-time can be presented by a convoluted product in which the embedding part D4 reads as follows

$$\mathrm{D4} = \int \left(\int_{d\mathrm{S}} d\vec{x} \cdot d\vec{y} \cdot d\vec{z} \right) * d\Psi(w) \tag{1.22}$$

where dS is an element of space-time and $d\Psi(w)$ is a function accounting for the extension of 3D coordinates to the fourth dimension through convolution ($*$) with the volume of space. The truth of the presentation (1.22) is proved below.

The set-distance $\Delta(A_i, A_j)$ considered above as a natural metric of topological spaces, which was initially established for two sets [44] and then generalized to manifold of sets [45], is able to clarify how two Poincaré sections are mapped. Let $\Delta(A_i, A_j, A_k, \ldots)$ be the generalized set-distance as the extended symmetric difference of a family of closed spaces:

$$\Delta(A_i)\big|_{i\in N} = \underset{\cup\{A_i\}}{\complement} \cup_{i\neq j} (A_i \cap A_j) \tag{1.23}$$

i.e., the complement of the intersection of these sets in their union. The complement of Δ, i.e., $\cup_{i \neq j}(A_i \cap A_j)$, in a closed space is closed in any case even if it involves open components with non-equal dimensions. Inside a timeless Poincaré section, the instans $m\langle\{A_i\}\rangle = \cup_{i \neq j}(A_i \cap A_j)$ characterizes the state of objects.

As distances Δ are the complements of objects, the system stands as a manifold of open and closed subparts. Mapping of these manifolds from one to another section, which preserve the topology, represents a reference frame. The topology permits one to characterize the eventual changes in the configuration of some components: if morphisms are observed, then such changes in the behavior of the components should be interpreted as a motion-like phenomenon when comparing the state of a section with the state of the mapped section (Figure 1.1).

All the spaces referred to above exist upon acceptance of the existence of the empty set as a primary axiom.

The set-distance Δ provides a set with the finer topology and the set-distance of non-identical parts and ensures a set with an ultrafilter [55]. Indeed, the set-distance is founded on a set $\{\cap \cup \ \}$ and it suffices to define a topology as union and intersection of set-distances including $\Delta(A, A) = \varnothing$. The latter case must be excluded from a filter (a special

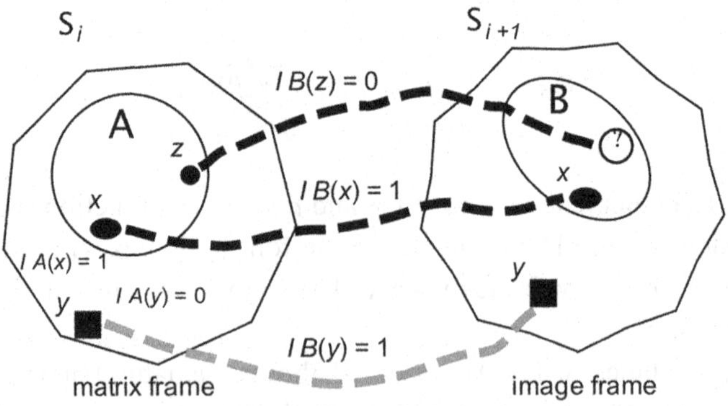

FIGURE 1.1 The mapping of the Poincaré section S_i into the section S_{i+1} places the conservation of the topologies of the general structure of the mapped spaces in such a way that the position of objects located inside these systems can be specified. A closed set is mapped into an equivalent closed set; an open set is mapped into an equivalent open set. Points $(x, y, \ldots)$ of the system with respect to the appropriate reference structures $(A, B, \ldots)$ are depicted by indicatrix functions $\mathrm{I}A(x)$, $\mathrm{I}B(x)$, …

subset of a partially ordered set), which is non-empty. Since any filter and topology are founded on a set $\{\cap \cup \in \}$, it is provided with the set-distance Δ. On the contrary, regarding a topology of a filter founded on any additional property (which can be denoted as ⊥), this property is not necessarily provided to a Δ-filter. Therefore, the topology and filter produced by Δ are, respectively, the finer topology and ultrafilter.

The mappings of both distances Δ and instans *m* from one into another section can be described by a function named the "moment of junction," because it has the global structure of a momentum. In the case of general topology of the system, the homeomorphic sequence of mappings provides a kind of reference frame, in which it becomes possible to assess the changes in the situation of points and sets of points that are present within these structures. Let points (*x*, *y*, *z*, ...) belong to either closed or open parts. For any *x* belonging to a set *E*(*i*) in a section *S*(*i*), an indicatrix function *I*(*x*) is defined by the correspondence of *x* with some *b*(*x*) in *S*(*i* + 1):

$$x \in E(i),\ I_{E(i)}(x) = 1 \text{ iff } x \in Ei;\ I_{E(i)}(x) = 0 \text{ iff } x \notin Ei$$
$$b(x) \in E(i+1),\ I_{E(i+1)}(x) = 1 \text{ iff } b(x) \in E(i+1);$$
$$I_{E(i+1)}(x) = 0 \text{ iff } b(x) \notin E(i+1) \tag{1.24}$$

Then a step function $f_{Ei,\ E(i+1)}$, which further is noted as f_E, can be introduced:

$$f_E = 1 \text{ iff: } I_{E(i)} = I_{E(i+1)};$$
$$f_E = 0 \text{ iff: } I_{E(i)} \neq I_{E(i+1)} \tag{1.25}$$

Summing over all points in the {*E*} results in the function $f_E(E)$ that describes a distribution of all indicatrix functions, which can reach 2^E. Then the expression of the proportion of points involved in the mapping of parts of *E*(*i*) into *E*(*i*+1) becomes

$$f^E(E) = f_E(E)/2^E,\ 0 < f^E(E) < 1 \tag{1.26}$$

Two species of the moment of junction are formed by means of the proportion $f_E^E(E)$ and the set-distance or the instans, because $E \supseteq \Delta(E) \cup m\langle E\rangle$, with the participation of the composition ⊥, and the distribution of points in the complement structures is not the complement of their distributions. Hence:

$$MJ_{\Delta} = \Delta(E) \perp f_E^E(E) \tag{1.27}$$

$$MJ_m = m\langle E \rangle \perp f_E^E(E) \tag{1.28}$$

and in the general case $MJ_{\Delta} \neq MJ_m$. As a composition of variables with their distribution, relations (1.27) and (1.28) actually represent a form of momentum.

The "moments of junction" (*MJ*) mapping one instans (a 3D section of the embedding 4D-space) to the next one apply to both the open objects (the distances) and their complements, which are the closed objects (the reference objects) in the embedding spaces. However, points standing for physical objects, which are able to move in a physical space, may be contained in both structures. This means that two kinds of mappings are composed with one another. Moreover, *MJ* is provided with reversibility, which accounts for the temporal reversibility of physical phenomena, which is discussed in literature [77].

The following theorem is held [55]: A space-time-like sequence of Poincaré sections is a nonlinear convolution of morphisms. The *proof* involves four steps.

(i) One kind of mapping ($\mathcal{M}$) connects a frame of reference to the next one: the same organization of the reference frame-spaces must be found in two consecutive instants of our space-time, otherwise, no change in the position of the contained objects could be correctly characterized. There may be some deformations of the sequence of reference frames on the condition that the general topology is conserved, and that each frame is homeomorphic to the previous one. Mappings ($\mathcal{M}$) will thus denote the corresponding category of morphisms.

(ii) The other kind of mapping ($\mathcal{J}$) connects the objects of one reference cell to the corresponding next one. Hence mappings ($\mathcal{J}$) behave as indicatrix functions of the situation of objects within the frames, and therefore, they are typically relevant from the condition "analysis situs" (the former name for topology, originally used by Poincaré himself). These morphisms thus belong to a complement category.

Then, each section, or timeless instant (that is a form of the above more general 'instans') of our space-time, is described by a composition (○) of these two kinds of morphisms:

$$\text{space-time instant} = \mathcal{M} \bigcirc \mathcal{J} \tag{1.29}$$

(iii) Stepping from one to the next instant is finally represented by a mapping **T**, such that the composition ($\mathcal{M} \bigcirc \mathcal{J}$) at iterate ($i$) is mapped into a composition ($\mathcal{M} \perp \mathcal{J}$) at iterate ($i + k$):

$$(\mathcal{M} \perp \mathcal{J})_{i+k} = \mathbf{T}^{\perp}(\mathcal{M} \bigcirc \mathcal{J})_i \tag{1.30}$$

So mapping ($\mathbf{T}^{\perp}$) emerges as a relation of the type $\mathcal{R}_{(i+k)}$ that maps a function F_{i+k} into F'_{j+k}:

$$F'_{j+k} = \boldsymbol{R}_{(k+j)} F_{i+k} \tag{1.31}$$

(iv) Relations (1.29)–(1.31) represents a case of the generalized convolution, namely, it is a nonlinear and multidimensional form of the convolution product, which has been first described by Bolivar-Toledo et al. [78]. We use this concept as an ideal tool for computing the behavior of visual perception.

Relation (1.31) is a form of convolution, which is obvious from the following example. Let $\alpha(j-k)$ be a particular form of $\mathcal{R}_{(i+k)}$ then Eq. (1.31) becomes:

$$F'_k = \sum \alpha(j-k) F_j \tag{1.32}$$

or for the case of an integrable space

$$F'(X) = \int a(X'-X) F'(X) d(X') \tag{1.33}$$

This relation represents a very close similarity with a distribution of functions in the sense of Schwartz [79]:

$$\langle f, \varphi \rangle = \sum \varphi(x) f(x) dx \tag{1.34}$$

or can be presented as a convolution product:

$$\int_E f(X-u)F(u)\,d(u) = (f * F)(X) \tag{1.35}$$

Thus, the connection between the abstract world of mathematical spaces and the physical universe of our observable space-time is provided by a convolution of morphisms, which supports the conjecture of relation (1.22).

1.5.2 SPACE-TIME AS A TOPOLOGICALLY DISCRETE STRUCTURE

A mathematical space can give rise to several topologies that range from coarser to finer forms, in an ordered relation [80]. In modern studies some researchers [81, 82] point to the existence of drastic topological fluctuations at Planck scales. Nevertheless, Bounias [54–56] noted that all contradictions vanish with properties of the empty hyperset providing discrete features at all scales, but also own the power of continuum that is physical "continuity" inside each fundamental cell.

Since the empty set is contained in itself, it is a non-well-founded set, or hyperset, or empty hyperset. Any parts of the empty hyperset are identical, whether a large part (∅) or a singleton {∅}; in this case the union of empty sets is also identical: $\varnothing \cup (\varnothing) \cup \{\varnothing\} \cup \{\varnothing, \{\varnothing\}\} \cup \{\varnothing, \{\varnothing\}, \{\varnothing, \{\varnothing, \{\varnothing\}\}\}\} \cup ... = \varnothing$. This is the major characteristic of a fractal structure, which means the self-similarity exists at all scales (in physical terms, from the elementary sub-atomic level to cosmic sizes). One empty set ∅ can be subdivided into two others; two empty sets generate something $(\varnothing) \cup (\varnothing)$ that is larger than the initial element. Consequently, the coefficient of similarity is $\rho \in\ ]½, 1[$. In other words, ρ realizes fragmentation when it falls within the interval]½, 1[and union when ρ is within the interval]0, ½[yields]0, 1[. The coefficient of similarity allows us to estimate the fractal dimension of the empty hyperset which, owing to the interval]0, 1[, becomes a fuzzy dimension.

In such a way time can be called nothing, because it is a singleton that does not have parts (otherwise it will be in contradiction to the definition of time as such). The nothingness singleton (∈) is absolutely unique. It is the lowest boundary of everything existing; this is the infimum of existence.

4D mathematical spaces have parts in common with 3D spaces, which yields 3D closed structures. There are then parts in common with 2D, 1D and zero dimension (points). General topology indicates the origin of time, which should be treated as an assembly of sections S_i of open sets, which are discussed above. In fact, fractality of space generates fuzzy dimensions, and hence a common part of a pair of open sets W_m and W_n with different dimensions m and n also accumulates points of the open space. If $m > n$, then those points, which belong to W_m and would not belong to the section of the given sets, cannot be included in a x-D object. The situation can be exemplified in the following way [83, 84]: "You cannot put a pot into a sheet without changing the shape of the 2D sheet into a 3D packet. Only a 2D slice of the pot can be a part of a sheet." Therefore, infinitely many slices, i.e., a new subset of sections with dimensionality from 0 to 3, ensure the raw universe in its timeless form.

In such a manner, a physical space is one that can be provided by closed intersections (timeless Poincaré sections) of abstract mathematical spaces. What happens to these sections S_i that all belong to an embedding 4D-space? A series of sections S_i, S_{i+1}, S_{i+2}, ... resembles the successive images of a movie, but nothing really moves. Therefore, the difference of distribution of objects within two corresponding sections will mean a detectable increment of time. Hence time will emerge from order relations holding on these sections.

As an example of such perception of time, a sequence of sections is shown in Figure 1.2. We can see that the volume of the central cell just exposes the ordered sequence of the Poincaré sections $\{S_i\}_i$, which the perceiver associates with time (though surrounding cells change their shapes, they preserve their own initial volume).

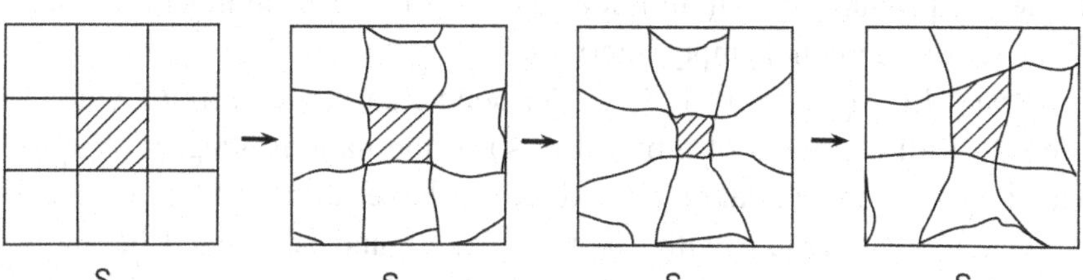

FIGURE 1.2 Sequence of sections perceived by the perceiver as time.

The moments of junction map a timeless Poincaré section representing one state of the involved spaces into another state. Each Poincaré section may present some relationship with what 't Hooft [85] called Cauchy surfaces of equal time. The moments of junction represent the interval between two successive states of a universe. Let $E(i)$ be a Poincaré section like $S(i)$ defined above: if it is an identity mapping, $MJ = \mathrm{Id}(S)$ where Id is the indicative function, then there is no time interval from $S(i)$ to $S(i+1)$. In all other cases, MJ represents two important parameters: first, it accounts for a differential time interval, and then attributes a differential element of the geometry of the corresponding space. In this sense, it has neither "thickness" nor duration. There is no "distance" in the Hausdorff sense between $S(i)$ and $S(i+1)$, but there is a change in the topological situation. Since the step from $S(i)$ to $S(i+1)$ is a discrete one, it follows that the corresponding space owns discrete/quantum properties, and these discrete properties are valid whatever the scales, as they are founded on the set difference that is not dependent on any scale nor size of the phenomena. It is noteworthy that these properties meet some requirement for space, time and matter, as suggested by 't Hooft [85].

1.6 HIERARCHY OF SCALES IN THE TESSELLATTICE

As has been demonstrated in previous sections of this chapter, the unfounded properties of the empty set provide existence to the tessellattice – a lattice that involves a tessellation of the corresponding abstract space with empty balls. In this section we continue the consideration of the subtle properties of the tessellattice [54].

Properties of the spaces whose members are empty set units exhibit very remarkable qualities. In particular, the following *lemma* is held: The Cartesian product of a finite initial section of integer numbers provides a variety of non-equal empty intervals.

Proof. Let $A_{(N+1)} = \{0, 1, 2, \ldots, N\}$ denote an initial section, which is the set of all members of a part $(M, \prec)$ of the natural integers (N) provided with an order relation $(\prec)$, which are lower than $(N+1)$. A set $\{\varnothing, a, b, \ldots\}$ is equipotent with $\{0, 1, 2, \ldots, n\}$ and is denoted below by (E_n). Then, since any set contains $\varnothing$, one has $(E_n \equiv A_{(N+1)})$. Consider now

$(E_n)^2 = (E_n) \times (E_n)$. The resulting set contains ordered pairs including the diagonal (*aa*, *bb*, *cc*, …) and the dissimilar pairs (*ab*, *ac*, *bd*, …) accounting for rational numbers ($aa \mapsto 1$, $ab \mapsto a/b$, $bb \mapsto 1$, $cd \mapsto c/d$, etc.). Since all the members of the diagonal are mapped into one, there remain $n^2 - (n - 1)$ distinct pairs. A rational number can be represented by a 2-simplex or facet, whose small sides are the corresponding integers. Jumping to 3D conditions with $(E_n)^3 = (E_n)^2 \times (E_n)$, each new rational is represented by a 3-simplex. This representation offers the advantage over the usual square to cube representation of avoiding several squares or cubes having to sharing common edges or facets. The number of 3-simplexes reflects the number $\mathbb{R}(E_n)^3$ of rational numbers available from any initial beginning segment $A_{(N+1)}$.

Between these rational numbers there should be intervals that are cuttings or segments. These segments represent all the available distances in the corresponding subpart of a 3D space. Such a subpart is involved in some part of an ordered sequence, which lies in the segment of an observable space-time.

Let us now consider a larger interval of (E_n) designating it as $[0, n]$. Denote by $m_E[0, n]$ a measure of the open part or interior $]0, n[$ of this interval in space (E_n): if E is a segment of N, one has

$$m_E[0, n] = \varnothing^N \tag{1.36}$$

This is also encountered in the Cartesian product $(E_n)^3$ when it includes $(\varnothing)$.

Consider any of the smaller intervals (σ) in $(E_n)^3$ and denote their measure by $m_E[\sigma(E_n)^3]$. By definition:

$$m_E[\sigma(E_n)^3] < m_E[0, n] \tag{1.37}$$

Consider $\varnothing^Q$ for any open or interior $]\sigma(E_n)^3[$. Then since $G \subset H \Rightarrow (g \subset H) \subset (H)$ and that the distance of the interior of a set to its frontier is naught [60], we get $\varnothing^Q \subset \varnothing^N$. This result imposes an order relation holding on empty sets that are constructed on various segments of a finite product, having finite initial sections of the set of natural integers, or that are equipotent to such a section.

An important *corollary* arises: A finite set of rational numbers inferred from a Cartesian product of a finite initial section of integer numbers establishes a discrete scale of relative sizes.

The *proof* consists of two steps.

(i) Intervals are constructed from mappings $\mathcal{G}$: $N^D \mapsto Q$ of $(N \times N \times N \times \ldots)$ in Q. For example, with D = 2, the smaller ratios available are $1/n$ and $1/(n-1)$, so that their distance is the smaller interval: $1/n(n-1)$). Consider now the few smaller intervals (σ) in $(E_n)^3$. Sizes of intervals ($\forall n > 1$) increase in the following order:

$$(\sigma_{(i)}) = 1/(n^2(n-1)) <$$
$$(\sigma_{(ii)}) = 1/(n(n-1)^2) <$$
$$(\sigma_{(iii)}) = n - 1/(n^2(n-1)^2) \tag{1.38}$$

The maximal value of the ratios of larger (n) to smaller ($1/(n^2(n-1))$) segments is

$$\max(\sigma)/\min(\sigma) = n^3(n-1) \tag{1.39}$$

One of the simplest possibilities of a scaling progression covering integer subdivisions (n) consists in dividing a fundamental segment (n = 1) by 2, then each subsegment by 3, etc. Thereby the size of structures is a function of iterations (n). At each next step (s_j) the ratio of size in dimension D is $(\prod s_j)^{\mathrm{D}}$. Then the maximal ratio ρ is defined by the Eq. (1.39):

$$\rho \propto \left\{ \left(\prod s_j\right)^{\mathrm{D}} \left(\prod s_j - 1\right) \right\}_{j=1 \to n} \tag{1.40}$$

The manifold $(\prod s_j)$ is a commutative Bourbaki-multipliable indexed on the integer section $I = [1, n]$. In practice, values can be presented as $\rho_j = a_j.10^{x_j}$, where in base 10 one takes a_j belonging to the (always existing) neighborhood of unity, i.e., $a_j \in]1[$, and consider the corresponding integer exponents x_j as the order of sizes of structures constructed from the lattice $\mathrm{L} = (\prod s_j)^{\mathrm{D}}$. Regarding extending distances (D = 1) to areas (D = 2) and volumes (D = 3), Eq. (1.40) consistently provides the predicted orders of scales listed in Table 1.1; the table represents quantum-like levels

TABLE 1.1 Scales of Sizes of Objects that Compose a Universe-Tessellattice as Prescribed by Relation (1.40)

	$\rho = \Pi_j (s_j)$		$\rho = \{\Pi_j (s_j) (\Pi_j (s_j) - 1)\}$		$\rho = \{\Pi_j (s_j)^2 (\Pi_j (s_j) - 1)\}$	
x	$(1 \pm 0.3) \times 10^X$	(s)	$(1 \pm 0.3) \times 10^X$	(s)	$(1 \pm 0.3) \times 10^X$	(s)
0	1.10^{0}	(1)	0	(1)	0	(1)
2	1.20×10^{2}	(5)				
10	0.87×10^{10}	(14)				
11					1.28×10^{11}	(7)
17	1.22×10^{17}	(19)				
21	1.12×10^{21}	(22)				
26					1.10×10^{26}	(12)
28	1.09×10^{28}	(27)				
29			1.27×10^{29}	(27)		
31	0.88×10^{31}	(29)				
33	0.87×10^{33}	(33)				
40	1.03×10^{40}	(35)			0.91×10^{40}	(16)
42			1.26×10^{42}			(22)
56	1.20×10^{56}	(45)	1.18×10^{56}	(27)		
59			0.93×10^{59}	(28)	1.33×10^{59}	(21)
61	1.24×10^{61}					
(*)						
82	0.83×10^{82}	(60)				
84					1.29×10^{84}	(27)
99			$1.12 \times 10''$			
100	1.20×10^{100}	(70)				
112			1.25×10^{112}	(45)		
115	1.13×10^{115}					
117	0.89×10^{117}	(79)				
120					1.10×10^{120}	(35)
(**)						
139					0.85×10^{139}	(39)
142	1.25×10^{142}	(92)				
150	1.00×10^{150}	(96)	*(***)*			
163	1.00×10^{163}	(103)				
171	1.24×10^{171}	(107)			0.92×10^{171}	(62)

of clusters of objects sharing successive orders of sizes having been constructed from one another.

Intervals constructed with powers of 10 on the neighborhood of unity (a]1[) are confronted through dimensions: (i) D = 1 with the simple multipliable set $\prod_j (s_j)$; (ii) D = 2 involving intervals $1/(n(n-1))$; (iii) D = 3 involving intervals $1/(n^2(n-1))$. Here the choice]0.7, 1.3[just reflects the case of a normal distribution quantile sufficiently close to unity as the mean. Notes: (*) and (**) – suggest further levels of higher scale universes; (***) a continued cluster from 10^{142} to 10^{171}, suggesting different ranking of cells of the tessellattice.

As can be seen from Table 1.1, predictable orders of size from $x = 1$ to $x = 60$ are clusters of space whose objects range from 1 (the Planck scale, i.e., the size of an elementary cell of the tessellattice), to $\sim 10^{10}$ elementary cells (roughly quark-like size), to about 10^{17} cells (atomic size), to 10^{21} cells (molecular size), to 10^{28} (human size), to 10^{40} cells (solar system size) up to 10^{56} cells (one of the largest structures, i.e., a galaxy or a cluster of galaxies, etc. up to the size of the whole universe). The universe offers quite a different organization of matter at different scales.

(ii) There is a finite number of segments or intervals, since their supremum is the number of 3-simplexes, which is finite, and contains redundant terms. This number is the number of pairwise combinations of distinct rationales. Therefore,

$$\sup(\mathcal{N}(E_n)^3) = \mathbf{C}^2_{n^2-n+1} = n(n-1)(n^2-n+1) \qquad (1.41)$$

So, there is a finite number of ratios of segment sizes that can impose upon a subpart of a space-time sequence having a limited number of relative scales for any of the objects represented by closed subspaces in $(E_n)^3$.

Corollary. The axiom of availability (stating that a rule applied to a set must be considered as explicitly applying to the set's members and parts) is necessary for an exploration of an unknown space, either mathematical or physical.

The simple *proof*. Suppose the axiom of availability is not stated: then, a complete subset of the rational numbers may not be provided in all bases by the Cartesian product of a segment of natural numbers. Let $E_n = \{1, \ldots, n\}$ with $n < 9$, and let two integers $p, q \in E_n$. Then, let the pair $(p, q) \in (E_n^2) = (E_n) \times (E_n)$. Usually (p, q) accounts for the ratio p/q, so that the set of pairs (p, q) is equipotent to the set of rational numbers noted as fractions: $(e_1.d_1d_2...d_i)$ where *e.* stands for the entire (integer) part and $d_1d_2...d_i$ for the decimal part. Let $n = 4$, and take $p = 1$, $q = 4$. Then the ordered pair (1, 4) stands for the ratio 1/4. However, writing 1/4 = 0.25 needs digit 5 to be available, whereas there are only 1, 2, 3, 4 available, not 5. Therefore, in this system, since digit 5 does not exist unless the additional axiom of the addition is introduced, the mapping of ordered pairs to the writing in base 10 of the corresponding rational numbers is not valid. The availability of the power set of parts, i.e., the infinitely iterated sets of parts of parts is enough to break this barrier.

1.6.1 TIGHT PACKING

Converging sequences of rational numbers are known to provide the set of real numbers, but this law is violated in the case of a space with finite dimension. On the contrary, infinitely descending sequences of pairs of the $\varnothing^Q$ and $\varnothing^N$ types can be found inside each part of $\{\varnothing\}$. Therefore, infinitely smaller intervals could always be found in the lower range of scales. These infinitely decreasing sizes are of a different nature in that they fill each discrete part $\{\varnothing\}$. Thus there should exist a set denoted by (¢), which has neither members nor parts [55]. Then the fundamental set of an embedding space U can be written as follows

$$F(U) = \{\cup(W)\} \cup ¢ \tag{1.42}$$

In particular, given a partition of $\cup(W)$ into W_X and W_Y, the separating distance between W_X and W_Y in $\cup(W)$ is naught iff it does not belong to the filter $\mathcal{F}$ holding on W, because $¢ \notin \mathcal{F}$:

$$\Lambda_{\cup(W)}(W_X, W_Y) = (¢) \tag{1.43}$$

The set with neither member nor parts is strictly unique. Set (¢) can be denoted as the "nothingness singleton" $\{¢\}$. Due to the uniqueness of (¢), the tessellattice is correctly tessellated since no gap can remain between any two or more of its empty tessellation balls. Besides, the set (¢) provides the tessellattice with an infimum, and thus with a partial order.

1.7 QUANTA OF FRACTALITY AND FRACTAL DECOMPOSITION

So space can be presented by the lattice $F(U) = \cup(W) \cup (¢)$, as required by relation (1.42). This lattice accounts for both relativistic space and quantum void, since: (i) the concept of distance and the concept of time have been defined on it, and (ii) this space holds for a quantum void since on the one hand, it provides a discrete topology, with quantum scales, and on the other hand, it is uniform.

The above relation (1.24) involves the mapping of a frame of reference into its image frame of reference in the next section of space-time. Such continuity represents the motion of an object in the perceived universe, which is exactly a case of "analysis situs" in the original meaning of topology used by Poincaré. Hence in the perception of a space-time continuity is provided iff the frames of references are conserved through homeomorphic mappings. This means that there is no need for exact replication; it is sufficient that the topological structures are conserved.

Therefore, the sequence of mappings of one into another structure of reference (e.g., elementary cells of the tessellattice) represents a variation of any cell volume along the arrow of physical time.

However, there is a case that precludes the conservation of homeomorphism. It occurs when a transformation of a cell involves some iterated internal similarity (Figure 1.3). Then, if $\mathcal{N}$ similar figures with similarity ratios $1/r$ are obtained, the Bouligand exponent (e) is given by an expression

$$\mathcal{N} \cdot (1/r)^{\mathrm{e}} = 1 \tag{1.44}$$

and the image cell acquires a dimensional change from d to $\mathrm{d}' = \ln(\mathcal{N}) / \ln r = \mathrm{e} > 1$.

Then, the presumptive homeomorphic part becomes deformed in such a way that the transformed cell no longer owns the property of the reference cell. This fractal transformation stands for the formation of a "solid," which can be associated with a "particled cell," or in terms of topology – more appropriately "particulate ball," since it is a new kind of topological ball,

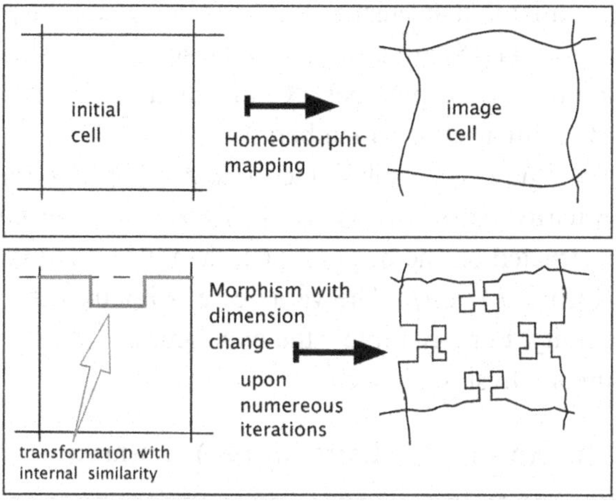

FIGURE 1.3 Continuity homeomorphic mappings (upper Figure) are broken when the deformation involves an iterated transformation with internal self-similarity (lower Figure). The mediator of transformations upon infinitely iteration is provided by empty set units Ø. In the lower Figure the first iterations are sketched, which creates the new figure in which the dimension D' > D where D is the initial dimension of a degenerate cell of the tessellattice.

$B[\varnothing, r(\varnothing)]$. Thus a particulate ball is represented by a non-homeomorphic transformation in a continuous deformation of space elementary cells.

Before examining the interaction of particulate balls with the degenerate space lattice and further with other particulate balls, it is necessary to demonstrate some mathematical preliminaries.

1.7.1 AN INITIATOR AND GENERATOR OF FRACTALITY

A minimum fractal structure is provided by a self-similar figure whose combination rule includes an initiator and a generator, and for which the similarity dimension exponent is higher than unity. Due to the self-similarity of $\geq \varnothing$, each time one considers the complement $\complement$ of itself in itself. This initiates the transformation of one ball in two identical balls, $\varnothing \mapsto \{(\varnothing), \varnothing\}$. This is continued into a sequence of {1/2, 1/4, …, l/(2n)} numbers at the nth iteration. Thus, the series

$$(I) = \sum_{(i=1\to\infty)} \{1/2^i\} \tag{1.45}$$

stands for the initiator that ensures the required iteration process. Any terms of expression (1.45) are indexed on the set of natural numbers providing infinite numbers of members. Coincidentally, 2^n also denotes the number of parts from a set of n members.

Let an initial figure (f) be subdivided into r subfigures at the first iteration. The similarity ratio is $\rho = 1/r$. If $N = r + f$ is the number of all subfigures constructed on the original one, then we obtain the Bouligand exponent $\mathrm{e} = \mathrm{Ln}(r + f)/\mathrm{Ln}\rho$. The value of e is bounded by unity if r is extended to infinity. For any finite r the exponent e is above unity, which typically occurs in physics. Hence

$$\{\min(\mathrm{e}) \mid \mathrm{e} > 1\} = \mathrm{Ln}(\max(r)+1)/\mathrm{Ln}(\max(r)) \tag{1.46}$$

These properties indicate a quantized fractality in the tessellattice.

1.7.2 DECOMPOSITION OF FRACTALS AND THE NOTION OF MASS

Let us consider a figure F. Let a fractal system Γ, which is developed on that figure, be denoted as $\Gamma = \{(\varnothing), (r + f)\}$. A more complex system is incorporated in several different subfigures that emerge in the following way. At the ith iteration the number of subfigures is $\mathcal{N}_i = (r + f)^i$ and the similarity ratio is $\rho_i = 1/r^i$. Corresponding subvolumes ($\mathcal{V}_i$) depend on the subvolumes of the previous iteration, which in the simplest case are related as $\mathcal{V}_i = \mathcal{V}_{i-1} \cdot (1/r)^3$. The total volume occupied by the subvolumes formed by the fractal iteration to infinity is the sum of the series

$$\mathcal{V}^{\text{fract}} = \sum_{(i=1 \to k)} \{(r+f)^i \cdot \mathcal{V}_{i-1} \cdot (1/r)^3\} \tag{1.47}$$

which means that a fractal decomposition consists in the distribution of the members of the set of fractal subfigures, $\Gamma \quad \mathcal{V}^{\text{fract}}$. These subfigures are constructed on the primary figure F and similar to it. But if the number of iteration k approaches infinity, then all subfigures of F become distributed and F is no longer a fractal. Such a decomposition process is depicted in Figure 1.4. Reciprocally, a fractal figure can be recomposed from an enumerable set of self-similar figures whose numbers and sizes are distributed as in relation (1.47).

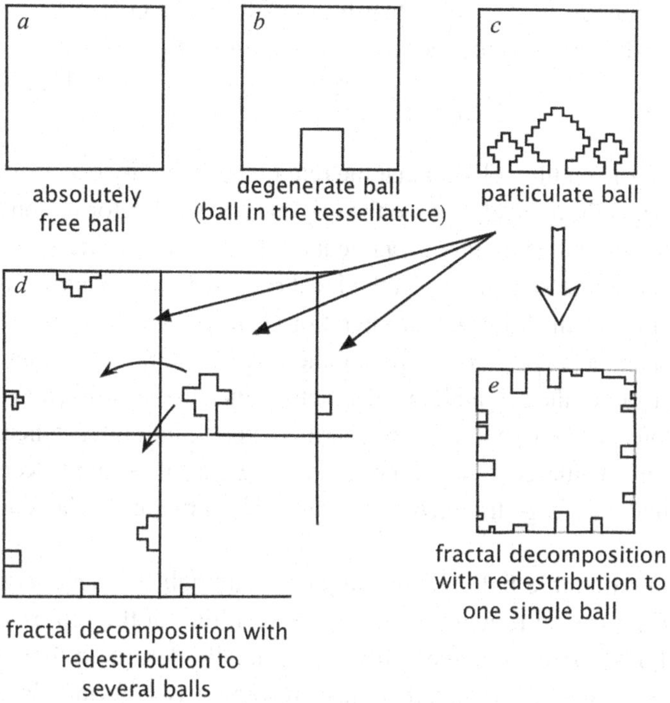

FIGURE 1.4 Topological ball is presented as a square. (*a*) – the absolutely free ball; (*b*), (*c*) – the ball inside of the tessellattice. Here, the state (*a*) corresponds to a degenerate ball; the state (*c*) shows the particled ball. (*d*) – fractal decomposition of the particled ball (*c*) with redistribution to several balls preserving fractality. (*e*) – fractal decomposition with redistribution in one single ball conserves the volume without keeping the fractal dimension, i.e., the quantum of fractality is not preserved.

The reciprocal decomposition of fractals points to the fact that the interaction of a particulate ball with surrounding cells of the tessellattice involves exchanges of structures, which resembles a kind of a reciprocal friction. Indeed, let a ball (*B*) contain a fractal subpart on it, as shown in Figures 1.3 and 1.4. Deformations can be transferred like fragments from one ball to another with conservation of the total volume of the full lattice (which is constituted by a higher scale empty set). If a fractal deformation is subjected to motion, it should collide with surrounding degenerate balls. Such collisions result in fractal decompositions at the expense of (*B*) whose exponent (e_B) decreases giving the profit to surrounding degenerate cells: if *k* is finite, one will have (e_1) > 1, (e_2) > 1, …, (e_k) > 1. The fractal decomposition gives rise to a distribution of coefficients η (e_k), whose most ordered form is a sequence of decreasing values:

$$\eta(e_k) = \{(e_i)_{(i \in]k,1])}\}$$

It follows from this relation that the remaining fractality decreases from the kernel (i.e., the area adjacent to the original particled deformation) to the edge of the distributed cloud of fragments (Figure 1.4*a*). At the edge, it can be conjectured that, depending on the local resistance of the lattice, the last decomposition (denoted as the *n*th iteration) can result in (e_n) = 1. Since in all cases one has (e_{n-1}) > 1, even though the corresponding remaining deformation is a part of the original fractal structure, then the resulting non-fractal deformations can be hypothetically distributed up to infinite distance.

Therefore, while central fragments carrying fractals exhibit decreasing higher boundaries, edge fragments are bounded by a rupture of the remaining fractality.

Thus, a particable ball becomes a particulate ball if a degenerate cell experiences a fractal reduction of volume, resulting in the representation of mass, but if the degenerate cell is non-fractally deformed then it does not experience mass. It seems that mass in some way is formed by fractal changes in an initially degenerate ball of the tessellattice. The value of mass is rather proportional to the number of iterations and a fractal completion would be only a theoretical (likely hyperbolic) limit. The number of iterations may provide an alternative kind of qualitative jump, where, for example, $n = 1$ iteration would stand for non-massive corpuscle, and $n \geq 2$ for massive particles.

A particulate ball, as described above in this subsection, provides a formalism describing the elementary particles proposed in Refs. [52, 53]. In this respect, mass is represented by a fractal reduction of volume of a ball, while just a reduction of volume as in degenerate cells is not sufficient to provide mass. Accordingly, if $\mathcal{V}^{\deg}$ is the volume of a cell in the tessellattice (the degenerate state of a ball), then the reduction of volume resulting from a fractal concavity is $\mathcal{V}^{\text{particle}} = \mathcal{V}^{\deg} - \mathcal{V}^{\text{fract}} = \Delta\mathcal{V} > 1$, or in line with expression (1.47)

$$\mathcal{V}^{\text{particle}} = \mathcal{V}^{\deg}\left(1 - \sum_i \{(r+f)^i \cdot \mathcal{V}_{i-1} \cdot (1/r)^3\}\right) \tag{1.48}$$

Since $(r+f)=(r)^{\mathrm{e}}$, expression (1.48) can be rewritten as follows

$$\mathcal{V}^{\text{particle}} = \mathcal{V}^{\deg}\left(1-\sum_{n}\left(\sum_{i}\left(\prod_{i}\left[\mathcal{V}_{i-1}^{\deg}\cdot(r_s)^{\mathrm{e}\,i}/(r_s)^{3i}\right]\right)\right)\right) \quad (1.49)$$

where in addition the sum by the s possible additional fractal concavities in the $(r+f)^i$ subfigures of the particulate ball is introduced.

Expression (1.49) relates the volume of a particulate ball to the fractal dimensional change (e), which can be expressed as the following important geometrical determination of the physical notion of mass: The mass m_B of a particled ball B is a function of the fractal-related decrease of the volume of the ball, i.e.,

$$m_B \propto \mathcal{V}^{\deg}/\mathcal{V}^{\text{particle}}\cdot(\mathrm{e}_V-1)_{\mathrm{e}_V\geq 1} \quad (1.50)$$

where (e) is the Bouligand exponent and (e_V-1) is the gain in dimensionality given by the fractal iteration; $\mathcal{V}^{\deg}$ is the volume of a degenerate ball and $\mathcal{V}^{\text{particle}}$ is the volume of the particle produced from this ball. Since $\mathcal{V}^{\deg}>\mathcal{V}^{\text{particle}}$ and $(\mathrm{e}_V-1)>0$, the right hand side of expression (1.50) is positive and greater than unity.

A decrease of volume alone is not sufficient for providing a ball with mass, since a dimensional increase is a necessary condition.

1.7.3 INCLUSIONS OF PHYSICAL PROPERTIES INTO THE TESSELLATTICE

The size of a degenerate cell in the tessellattice can be identified with the Planck length $\ell_{\mathrm{P}}=\sqrt{\hbar G/c^3}\cong 1.616\times 10^{-35}$ m. Since from a physical point of view any lattice should have elasticity, we have to ascribe this property also to the tessellattice.

The elasticity of the tessellattice favors an exchange of fragments of the fractal structure between the particulate ball and the surrounding degenerate balls, or cells of the tessellattice. Thus due to the interaction with the elastic environment, the particled ball should experience a rather periodical decomposition of its fractal state. The ball's fractal fragments spread out of the ball to a distance R_{bound}, which is the boundary for spreadable fragments, and then come back to the particled ball. It is reasonable to relate the velocity of these fragments with the fundamental velocity of the tessellattice (which preliminary we can conjugate with the speed of light $c = 3\times10^8$ m/s). The distance R_{bound} can be associated with the Compton wavelength λ_{Com} of the particled ball, which represents a radius of hardness of any canonical particle [4]. As is known, for the electron $\lambda_{Com} = 2.426\times10^{-12}$ m and for the proton $\lambda_{Com} = 1.321\times10^{-15}$ m.

The fractality of particle-giving deformations gathers its space parameters $(\varphi_i)_i$ and velocities (υ) into a self-similarity expression, which provides a space-to-time connection. Indeed, let (φ_0) and (υ_0) be the reference values. Then the similarity ratios are $\rho(\varphi) = (\varphi_i / \varphi_0)$ and $\rho(\upsilon) = (\upsilon / \upsilon_0)$; as the result

$$\rho(\varphi)^{e} + \rho(\upsilon)^{e} = 1 \tag{1.51}$$

Since $(\varphi_i)_i = \{\text{distances}\,(l), \text{and masses}\,(m)\}$, one can write $m_0 / m = l / l_0$, so that

$$(m_0 / m)^{e} + (\upsilon_0 / \upsilon)^{e} = (l / l_0)^{e} + (\upsilon_0 / \upsilon)^{e} = 1 \tag{1.52}$$

While coefficient (e) gets a value above unity, which is the requirement of fractal decomposition, the geometry outside e = 1 escapes the usual (3D+t) space-time and, owing to the previously demonstrated necessity of a embedding 4D (timeless) space (see Section 1.5 and Ref. [55]), the coefficient (e) must reach e = 2. Hence, the boundary conditions provide the following results:

$$(m_0 / m)^2 + (\upsilon_0 / \upsilon)^2 = 1 \quad \Leftrightarrow \quad m = m_0 / [1 - (\upsilon / \upsilon_0)^2]^{1/2} \tag{1.53}$$

$$(l / l_0)^2 + (\upsilon / \upsilon_0)^2 = 1 \quad \Leftrightarrow \quad l = l_0 / [1 - (\upsilon / \upsilon_0)^2]^{1/2} \tag{1.54}$$

The Lagrangian (L) should obey a similar law and (L/L_0) should fulfill relation (1.52) as a form of $\rho(\varphi)^{e}$. Then $(L / L_0)^2 + (\upsilon / \upsilon_0)^2 = 1$ and putting $L_0 = -m_0 \upsilon_0^2$ we derive

$$L = -m_0 \upsilon_0^2 [1 - (\upsilon / \upsilon_0)^2]^{1/2} \tag{1.55}$$

As can be seen, by analogy with special relativity all used parameters, namely, m_0, m, l and υ are the parameters of a moving object, while $\upsilon_0 \equiv c$ is the speed of light.

1.8 MATHEMATICAL PECULIARITIES OF BALLS IN THE TESSELLATTICE

The previous sections have demonstrated that the fundamental space is based on the empty set along with the set theory and topology extended to non-well-founded sets, as the combination of those rules are necessary and sufficient conditions for the existence of a physical-like world. Thus the physical world has to be constituted as the tessellattice, i.e., a lattice of empty elements that are balls exhibiting self-similarity and fractal properties. Elementary balls within a given range of size are attributed to a virtual volume at the free state.

These volumes are reference frames in which an operator named the "moment of junction" assesses the position of objects. The moment of junction connects one Poincaré section to the next and owns the structure of a moment. These volumes belong to the space of the topologically open distances, as they play the role of the topological complement of the space of objects, which are topologically closed. So such degenerate balls characterize the position and distances between objects, because their morphisms are homeomorphic.

On the other hand, balls exhibiting dimensional changes (through fractal shaping) no longer fulfill the condition of homeomorphism. Because of that, they have been attributed to the class of objects. In the present section we provide a rationale for the existence of fractal-changed balls and examine possible shapes of such fractal changed balls.

1.8.1 PRELIMINARIES: THE DISTRIBUTION OF VARIABLES

The distribution of variables x, y is a density function $h(x, y)$ that admits margin densities for each variable [86, 56].

$$f(x) = \int h(x, y)\,dx \text{ and } g(x) = \int h(x, y)\,dx \tag{1.56}$$

If E is a probabilized space, which is the case of the topological spaces, and the variables are continued, then the probability that x, y belong to E is:

$$P(x, y) \in E = P(E) = \iint_E h(x, y)\, dx dy \tag{1.57}$$

In the case of discrete variables the integral is replaced by a union or a sum. The repartition function (i.e., distribution function) is represented by a summation in either sense with boundaries:

$$H(x, y) = P\{(X \leq x) \cap (Y \leq y)\} \tag{1.58}$$

More generally, one may consider $x \in [u, \upsilon]$ within a domain $[a, b]$ of E. Then the probability of finding x in the closed segment $[u, v]$ is $P(u \leq x \leq \upsilon) = (\upsilon - u) / (b - a)$ if E is totally ordered. In other cases, the process is extended to $y \in [q, r]$ and so on; alternative solutions have been given in previous sections. In a discrete space probabilities are not multiplicative, while in a continuous space the repartition functions are multiplicative: $H(x, y, \ldots) = F(x) \cdot G(y) \ldots$

Now, let X, Y, ... be random objects defined in the same probabilized space, and $Z = \Phi(X, Y, \ldots)$ a real function in E. Then the moment, including the expected value, has the form

$$E(Z) = \iint_{\ldots} \Phi(X, Y, \ldots) h(x, y)\, dx dy d(\ldots) \tag{1.59}$$

Whatever the form of a distribution, it owns a family of moments m_k^c of order k and centered on c; the expected value is $E = m_1^0$, and the variance Var $= m_2^E$.

One particular case is the covariance $\mathrm{Cov}(X, Y) = E(X, Y) - E(X) \cdot E(Y)$, so that if one has $\Phi(X, Y, \ldots) = X + Y$, then

$$\mathrm{Var}(X + Y) = \mathrm{Var}(X) + \mathrm{Var}(Y) + 2\mathrm{Cov}(X, Y) \tag{1.60}$$

If variables X and Y are independent, then $\mathrm{Cov}(X, Y)$ is bounded by zero; if X and Y are completely self-similar, like in any subpart of a fractal structure, then $\mathrm{Cov}(X, Y)$ is a maximum.

In the case of the distribution $K(z)$ as the probability to get the sum $(X + Y + \ldots)$ ″z is given by the derivative of the repartition

$$P(X+Y+\ldots \le z) = \iint_{x+y+\ldots \le z} \ldots h(x,y,\ldots)\,dx\,dy\,d(\ldots) \tag{1.61}$$

The summation on one variable, e.g., y, is bounded by $z-(x+\ldots)$ and has the form

$$P(X+Y+\ldots \le z) = K(z) = \int_{-\infty}^{\infty} \ldots (f(x)) \left(\int_{-\infty}^{z-(x+\ldots)} g(y)dy \right) \tag{1.62}$$

or in terms of distribution

$$k(z) = \int_{-\infty}^{\infty} \ldots (f(x)\ldots)(g(z-(x+\ldots)))\,dx \tag{1.63}$$

which is a typical convolution function.

However, the morphisms of distances and objects (1.30) fulfill a non-linear form of generalized convolution, namely, $(\mathcal{M} \perp \mathcal{J})_{i+k} = \mathbf{T}^{\perp}(\mathcal{M} \bigcirc \mathcal{J})_i$. Here, $\mathcal{M}$ and $\mathcal{J}$ are morphisms of distances and objects, respectively; $\mathbf{T}^{\perp}$ is an operator mapping a Poincaré section (S_i) into (S_{i+1}) on the basis of the moment of junction *MJ* that is a composition function of either the set distances or their complements (the 'instans') with a distribution function. The operator $\mathbf{T}^{\perp}$ translates a composition rule (here: $\bigcirc$) into ($\perp$).

Redundancy will be considered in either active or commutation forms. The latter involves multiple convolutions of densities:

$$f^{n*} = f_1 * f_2 * \ldots f_n * \ldots * f_n \tag{1.64}$$

1.8.2 ON THE LAW DETERMINING THE SEQUENCE OF POINCARЙ SECTIONS

Let S_i be a closed intersection of topological dimension n produced by the intersection of an n-subspace with an m subspace ($m > n$) belonging to the set of parts of the embedding ω-space (W^{ω}).

A universe can be constructed in a space (W^{ω}) = $\{X, \perp\}$ where $X \in \{X^3 \cap X^4\}$ is a set and ($\perp$) is a combination rule determining the choice of S_{i+1} from S_i.

Dimensions $\omega = 4$ and $n = 3$ provide an optimal situation, in terms of mathematical organizational properties, which would place our space-time among the most efficient universe configurations [56]. Hence it is sufficient to consider $n = 3$ coming from $\omega = 4$.

Of course, there may exist as many universes as there are laws ($\perp$). However, one particular case deserves a special attention. Let S_i denote a 3D Poincaré section of W^4. Continuity in mappings of members of S_i in the sequence $\{S_i\}_i$ is favored if the successors S_{i+1} are such that

$$\{S_i\}_i = \left(\forall S_i,\ S_i \cap S_{i+1} = \max\{S_i \cap S_{i+k}\}\ k > i\right) \tag{1.65}$$

Below we will see that set distance functions possess continuity and continuity also takes place in ordered Poincaré sections of space [56].

The following definition recalls the generalized distance provided by topologies as it has been presented in Section 1.2.3. If E is a topological space $E = \{X, T\}$, and $A, B, C, \ldots, G, \ldots$ are subspaces constituted from the set of parts of set X composing E, then the separating set-distance between A and B within E is denoted by $\Lambda_E(A, B)$ and identified by

$$\Lambda_E(A, B) = \min\{(G \in E),\ (A \cap G \neq \varnothing,\ B \cap G \neq \varnothing) : \Delta(A, G) \cap \Delta(B, G)\} \tag{1.66}$$

where A denotes the simple set-distance (1.10) as the symmetric difference

$$\Delta(A, B) = \complement_{\{A \cup B\}}(A \cap B) \tag{1.67}$$

The generalized set-distance is given by the relation

$$\Delta_E(A, B) = \min\{(G \in E), (A \cap G \neq \varnothing,\ B \cap G \neq \varnothing) : \Delta(A, B, G)\} \tag{1.68}$$

with the symmetric distance

$$\Delta(A, B, G) = \complement_{\{A \cup B \cup G\}}\left((A \cap B) \cap (A \cap G)(B \cap G)\right) \tag{1.69}$$

If $G = \varnothing$, then relation (1.68) reduces to (1.66) and Δ_E reduces to Δ.

These considerations allow one to conclude that the mapping $f: \mapsto \mathbb{R}$ of the set distance (Δ) on the real numbers ($\mathbb{R}$) is continuous.

Besides, let A, B in E and $f(A)=a$, $f(B)=b$, $f(G)=g$ in $\mathbb{R}$. Then the mapping ϕ: $\Lambda_E \mapsto \mathbb{R}$ of the separating distance on the set of real numbers is continuous if a, b, g are cuttings. In the case a, b, g are initial segments, the mapping remains continuous if E is totally ordered, while if E is only partly ordered by inclusion or intersection, then the mapping ϕ is continuous for any $e < a$ or $e > b$.

Now let (S_i) be one 3D timeless section in W^4, a_i be a member or a part of (S_i) and $\Upsilon(a_i)$ is a neighborhood of (a_i) in (S_i). Call $(\underline{a}_i)_{i+k}$ and $(\underline{\Upsilon}(\underline{a}_i)_{i+k})$ the homeomorphic projections of a_i and $\Upsilon(a_i)$ on (S_{i+k}). $\Delta((S_i), (S_{i+1}))$ is minimal and that for the same reason, $\Delta((\underline{\Upsilon}(\underline{a}_i)_{i+1}, (\Upsilon(a_{i+1})))$ is minimal, which is consistent with the clause of continuity. If, in contrast, there exists a section (S_{i+h}) whose distance with (S_i) is smaller than $\Delta((S_i), (S_{i+1}))$, then the neighborhood $(\Upsilon(a_{i+1}))$ may be contained in $\Delta((S_i) \cap (S_{i+h}), (S_i))$. This depicts continuity in ordered Poincaré sections of space.

1.8.3 DISTRIBUTION OF DEFORMATIONS OF THE TESSELLATTICE BALLS

Poincaré sections $\{S_i\}_i$ are composed of distance (Δ and Λ) and objects ($m\langle\rangle$), which are open and closed, respectively. The space of distances ensures the reference frame from which the topological changes of the localization of objects are observed. This space has been shown in Section 1.3 to be basically composed of elementary cells represented by free forms $C^{(o)}$ and degenerate forms C^{deg}. A putative volume $\mathcal{V}^{(o)}$ is attributed to free balls that are devoid of any deformations and thus described by the identity mapping (Id) from (S_i) to (S_{i+1}). In contrast, in the tessellattice, degenerate cells result from homeomorphic transformations; these transformations involve some change in their volumes (in 3D sections) without dimensional alteration. Then, if $\delta\mathcal{V}^{(o)} \simeq \eta \cdot \mathcal{V}^{(o)}$ canonically denotes such a change in the volume, then $\mathcal{V}^{deg} = \mathcal{V}^{(o)} \cdot (1-\eta)$. From sections (S_i) to (S_{i+1}) one has $\delta\mathcal{V}_i^{(o)}$ mapped into $\delta\mathcal{V}_{i+1}^{(o)}$. Within each section, the set of all such deformations becomes $\cup_i \{C_i^{deg}\}_i$ and $\cup_{i+1} \{C_{i+1}^{deg}\}_{i+1}$, respectively.

The distribution of these cells within each section involve a number of variables and can be described by a multiple convolution as in relation (1.64).

Now we shall consider the behavior of the homeomorphic projection $(\underline{C}^{\deg})_{i+1}$ and the non-homeomorphic one $(C^{\deg})_{i+1}$. According to relation (1.60), one has

$$\mathrm{Var}\left((\eta_i^{\deg})+(\eta_i^{\deg})\right)_{i+1}=\mathrm{Var}\left((\underline{\eta}_i^{\deg})\right)_{i+1}+\mathrm{Var}(\eta_{i+1}^{\deg})$$
$$+2\mathrm{Cov}\left((\underline{\eta}_i^{\deg})_{i+1},\ (\eta_{i+1}^{\deg})\right) \qquad (1.70)$$

and with $N^{\deg}$ the cardinal (Card) of set $\{C_i^{\deg}\}$:

$$\mathrm{Var}\left(\cup_i(\eta_i^{\deg})\right)=\cup_i\ \mathrm{Var}(\eta_i^{\deg})+N^{\deg}\,\mathrm{Cov}\left(\{\eta_i^{\deg}\}_i\right) \qquad (1.71)$$

Gathering Eqs. (1.70) and (1.71) we obtain

$$\mathrm{Var}(\cup(\eta))=\cup\mathrm{Var}(\eta)+\mathrm{Card}(\cup)\mathrm{Cov}(\{\eta,\ \})\,; \qquad (1.72)$$

these variances are subjected to boundary conditions depending on the level of dependence or independence of $(\underline{\eta}_i^{\deg})$ and $(\eta_i^{\deg})$.

In this case if the components $\{\delta\,\mathcal{V}_i^{(\mathrm{o})}\}$ exhibit the maximum of similarity, then $\mathrm{Cov}(\{\eta,\ \})$ in Eq. (1.72) reaches $\max\mathrm{Cov}(\{\eta,\ \})$. This is achieved through the fractal properties of the lattice whose cells are self-similar balls composed with the empty hyperset $\{\varnothing^{\varnothing}\}$.

However, if the variables are totally independent, as in a completely random space, one gets $\mathrm{Cov}(\{\eta,\ \})=0$ in expression (1.72). In this case the variance of the sum is minimal, which supports the proposition claimed in relation (1.65) for the selection of (S_{i+1}) from (S_i).

Therefore, the degenerate tessellattice contains a non-denumerable infinity of sub deformations.

Indeed, the empty hyperset provides existence of an n-space, where n is as great as required, endowed with the power of continuum [47]. In itself, each empty set unit provides an empty complement, so that each unit provides a sequence of structures fitted one into the other, which can be indexed on a sequence of the $\{1/2^{ni}\}_i$ type (see Section 1.7). Thus, the distribution of volumes $\{\delta\,\mathcal{V}_i^{(\mathrm{o})}\}$ in the degenerate space contains infinitely many times the collections of deformations required for constituting a quantum of fractality. These quanta are constantly available in the topological neighborhood of a cell in (S_i), so the law of selection of the next section will select $\{S_{i+1}\}$ in (W^4), such that the same set of deformation is organized into one single structure, that is a fractal.

This consideration allows the following founding statement: The combination rule of a continued space-time-like sequence of Poincaré sections fulfilling the option (1.65) exhibits a trend to collapse random distributions of degenerate cells into massive objects.

Continuity associated with the condition of maximum intersection (1.65) favors the collapse of scattered deformations into one single aggregate forming a fractal structure. The formation of fractals results in a dimensional change of the affected cells. So fractally deformed cells are no longer homeomorphic images, and therefore, they acquire mass, in the sense defined in relation (1.51). Consequently, these cells escape the class of 'reference frame' or distances, and fall into the class of 'objects'. They become 'particulate balls' with volumes $\mathcal{V}^{\text{part}}$ as described in Section 1.7.

1.8.4 STRUCTURAL CLASSES OF PARTICABLE CELLS: LEPTONS AND QUARKS

The combination of parameters of cells denoted as $\vartheta = \{\rho,\ f, I\}$ may be called a quantum of fractality where I is the initiator that provides the iteration process (1.45), ρ the self-similarity ratio (that describes a number of subfigures r to which the object is subdivided, $\rho = 1/r$) and f the additional number of subfigures involved in the $(1/\rho)$ fragments of the initial figure.

The appropriate fractal structure denoted (Γ) can be decomposed in a sequence of elementary components $\{C_1, C_2, \ldots, C_k, \ldots\}$, see Subsection 1.7.2. If these elementary deformations, which do not represent fractal iterations, are collected in one cell, then the cell contains all the quanta of fractality, though its dimension is not changed (Figure 1.4, *e*). The cell remains non-massive as it stands and its possible motion is determined by the velocity of transfer of non-massive deformations permitted by the elasticity of the space tessellattice as a whole. Since the deformations are ordered and distributed in one particular structural block, it owns a stability through mappings of Poincaré sections. Such particles would correspond to the boson family, namely, pseudo-particles representing just transfer of packs of deformations in an isolated form.

However, if any single ball carries a group $\{\Gamma_i\}_i$ of quantum fractals, it will represent a class of massive particles. Namely, the class of leptons: electron, muon and τ-lepton (and also their appropriate antiparticles). The mass of a particle is determined by relation (1.50); therefore for leptons

$$m_\kappa = \mathcal{V}^{\deg} / \mathcal{V}_\kappa^{\text{lepton}} \cdot (\mathrm{e}_\mathcal{V} - 1)_{\mathrm{e}_\mathcal{V} \ge 1} > 1 \tag{1.73}$$

where $\kappa = 1, 2, 3$ (three leptons). The muon possesses deeper quantum fractals and a smaller volume than the electron; the τ-lepton has even deeper quantum fractals and less volume than the muon, because of the known inequality of their masses: $m_e < m_\mu < m_\tau$. The lepton-like structure (Figure 1.4, *c*) would be presented as below:

$$\ell_{i,k} = (\{\rho_i,\ f_i,\ \mathrm{e}_i\}_{i,k}) \tag{1.74}$$

A quark-like family can be represented by such common structures:

$$q_{i,k} = (\{\rho_i^{-1},\ f_i,\ \mathrm{e}_i\}_{i,k}) \tag{1.75}$$

namely, quarks have a structure opposed to leptons, which is schematically shown in Figure 1.5. Indeed, balls in the tessellattice must be at least slightly contracted (the so-called degenerate state) in comparison with an absolutely free ball. So, if in the case of leptons a particable ball is further contracted through a fractal iteration, in the case of quarks we may anticipate the opposite phenomenon: a degenerate ball straightens its contracted state at least partly by means of a fractal iteration. Hence, for quarks the following relationship holds

$$\mathcal{V}^{\deg} / \mathcal{V}_\beta^{\text{quark}} \cdot (\mathrm{e}_\mathcal{V} - 1)_{\mathrm{e}_\mathcal{V} \ge 1} < 1 \tag{1.76}$$

where $\beta = 1, 2, \ldots, 6$ (six quarks).

The inequality (1.76) shows that quarks do not have mass in the sense of lepton masses (expressions (1.73) and (1.51)): leptons are fractal volumetrically contracted objects, but quarks are fractal volumetrically inflated

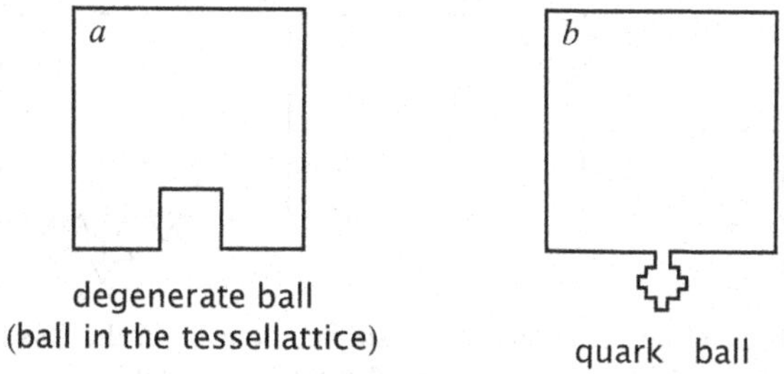

FIGURE 1.5 Ball in the degenerate tessellattice (*a*) and the same ball that is partly straightened under a fractal volumetric iteration, which results in the formation of the quark (*b*). The volume $\delta \mathcal{V}^{\text{qurk}}$[Figure (*b*)] added to the ball due to the fractal inflation shall not exceed the volume $\delta \mathcal{V}^{\text{deg}}$ [Figure (*a*)], which each free ball lost when combining tessellattice.

objects. That is why these two classes of particles exhibit completely different mechanics in the tessellattice, which will be considered in the next chapters. However, in any case the dimensionality of a particulate ball is above 3D and approaching 4D.

One more peculiarity of an elementary particle is the electric charge. In considering the volumetric characteristics of cells of the tessellattice, we have concluded that a particulate ball should obey some convexity/concavity trends and symmetry properties. Would it be associated with the notion of the charge?

1.8.5 THE SURFACE OF A PARTICULATE BALL

Can the surface of a particulate ball, i.e., a cell that has experienced volumetric fractal iterations, be additionally processed with fractals? If such a processing takes place, then the surface will change the dimensionality from 2D to a value 2D' that satisfies inequalities 2D < 2D' < 3D. Figure 1.6*a* depicts a pseudo-neutral interface between two adjacent cells, which have both cavities and protuberances. These concavities and convexities can be associated with the origin of the negative and positive charge of a lepton or quark.

Thus, the electric charge can appear as the quantum surface fractal of a particulate ball, i.e., when the surface has undergone fractal iterations. Figures 1.6*b* and 1.6*c* illustrate the fractal deformation of the surface of the ball; the

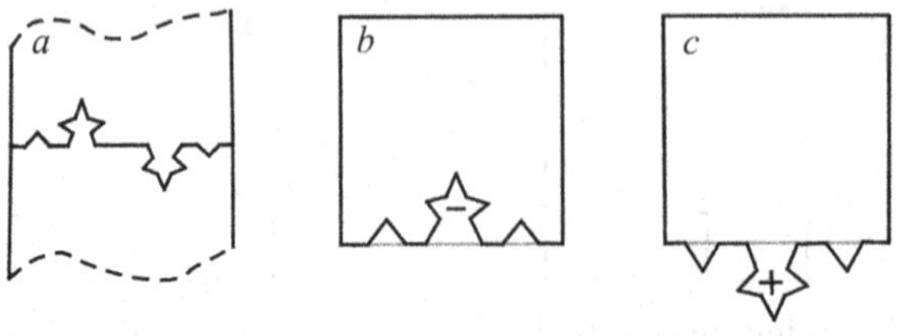

FIGURE 1.6 Fractal deformations of the surface of cells in the tessellattice. *a* – the interface between adjacent cells that essay the surface fractal iteration; *b*, *c* – electric charge "–" and "+", respectively, as the quantum surface fractal of a particulate ball.

two forms of the electric charge are shown – negative (the fractal iteration is orientated inside the particle) and positive (the fractal iteration is orientated outside the particle). Thereby the positive charge of a canonical particle is represented by the particle's surface that is covered with spikes projecting away from the particle; respectively, the source of the negative charge is the particle's surface covered with spikes projecting into the particle.

We anticipate that a canonical particle, i.e., a lepton or quark, has the shape of a sphere, which is logical. Then fractal images of iteration functions on the Riemann sphere can be the fractal image of an expression that is the stereographic projection of a point expressed in spherical coordinates onto the plane.

It should be noted that such a fractal sphere was described by Barnsley [87] as follows. The technique used to generate the Riemann sphere is to map each point to a point on the surface where a line connecting the north pole to the plane point intersects the sphere, and we choose the angle ϕ between this line and the diameter of the sphere connecting the poles to generate a latitude coordinate. For a longitude coordinate we take the polar coordinate angle of the plane point with respect to the positive real line. Let the point z in the plane be expressed as $z = re^{i\theta}$, which makes it possible to choose θ. Now the coordinate transformation for ϕ is needed. The line from the north pole to the point z forms the hypotenuse of a right triangle, the side adjacent is the radius of the sphere to the north pole, and the side opposite is the plane vector to z. Consequently, $r = |z| = \tan\phi$. The coordinate transformation can be expressed as

$$z = z_1 + iz_2 = \tan\phi\,\cos\theta + i\tan\phi\,\sin\theta = \tan\left(\phi\, e^{i\theta}\right) \qquad (1.77)$$

How do these coordinates change when one rotates the sphere? As with the equator, the θ coordinate becomes the negative angle with the positive real line, $\theta \mapsto -\theta$. For ϕ we need some plane geometry. If we have a diameter of a circle in the plane, and a point that is not on the diameter, then the triangle formed by the chords between the endpoints and the third point is a right triangle. Then we take the same planar circle, and draw the lines connecting a point on the perpendicular bisector of the diameter to the endpoints. Drawing a line from one of the ends of the diameter to the point at which the line from the other end to the bisector intersects the circle, the triangle so formed is similar to the two triangles formed by the lines previously drawn to the bisector. The tangent of the angle from the north pole to a point on the sphere is $\tan\phi$ and the tangent of the angle formed by the south pole is $\cot\phi$.

When one rotates the Riemann sphere around the real line, one has exactly this situation: the angle ϕ of the original point on the sphere is now the angle measured from the south pole. Consequently, the tangent of the north pole is $\cot\phi$. Now looking at the action of $f(z)$ on $z = \tan(\phi e^{i\theta})$, we have $f(z) = \cot(\phi e^{i(-\theta)})$. So the spherical coordinates are transformed in the following way: $\theta \mapsto -\theta$ and $\tan\theta \mapsto \cot\theta$. As a result, the corresponding points on the sphere are expressly distributed, which is shown in Figure 1.7. The number of points inside and outside of the unit circle (i.e., on the inner and outer surface of the same sphere) are the same.

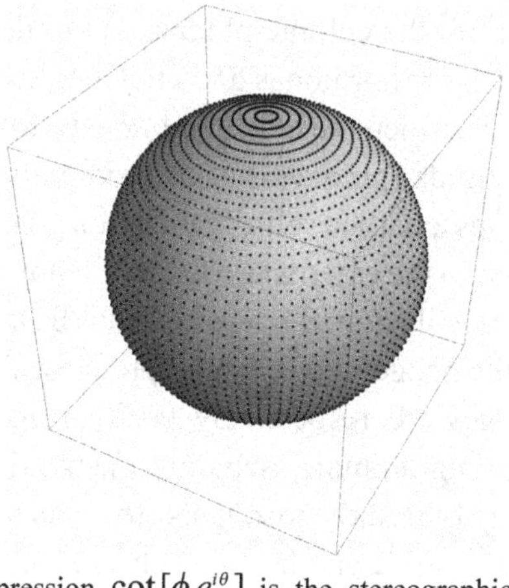

FIGURE 1.7 The expression $\cot[\phi e^{i\theta}]$ is the stereographic projection of a point expressed in spherical coordinates onto the plane. At the calculation of the fractal sphere, i.e., point interaction functions on the sphere, the algorithm [88] for the *Mathematica* software has been used (where the angle is divided into 48 steps).

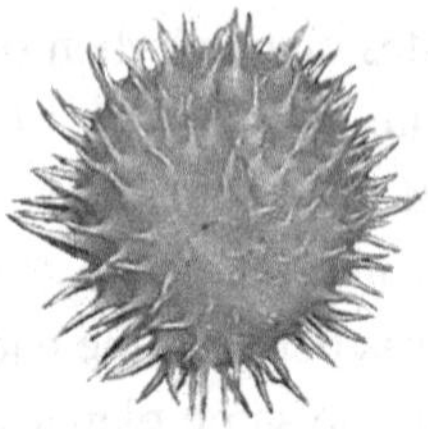

FIGURE 1.8 The surface of the horse-chestnut fruit is a macroscopic manifestation of the elementary positive charge in the plant world. In the animal world a similar shape is seen in a hedgehog and among fish – a hedgehog-fish.

Figure 1.8 depicts one of the manifestations of the electric charge, as the surface phenomenon, in the macroscopic living world. This is an expression of the property of the tessellattice that allows self-organization at different scales, which in general aspects must repeat each other. In fact, Table 1.1 predicts clusters of space at a set of scales starting from the Planck length and continuing up to the size of the universe. Thus, we encounter similar manifestations of this positivity and negativity in the biological world. The boundary of the nutrition system of a living organism appears as a kind of a charge structure as well: in flora it is a root (positive charge) and in fauna it is a stomach (negative charge). The topology of leaves in plants may also be treated as negativity, because leaves synthesize food by photosynthesis. At the cellular level this is most clearly seen in the exterior membrane (the plasma membrane) of the cell that either: (i) represents positivity in the form of protrusions from the cell membrane, as motile cilia/flagella, or as secreted vesicles that carry hormones and enzymes away from the cell, or (ii) represents negativity when the cell membrane in-folds to form vacuoles containing absorbed fluids (pinocytosis) or food particles (phagocytosis). These expressions of tessellattice self-organization continue to be seen, like a typical charge stutter, in the external surfaces of many multicellular organisms, for example in the illustrated horse-chestnut fruit and species of cacti (amongst flora) and the hedgehog fish and reptilian scales (amongst fauna). Male and female sexes are respectively positive and negative charges. In the social area among animals including mankind we can distinguish vowel sounds (coming out) and consonants (that sound inside), which can also be considered as positive and negative charges, correspondingly.

Here are some examples: (i) *Paramecium* is a common single-celled organism that swims using its beating cilia and ingests bacteria (food) into food vacuoles, where they are digested and absorbed. (ii) A living cell secretes enzymes out of the cell. (iii) *Amoebae* feeds by phagocytosis (in the same manner as white blood cells). (iv) A living cell membrane infolds to take fluids into the body of the cell.

So we can now present a rough portrait of both a lepton and quark. In Figure 1.9 we can see a ball in a completely free state (somewhere outside of our universe), or maybe in the interface between a set of universes, which is basically allowed in the tessellattice through its spontaneous clustering not only on a small scale, but also at the cosmic size (see expression (1.40) and Table 1.1). A ball in the universe being in the degenerate state among similar degenerate balls has to have a smaller size, $d_{\text{in universe}}$, because it must allow not only a fractal contraction but also a fractal expansion. The fractal contracted state, which is specified with the loss of volume, we associate with the creation of a lepton. In line with the definition (1.51) and (1.73), the stronger the contraction, the heavier a lepton, i.e., sequentially: electron, muon and τ-lepton.

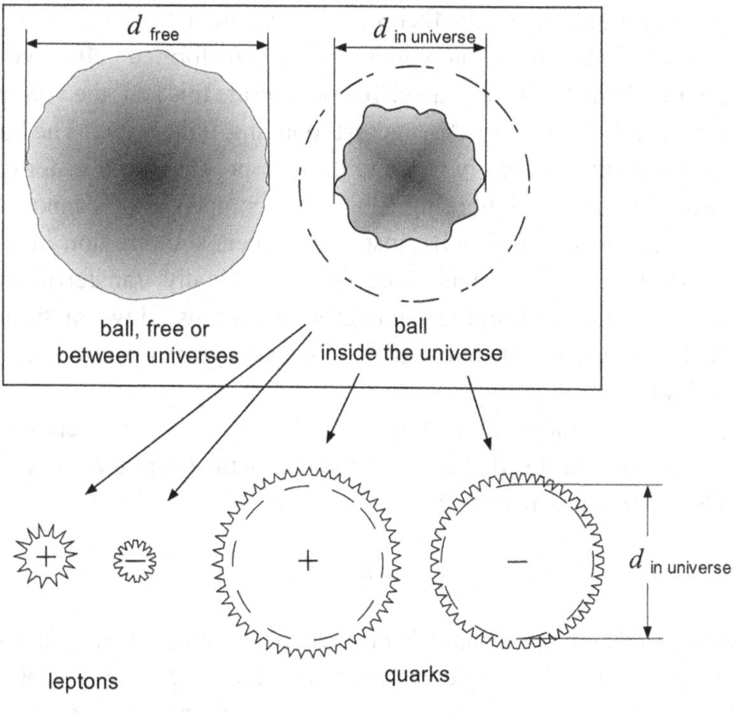

FIGURE 1.9 Topological ball, a primary entity of Nature, which in the physics language can be called 'superparticle.' Due to fractal deformations the ball shrinks to the state of lepton or quark. Positive and negative charged particles are shown.

The fractal-inflated state, which is characterized by an increase in volume of the appropriate particulate ball, is associated with the creation of a quark. The family of quarks includes 6 flavors whose fractal volume discretely grows from the smallest to the largest (namely, $d_{\text{in universe}}$ approaches d_{free}): up, down, strange, charm, bottom and top.

Leptons and quarks are characterized by the electric charge, plus or minus, which is given by the quantum of surface fractality of the appropriate particulate ball.

1.8.6 THE LATTICE DISTORTION AROUND A PARTICULATE BALL

A cell in the tessellattice, which underwent fractal transformations to the state of a particle, must induce some geometric changes to surrounding cells just to compensate lost volumes. While the area of the iterated self-similar transformation may be theoretically infinite, its volume in 3D sections is not infinite. Hence the volume of the transformed cell is reduced owing to iterations to a finite value. To compensate this lost, the volume of surrounding cells is increased by a corresponding finite value. The homogeneity of the lattice is partly restored through progressive transfer of the additional volume to neighboring cells. Such compensation cannot occur only in a few surrounding cells, instead it covers some region in which cells reside in an intermediate state that is gradually transferring from semi-fractal to homeomorphic volumetric alterations. How far from the particle, i.e., its particulate ball, can the distortion spread into the degenerate tessellattice?

Experimental studies give us the unambiguous answer: such a deformation coat has a radius that is characterized by the Compton wavelength λ_{Com}. The Compton wavelength

$$\lambda_{\text{Com}} = h/(mc) \tag{1.78}$$

is treated as a quantum mechanical property of a particle; it was introduced by Compton [4] in his explanation of the scattering of X-ray photons by electrons (a process known as Compton scattering). For the electron $\lambda_{\text{Com}} = 2.4263102367(11) \times 10^{-12}$ m [89].

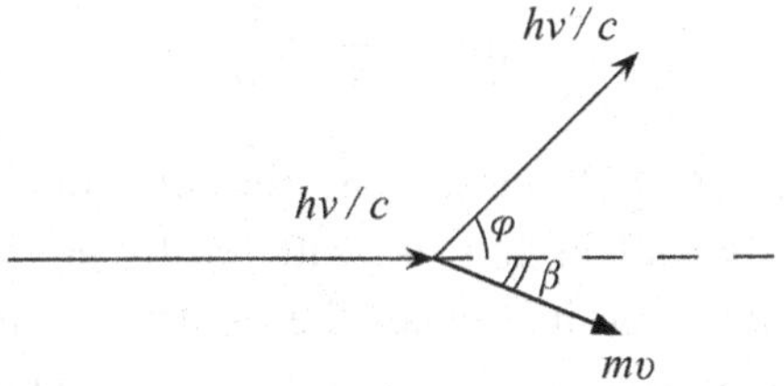

FIGURE 1.10 Compton scattering.

The theoretical consideration of collisions of quanta of light with electrons was remarkably presented by Born [90] (Figure 1.10). The equation characterizing the law of energy conservation and two appropriate equations for the law of the momentum conservation are

$$h\nu + m_0 c^2 = h\nu' + m c^2 \tag{1.79}$$

$$h\nu / c = h\nu' / c \cdot \cos\varphi + m\upsilon \cos\beta \tag{1.80}$$

$$0 = h\nu' / c \cdot \sin\varphi - m\upsilon \sin\beta \tag{1.81}$$

Here, ν is the frequency of quantum of light, m_0 and $m = m_0 / \sqrt{1-\upsilon^2/c^2}$ are the electron's rest mass and the total mass, respectively; φ is the angle of deviation of the quantum of light after the scattering with the electron and β is the angle of deviation of the electron after collision with the quantum of light. The final expression, Compton's formula for the change of the wavelength of the quantum of light due to the collision with the electron, can easily be derived from Eqs. (1.79)–(1.81):

$$\Delta\lambda = \lambda' - \lambda = c \cdot \left(\frac{1}{\nu'} - \frac{1}{\nu} \right) = (1 - \cos\varphi) \frac{h}{m_0 c} \tag{1.82}$$

where in the right hand side we can see the Compton wavelength of the electron. This length, λ_{Com}, is not formal. We can rewrite equations (1.79)–(1.81) in the form that explicitly shows how spatial intervals characterizing the scattering objects behave,

$$\frac{1}{\lambda} + \frac{1}{\lambda_{\text{Com}}} = \frac{1}{\lambda} + \frac{1}{\lambda_{\text{Com}}} \frac{1}{\sqrt{1-\upsilon^2/c^2}} \tag{1.83}$$

$$\frac{1}{\lambda} = \frac{1}{\lambda'} \cos\varphi + \frac{1}{\lambda_{\text{Com}}} \frac{\upsilon}{c} \cos\beta \tag{1.84}$$

$$0 = \frac{1}{\lambda'}\sin\varphi - \frac{1}{\lambda_{\text{Com}}}\frac{\upsilon}{c}\sin\beta \tag{1.85}$$

Equations (1.83)–(1.85) show that the quantum of light, which is described by the wavelength λ, is scattered by an object that has the characteristic size λ_{Com} that influences the light, such that its wavelength changes. Hence, the object's wavelength λ_{Com} is much more rigid, than the wavelength λ of the running quantum of light. Eqs. (1.83) to (1.85) result in the same Compton's expression (1.82).

Thus, canonical particles possess an actual radius of hardness, which is determined by the Compton wavelength (1.78). Notwithstanding this, in orthodox quantum mechanics the size of a canonical particle does not play a part in the theory and the same takes place in high-energy physics where a particle is regarded as point-like.

A smaller or "reduced" Compton wavelength $\lambda\!\!\!^{-}_{\text{Com}} = \lambda_{\text{Com}} / (2\pi)$ naturally represents the mass of a particle on the quantum scale, because it manifests itself as an alternate in the major fundamental equations of quantum mechanics. In fact the Schrödinger equation of an electron in a hydrogen-like atom can be rewritten by using this parameter:

$$\frac{i}{c}\frac{\partial}{\partial t}\psi = -\tfrac{1}{2}\lambda\!\!\!^{-}_{\text{Com}}\nabla^2\psi - \frac{\alpha Z}{r}\psi \tag{1.86}$$

where the dimensionless value $\alpha \approx 1/137$ is the fine structure constant.

The reduced Compton wavelength appears also in the Dirac and Klein–Gordon equation for a free particle, respectively:

$$i\gamma^{\mu}\partial_{\mu}\psi + \lambda\!\!\!^{-}_{\text{Com}}\psi = 0 \tag{1.87}$$

$$\nabla^2\psi - \frac{1}{c^2}\frac{\partial^2}{\partial t^2}\psi = \lambda\!\!\!^{-2}_{\text{Com}}\psi \tag{1.88}$$

Not only is the Compton wavelength manifested in experiments studying light scattering by charged particles. Thomson [91] scattering at which a free charged particle elastically scatters electromagnetic radiation, as described by classical electromagnetism, exhibits the classical radius of the electron r_e. The Thomson scattering can be considered as the low-energy limit of Compton scattering: the particle kinetic energy and photon frequency are the same before and after the scattering.

The Thompson scattering cross-section is

$$\sigma_{\text{Thom}} = 8\pi r_e^2 / 3 \tag{1.89}$$

where the classical electron radius reads

$$r_e = \frac{e^2}{4\pi\varepsilon_0 m_0 c^2} = 2.8179403227(19)\times10^{-15}\ \text{m [89]} \tag{1.90}$$

The cross-section (1.89) was verified experimentally. It directly demonstrates that the classical electron radius (1.90) is also a real parameter of the electron. Nevertheless, neither the Compton wavelength, nor the classical electron radius is included in any modern model of elementary particles; these notions do not appear in the Standard Model of particle physics. Is that not strange, dear researchers of high-energy physics?

For a charged particle considered as a particulate ball being in the tessellattice, the appearance of two characteristic lengths, λ_{Com} and r_e, are quite natural. λ_{Com} determines the scale of the deformation coat that compensates the lost volume of the particulate ball due to its volumetric fractal iterations. r_e defines the scale of the electrically polarized part of the deformation coat, which is responsible for the compensation of the loss of area of the particulate ball owing to its surface fractal iterations. Figure 1.11 depicts the particle surrounded with its deformation coat that in addition is also electrically polarized.

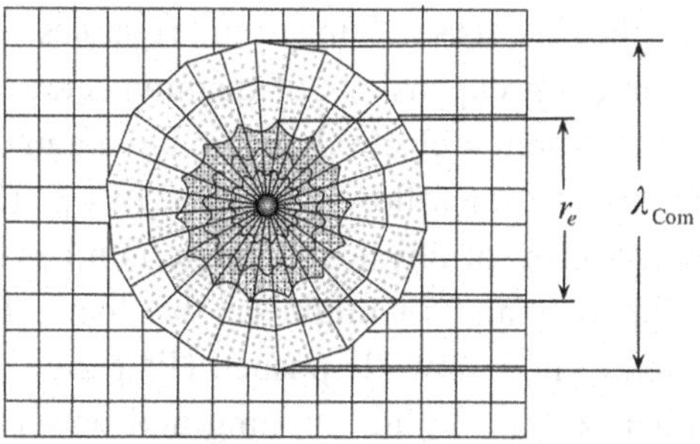

FIGURE 1.11 Deformation coat around the particled ball (a lepton) in the tessellattice. Two kinds of coats, which are superimposed, are shown: one is associated with the compensation of the lost volume by the particled ball (its size is specified by the particle's Compton wavelength λ_{Com}) and the second owing to the compensation of the lost area of the surface of the particled ball (its size is defined by the particle's classical radius, which for the electron is the classical electron radius r_e).

1.8.7 THE ORIGIN OF THE FINE STRUCTURE CONSTANT

Sommerfeld [92] introduced the fine structure constant α as the ratio of the speed of the electron in the first orbit of the Bohr atom to the speed of light. He used this dimensionless parameter to calculate the fine splitting of the spectral lines of hydrogen atom.

Many major characteristics in atomic physics also include the fine structure constant. The Thompson scattering cross-section (1.89) can be rewritten in terms of the Compton wavelength and the fine structure constant, $\sigma_{\text{Thom}} = 8\pi / 3 \cdot (\alpha \lambda\!\!\!^{-}_{\text{Com}})^2$. Besides, the Bohr radius is also related to the electron's reduced Compton wavelength and the fine structure constant, $a_0 = \lambda\!\!\!^{-}_{\text{Com},\, e} / \alpha$, as well as the Rydberg constant, $R_\infty = \alpha^2 / (4\pi \lambda\!\!\!^{-}_{\text{Com},\, e})$. A review article on the study of the origin of α was written by Kragh [93].

In quantum theories the fine structure constant is considered as a value that conglomerates the electron charge e, the Planck constant $\hbar$, and the speed of light c: $\alpha = e^2 / (\hbar c)$ in electrostatic cgs units or $\alpha = e^2 / (4\pi \varepsilon_0 \hbar c)$ in SI units. The dimensionless quantity $\alpha \simeq 1/137$ has the same numerical value in all systems of units. With increase in energy of interacting charged particles, when they approach closer than $2r_e$, the value of α increases. It is believed that this value cannot be predicted theoretically and is based only on experimental data.

In quantum electrodynamics and quantum field theory the fine structure constant has the value of the interaction constant that characterizes the strength of the electromagnetic interaction, or the coupling constant, between electric charges and photons. Its small magnitude enables very accurate predictions in the perturbation expansions of quantum electrodynamics; α is used in the analysis of quantum electrodynamic Feynman diagrams.

Feynman wrote about α as follows (see Ref. [5], p. 129): "It's one of the greatest damn mysteries of physics: a magic number that comes to us with no understanding by man. You might say the 'hand of God' wrote that number, and 'we don't know how He pushed His pencil'." Studies of the origin of the number $\alpha \simeq 1/137$ are ongoing and new researchers bring their new ideas and hypotheses regarding its origin, which already looks like some kind of a science of alphaology (α is considered as a unification of first prime numbers, a unification of next prime numbers incorporated in some formulas with the number π, or an incorporation of some numbers with the dimensionless constant of dynamic chaos, etc.).

In the past Lunn [94] expressed an interesting suggestion; he connected the fine structure constant with the nuclear mass defect, and considered its possible relationship with gravity through the relation $Gm^2 / e^2 = \alpha^{17} / (2048\pi^6)$ where G is Newton's gravitational constant. Recently Oldershaw [95, 96] has taken the next step – he unifies the electric charge with the Planck mass through the constant α. Namely, since the Planck mass $m_P = \sqrt{\hbar c / G}$ and $\alpha = e^2 / (4\pi\varepsilon_0 \hbar c)$, he combines these relations to obtain:

$$\alpha = \frac{e^2 / (4\pi\varepsilon_0)}{Gm_P^2} \tag{1.91}$$

Relation (1.91) shows that the fine structure constant represents the ratio of the strengths of the fundamental unit of electric and gravitational interactions.

Oldershaw's result (1.91) is very interesting with a geometric perspective: α describes the ratio between dimensions of surface and volumetric fractals located on the particulate ball, because above we have related, respectively, mass with the volume and charge with the surface of the particulate ball studied. In other words, α determines the difference in convolutions of the particle's surface dimension $2\text{D} + \delta(2\text{D})$ and its volume dimension $3\text{D} + \delta(3\text{D})$; that is, α shows how $\delta(2\text{D})$ is different from $\delta(3\text{D})$:

$$\alpha = \delta(2\text{D}) / \delta(3\text{D}) \tag{1.92}$$

The spatial polyhedrons (Figure 1.11) can be characterized by the radii of the appropriate described spheres. These radii are the classical electron radius r_e and the Compton wavelength λ_{Com}, which set the numerical value of the parameter (1.92); namely, complementing the ratio r_e/λ_{Com} by a factor 2π, we obtain the explicit definition of the fine structure constant:

$$\alpha \equiv 2\pi \cdot \frac{r_e}{\lambda_{\text{Com}}} = \frac{r_e}{\lambda\!\!\!^{-}_{\text{Com}}} = 7.2973525662(16) \times 10^{-3} \tag{1.93}$$

Recent experimental data confirmed $\alpha = 7.2973525664(17) \times 10^{-3}$ [89]. Therefore the expression (1.93) should be treated as the natural definition of the fine structure constant. It should be noted that Michaud [97] previously already demonstrated that the radius r_e is obtained by multiplication of λ_{Com} and α divided by 2π.

We have considered volumetric (3D $\rightarrow$ 3D + δ(3D)) and surface (2D $\rightarrow$ 2D + δ(2D)) fractal changes of a ball, which are associated with the emergence of physical entities in the ball – the particle mass and the particle charge, respectively.

One more kind of fractal iteration has not been examined yet: 1D $\rightarrow$ 1D + δ(1D). It is obvious that it is the vector velocity of a particle, which sets 1D-structure in the system associated with the quantum system studied. During motion the velocity vector will be disintegrated coming on to excitations generated by the particle. In such a manner 1D will be increased to 1D + δ(1D).

The fine structure constant introduced in relations (1.92) and (1.93), which represents local changes in space caused by different topologies – volumetric and surface – in the deformation coat of a particle, must affect the flat space-time interval $ds^2 = c^2 dt^2 - d\vec{r}^{\,2}$ on the atomic scale changing it to the form $ds^2 = \alpha^2 c^2 dt^2 - d\vec{r}^{\,2}$. From this new equation in the limit $ds \rightarrow 0$ we get the relation $\alpha c = \upsilon$ where υ is the velocity of the particle. This relation for the space-time interval was suggested by Shpenkov [98], though the introduction of the ratio $\upsilon / c = \alpha$ was already done by Sommerfeld [92] a hundred years ago.

1.8.8 QUASI-PARTICLES OF THE TESSELLATTICE

When a particulate ball, i.e., a canonical particle, starts to move, it begins to contact new and new oncoming cells of the tessellattice (Figure 1.12). Such contacts activate the process of distribution of fractal fragments from the particulate ball to nearest cells, which is schematically shown in Figures 1.4*c* and 1.4*d*. Besides, the passing of fractal fragments from the particle to cells of the tessellattice is accompanied with the transfer of the particle's velocity. These fractal fragments are able to hop from cell to cell by a relay mechanism. Such excitations of the tessellattice can be associated with physical notions of quasi-particles, or field particles.

If the particle is neutral (no charge), then these excitations carry only volumetric fractal fragments, i.e., in terms of physics, the excitations carry fragments of the particle mass.

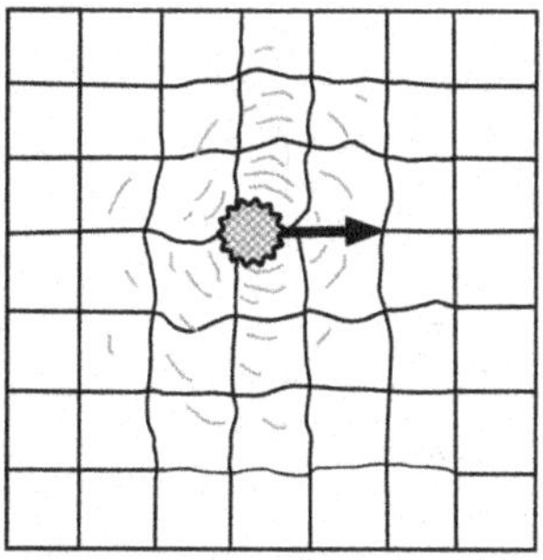

FIGURE 1.12 Interaction of a moving particulate ball with surrounding cells of the tessellattice.

These excitations were named 'inertons' by the author [52, 53], as they behave as carriers of the force of *inertia*, which by general definition is the resistance of any physical object on the side of space to any change in its state of motion (i.e., inertia is the result of the interaction of the moving object with the ordinary physical space). Inertons are very widespread in nature. The study of the motion of a particle and its cloud of inertons is the subject of the next chapters.

If a moving particle is charged, i.e., includes also surface fractals, it should be accompanied by additional excitations. Each cell can take on itself two different kinds of fractals – volumetric and surface, which can coexist simultaneously in the same cell because the particulate ball is surrounded with the deformation coat in which volumetric and fractal distortions of cells can interfere up to the distance of r_e (Figure 1.11). Hence the moving particle has also to be accompanied with excitations of space, which collect both volumetric and surface fractal fragments. Surface fractal excitations are nothing but photons and they represent the polar state of a cell; such a state migrates in space by hopping from cell to cell. But photons can escape from the particle's cloud and in such a case they become free photons whose behavior is described by conventional electrodynamics; free photons move in the tessellattice by a relay mechanism hopping from cell to cell with the speed of light c, which is the fundamental velocity of the tessellattice. Figure 1.13 demonstrates a degenerate cell of space and the major excitations of the tessellattice: a photon (Figure 1.13*a*), inerton (Figure 1.13*d*) and a mixed quasi-particle, an inerton-photon (Figure

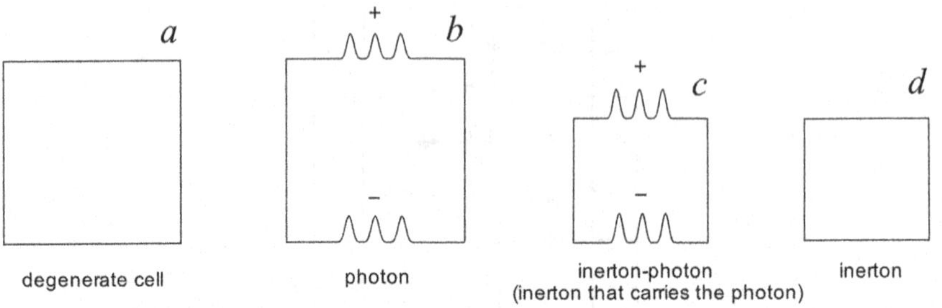

FIGURE 1.13 Degenerate cell (*a*) of the tessellattice and its possible excitations: *b* – the photon as the cell with a polarized surface; it is characterized by pure surface fractal fragments – convexities and concavities; *c* – the photon formed on the inerton, i.e., this is the inerton-photon that is characterized by both surface and volumetric fractal fragments (this excitation has a rest mass and the appropriate cell is more deformed than is the case *b*); *d* – the inerton that is specified with a volumetric fractal contraction of the kernel cell (this excitation carries a fragment of mass).

1.13*c*) that represents a substrate of the matter wave of any moving charged particle, such as an electron, proton, etc.

The third fundamental excitation is an excitation that emerges at the motion of a quark. Such kinds of excitations are analogous to inertons, but represented by inverted volumetric fractals. They were named 'qinertons' (inertons of quarks) [99]. A whole cloud of qinertons might be identified with a gluon of quantum chromodynamics. However, qinertons are excitations that can exist only in the interior of hadrons. The behavior of quarks and the structure of hadrons are discussed in Chapter 6.

Although spatial excitations – an inerton, qinerton and photon – they move by hopping from cell to cell of the tessellattice; when migrating each of them is not entirely localized in one cell of the tessellattice. A kernel cell of the excitation is recognized as the Planck length ~10^{-35} m, but the fractal perturbation vanishes only at some boundary distance r_{bound} that might be put ~ 10^{-31} m.

We agreed above that the Planck length $\ell_{\text{P}} \approx 10^{-35}$ m defines the size of the particle itself, i.e., its kernel cell. Studies in high-energy physics point out that at the distance around 10^{-31} m all fundamental forces should be unified (Figure 1.14) and the major concept regarding such a conclusion has not changed for at least 25 last years [100–102]. So the scale from 10^{-35} to 10^{-31} m may be treated as a transition region around a cell deformed with fractal iterations.

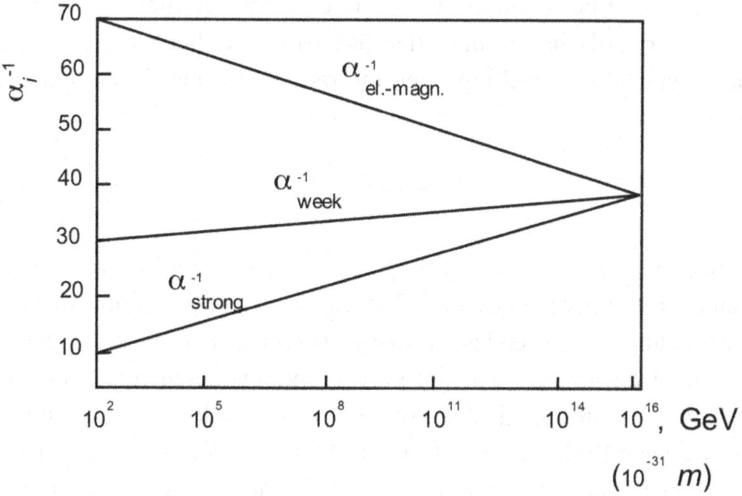

FIGURE 1.14 Gauge couplings unification.

1.8.9 STACKING OF CELLS IN THE UNIVERSE

In Section 1.6 we have shown that the tessellattice permits a spontaneous formation of clusters that may embrace an enormous number of cells. The highest scales may include the greatest number of cells, from 10^{82} to 10^{120}, then 10^{139} and 10^{142} to 10^{171} [55]. Expression (1.40), which demonstrates such union of cells, predicts the existence of galaxies, their unification into clusters and also the formation of a set of universes as well. In any case, the expression predicts the existence of our universe having formed as a cluster of primary topological balls that can be considered as cells of the total tessellattice.

The tessellattice and space-time introduced in it (Section 1.5) allow us to conclude that in any one Poincaré section representing a timeless instantaneous state (an instans) of the universe, the lattice of space is represented by a stacking of balls with nonidentical shape. From astrophysical observations we know that large astronomical bodies (planets or stars) are characterized by a larger concentration in their central part.

So in the central part of the tessellattice, the volume that is available is limited by the concentration of the stacking, and this limit is likely to be decreasing while going inward from the outer coats of the tessellattice. In a simple estimation, we denote by (a) the radius of the canonical (smallest) volume that can be transferred from one ball to another. Assuming that each cell forwards its volume (the size *a*) to another situated closer to the periphery, in the stacking, then the radius of a ball in the *n*th coat is approximated by:

$$\ell_n = \ell_1 + (n-1)a \tag{1.94}$$

Hence in the universe elementary cells exhibit increasing volumes from the center to the periphery of a 3D stacking. Transfers of non-fractal elementary volumes between balls are operated without dimensional increase. Indeed, at each given scale, the corresponding increments *a* are represented by similar topological features. In effect: following relation (1.66)-(1.69), we have for $n = 2$: $\ell_2 = \ell_1 + a$ and for $n = 1$: $\ell_1 = \ell_0 + (n-1)a = \ell_0$. From the last relation we derive $\ell_0 = \ell_2 - a$ where ℓ_0 stands for the radius of a founding ball located in the middle part of the tessellattice. Let $\ell_0 = (\varnothing)$ be an empty set. Then $\ell_2 = \ell_1 + a$ can be represented by a pair ($\varnothing$, $\{\varnothing\}$), where $\{\varnothing\}$ is the frontier of the ball ℓ_2. The element $\{\varnothing\}$ is what is exchangeable, and since it is a frontier, it has a dimension lower than the dimension of the interior, dim (a) < dim (ℓ). Thus exchanges do not modify the dimensionality of involved balls.

While the above considerations are valid for a particleless lattice, if the lattice is filled with particled balls, then the emerged matter due to the inerton clouds induces a kind of pressure, which may also result in proposed relation (1.94).

Consequently, it may be considered that these exchanges apply to the frontiers of the balls, which will result in changes in the concentration of their internal structures. It is noteworthy that the concentration has been used as a probe for the identification of the packing of balls. Otherwise, the adjunction of *a* to a radius r_n may result in the reunion of two spaces having nonequal dimensions, which can result in a structure of the "beaver space" type as described by Bounias [55].

A measure on such a lattice space by using a scanning function will not scan the same components in elementary balls that are situated at various distances from their origin [55]. Since it is likely that there occurs an increase of the composition of balls from this origin, then the scale will decrease with increasing distances: in effect, a larger set of scanned structures will appear at farther distances. Then, the measure of remote distances will be overestimated when using a local scale. This might account for the phenomena known as the Doppler effect, in turn usually involving the Hubble constant, and the acceleration of the universe, which will be considered in detail in Chapter 10.

1.8.10 BIG BANG

Among the major results of Bounias [54–57] is the proof that the lattice existing from empty hyperset units provides a manifold of quantized scales represented by a set of defined integer ratios and there exist empty set units of various size, with integer versus rational similarity ratios. One universe is represented by one particular sequence of Poincaré sections selected through a particular combination rule, $U_i = \{\{S_l\}_l, \perp\}_i$. Such a universe represents a manifold of organized empty set units, and since the lattice in which it is embedded is strictly fractal, the reunion of these empty set units is a higher scale empty set. Therefore $U_i = \varnothing_i$.

Now let $\varnothing_i$ be the size of a free ball in a part of the embedding lattice. This part of the whole embedding lattice plays a role of an "over universe" with a set-distance defined in relation (1.66) and it can be denoted $\varnothing_{+i}$. The distribution collapse of degenerate balls of $\varnothing_{+i}$ results in the formation of a particle whose subparts contain potentially as many quanta of fractality as U_i contains massive objects. The creation of a particle inside U_i is a primordial condensation of a ball into U_i inside U_{+i}.

Such a consideration makes possible the existence of "big-bangs": they occurred, occur and will occur in at least countless balls of the embedding lattice, without need for an external provision of energy.

However, these "big-bangs" fulfill some conditions, namely, the combination rule ($\perp$) determining the choice of S_{i+1} from S_i. Relations between different universes and past-to-present successive universes can exist only through the same law ($\perp$).

It does not seem likely that a big-bang could happen in the whole universe; it may occur rather only in some small part(s) of it. Figure 1.15 depicts a scenario that is not consistent with the mathematical laws.

FIGURE 1.15 The Big Bang (a cartoon by Michel Bounias from his letter dated 28 March 2000).

KEYWORDS

- **charge**
- **distance**
- **founding element**
- **founding lattice**
- **fractal**
- **inerton**
- **lepton**
- **mass**
- **measure**
- **photon**
- **quark**
- **qinerton**
- **set**
- **space**
- **tessellattice**
- **topological ball**

CHAPTER 2

SUBMICROSCOPIC MECHANICS

CONTENTS

2.1 CONCEPTUAL DIFFICULTIES OF ORTHODOX QUANTUM MECHANICS

Fundamental phenomena of quantum mechanics are now studied by using very precise techniques. In particular, by using neutron interferometry and polarimetry with thermal neutrons or holographic-grating interferometry researchers [103] have presented a survey of neutron-optical experiments that test a variety of peculiarities of quantum theory, such as quantum contextuality, multi-partite entanglement of single-neutrons, topological phases or decoherence effects, and violation of Heisenberg's error-disturbance uncertainty relation. Work has been performed to test the foundations of quantum mechanics with photons [104] and recent photonic experiments to test two foundational themes in quantum mechanics have been reviewed and examined: wave-particle duality, central to recent complementarity and delayed-choice experiments, and Bell nonlocality.

These kinds of studies somewhat repeat previous experiments albeit now at a more accurate level but at the same time exclude any option for the appearance of a new paradigm.

On the other hand, it is a matter of fact that quantum mechanics operates with notions that are not determined in the frame of formal mechanics. Among them we can cite the notion of "wave-particle" duality (that is not apparent in classical and high energy physics); the probabilistic interpretation of the Schrödinger wave ψ-function and hence the probability amplitude and its phase; long-range action of the quantum mechanical interaction; Heisenberg's uncertainty principle; the passage to the so-called operators of physical values, etc. For example, *Quantum Mechanics* by Landau and Lifshitz [105] begins with the words that in quantum mechanics all physical values are described by operators. So from the very beginning any physical description and physical interpretation of the processes examined by the formalism of quantum mechanics are rejected; only a set of initial and final energy states of the particle under consideration are the subject of study.

Modern views on an interpretation of quantum mechanics are reduced to a set of statements that allow one to explain the result of a measurement by avoiding the intrusion of the observer into the process of that measurement. Thus, an interpretation of quantum mechanics is often considered as an interpretation of its mathematical formalism when some important mathematical parameters are related to certain physical meanings.

Within physics education researchers identify difficulties for students in learning the conceptual and mathematical bases of quantum physics [106]: At undergraduate level, it is difficult to learn the concepts of quantum theory for many reasons, such as the probabilistic approach for determining the position of a particle, the uncertainty relations between observables, the non-physical abstract wave function that carries all the information about the real particle, the collapse of the wave function, and the use of advanced mathematics and different notations.

Let us list some of the most obvious unresolved issues, or conflicts that are presented in quantum mechanics [107, 108, 13]:

- the Schrödinger equation describes the quantum mechanics of particles but the equation cannot explain the reason of long-range action and wave behavior of the particles. Besides, a classical parameter – mass – enters the quantum equation;
- what is the wave ψ-function? This problem still thrills the curiosity of researchers;
- where is the particle mass located when the particle as the whole is fuzzy in an undetermined volume as prescribed by the wave ψ-function?;
- in the modern interpretation, the wave ψ-function is quite abstract and non-physical;
- in the process of measurement the abstract ψ-function suddenly collapses into a measurable actual point particle;
- all correct theories should be Lorentz invariant, i.e., they and Einstein's special relativity should be in agreement. Nevertheless, although the Schrödinger equation is not Lorentz invariant, it perfectly describes quantum phenomena. How is that possible?;
- there is no correct determination of values E and v in the expression $E = h\nu$ applied to a moving canonical particle. In one case $E = \frac{1}{2}m\upsilon^2$ and in the other one $E = mc^2 / \sqrt{1-\upsilon^2/c^2}$. Which one is true?;
- what does the de Broglie wavelength λ of a particle mean?;
- what is the nature of the phase transition in the quantum system being studied when we pass from the description based on the Schrödinger equation to that based on the Dirac equation?;
- what is spin? It is one more mystery of the microworld. Quantum field theories define it as an "inseparable and invariable property of a particle" and that is all.

High-energy physicists gave up on all these internal problems of quantum mechanics. Particle physicists study really fundamental things, but somehow forget that their studies are precisely rooted in quantum mechanical formalism. For example, recently Professor J. Gao [109] has written an inadequate article about the Large Hadron Collider at CERN, saying: "The Standard Model theory of particle physics is now gloriously complete … and now is the time to nail it down with precision and to match to new theories to cover the unknown components of the Universe, such as Dark Matter and Dark Energy, through Higgs with its field stretched out to the whole Universe space. The Higgs couples not only to known fundamental particles in the Standard Model but might also couple to unknown parts of the Universe," and so on... High-energy physicists are modestly silent about the fact that the LHC did not find any particle of the long predicted chain of SUSY particles and nothing has been found that would support the idea of 1D strings of many dimensions, etc. Besides, they are in no hurry to answer such simple questions as: what is mass?, what is electric charge?, what is a photon?, what is the principal difference between leptons and quarks and why do they interact so improbably?, what is the mechanism of the conferment of quarks?, and so on. So, it seems particle physicists have even more conceptual difficulties in their theories than those we can see in their mother-theory, quantum mechanics.

But let us come back from this brief lyrical digression to our main topic – quantum mechanics and its conceptual problems. In 1952 Bohm [12] published two papers on the pilot wave. Bohm's papers were an extended version of de Broglie's idea that was presented at the Solvay conference in 1927, but at that time due to criticism from the audience de Broglie reverted to the statistical interpretation of quantum mechanics.

Bohm's [12] papers inspired many researchers to start to examine approaches with so-called "hidden" variables. Initial approaches were identified with the theory of the pilot wave, or the de-Broglie-Bohm description of quantum mechanics. In those studies the wave ψ-function is subdivided into two multipliers and one of the multipliers is called the quantum potential. Nevertheless, de Broglie himself did not consider the approach as fundamental, because these two multiplier functions still continue to be determined in the framework of an abstract phase space in which the formalism of quantum mechanics has been developed. Quantum mechanics is based on postulates, above all its formalism prescribes that the state of a quantum system either pure or mixed is given by a vector in a complex vector space, namely, a Hilbert space.

Since 1952, when under the influence of Bohm's [12] publication, de Broglie changed his views on the foundations of quantum mechanics and in searching for a deterministic theory began to move away from its probabilistic interpretation, sadly causing many great physicists to turn away from him in their belief that he had become erratic. Nevertheless, de Broglie published a series of remarkable works on the foundations of quantum mechanics [11, 110, 111]. Especially valuable is his finally widely recognized book [111], which very critically considers the conventional concept of quantum mechanics. De Broglie was searching for the double solution theory in which a quantum system should be determined in the real physical space and his main idea was not dualism, i.e., a merger of the wave and particle into a wave-particle, but a coexistence between waves and particles. De Broglie believed that there would be hidden laws that provide the basis for motion of a particle and that the description of phenomena should also be the goal of physics, not only their prediction. As noted by Valentini [112], in de Broglie's theory, quantum theory emerges as a special subset of a wider physics, which allows non-local signals and violation of the uncertainty principle.

De Broglie [11] noted that any particle, even isolated, has to be imagined as being in continuous "energetic contact" with a hidden medium that constitutes a concealed regulator. This hypothesis belongs to Bohm and Vigier [113], who named this invisible regulator the "subquantum medium." Hence by Bohm, Vigier and de Broglie the particle has to be considered as continuously exchanging energy and momentum with such a hidden regulator: "If a hidden subquantum medium is assumed, knowledge of its nature would seem desirable. It certainly is of quite complex character" [11]. Nevertheless, de Broglie did not consider a subquantum medium needed for the future quantum theory as a universal reference medium, because in his belief this would be contrary to relativity theory.

However, on the other hand, why should we refer to relativity theory? It has no fewer conceptual difficulties than quantum mechanics. We have a rigorous theory of space presented in Chapter 1, which is able to become that hidden subquantum medium needed for the construction of particles, fields, subquantum mechanics and gravity. So, let us proceed to develop the mechanics of the tessellattice, which so far has been veiled by the crude formalism of orthodox quantum physics. Submicroscopic mechanics will also help us to cope with difficulties associated with the necessity for uniting quantum theory with gravity.

2.2 A NON-RELATIVISTIC PARTICLE IN THE TESSELLATTICE

As is known from solid-state physics the motion of a large polaron comprises the motion of the charge carrier and the carrier's polarization "coat," namely, the moving charge encapsulates its polarization coat. The same occurs in a liquid: an ion moves together with its solvation shell.

Similarly, in the tessellattice a moving particle has to pull its deformation coat, which is characterized by its radius being equal to the Compton wavelength (Figure 1.10). The particle's deformation coat should remain invariable at any point of the particle's path, which means that cells that form a deformation coat are adjusted to the particle at each point of the particle's path. Therefore the rapidity at which cells adapt to the particle much exceeds the particle's velocity. It is reasonable to assume that the speed of adjustment of the cells is equal to the speed of light c, the fundamental velocity of the tessellattice.

The particle moving in the tightly packed tessellattice cannot move freely; it certainly should experience frictional loss of velocity at each act of collisions with oncoming cells. As a result of such interaction with the tessellattice, the particle will emit elementary excitations that have been named above as 'inertons' (Figure 2.1). Thus, the initial speed of the particle's inertons also has to be associated with the speed of light.

One can say that the particle moving in such a way will quickly stop. However, if the tessellattice possesses elasticity, inertons emitted by the particle will be bounced off the lattice and come back to the particle. So the particle induces a deformation potential in its environment, i.e., the potential depends on the particle's velocity. This kind of potential, which depends on velocity, is known in classical mechanics [114]. This means that inertons that are emitted ahead by the particle at an angle to the particle's path will return to it again from the rear, as Figure 2.2 depicts. Hence, the movement in the tessellattice is possible without loss of energy.

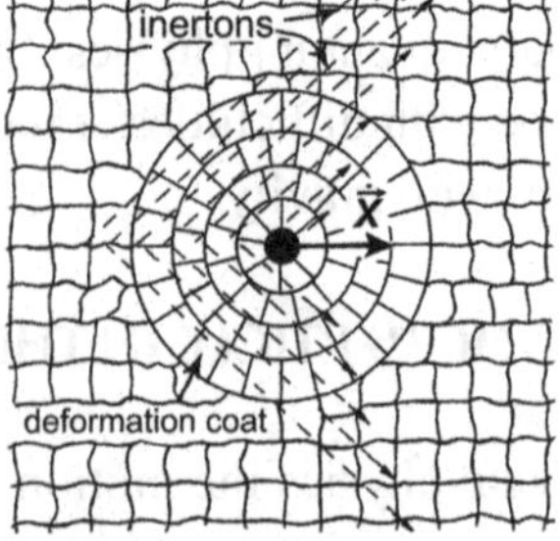

FIGURE 2.1 Particle moving in the tessellattice surrounded by the particle's deformation coat and accompanied by emitted inertons.

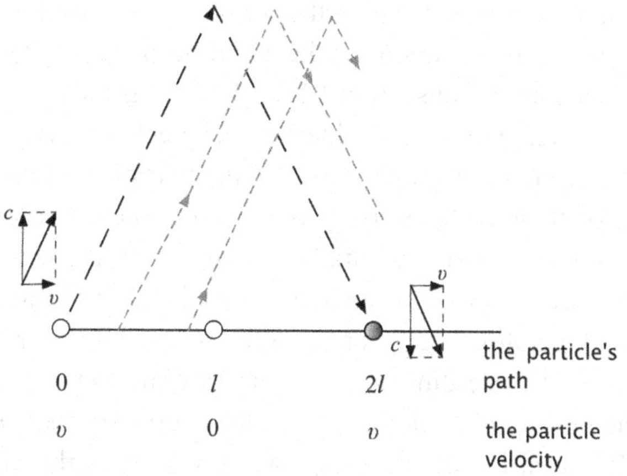

FIGURE 2.2 Particle emits inertons that are reflected back from the tessellattice to the particle. In the section from 0 to l the particle loses its velocity. In the moment of contact with returned inertons the particle reacquires the lost velocity, which occurs in the section from l to $2l$.

2.2.1 THE LAGRANGIAN AND THE EQUATIONS OF MOTION

The discussed motion of a particle, which emits and then absorbs its inertons, can be described by the following Lagrangian [52]

$$L = \tfrac{1}{2} g_{ij} \dot{x}_i(t) \dot{x}_j(t) + \tfrac{1}{2} \sum_{n=1}^{N} \tilde{g}_{ij}^{(n)} \dot{x}_{(n)}^{i}(t_{(n)}) \dot{x}_{(n)}^{j}(t_{(n)}) - \sum_{n=0}^{N} \frac{2}{T_n} \delta_{t-\Delta t_{(n)},\, t_{(n)}}$$
$$\times \left\{ x_i(t) \sqrt{g_{is} (\hat{A}^{-1} \tilde{g}_{sj}^{(n)})_0}\; \dot{\chi}_{(n)}^{j}(t_{(n)}) + x_i(t)\big|_{t=0} \sqrt{g_{is} (\hat{A}^{-1} \tilde{g}_{sj}^{(n)})_0}\; \chi_{(n)}^{j}(t_{(n)}) \right\} \tag{2.1}$$

Here, the first term characterizes the particle, the second term specifies the ensemble of N inertons, emitted by the particle, and the third term is attributed to the contact interaction between the particle and the ensemble of inertons. x^i and $\dot{x}^i$ are the ith components of the position and the velocity of the particle, respectively; g_{ij} is a metric tensor generated by the particle in its vicinity in three-dimensional space. Regarding the above indices, n corresponds to the number of the respective inerton; they are enclosed in parentheses to distinguish them from the other indices [i, j, and k in Eq. (2.1)] describing components of vector and tensor quantities; $\chi_{(n)}^{i}$ and $\dot{\chi}_{(n)}^{i}$ are components of position and velocity of the nth inerton; $\tilde{g}_{ij}^{(n)}$ is the metric tensor generated by the *n*th inerton in three-dimensional space (it describes local deformation in the neighborhood of the excitation, as has been discussed in Subsection 1.8.7); $1/(T_n/2)$ is the frequency of collisions of the particle with the *n*th inerton; Kronecker's symbol $\delta_{t-\Delta t_{(n)},\, t_{(n)}}$ provides the agreement of proper times of the particle and the *n*th inerton at the instant of their collision ($\Delta t_{(n)}$ is the time interval after the expiry of which, measuring from the initial moment $t = 0$, the moving particle emits the *n*th inerton); $\langle \dot{x}_i(t)\rangle$ is an average velocity of the particle during the period $T_{(n)}$. Operator $\hat{A}^{-1}$ appearing in the interaction energy characterizes the rotation of three-dimensional space around the X^i axis; matrix A belongs to the group of rotation $SO(3)$ of three-dimensional space around the x^i axis. This rotation of space eventually results in the motion of the inerton metric $\tilde{g}_{ij}^{(n)} \rightarrow \tilde{g}_{k+\alpha,j}^{(n)}$ (with regard to the cyclic permutation of the index α in the same manner as indices k and j take on values 1, 2, 3), i.e., operator $\hat{A}^{-1}$ acts according to the following rule:

$$g_{ik}(\hat{A}^{-1}\tilde{g}_{kj}^{(n)})_0 = g_{ik}\left(\frac{1}{A}\frac{\partial\varsigma_{(n)}^k}{\partial\chi_{(n)}^k}\frac{\partial\varsigma_{(n)}^k}{\partial\chi_{(n)}^j}\right)\Bigg|_{x_{(n)}=0}$$

$$= g_{ik}\left(\frac{\partial\varsigma_{(n)}^k}{\partial A\chi_{(n)}^k}\frac{\partial\varsigma_{(n)}^k}{\partial\chi_{(n)}^j}\right)\Bigg|_{x_{(n)}=0} = g_{ik}(\tilde{g}_{k+\alpha}^k)_0 \tag{2.2}$$

The radius vector $x_{(n)}(x_{(n)}^j)$ of the *n*th inerton is measured from the radius vector of the particle x(x^t). For example, if rotation takes place in the Euclidean space around the X^1 axis, then [115]

$$A = \begin{pmatrix} \cos\varphi & \sin\varphi & 0 \\ -\sin\varphi & \cos\varphi & 0 \\ 0 & 0 & 1 \end{pmatrix} \tag{2.3}$$

where φ is the angle of rotation. Thus the operator $\hat{A}^{-1}$ acting according to the rule (2.3) provides the transfer from the space of the particle's motion to the space of the *n*th inerton's motion, i.e., it shifts the inerton to the trajectory different from that of the particle (Figure 2.2). For simplicity, we will use the notation below

$$\widehat{B}_{ij}^{(n)} \equiv \sqrt{g_{ij}(\widehat{A}^{-1} g_{kj}^{(n)})}\,|_0 \tag{2.4}$$

Let us now derive the Euler-Lagrange equations

$$d(\partial L / \partial \dot{Q}^k) / dt - \partial L / \partial Q^k = 0 \tag{2.5}$$

for the particle

$$Q^k = x^k(t_{(n)} + \Delta t_{(n)}) \equiv x_{(n)}^k(t_{(n)} + \Delta t_{(n)}) \tag{2.6}$$

and for the inerton

$$Q^k \equiv \chi_{(n)}^k(t_{(n)}) \tag{2.7}$$

The Euler-Lagrange equations for extremals read:

$$\ddot{x}_{(n)}^s + \Gamma_{ij}^s \dot{x}_{(n)}^i \dot{x}_{(n)}^j + \frac{2g^{ks}}{T_{(n)}}\left(\frac{\partial \widehat{B}_{ij}^{(n)}}{\partial x_{(n)}^k}\right)\left(\dot{x}_{(n)}^i \chi_{(n)}^j + x_i(t)\big|_{t=0}\, \chi_{(n)}^j\right) + \frac{2g^{ks}}{T_{(n)}} \widehat{B}_{kj}^{(n)} \dot{\chi}_{(n)}^j = 0 \tag{2.8}$$

$$\ddot{\chi}_{(n)}^s + \tilde{\Gamma}_{ij}^s \dot{\chi}_{(n)}^i \dot{\chi}_{(n)}^j + \frac{2\tilde{g}^{ks}}{T_{(n)}}\left(\frac{\partial \widehat{B}_{ij}^{(n)}}{\partial x_{(n)}^k} - \frac{\partial \widehat{B}_{ik}^{(n)}}{\partial x_{(n)}^j}\right) x_{(n)}^i \dot{\chi}_{(n)}^j$$
$$- \frac{2\tilde{g}^{ks}}{T_{(n)}} \widehat{B}_{ki}^{(n)} \left(\dot{x}_{(n)}^j - x_i(t)\big|_{t=0}\right) = 0 \tag{2.9}$$

Here, Γ_{ij}^s and $\tilde{\Gamma}_{ij}^{(n)s}$ are symmetrical connections (see, e.g., Ref. [115], p. 293) for the particle and for the *n*th inerton, respectively; indices *i*, *j* and *s* take the values 1, 2, 3.

Let us analyze Eqs. (2.8) and (2.9). We shall specify the form of tensors g_{ij} and $\tilde{g}_{ij}^{(n)}$ that are determined in 3D space. We have noted that the deformation of the tessellattice in the neighborhood of the particle with a mass m_0 can be related to the induction of a deformation potential *V*(*r*). If we relate the potential to self-energy of the particle, then a non-parameterized metric tensor in this region of the space will take the form

$$g_{ij} = \text{const } m_0 \, \delta_{ij} \cdot \left(1 - V(m_0;\ r)\right) \tag{2.10}$$

where the first term, which does not include V, corresponds to the metric of flat space, i.e., the degenerate tessellattice in which we have $g_{ij} \to \delta_{ij}$. Similarly for the nth inerton

$$\tilde{g}_{ij}^{(n)} = \text{const } \mu_{0(n)} \delta_{ij} \cdot \left(1 - W_{(n)}(\mu_{0(n)};\ r)\right) \tag{2.11}$$

where $\mu_{0(n)}$ is the mass of the nth inerton. Because the inequality $\mu_{0(n)} << m_0$ holds, the potential $W_{(n)}$ is local but identical by its nature to the potential V, which we have considered in Subsection 1.8.7. The potentials are proportional to the masses that induced them, namely: $V \propto m_0$ and $W_{(n)} \propto \mu_{0\,(n)}$.

The transformation of the metric $\tilde{g}_{ij}^{(n)}$ to $\tilde{g}_{i+\alpha,j}^{(n)}$ under the effect of the operator $\hat{A}^{-1}$ occurs in three steps: (i) the particle sticks together with an oncoming nth cell of the tessellattice, (ii) the intensification of inner fractal changes in both the particle and the cell and (iii) the separation of a fractal fragment, which is specified by the metric $\tilde{g}_{i+\alpha,j}^{(n)}$, from the particle and the travelling of this nth inerton by its own trajectory as Figure 2.2 exhibits. In the first stage the interaction operator $\widehat{B}_{ij}^{(n)} = 0$ and hence the difference of term between equations (2.8) and (2.9) is reduced to the equation

$$(\ddot{x}_{(n)}^{s} - \ddot{\chi}_{(n)}^{s}) + \left(\Gamma_{ij}^{s}\, \dot{x}_{(n)}^{i}\, \dot{x}_{(n)}^{j} - \tilde{\Gamma}_{ij}^{(n)s}\, \dot{\chi}_{(n)}^{i}\, \dot{\chi}_{(n)}^{j}\right) = 0 \tag{2.12}$$

Eq. (2.12) can be treated as the equation that determines the point of intersection of the geodesic lines of the particle and the nth inerton. At this point, the particle and the nth inerton are united representing a common system and because of that the acceleration that one of the partners is experiencing coincides with that of the other partner. Therefore, the difference in the first set of parentheses in Eq. (2.12) is equal to zero, and we obtain the relation

$$\Gamma_{ij}^{s}\, \dot{x}_{(n)}^{i}\, \dot{x}_{(n)}^{j} = \tilde{\Gamma}_{ij}^{(n)s}\, \dot{\chi}_{(n)}^{i}\, \dot{\chi}_{(n)}^{j} \tag{2.13}$$

The structures of fields Γ_{ij}^{s} and $\tilde{\Gamma}_{ij}^{(n)s}$ in the point of cross-geodesics are identical. However, the values of intensity of the fields are different: Γ_{ij}^{s} is generated by the mass m but $\tilde{\Gamma}_{ij}^{(n)s}$ is generated by the mass $\mu_{0(n)}$. Therefore the relation $\Gamma_{ij}^{s} / \tilde{\Gamma}_{ij}^{(n)s} = m_0 / \mu_{0(n)}$ applies. Putting the initial velocity of the particle equal to υ_0 and the velocity of the particle after the nth collision equals υ_{0n} and taking into account that the initial velocity of the nth inerton $|\dot{x}_{(n)}|_0 = c$, we obtain

$$m_0 \upsilon_0^2 = \mu_{0(n)} c^2 \tag{2.14}$$

The nth inerton is transferred by the lattice vibrations deep into the tessellattice in a transverse direction to the vector velocity of the particle (Figure 2.3), which occurs under the influence of the elastic forces acting in the tessellattice. For the realization of the described mechanism the operator $\widehat{B}_{ij}^{(n)}$ in Eqs. (2.8) and (2.9) should provide a rotation of the space on the angle $\varphi = \pi / 2$. In the new coordinate system related to the nth inerton the axes are redesignated and then the direction cosine in expression (2.3) becomes equal to unit.

Besides, at each collision the particle passes a momentum on to the nth inerton, such that the nth inerton additionally also has a longitudinal component of velocity $\upsilon_{(n)}$, which is less than the initial velocity υ_0 of the particle. Thus after the nth collision the particle and the nth inerton scatter following their own paths (Figure 2.2).

The particle is described by the metric tensor $g_{ij} = \text{const}\ m_0 \delta_{ij}$ and the nth inerton is specified by the metric tensor $\tilde{g}_{ij}^{(n)} = \text{const}\ \mu_{0(n)}\, \delta_{ij}$. These two tensors are constants; therefore, all their deviations along the respective trajectories are equal to zero. As a result, only two terms remain in Eqs. (2.8) and (2.9), namely, the first and the last ones. They are transformed as follows:

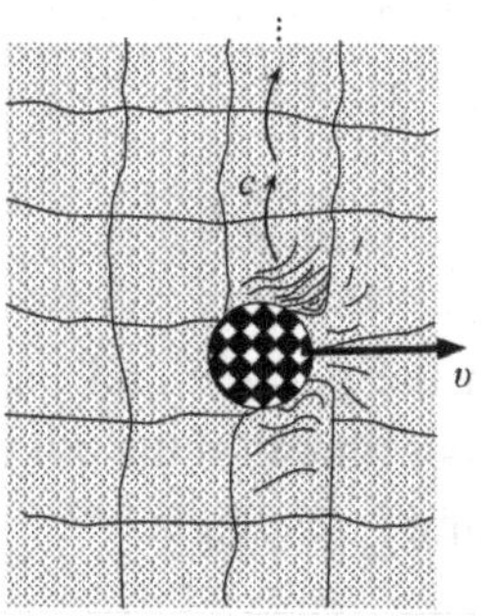

FIGURE 2.3 Tessellattice guides the inerton in the direction perpendicular to the particle's path. The initial velocity of the inerton equals the velocity of light (though gradually it loses velocity owing to the deformation potential of the particle).

$$g^{\cdots}\widehat{B}^{(n)} \rightarrow \sqrt{\mu_{0(n)} / m_0} = \upsilon_{0(n)} / c \tag{2.15}$$

$$\tilde{g}^{\cdots}\widehat{B}^{(n)} \rightarrow \sqrt{m_0 / \mu_{0(n)}} = c / \upsilon_{0(n)} \tag{2.16}$$

Here we have taken into consideration relation (2.14) and the fact that by definition $g^{sk} g_{kj} = \delta_j^s$. Below we omit the parentheses for the index n.

2.2.2 SOLUTION TO THE EQUATIONS OF MOTION FOR THE NON-RELATIVISTIC PARTICLE

Let us assume that the particle moves along the axis $X^1 \equiv X$ in a Cartesian coordinate system and the coordinate x describes a position of the particle. The reference point of the inerton radius vector $\chi_n(\chi_n^i)$ is associated with the particle. Let projections of these vectors on axes X^1, X^2 and X^3 be χ_n^1, χ_n^2 and χ_n^3, respectively. If we introduce a generalized coordinate $\chi_n^{\perp} = \sqrt{(\chi_n^2)^2 + (\chi_n^3)^2}$ then the problem is reduced to a two-dimensional one: the particle moves along axis X together with an ensemble of inertons that are characterized by projections $\chi_n^{\|}$ and $\chi_n^{\perp}$ on the axis X^1 and X^2, respectively. Thus equations (2.8) and (2.9) in the Euclidean space with regard for relations (2.15), (2.16) take the form

$$\ddot{x}_n + \frac{2}{T_n} \frac{\upsilon_{0n}}{c} \dot{\chi}_n^{\perp} = 0 \tag{2.17}$$

$$\ddot{\chi}_n^{\perp} - \frac{2}{T_n} \frac{c}{\upsilon_{0n}^{\|}} (\dot{x}_n - \upsilon_0) = 0 \tag{2.18}$$

$$\dot{\chi}_n^{\|} = 0 \tag{2.19}$$

Recall that indices n on the particle and inerton parameters point to their dependence on the proper time t_n of the nth inerton. Differentiation of Eq. (2.18) with respect to time t_n results in

$$\dddot{\chi}_n^{\perp} - \frac{2}{T_n} \frac{c}{\upsilon_0} \ddot{x}_n = 0 \tag{2.20}$$

Let us express $\ddot{x}_n$ in terms of Eq. (2.20) and substitute it into Eq. (2.17). We get the equation of a harmonic oscillator for $\chi_n^{\perp}$:

$$\ddot{\chi}_n^{\perp} + (2/T_n)^2 \chi_n^{\perp} = C_{1n} \tag{2.21}$$

From Eq. (2.21) we can see that the *n*th inerton participates in harmonic oscillations along the axis perpendicular to the trajectory of the particle. Let us choose the following initial conditions:

$$\chi_n^{\perp}\Big|_{t_n=0} = 0, \ \dot{\chi}_n^{\perp}\Big|_{t_n=0} = c \tag{2.22}$$

Then the solution to Eq. (2.21) takes the form:

$$\chi_n^{\perp} = \Lambda_n \sin[t_n / (T_n / 2)] \tag{2.23}$$

$$\dot{\chi}_n^{\perp} = c \cos[t_n / (T_n / 2)] \tag{2.24}$$

where Λ_n is the amplitude of oscillations of the *n*th inerton, i.e., the maximum distance which the inerton reaches when it is separated from the particle. In the present case the following relation holds:

$$\Lambda_n / (T_n / 2) = c \tag{2.25}$$

Integrating Eq. (2.17) we get

$$\dot{x}_n = C_{2n} - \frac{2}{T_n} \frac{\upsilon_{0n}}{c} \chi_n^{\perp} \tag{2.26}$$

Working from the initial conditions for the particle

$$\dot{x}_n(t_n + \Delta t_n)\,|_{t_n=0} = \dot{x}(\Delta t_n) = \upsilon_{0n} \tag{2.27}$$

which allows us to obtain from Eq. (2.26) with regards to Eqs. (2.23) and (2.27)

$$\dot{x}_n = \upsilon_{0n}\{1 - \sin[t_n / (T_n / 2)]\} \tag{2.28}$$

$$x_n = \upsilon_{0n} t_n + \lambda_n \{\cos[t_n / (T_n / 2)] - 1\} \tag{2.29}$$

where the following notation has been introduced:

$$\lambda_n = \upsilon_{0n} T_n / 2 \tag{2.30}$$

The solutions (2.28)–(2.30) show that in the interval from $t_n = 0$ to $t = T_n$ the particle velocity oscillates between $\dot{x}_n = \upsilon_{0n}$ and $\dot{x}_n = 0$.

Let us find the relationship between the initial velocity υ_{0n} of the particle after the emission of the *n*th inerton (at the moment $t_n = 0$ where $n \neq 0$) and the initial velocity υ_0 (at the moment $t = t_n = 0$, i.e., when $n = 0$). Putting $n = 0$ in Eq. (2.28) we come from the proper time of the *n*th inerton t_n to the proper time t of the particle:

$$\dot{x}_n(t_n + \Delta t_n) = \dot{x}(t) = \upsilon_0 \{1 - \sin[\Delta t_n / (T_n / 2)]\} \tag{2.31}$$

Then at the moment $t_n = \Delta t_n$

$$\dot{x}(\Delta t_n) = \upsilon_0 \cdot \{1 - \sin[\Delta t_n / (T_n / 2)]\} \tag{2.32}$$

or since $\Delta t_n = nT / 2N$ (see Figure 2.4), we get instead of Eq. (2.32)

$$\dot{x}(\Delta t_n) = \upsilon_0 \cdot \left[1 - \sin\left(n / (2N)\right)\right] \tag{2.33}$$

But it is at this instant of time when $t = \Delta t_n$ that the *n*th inerton is emitted. Therefore solution (2.33) should be compared with expression (2.27). As a result, we acquire for the initial velocity of the *n*th inerton

$$\upsilon_{0n} = \upsilon_0 \cdot \left[1 - \sin\left(n / (2N)\right)\right] \tag{2.34}$$

Thus we have obtained harmonic solutions for the velocity and the coordinate of the particle, and in this case the motion of the particle is characterized not only by the time half period T_n of the cycle, but also by the spatial period λ_n. However, there is a time delay $2\Delta t_n$ (and also a spatial distance) between the moment of absorption of the *n*th inerton and the moment of its next emission (see Figure 2.4). Therefore, to extend solutions (2.28) and (2.29), as well as (2.23) and (2.24), to the subsequent oscillations of the particle, we have to add into these expressions a jump that describes the transition of the particle from the *q*st to the next

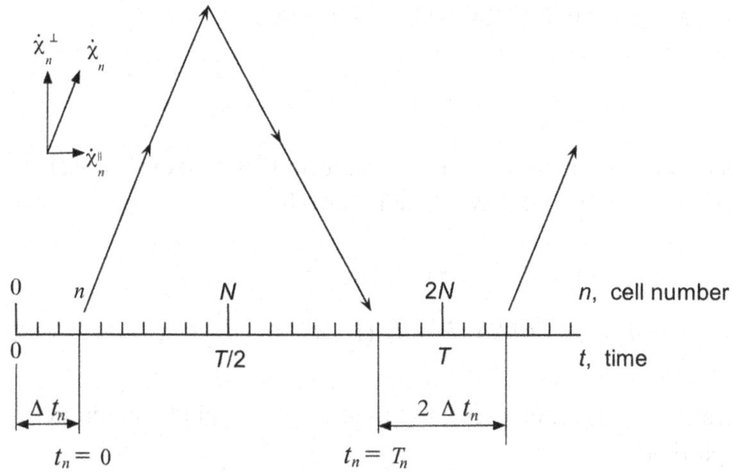

FIGURE 2.4 The trajectory of motion of the nth inerton. It is emitted in the moment of time Δt_n (where t and t_n are the proper times of the particle and the nth inerton, respectively, with $t = t_n + \Delta t_n$ where $\Delta t_n = nt / (2N)$, $T_n = T - 2\Delta t_n \equiv \mathrm{T} \cdot (1 - n / N)$ and $n = \overline{0,\ N-1}$).

(q+1)th oscillation. In other words, we have to include a quasi cyclicity into the parameter t_n. Substitution with $t_n \to t_{qn}$ meets this requirement, i.e., in the expressions from (2.23) to (2.33) the role of proper time will play the quasi-continuous parameter t_{qn} defined as follows

$$t_{qn} = t_n + 2(q-1)T_n,\ \ n = 1, 2, 3, \ldots;$$

$$T_n = T \cdot (1 - n / N),\ \ 0 \le t_n \le T_n,\ \ n = \overline{0,\ N-1} \qquad (2.35)$$

The solution to Eq. (2.19) is $\dot{\chi}_n^{\parallel} = \text{const}$. We cannot use the law of the conservation of momentum to find $\dot{\chi}_n^{\parallel}$, as collisions of the particle with oncoming cells of the tessellattice are inelastic (after each collision an inerton is either created from the particle or adsorbed by the particle). However, in the time $t_n = T_n$ that passed since the moment of emission of the nth inerton, the latter has to be absorbed by the particle again. The path traversed by the particle in time $t_n = T_n$, according to Eq. (2.29) is equal to

$$x_n(T_n) = 3\upsilon_{0n} T_n / 2 \qquad (2.36)$$

Comparing relation (2.36) and $\dot{\chi}^{\parallel}_{n}$, we obtain

$$\dot{\chi}^{\parallel}_{n}(T_n) = 3\upsilon_{0n} / 2. \tag{2.37}$$

This solution can be rewritten in the form that includes a quasi-continuous time parameter; namely, we may introduce

$$\tau_{qn} = (2q-1)\Delta t_n + (q-1)T_n, \; n = 1, 2, 3, \ldots;$$
$$\Delta t_n = nT / (2N), \; T_n = T \cdot (1 - n / N), \; n = \overline{0, \; N-1}. \tag{2.38}$$

Then Eq. (37) can be presented in terms that include the proper time t of the particle

$$\dot{\chi}^{\parallel}_{qn} = \tfrac{3}{2}\upsilon_{0n}\, \vartheta(t - \tau_{qn})\, \vartheta(\tau_{qn} + T_n - t) \tag{2.39}$$

where the Heaviside step function $\vartheta(t) = 1$ when $t \geq 0$ and $\vartheta(t) = 0$ when $t < 0$.

Figure 2.5 exhibits the principle of the motion of a particle with its inertons. The motion is characterized by periodicity in both the particle and inertons. The particle moves by interacting with its inertons – the frequency of collisions with the nth inerton is $1/(T_n/2)$. After going through the section λ_n the particle stops, as it has emitted all its N inertons. But from the next section λ_n the particle starts to gradually absorb these inertons, such that we can say that inertons guide the particle. Then the described situation is repeated again and again, which means that the moving particle undergoes spatial oscillations (the number of cycles q = 1, 2, 3, …). These oscillations are characterized by the initial parameters of the particle – the amplitude λ and the period of collisions $T/2$. The value λ can be treated as a spatial period of the particle's spatial oscillations. Similar parameters also specify the particle's inertons; the inerton envelope possesses the amplitude Λ and the same period of collision T. Initial velocity of the particle and inertons are υ_0 and c, respectively.

The proper time t of the particle can be considered as a natural parameter proportional to the particle's path length l, $t = l/\upsilon_0$. This parameter should necessarily be continuous. Such a requirement may be readily satisfied, if to the initial condition of $\dot{x}\big|_{t=0} = \upsilon_0$ (2.27) we add another

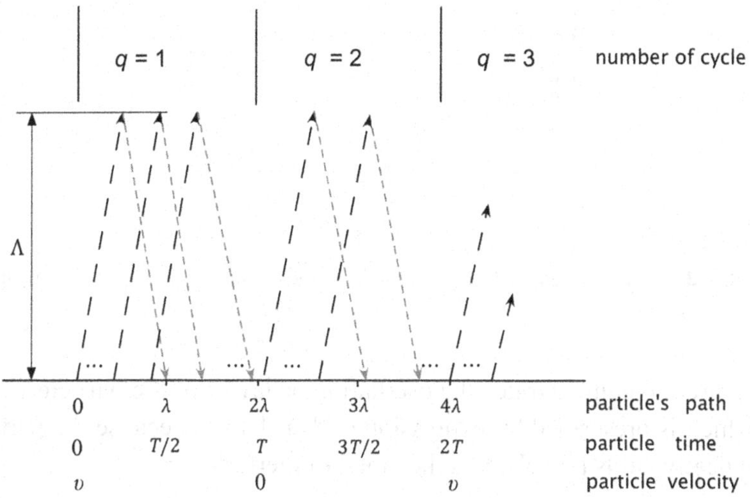

FIGURE 2.5 Trajectories of motion of the particle and its inertons. The oscillating movement of the particle and inertons, which accompany it, clearly manifest themselves.

requirement – the periodic recurrence of the value of the initial velocity with time (see Figure 2.5):

$$\dot{x}\big|_{t=0} = \dot{x}\big|_{t=qT} = \upsilon_0, \; q = 1, 2, 3, \ldots \tag{2.40}$$

[with arbitrary t, variable $\dot{x}(t) \geq 0$]. At the condition (2.40) the solution to the particle, expressions (2.28) to (2.30), as a function of the particle's proper time t become continuous:

$$\dot{x}(t) = \upsilon_0 \cdot \left(1 - \left|\sin[t/(T/2)]\right|\right) \tag{2.41}$$

$$x(t) = \upsilon_0 t + \lambda \cos[t/(T/2)] \frac{\sin[t/(T/2)]}{|\sin[t/(T/2)]|} - \lambda \tag{2.42}$$

$$\lambda = \upsilon_0 T/2 \tag{2.43}$$

Here, the position $x(t)$ of the particle as a function of t is continuous with the initial condition $x(0) = 0$. The value T is the natural parameter at the length of the spatial period λ of the particle, $T/2 = \lambda/\upsilon_0$ as well as the particle's proper time t being the natural parameter for the entire particle's path, $t = l/\upsilon_0$

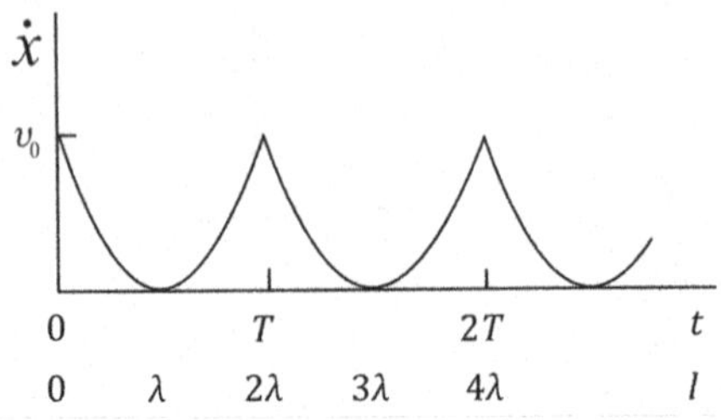

FIGURE 2.6 Oscillations of the particle velocity along the particle path by expression (2.41).

Figure 2.6 demonstrates the oscillation behavior of the particle velocity, which is prescribed by expression (2.41). This is because the particle periodically emits and absorbs the cloud of inertons.

2.3 A RELATIVISTIC PARTICLE IN THE TESSELLATTICE

In Section 2.2, we have proceeded from the modernized classical nonrelativistic Lagrangian of a particle

$$L_{\text{nonrel.}} = \tfrac{1}{2} g_{ij} x^i x^j + U(x, \chi, \dot{\chi}) \tag{2.44}$$

where the function U includes the potential energy of the interaction between the particle and its inertons and the kinetic energy of the ensemble of inertons (in this Section the vectors x and $\dot{x}$ pertain to the particle, and the vectors χ and $\dot{\chi}$ pertain to inertons).

Studying a relativistic particle we shall proceed in a similar fashion, i.e., in the classical relativistic Lagrangian

$$L_{\text{rel.}} = -m_0 c^2 \sqrt{1 - \upsilon_0^2 / c^2} \tag{2.45}$$

we shall substitute

$$\upsilon_0^2 \rightarrow [g_{ij} x^i x^j + U(x, \chi, \dot{\chi})] / g \tag{2.46}$$

The modernized relativistic Lagrangian (2.45) of a moving particle, in view of the interaction with inertons, can be written in an explicit form as follows [49]:

$$L_{\text{rel.}} = -gc^2 \left\{ 1 - \frac{1}{gc^2} \left[g_{ij}\, \dot{x}_i(t)\, \dot{x}_j(t) + \sum_{n=1}^{N} \hat{g}_{ij}^{(n)}\, \dot{x}_{(n)}^{i}(t_{(n)})\, \dot{x}_{(n)}^{j}(t_{(n)}) \right. \right.$$

$$- \sum_{n=0}^{N} \frac{2}{T} \delta_{t-\Delta t_{(n)},\, t_{(n)}} \left(x_i(t) \sqrt{g_{is} (\hat{A}^{-1} \tilde{g}_{sj}^{(n)})_0}\, \dot{\chi}_{(n)}^{j}(t_{(n)}) \right)$$

$$\left. \left. + \langle \dot{x}^i(t) \rangle \sqrt{g_{is} (\hat{A}^{-1} \tilde{g}_{sj}^{(n)})_0}\, \chi_{(n)}^{j}(t_{(n)}) \right) \right] \right\}^{1/2} \qquad (2.47)$$

Here, g_{ij} are components of the metric tensor produced by the particle in the 3D space; along the particle's path $g_{ij} = \text{const}\ \delta_{ij}$ and $g = g_{ij}\ \delta^{ij}$ is the convolution of the tensor; $\tilde{g}_{ij}^{(n)}(x_{(n)})$ are components of the metric tensor of the nth inerton and along its trajectory the tensor $\tilde{g}_{ij}^{(n)}$ is supposed to be equal to const δ_{ij} (index n is enclosed in parentheses to distinguish it from the indices i, j, s that describe tensor and vector quantities); $1/T_{(n)}$ is the frequency of collisions of the particle with the nth inerton, N is the number of inertons emitted by the particle; $\delta_{t-\Delta t_{(n)},\, t_{(n)}}$ is Kroneker's symbol that provides the agreement between the proper time t of the particle and the proper time $t_{(n)}$ of the nth inerton at the moment of their collision ($\Delta t_{(n)}$ is the time interval measured since the initial moment $t = 0$ after which the moving particle emits the nth inerton). The operator $\hat{A}^{-1}$ sets the separation/unification of the nth inerton at moments of its emission/absorption by the particle.

The proper time t of the particle is expressed through the proper time $t_{(n)}$ of the nth inerton, $t = t_{(n)} + \Delta t_{(n)}$, which allows us to write the Euler-Lagrange equations for the particle and the nth inerton in terms of one of these two time parameters, for instance, via $t_{(n)}$:

$$d(\partial L_{\text{rel.}} / \partial \dot{Q}^k) / dt_{(n)} - \partial L_{\text{rel.}} / \partial Q^k = 0 \qquad (2.48)$$

where for the particle and the nth inerton we put $Q^k \equiv x^k(t_{(n)} + \Delta t_{(n)})$ and $Q^k \equiv \chi^k(t_{(n)})$, respectively.

If the particle moves along the axis $X^1 \equiv X$, then the motion of inertons is defined by the coordinates transversal and longitudinal to the axis X, that is, $\chi_{(n)}^{\perp} = \sqrt{(\chi_{(n)}^{2})^2 + (\chi_{(n)}^{3})^2}$ and $\chi_{(n)}^{\parallel}$, respectively. The solution of the Euler-Lagrange equations (2.48) coincide with those derived for the non-relativistic particle and its inertons (2.28)–(2.30) and (2.23)–(2.25), respectively (where the parentheses for index n are omitted).

Considering the proper time of the particle as a natural parameter proportional to the particle's path length l, $t = l/\upsilon_0$, we arrive at the continuous solution (2.41)–(2.43). In each section of 2λ the relation $T = 2\lambda/\upsilon_0$ holds. In the case of inertons, the period of collisions T_n may be viewed as the natural parameter at the length of the spatial period Λ_n, $T_n = \Lambda_n / c$.

2.4 WAVE NON-RELATIVISTIC MECHANICS

Let us treat an ensemble of inertons, which accompany the particle under consideration whose velocity satisfies the inequality $\upsilon_0 << c$, as one integral object, a cloud of inertons. Then our study is significantly simplified and can be reduced to the consideration of a system of two objects: the particle and its cloud of inertons, which the particle periodically emits and adsorbs when moving along its path. In this case the Lagrangian (2.1) is transformed to the following one written in two-dimensional Euclidean space

$$L = \tfrac{1}{2} m_0 \dot{x}^2 + \tfrac{1}{2} \mu_0 \cdot \left[(\dot{\chi}^{\parallel})^2 + (\dot{\chi}^{\perp})^2 \right] - \frac{2\pi}{T} \sqrt{m_0 \mu_0}\, x \dot{\chi}^{\perp} \qquad (2.49)$$

In the Lagrangian (2.49) the first term describes the kinetic energy of the particle with the mass m and the velocity $\dot{x}$, which moves along the axis X; the second term depicts the kinetic energy of the whole inerton cloud whose mass is μ_0 and its center-of-mass has the coordinate $\chi^{\parallel}$ along the particle's path and $\chi^{\perp}$ is the transverse coordinate; the third term is the interaction energy between the particle and the inerton cloud where $1/T$ is the frequency of their collisions.

By using the substitution

$$\dot{x}^{\perp} = \dot{\tilde{\chi}} + 2\pi \sqrt{m_0 / \mu_0}\; x / T \qquad (2.50)$$

we carry out a kind of a canonical transformation that leads to the following Lagrangian

$$\tilde{L} = \tfrac{1}{2} m_0 \dot{x}^2 - \tfrac{1}{2} (2\pi / T)^2 m_0 x^2 + \tfrac{1}{2} \mu_0 \cdot \left(\dot{\tilde{\chi}}^2 + (\dot{\chi}^{\parallel})^2 \right) \qquad (2.51)$$

We can see from the effective Lagrangian (2.51) that in such a presentation the particle's behavior is described as a classical harmonic oscillator and the accompanying inerton cloud moves by its own hidden principle (though it does not disturb the particle).

The Hamiltonian function according to the definition

$$H = \sum_i \dot{Q}_i \partial L / \partial \dot{Q}_i - L$$

In our case the Hamiltonian is

$$H = \dot{x} \partial L / \partial \dot{x} + \dot{\tilde{\chi}} \partial L / \partial \dot{\tilde{\chi}} - \tilde{L} \tag{2.52}$$

The effective Hamiltonian based on the Lagrangian (2.51) of the oscillating particle in the system of the center-of-mass of the particle and its inerton cloud in the explicit form becomes

$$H = p^2 / (2m_0) + m_0 (2\pi / T)^2 x^2 / 2 \tag{2.53}$$

Solutions of the equations of motion given by the Hamiltonian (2.53) are well known for different presentations. In particular, the function (2.53) allows one to derive the Hamilton-Jacobi equation

$$(\partial S_1 / \partial x)^2 / (2m_0) + m_0 (2\pi / T)^2 x^2 / 2 = E \tag{2.54}$$

from which we obtain the equation for a shortened action:

$$S_1 = \int_{x_0}^{x} p\, dx = \int_{x_0}^{x} \sqrt{2m_0 [E - (2\pi / T)^2 x^2 / 2]}\, dx \tag{2.55}$$

The function (2.55) enables the solution x as a function of t in the form

$$x = \frac{\sqrt{2E / m_0}}{2\pi / T} \sin(2\pi t / T) \tag{2.56}$$

Now we can calculate the increment ΔS_1 of the action (2.55) of the particle during the period T; in terms of the action-angle variables

$$\Delta S_1 = \oint p\, dx = \oint \sqrt{2m_0 \left(E - m_0 (2\pi / T)^2 x^2\right)}\, dx$$
$$= \oint \sqrt{2m_0 \left(E - E \sin^2(2\pi t / T)\right)} \sqrt{2E / m_0} \cos(2\pi t / T)\, dt$$

$$= 2E\int_0^T \cos^2(2\pi t/T)\,dt = 2E\left(\frac{t}{2} + \frac{\sin(4\pi t/T)}{4(2\pi/T)}\right)\Bigg|_{t=0}^{t=T}$$

$$= ET \qquad (2.57)$$

The final result (2.57) can be rewritten as follows

$$\Delta S_1 = E \cdot T = E/\nu \qquad (2.58)$$

where the notation ν = 1/T is entered.

Since the constant E is the initial energy of the particle, i.e., $E = \frac{1}{2}m_0\upsilon_0^2$, the increment of action (2.58) can also be presented in the form

$$\Delta S_1 = \tfrac{1}{2}m_0\upsilon_0^2 \cdot T = m_0\upsilon_0 \cdot \tfrac{1}{2}\upsilon_0 T = m_0\upsilon_0 \cdot \lambda \qquad (2.59)$$

where the parameter λ is the spatial amplitude of oscillations of the particle along its path, which has been discussed in Subsection 2.2.2 (see also Figure 2.5).

If we equate the increment of the action ΔS_1 to the Planck constant h, we immediately arrive at the two major relationships of quantum mechanics introduced by de Broglie [10] for a particle:

$$E = h\nu,\ \lambda = h/(m_0\,\upsilon_0) \qquad (2.60)$$

Thus the amplitude of special oscillation of a particle is exactly the particle's de Broglie wavelength.

Having obtained the relationships (2.60), we can present the complete action for a particle

$$S = S_1 - Et = \int^x p\,dx - Et \qquad (2.61)$$

in two equivalent forms:

$$S = m_0\upsilon_0 x\ - Et \qquad (2.62)$$

and

$$S = h\cdot(x/\lambda\ - \nu t) \qquad (2.63)$$

In relations (2.62) and (2.63) all parameters are completely determined:

- λ, which is known as the particle's de Broglie wavelength; this characteristic equals the spatial amplitude of oscillations of the particle $\lambda = \upsilon_0 T/2$ in the reference frame associated with the center-of-mass of the system {particle + its inerton cloud}. Thus

$$\lambda_{\text{dB}} \equiv \lambda = \tfrac{1}{2}\upsilon_0 T = \upsilon_0/(2\nu); \tag{2.64}$$

- $\nu = 1/T$ is the frequency of the particle oscillations along its path; $1/(2T) = 2\nu$ is the frequency of collisions of the particle with its inertons;
- t is the proper time of the particle, which means that the system studied is Lorentz invariant;
- h is the minimum increment of the action of the particle per cyclic period, which occurs in the case when the particle is freely guided by the degenerate space, i.e., the tessellattice without any intervention of outside factors (fields, particles, obstacles, etc.). In mechanics the mass m of the particle and elasticity constant γ of the medium/bonds set the frequency of the particle oscillations: $\nu = \sqrt{\gamma/m_0}/(2\pi)$. Parameters m and E are the characteristics of the particle. Hence, assuming $\Delta S_1 = h$ we automatically determine the Planck constant as an adiabatic invariant of the oscillator whose elasticity constant is defined by elastic properties of the tessellattice. Besides, the action (2.63) shows that the system {particle + its inerton cloud} behaves as a wave that is harmonized with the particle by the de Broglie relationships (2.60).

Both the relation (2.62) and (2.63) are obtained from the equation (2.61) but in different presentations: the first one in conventional terms of classical mechanics – the momentum $p_0 = m_0\upsilon_0$ and the kinetic energy $E = p_0^2/(2m_0)$ and the second one in terms of the amplitude λ and the frequency ν of the particle oscillations that occur along the particle path.

The relationships (2.62), (2.63) and (2.60) allow the derivation of the Schrödinger equation (comp. with de Broglie [106]). If in a conventional wave equation

$$\Delta\psi - \frac{1}{(\upsilon_0/2)^2}\frac{\partial^2\psi}{\partial t^2} = 0 \tag{2.65}$$

(where $\frac{1}{2}\upsilon_0$ is the average velocity of the particle in the spatial period λ) we insert a wave function, whose phase is based on the action (2.63),

$$\psi = a \exp\{i 2\pi [x/\lambda - \nu t]\} \tag{2.66}$$

and set $\upsilon_0 = \lambda \cdot 2\nu$, we get the wave equation in the following presentation:

$$\Delta\psi + \left(\frac{2\pi}{\lambda}\right)^2 \psi = 0 \tag{2.67}$$

Then putting $\lambda = h/p$ and extracting the momentum p from the function (2.55) (i.e., $p^2 = 2m[E - U]$ where U is a potential) we finally obtain a conventional time-independent Schrödinger equation

$$\Delta\psi + \frac{2m_0 [E - U]}{\hbar^2} \psi = 0 \tag{2.68}$$

In the end of this section, we shall consider how the amplitudes of the particle and its inerton cloud are related. The frequency of collisions of the particle with the cloud can be written in two ways. On the one hand, the frequency of the particle's collisions with the inerton cloud is $1/(T/2) = \upsilon_0/\lambda$, according to relation (2.30). On the other hand, according to relation (2.25), the frequency of collisions of the inerton cloud with the particle is $1/(T/2) = c/\Lambda$. Combining these two relations, we get an interesting relationship

$$\Lambda = \lambda \frac{c}{\upsilon_0} \equiv \lambda_{\text{dB}} \frac{c}{\upsilon_0} \tag{2.69}$$

which binds the amplitudes of the particle (λ) and its inerton cloud (Λ). For example, in a hydrogen atom the Bohr radius $a_0 = 0.59177 \times 10^{-11}$ m, which means that the electron's de Broglie wavelength $\lambda = 2\pi a_0 = 3.3249 \times 10^{-10}$ m. The speed of the electron is $\upsilon_0 = 2.1877 \times 10^6$ m·s^{-1}. Therefore relationship (2.69) allows the calculation of the amplitude of the electron's inerton cloud: $\Lambda = \lambda c/\upsilon_0 = 1.45 \times 10^{-8}$ m. Thus, the electron's inertons reach a distance that exceeds the size of the hydrogen atom by about two orders of magnitude.

2.5 WAVE RELATIVISTIC MECHANICS

When the velocity υ_0 of a moving particle is close to the speed of light c, then the system under consideration is characterized by the relativistic Lagrangian (2.47). If the ensemble of inertons is treated as an entire object, namely, the particle's inerton cloud with an effective mass at rest μ_0, then the Lagrangian (2.47) in the Euclidean space becomes

$$L_{\text{rel.}} = -m_0 c^2 \left\{ 1 - \frac{1}{m_0 c^2} \left[m_0 \dot{x}^2 + \mu_0 \dot{\chi}^2 - \frac{4}{T} \sqrt{m_0 \mu_0} (x \dot{\chi} + \upsilon_0 \chi) \right] \right\}^{1/2} \tag{2.70}$$

The function (2.70) describes the particle with the mass at rest m_0 that moves along the X-axis with the velocity $\dot{x}$ (υ_0 is the initial velocity); χ is the distance between the inerton cloud and the particle, $\dot{\chi}$ is the velocity of the inerton cloud in the frame of reference connected with the particle, and $2/T$ is the frequency of collisions between the particle and the inerton cloud. x and χ are functions of the proper time t of the particle, which is treated as a natural parameter, i.e., $t = l/\upsilon_0$ where l is the length of the particle path.

The Euler-Lagrange equations for the Lagrangian (2.70) are reduced to the following system (where the relationship (2.14) is used)

$$\ddot{x} + \frac{2}{T}\frac{\upsilon_0}{c}\dot{\chi} = 0 \tag{2.71}$$

$$\ddot{\chi} - \frac{2}{T}\frac{c}{\upsilon_0}(\dot{x} - \upsilon_0) = 0 \tag{2.72}$$

The solution for the inerton cloud obtained from Eqs. (2.71) and (2.72) has the form

$$\chi = \Lambda \left| \sin[t/(T/2)] \right| \tag{2.73}$$

$$\dot{\chi} = c \frac{|\sin[t/(T/2)]|}{\sin[t/(T/2)]} \cos[t/(T/2)] = c(-1)^{\lfloor t/T \rfloor} \cos[t/(T/2)] \tag{2.74}$$

The solution for the particle ($\dot{x}$ and x) complies with expressions (2.41)–(2.42). The solutions for the particle and its inerton cloud satisfy the initial conditions $x(0) = \chi(0) = 0$ in relation to (2.40) and the inequalities $x,\ \dot{x},\ \chi \geq 0$ and $c \geq \dot{\chi} \geq -c$.
If we introduce a new variable $\tilde{\chi}$ by the rule

$$\dot{\chi} = \dot{\tilde{\chi}} + 2\sqrt{m_0 / \mu_0}\, x / T \tag{2.75}$$

into the Lagrangian (2.70), we obtain

$$L_{\text{rel.}} = -m_0 c^2 \left\{ 1 - \frac{1}{m_0 c^2} \left[m_0 \dot{x}^2 - m_0 (1/T)^2 x^2 + \mu_0 \dot{\tilde{x}}^2 - \frac{4}{T} \sqrt{m_0 \mu_0}\, \upsilon_0 x \right] \right\}^{1/2} \tag{2.76}$$

Here, in the Lagrangian (2.76) the first two terms in the brackets under the radical describe the particle in the harmonic potential; the last two terms exhibit the effective kinetic and potential energy of the inerton cloud.

Let us write the Hamiltonian (2.52) that corresponds to the Lagrangian (2.76) introducing the cyclic frequency:

$$H_{\text{rel.}} = m\,(2\pi / T)^2 x^2 + m c^2 + 2\pi \sqrt{m \mu_0}\, \upsilon_0 x / T \tag{2.77}$$

where $m = m_0 / \sqrt{1 - \upsilon_0^2 / c^2}$. When deriving the Hamiltonian (2.77), we have taken into account that in the Lagrangian (2.76) the radical $\{\ldots\}^{1/2}$ is a constant equal to $\sqrt{1 - \upsilon_0^2 / c^2}$.

On the other hand, in accordance with the derivation of momenta for the particle (p) and the inerton cloud ($\tilde{p}$), we have

$$p = \partial L_{\text{rel.}} / \partial \dot{x} = m\dot{x},\ \tilde{p} = \partial L_{\text{rel.}} / \partial \dot{\tilde{x}} = m\dot{\tilde{x}} \tag{2.78}$$

Relations (2.78) and the Lagrangian (2.76) allow the presentation of the Hamiltonian (2.77) as below

$$H_{\text{rel.}} = p^2 / m + \tilde{p}^2 / \mu + (m_0 c^2) / m \tag{2.79}$$

where $\mu = \mu_0 / \sqrt{1-\upsilon_0^2/c^2}$.

Both presentations (2.77) and (2.79) of the relativistic Hamiltonian are equivalent. By combining these expressions, we get

$$H_{\text{rel.}} = p^2/(2m) + m(2\pi/T)^2 x^2/2 + (m^2 + m_0^2)c^2/ \\ (2m) + \tilde{p}^2/(2m) + 2\pi\sqrt{m_0\mu}\,\upsilon_0\chi/T \qquad (2.80)$$

The first two terms in expression (2.80) describe the particle and represent the Hamiltonian of the harmonic oscillator; the third term is the renormalized energy of the particle at rest, and the last two terms are the kinetic and renormalized potential energy of the inerton cloud.

Let us separate out of the Hamiltonian (2.80) the effective Hamiltonian of the particle that describes its behavior relative to the center of inertia of the {particle + its inerton cloud} system:

$$H_{\text{rel.}}^{(\text{effect.})} = p^2/(2m) + m(2\pi/T)^2 x^2/2 \qquad (2.81)$$

Thus we can write the Hamilton-Jacobi equation

$$(\partial S_1/\partial x)^2/(2m) + m(2\pi/T)^2 x^2/2 = E \qquad (2.82)$$

and then following the calculations carried out in Section 2.2 (namely, see expressions (2.55)-(2.63)) we finally arrive at the de Broglie relationships

$$E = h\nu, \ \lambda = h/(m\upsilon_0) \qquad (2.83)$$

where the particle's frequency ν = 1/*T*, the kinetic energy $E = m\upsilon_0^2/2$ and the mass $m = m_0/\sqrt{1-\upsilon_0^2/c^2}$.

These relationships (2.83) admit the derivation of the Schrödinger wave equation (2.68) in such a way that it is Lorentz-invariant, because it includes the contracting factor $\sqrt{1-\upsilon_0^2/c^2}$ and includes the time t as a natural parameter.

By definition (1.51) the mass of the particle as a local volumetric fractal deformation is the ratio between the volume $\mathcal{V}^{\text{deg.}}$ of a degenerate cell and the volume $\mathcal{V}_0$ of the particulate cell: $m_0 = \text{const}\ \mathcal{V}^{\text{deg.}}/\mathcal{V}_0$. As the particle moves along the *X*-axis, the contraction of its size by a factor of $\sqrt{1-\upsilon_0^2/c^2}$

in this direction automatically leads to an increase in its mass by a factor of $1/\sqrt{1-\upsilon_0^2/c^2}$. Indeed, as $m_0 \propto \mathcal{V}_0^{-1} = (R_{0x}R_{0y}R_{0z})^{-1}$, then

$$m = m_0 / \sqrt{1-\upsilon_0^2/c^2} \propto \left[\left(R_{0x}\sqrt{1-\upsilon_0^2/c^2}\right)R_{0y}R_{0z}\right]^{-1} \tag{2.84}$$

where R_{0x} is the typical size of the particle in the state of rest along the axes $i = \{X, Y, Z\}$.

The result (2.84) can also be obtained in another way, namely, from the consideration of the particle travelling together with its deformation coat as a fluid element of a liquid in hydrodynamics. We may start from the known equation (see e.g., Ref. [116])

$$\rho\, d\vec{\upsilon}/dt = -\nabla P \tag{2.85}$$

where ρ is the density of the chosen fluid element and P is the pressure in the liquid in the place of the element. We assume that the motion is adiabatic, i.e., the change in pressure on the element caused by the whole liquid is proportional to the variation in density of this element (Ref. [116], p. 351):

$$(\partial P/\partial\rho)_{\text{entropy}} = c^2 \tag{2.86}$$

where c is the maximum velocity for this liquid (sound velocity). Equation (2.85) is nonlinear and consequently allows for multiple solutions. However, there is only one possibility when the equation becomes linear, and, therefore, it has a unique solution.

In hydrodynamics, the point is limited by the proper size of the element of the liquid. This size is enormous compared with the particle size. Hence the character of change of the substantial derivative $d\vec{\upsilon}/dt$ in Eq. (2.85) should be specified by a special condition. The main peculiarity is the non-stationary motion of the element: its velocity changes from υ_0 to 0, then from 0 to υ_0, etc., with each section λ and the time interval $T/2$. Hence the substantial derivative may be defined as

$$\frac{df(q)}{dq} = \lim_{\Delta q \to \left\{\begin{smallmatrix}\lambda/2\\ T/2\end{smallmatrix}\right.} \frac{f(q+\Delta q)}{\Delta q} = \lim_{\Delta q \to \left\{\begin{smallmatrix}\lambda/2\\ T/2\end{smallmatrix}\right.} \frac{\Delta f}{\Delta q} \tag{2.87}$$

Then Eq. (2.85) can be substituted by its discrete analogue:

$$\rho \Delta \upsilon / \Delta t = -c^2 \rho \Delta / \Delta l \tag{2.88}$$

where $\Delta l = \lambda$ and $\Delta t = T/2$ are the spatial and time intervals within which the velocity of the element changes from υ_0 to 0, i.e. $\Delta \upsilon = -\upsilon_0$.

When the velocity changes from υ_0 to 0, the pressure changes inversely – from the minimum value P to the maximum P_0; therefore $\Delta P = P - P_0$. Then, from the discretely analogous formula (2.86) $\Delta\rho = \rho - \rho_0$ $\Delta P / \Delta\rho = c^2$ we derive that $\Delta\rho = \rho - \rho_0$ where ρ and ρ_0 are the density of the liquid element at the moment of motion and in the state of rest, respectively. Hence, from Eq. (2.88), with regard to the relationship of $2\lambda/T = \upsilon_0$ we gain

$$\rho = \rho_0 / (1 - \upsilon_0^2 / c^2) \tag{2.89}$$

At the next stage of the motion of the element, from λ to 2λ, the parameters of Eq. (2.88) are: $\Delta l = \lambda$, $\Delta t = T/2$, $\Delta \upsilon = \upsilon_0$ and $\Delta\rho = \rho_0 - \rho$, which again results in expression (2.89). Expression (2.89) is true even at $\upsilon_0 \to c$.

The liquid element studied can be associated with the particle's cloud of inertons. Inertons should carry the same mass as the particle does, as they carry fractal volumetric fragments of the particle. Hence the value of the particle's volume $\mathcal{V}_0$ plays the role of an effective volume of the liquid element studied. So for the liquid element $m_0 \propto 1/\mathcal{V}_0$ and hence we have $\rho_0 \propto 1/\mathcal{V}_0^2$. As relationship (2.89) demonstrates, the motion of the element with the velocity υ_0 results in a decrease of the total volume of the element along the direction of its motion: $\rho_0 \to \rho \propto 1/(\mathcal{V}_0\sqrt{1 - \upsilon_0^2/c^2})^2$. Then the mass of the element and also the mass of the particle (due to the interdependence of these masses) changes:

$$m_0 \propto 1/\mathcal{V}_0 \to 1/(\mathcal{V}_0\sqrt{1 - \upsilon_0^2/c^2})^2 \tag{2.90}$$

Thus the mass increases in the direction of the velocity vector by a factor of $1/\sqrt{1 - \upsilon_0^2/c^2}$, which is in agreement with the experimental fact and the formalism of special relativity.

KEYWORDS

- **conceptual difficulties**
- **inertons**
- **quantum mechanics**
- **submicroscopic mechanics**
- **tessellattice**
- **wave mechanics**

CHAPTER 3

INERTONS UNVEILED

CONTENTS

As we have seen in Chapter 2 concerning Poincaré's idea [9] about the motion of a particle being surrounded with excitations of the ether and de Broglie's idea [11] for a double solution theory to explain that a real wave guides the particle, both of these ideas have now been realized. The development of the submicroscopic theory became possible owing to the theory of physical space described in Chapter 1.

In the present chapter, we will consider important peculiarities of quantum systems by using the submicroscopic concept, which cannot be investigated in the framework of the usual formalism of quantum mechanics. We also will return to the derivation of the Schrödinger equation, the physical interpretation of the wave ψ-function and will look into the interior of other well-established notions of quantum mechanics.

Therefore, to begin, we will consider the basic notions of quantum mechanics, such as tunneling and Heisenberg's uncertainty principle.

3.1 THE PHENOMENON OF TUNNELING

Quantum tunneling is a quantum mechanical phenomenon where a particle tunnels through an energy barrier that classically cannot be overcome. Tunneling is often explained using the wave–particle duality of matter and somehow through the Heisenberg uncertainty principle.

The first detailed theoretical consideration of the phenomenon of the tunneling effect was published in 1928 by Mandelstam and Leontowitsch [117]. In the same year, the tunneling effect was also independently suggested by other researchers [118–120]; they described alpha decay of a series of unstable nuclei, alpha-emitters, by using the WKB approximation [121–123]. Since then, a number of modern methods in application to the problem of tunneling have been described [124]. Moreover, Turok [125] performed a description of quantum tunneling in real time using complex classical trajectories.

Of course the starting point for the description of the phenomenon is the one-dimensional time independent Schrödinger equation

$$-\frac{\hbar}{2m}\frac{d^2\psi}{dx^2} + U(x)\psi = E\psi \tag{3.1}$$

where $U(x)$ is the potential energy of the barrier and E is the energy of the particle with a mass m. Figure 3.1 depicts the simplest rectangular form of the barrier. The barrier divides Eq. (3.1) into three equations: before the barrier, inside the barrier and after the barrier.

For each of the three zones one can write the appropriate wave ψ_i-function, where $i = 1, 2, 3$. Knowing all three functions, we then can calculate the density of the particle flux. The barrier transmission coefficient is defined by a ratio of the flux of passing particles to the flux of incident particles

$$\mathcal{T}(E) = C\exp\left\{-2\int_{x_1}^{x_2}\sqrt{2m[U(x)-E]/\hbar^2}\,dx\right\} \tag{3.2}$$

where C is a constant close to unit. In the case of the rectangular barrier (Figure 3.1) the transmission coefficient is $\mathcal{T}(E) \approx \exp\{-2\sqrt{2m[U(x)-E]/\hbar^2}\, l\}$.

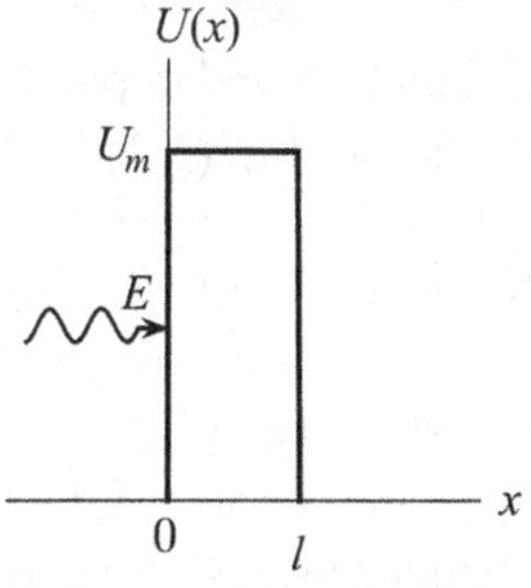

FIGURE 3.1 The simplest potential barrier whose height is U_m and the width is l.

However, the real distance that an electron passes inside the barrier cannot be predicted and evaluated in the framework of the quantum mechanical formalism. Only an experimental study utilizing a highly precise scanning tunneling microscope permits an estimation of such distance. In that case, the barrier is due to the gap between the tip of the microscope and the surface of the object studied. Since the tunnel current following the transmission coefficient depends exponentially on the barrier width, this device is extremely sensitive to height variations on the examined sample.

How is the problem of the deep penetration of a particle seen in submicroscopic mechanics? The particle falls on the barrier with the energy $E = m\upsilon_0^2 / 2$ (Figure 3.1) and its velocity and position are described by expressions (2.41) and (2.42), respectively. The particle cannot stop immediately when it suddenly meets an obstacle at the position $x = 0$. The particle continues to move inside the barrier until it passes the distance λ as prescribed by solution (2.42), because the particle is guided by inertons. However, the obstacle with the potential $U(x)$ of course affects the particle because the potential tries to slow down the particle's inertons.

The problem is easy to solve by considering the oscillations of the particle near the center-of-mass of the system {particle + its inerton cloud}, which is sketched by the Hamiltonian (2.53). When the particle enters the barrier, the energy of oscillations decreases and the particle starts to obey the equation of the damped harmonic oscillator

$$\ddot{x} + 2\beta\dot{x} + (2\pi / \bar{T})^2 x = 0 \tag{3.3}$$

where β is a constant with the dimension of angular frequency, which specifies the strength of the damping and depends on the potential *U(x)*; $\overline{T}$ is a peculiar period. The solution to Eq. (3.3) is

$$x(t) = \overline{\lambda}\, e^{-\beta t} \sin(2\pi t / \overline{T}) \tag{3.4}$$

where the amplitude of the particle is $\overline{\lambda}$ (2.64). It is obvious $\overline{\lambda}$ can spread up to the value of the double spatial period of the particle, i.e., $2\lambda_{\text{dB}}$. Hence during the time $t = \overline{T}/2$ until inertons push the particle, the particle is wedging into the barrier up to a distance $x_{\text{max}} = 2\lambda_{\text{dB}} e^{-\beta T/2} \sin(\pi/2)$. Putting the damping constant $\beta = 1/(T/2)$ we can estimate the maximum distance $x_{\text{max}} \approx 2\lambda_{\text{dB}} e^{-1} = 0.73576\,\lambda_{\text{dB}}$. For example, take the case of an electron with a speed $\upsilon_0 = 10^6$ m/s (which corresponds to the energy 2.8 eV) in a metal and with a de Broglie wavelength $\lambda_{\text{dB}} = 7.27 \times 10^{-10}$ m. Then the maximum distance, over which the electron seeps through the barrier will be equal to $x_{\text{max}} \approx 5.35 \times 10^{-10}$ m. The time needed for the electron to overcome this distance in the linear approximation is $t = \lambda_{\text{dB}} / \upsilon_0 = T/2 \approx 2.3 \times 10^{-16}$ s. Note that many researchers studied a time associated with the passage of a particle under tunneling barrier; however, all those considerations were based solely on the hypothesis of an abstract wave packet that tunneled through the barrier studied (see, e.g., Ref. [126]).

3.2 THE UNCERTAINTY PRINCIPLE

The uncertainty principle, which is established in quantum mechanics, declares that particles cannot have separate well-defined meanings of positions and velocities, but only a quantum state – a peculiar combination of position and velocity. Properties, which are not known, have to be delineated by probabilities based on the wave ψ-functions.

Heisenberg [127] inferred his formulation of the uncertainty principle via a thought experiment in which the position of an electron was measured using a gamma-ray microscope. His formula is $\varepsilon(q)\,\eta(p) \geq h/(4\pi)$: when one measures the position of an electron with an error $\varepsilon(q)$, the momentum of the electron also acquires the additional amount $\eta(p)$. An experimenter cannot know both the position and the momentum at the same moment in time. Heisenberg noted that a selection from an abundance of possibilities takes place in quantum systems, which also puts a limitation on future outcomes. Heisenberg worked with a broadened wave packet with calculable probability – a typical approach to the description of quantum systems.

Later on Kennard [128] derived a different formulation of the uncertainty principle, which was later generalized by Robertson [129]: $\sigma(q)\,\sigma(p) \geq h/(4\pi)$, i.e., one cannot suppress quantum fluctuations of both position $\sigma(q)$ and momentum $\sigma(q)$ lower than a certain limit simultaneously. The fluctuations exist within themselves without respect to whether the position q and the momentum p of the quantum system are measured or not. This inequality cannot foresee any behavior of the parameters q and p at the time of measurement. Nowadays it seems that Kennard's formulation of the uncertainty principle is more often used.

Other authors deriving the uncertainty relation

$$\Delta x \Delta p \geq \tfrac{1}{2}\hbar \tag{3.5}$$

based their consideration on the behavior of a wave packet of finite length (see, e.g., Fermi [130] who referred to the proof conducted by E. Persico – see Engl. translation: Persico, E. (1950), *Fundamentals of Quantum Mechanics*, Prentice-Hall, New York) or the representation of the wave ψ-function as a superposition of plane waves corresponding to the discrete spectrum (Born [90], p. 383), and so on.

In such a manner the uncertainty principle is considered as a corollary of the wave-particle duality of nature when a canonical particle is called a wave-particle and then all the characteristics of classical waves are automatically attributed to the particle. However, a wave packet is not stable and dissipates over time. Hence the description of a particle by using a wave packet is an approximate description. De Broglie [111] in his book devoted to the uncertainty principle analyzed this topic as well as the principle of spectral expansion in detail. He showed that the uncertainty relation (3.5) is due to: (i) the attribution of a wave function ψ to a particle, (ii) the reliance on the properties of Fourier series and Fourier integrals, and (iii) the recognition of the commutation relation for the operators of the particle's position and momentum

$$[\hat{x},\ \hat{p}] = ih, \text{ or } \hat{x}\hat{p} - \hat{p}\hat{x} = ih \tag{3.6}$$

Examining a forth uncertainty relation, between energy and time, $\delta E \cdot \delta t \sim h$, de Broglie noted that it is also a consequence of the Fourier expansion analysis.

Therefore, in conventional quantum mechanics the uncertainty relations emerge only owing to the consideration of the wave part of the wave-particle duality (a superposition of monochromatic waves or a wave package). It is this approximation that gives rise to the inequality $\Delta x\, \Delta k > 1$ for changes of the position Δx and the wave number Δk of the wave package studied; multiplying this inequality by the Planck's constant, we derive the uncertainty relation (3.5) [11].

However, in spite of this way of analysis we have seen in Section 2.2 that a subquantum system {particle + its inerton cloud} is characterized by the action S that can be written in two presentations: through the particle's classical parameters – the kinetic energy E, mass m and velocity υ_0 (2.62) and via a position x of the particle, the amplitude $\lambda \equiv \lambda_{\mathrm{dB}}$ of the particle spatial oscillations (the de Broglie wavelength), and the frequency ν of collision of the particle with its inerton cloud (2.63). The second of these two presentations exactly copies the presentation of a typical classical wave, but the first one also requires taking the properties of the particle strictly into account (2.62). Considering the two presentations of the action together call for the submicroscopic mechanics stated in Chapter 2, which makes it possible to predict both the particle's position x, its velocity υ_0 and hence the momentum $p=m\upsilon_0$. This particularly means that the origin of the commutator (3.6) is not in the uncertainties between Δx and Δk, but in the relationships (2.60)–(2.64). These relationships are due to the oscillating motion of the particle and its inerton cloud in each section 2λ of the particle's path, which leads to the quantum increment in the action by the value h.

Just to show that the uncertainty principle is not so fundamental, the reader is referred to the studies of Nobel Prize winner Hans Georg Dehmelt and his collaborators [131–133]; in a series of high precision measurements they were able to gauge a number of parameters of an electron practically at rest.

Hofer [134, 135] revealed in high-resolution scanning tunneling microscopy experiments that the interpretation of the density of electron charge as a statistical quantity leads to a conflict with the Heisenberg uncertainty principle; the uncertainty principle was violated by close to two orders of magnitude. Moreover, in a more recent work Hofer [136] concluded: "measured with errors of less than 0.1 pm, energies with about 0.1 meV, and temperature levels below 20 mK – it can be inferred that standard quantum mechanics, with its inherent uncertainties, is a model at the end of its natural lifetime."

Boyd [137] has stated that, concerning the refutation of Heisenberg's uncertainty regarding photons: "NASA has developed a timer system accurate to 10 femtoseconds, with projections of improvements into the .001 femtosecond regime. Emission time is then not an issue over a premeasured course and thus location of the photon is known to within the accuracy limits of the timer. We can know momentum with certainty because the momentum of a photon is directly related to its frequency. If you know the frequency, you know the momentum. The other parameters mentioned above follow along similar lines. Thus, contrary to Heisenberg, I can know both the momentum and the position of the photon simultaneously, with absolute accuracy."

Dumitru [138] examined a number of possible applications of uncertainty relations and came to the conclusion that they do not have any crucial significance for physics. Moreover, he finished his paper with words by Dirac [139]: "uncertainty relations in their present form will not survive in the physics of the future."

Besides, we have to mention the phenomenon of quantum entanglement: a pair or particles interacts in such a way that the quantum state of each particle cannot be described independently—instead, a quantum state is given for the system as a whole. Measurements of physical properties of entangled particles (position, momentum, spin, polarization, etc.) show the whole correlation. It would seem pointless to qualify these words here with source references as they have already become abundant in academic publications. At last, on the web site of CODATA (http://physics.nist.gov/cuu/Constants/index.html), internationally recommended values of the fundamental physical constants have been defined with extreme accuracy and the uncertainty principle did not in any way affect these final definitions.

All the arguments above allow us to state that the uncertainty principle must be refuted forever.

3.3 THE PHASE TRANSITION FROM SCHRÖDINGER'S TO DIRAC'S

The Schrödinger wave equation considered in Chapter 2 is specified with the frequency $\nu = m\upsilon_0^2/(2h)$, which is given by a particle having the kinetic energy $m\upsilon_0^2/2$, where $m = m_0/\sqrt{1-\upsilon_0^2/c^2}$. In the relativistic quantum theory the frequency

$$\nu_{\text{rel.}} = \sqrt{p^2c^2 + m_0^2c^4}\,/\,h \tag{3.7}$$

characterizes the spectrum of a wave-particle in the Dirac wave equation

$$(i\hbar\partial/\partial t - \widehat{H}_{\text{Dirac}})\psi_{\text{Dirac}} = 0 \tag{3.8}$$

In as much as the energy of the particle at rest m_0c^2 is a peculiar intrinsic potential energy, it is reasonable for our model described in Chapter 2 that this energy does not displace itself in the particle spectrum. So in what manner can the hidden energy m_0c^2 be evident in an explicit form (3.7) in the Dirac formalism? The answer is concealed in the tessellattice that surrounds the particle. The deformation coat emerging around the particled cell is a linear response to this local defect in the tessellattice, which screens the defect. The radius of the deformation coat is the particle's Compton wavelength $\lambda_{\text{Com}} = h/(m_0c)$. Cells in the deformation coat are in a tensioned state, which aspires to decompose the fractal state of the particulate cell. This means that fractal fragments under such a tension may indeed dance in the framework of the deformation coat. Harmonic oscillations of fractal fragments between the kernel cell and the whole deformation coat are able to support the balance in the system. So the situation looks like as though the fractals are partly distributed by the coat, i.e., the kernel cell periodically throws off its mass into the cells of the coat and the mass is then returned to the particle by the elastic tessellattice.

Therefore, the deformation coat can be considered as a crystallite with N entities that have been named "superparticles" [140]. Each entity can be described by the lattice vector $\mathbf{n}$. The $\mathbf{n}$-entity is characterized by a mass m_{n} and a displacement from the equilibrium position whose components are $\zeta_{\mathbf{n}\beta}$ (β =1, 2, 3). Entities interact only with the nearest neighbors and consequently, if the position of a certain cell depends on the vector **n**, then its nearest neighbor can be described by the vector **n** + **a**, where **a** is the crystallite structure constant. The Lagrangian of the crystallite in the harmonic approximation appears as

$$L = \frac{1}{2}\sum_{\mathbf{n},\beta} m_{\mathbf{n}} \dot{\zeta}^{2}_{\mathbf{n}\beta} - \frac{1}{2}\sum_{\mathbf{n},\beta\beta'} \gamma_{\beta\beta'} (\zeta_{\mathbf{n}\beta} - \zeta_{\mathbf{n}-\mathbf{a}\,\beta})^2 \tag{3.9}$$

where $\gamma_{\beta\beta'}$ is the crystallite elasticity tensor.

In the case of a solid in which all masses are the same, i.e., $m_{\mathrm{n}} = m$, it is more convenient to apply the Lagrangian (3.9) to the collective variables $A_k = (A_{-k})^*$ by using the canonical transformation

$$\zeta_{\mathbf{n}\beta} = \frac{1}{\sqrt{N}}\sum_{\mathbf{k}} e_\beta(\mathbf{k}) A_{\mathbf{k}} \exp(i\,\mathbf{k}\,\mathbf{n}) \tag{3.10}$$

where the quantities $e_\beta(\mathbf{k})$ represent the three vibration branches (one longitudinal and two transverse). The transformation allows one to obtain the Hamiltonian function

$$H = \frac{1}{2}\sum_{\mathbf{k},\beta}\left[\frac{1}{m} P_{\mathbf{k}\beta} P_{-\mathbf{k}\beta} + m\,\Omega^2_\beta(\mathbf{k})\, A_{\mathbf{k}\beta} A_{-\mathbf{k}\beta}\right] \tag{3.11}$$

and then the Hamiltonian operator

$$\widehat{H} = \frac{1}{2}\sum_{\mathbf{k},\beta} \hbar\Omega_\beta(\mathbf{k}) \left[\hat{b}^{+}_{\mathbf{k}\beta} \hat{b}_{\mathbf{k}\beta} + \frac{1}{2}\right] \tag{3.12}$$

The energy E of vibrations of the solid crystal is determined as

$$E = \sum_{\mathbf{k},\beta} \hbar\Omega_\beta(\mathbf{k}) f_{\mathbf{k}} \tag{3.13}$$

where f_{k} is the Planck distribution function for phonons. Expression (3.13) shows that the energy spectrum of the solid crystal is formed by the sum of the whole set of its vibrations.

However, in the present case of the crystallite formed around a particle in the tessellattice the situation is different. When the particle is born, the crystallite is quickly formed adiabatically around it, with the velocity c. Therefore, when the crystallite is formed, the entities involved in its formation are coherently excited and, as a result, the whole crystallite, being at a zero temperature, appears to be in only one, the lowest excited state. Hence the transition to the collective variables (3.10) does not incorporate the whole set of states. The transformation to the variable A_{k} should rather be chosen in the following form

$$\zeta_{\mathbf{n}\beta} = e_{\beta}(\mathbf{k})\, A_{\mathbf{k}} \exp(i\,\mathbf{k}\,\mathbf{n}_{\beta}) \tag{3.14}$$

which reflects the specific initial conditions of the crystallite formation. Substituting the relation (3.14) into the Lagrangian (3.9) we gain the crystallite Lagrangian written in the form of the collective variables and their derivatives

$$L_{\mathrm{cr.}} = \tfrac{1}{2}\sum_{\mathbf{n}} (m_{\mathbf{n}} \dot{A}_{\mathbf{k}} \dot{A}^{*}_{\mathbf{k}} - m_{\mathbf{n}} \omega^{2}_{\mathbf{k}} A_{\mathbf{k}} A^{*}_{\mathbf{k}}) \tag{3.15}$$

where the notation

$$\sum_{\mathbf{n}} m_{\mathbf{n}} \omega^{2}_{\mathbf{k}} = \sum_{\mathbf{n}} \delta_{\mathbf{nn}}\, 4 \sum_{\beta\beta'} \gamma_{\beta\beta'} \sin^{2}(\mathbf{k}\mathbf{n}_{\beta'}) \tag{3.16}$$

is introduced. In the long-wave approximation, $a\,k << 1$, the relation (3.16) becomes

$$\omega^{2}_{\mathbf{k}} \cong N \sum_{\beta\beta'} \gamma_{\beta\beta'} (k a_{\beta'})^{2} / m_{\Sigma} \tag{3.17}$$

where $m_{\Sigma} = \sum_{\mathbf{n}} m_{\mathbf{n}}$. From the relation (3.17) we derive the expression for the sound velocity of the collective vibrations of the crystallite, which is identified with the fundamental speed, i.e., the velocity of light,

$$c = \omega_{\mathbf{k}} / k_{N} = a \sqrt{N \sum_{\beta\beta'} \gamma_{\beta\beta'} / m_{\Sigma}} \tag{3.18}$$

As it follows from the Lagrangian (3.15), the generalized momentum of the crystallite vibration mode is

$$P_{\mathbf{k}} = \partial L_{\text{cr.}} / \partial \dot{A}_{\mathbf{k}} = m_{\Sigma} \dot{A}^{*}_{\mathbf{k}} \tag{3.19}$$

Then the spectrum of the crystallite becomes

$$E_{\mathbf{k}} = \tfrac{1}{2} | P_{\mathbf{k}} |^2 / m_{\Sigma} + \tfrac{1}{2} m_{\Sigma} \omega_{\mathbf{k}}^2 | A_{\mathbf{k}} |^2 \tag{3.20}$$

Having the energy (3.20), we can construct the action and then can write the Hamilton-Jacobi equation

$$\tfrac{1}{2} (\partial S_{\mathbf{k}} / \partial | A_{\mathbf{k}} |)^2 / m_{\Sigma} + \tfrac{1}{2} m_{\Sigma} \omega_{\mathbf{k}}^2 | A_{\mathbf{k}} |^2 = E_{\mathbf{k}} \tag{3.21}$$

Using the action-angle variables equation (3.21) allows us to derive an increment of the action per the period

$$J = E_{\mathbf{k}} / [\omega_{\mathbf{k}} / (2\pi)] \tag{3.22}$$

Equating $J / (2\pi) = \hbar$ we get from expression (3.22)

$$E_{\mathbf{k}_0} = \hbar \omega_{\mathbf{k}_0} \tag{3.23}$$

Here, $\omega_{\mathbf{k}_0}$ is the cyclic frequency that describes standing spherical oscillations of the deformation coat.

The energy of the particle at rest $m_0 c^2$ should be equal to the energy of the deformation coat (3.23), which results in the relation

$$\hbar \omega_{\mathbf{k}_0} = m_0 c^2 \tag{3.24}$$

where $\omega_{\mathbf{k}_0}$ is the cyclic frequency and $| \mathbf{k}_0 | = k_0 = 2\pi / \lambda_0$ is the wave number of the oscillating crystallite in the **k**-space; λ_0 is the amplitude of oscillations, which is given by the crystallite size. From the relation (3.24) we immediately derive

$$\lambda_0 = m_0 c / h \tag{3.25}$$

which shows that λ_0 is nothing but the Compton wavelength λ_{Com} of the particle located in the center of the crystallite, i.e., the deformation coat.

The moving particle possesses the total energy mc^2, which also changes the energy of the crystallite mode to

$$\hbar\omega_{\mathbf{k}(\mathbf{v})} = mc^2 \quad (3.26)$$

and then the amplitude of oscillation along the velocity vector $\vec{\upsilon}_0$ shrinks in size:

$$\lambda_{\text{Com},\,\upsilon} = \lambda_{\text{Com}}\sqrt{1-\upsilon_0^2/c^2} \quad (3.27)$$

Thereby,

$$\lambda_{\text{Com},\,\upsilon} = h/(mc) \quad (3.28)$$

The results (3.27) and (3.28) allow us to conclude that the frequency (3.7), which supposedly describes the wave behavior of the particle in the Dirac theory, in reality features the collective vibrations of the deformation coat (or the crystallite):

$$\nu_{\text{rel.}} = \omega_{\mathbf{k}(\mathbf{v})}/(2\pi) = mc^2/h \quad (3.29)$$

Let us now turn our consideration to inertons emitted by the particle at its motion. The inerton cloud spreads out to the distance Λ from the particle in the transverse directions. The amplitude Λ of oscillations of the inerton cloud is connected to the amplitude λ of spatial oscillations of the particle by the relationship (2.69). Note a similar relationship exists between λ and the Compton wavelength: when comparing the formulas $\lambda_{\text{dB}} = h/(m\upsilon_0)$ and $\lambda_{\text{Com},\upsilon} = h/(mc)$, we derive

$$\lambda_{\text{dB}} = \lambda_{\text{Com},\,\upsilon}\, c/\upsilon_0 \quad (3.30)$$

From relations (3.30) and (2.69) we deduce a very interesting relationship

$$\Lambda = \lambda_{\text{Com},\,\upsilon}\, c^2/\upsilon_0^2 \quad (3.31)$$

It follows from relation (3.31) that when a particle's motion is slow ($\upsilon_0^2 << c^2$) the amplitude of the inerton cloud significantly exceeds the size of the deformation coat, i.e., $\Lambda >> \lambda_{\text{Com},\,\upsilon}$ (Figure 2.1). This means that the inerton cloud guides the particle and already probes the space to a distance about Λ from the particle. The inerton cloud is able to interact with obstacles and transfers the appropriate information to the particle. Such motion is close to the motion that de Broglie called "motion by guidance," which he related to a constant intervention from a subquantum medium. Hence in this case while measuring the coordinate and/or the momentum of the particle along the direction of its motion, an instrument will measure the inerton cloud that carries the same kinetic energy along the particle's path as the particle itself, $m\upsilon_0^2/2$. Thus the measurement of the deformation coat is apparently not displayed in the so-called nonrelativistic approximation. Therefore, in the present case the Schrödinger and Pauli equations can be used in order to analyze the particle behavior.

In the so-called relativistic limit, $\upsilon_0 \to c$, it is evident from the relation (3.31) that $\Lambda \approx \lambda_{\text{Com},\,\upsilon}$, which implies that the inerton cloud is practically completely closed in the deformation coat. Because of this, when measuring the coordinate or the energy of the particle along the direction of the particle's motion, the instrument will register the entire moving region, i.e., both the inerton cloud and the deformation coat. But the total energy $\hbar\omega_{\mathbf{k}_\upsilon}$ of the latter exceeds the kinetic energy $m\upsilon_0^2/2$ of the particle (as well as the inerton cloud) and, consequently, in this case the energy of the particle at rest m_0c^2 will explicitly be manifested. This approximation falls within the Dirac equation formalism.

3.4 SPIN AND THE PAULI EXCLUSION PRINCIPLE

A number of papers have been devoted to the study of different aspects associated with the notion of spin (see, e.g., Refs. [140–142]), which usually is treated as an intrinsic form of angular momentum carried by an elementary particle. Spin is treated like a vector quantity, which is an inner property of a particle and this characteristic is rather isotropic for a free particle; nevertheless spin has a definite magnitude and orientation, but quantization makes this orientation different from that of an ordinary vector.

Recently Hofer [140, 135] has proposed a theory of extended electrons in which their wave properties are related to some form of density oscillation. Namely, according to his theory a free electron traveling along the z-axis with a constant velocity υ has to undergo a density oscillation, which is described by a plane wave: $\rho(z,\ t) = \rho_0 \cdot [1 + \cos(4\pi z / \lambda - 4\pi v\, t)]/2$. In comparison, the Schrödinger wave function is determined as

$$\psi = \sqrt{\rho_0}\, \exp\left[i\left(2\pi z / \lambda - 2\pi v t\right)\right] \tag{3.32}$$

The spin of an electron is defined as

$$\psi \mathbf{e}_2 \psi^+ = \rho_0\left[\mathbf{e}_2 + \sin\left(2\pi z / \lambda - 2\pi v t\right)\mathbf{e}_1\right] \tag{3.33}$$

where the directions of the reference vectors $\mathbf{e}_1$ and $\mathbf{e}_2$ are perpendicular to the direction of electron motion. Thereby the spin vector has the form $\mathbf{s} = \frac{1}{2}\psi \mathbf{e}_2 \psi^+$ and is oriented under an angle of $\pi/4$ to the direction of the electron velocity vector. Thus Hofer, using geometric algebra (also known as Clifford algebra) determined the spin of an electron with respect to the velocity vector of the electron, i.e., as a property of the electron itself, but not with respect to the external magnetic field. Spin-properties of the electron are referred to intrinsic field components and such description satisfies the measurements of spin in an external field yielding the two possible opposite orientations. Hofer's theory allows the consideration of spin-dynamics of single electrons in terms of a modified Landau-Lifshitz equation, which is in agreement with experimental manifestations of spin.

Hofer's approach demonstrates the importance of the internal structure of the electron in understanding the notion of spin. In paper [140] an additional degree of freedom was introduced to the Lagrangian (3.70) of a relativistic particle to describe a kind of an intrinsic motion. When an external electromagnetic field is superimposed to the system, the intrinsic variables may engage with the field and, as a consequence, they will explicitly appear in the Pauli equation.

In the model [140], those intrinsic variables were associated with the particle's oscillations having a shape between a bean-like and spherical form (though in principal it could be any kind of an oscillation motion: pulsations, twisting, etc.), the appropriate Hamiltonian looked as follows

$$H_{\uparrow(\downarrow)} = \sqrt{c^2\mathbf{p}^2 + c^2\boldsymbol{\pi}_{\uparrow(\downarrow)}^2 + m_0^2c^4} \tag{3.34}$$

i.e., the canonical form $\sqrt{c^2\mathbf{p}^2 + m_0^2c^4}$ has been modernized by introducing an additional term $c^2\boldsymbol{\pi}_{\uparrow(\downarrow)}^2$ where the indexes ↑ and ↓ mean the direction of pulsation of the particle (forwards or backwards, respectively). The term $c^2\boldsymbol{\pi}_{\uparrow(\downarrow)}^2$ exhibits the energy of pulsations, which allows the linearization of the Hamiltonian $H_{\uparrow(\downarrow)}$, because logically, in this case, one may split the square root only when the expression under the radical contains matrix components. So in the function (3.34) we can move to the operators and then linearize the radical over $\hat{\mathbf{p}}$ with the exclusion of the intrinsic momentum operator $\hat{\boldsymbol{\pi}}_{\uparrow(\downarrow)}$, which results in the Dirac Hamiltonian operator

$$\hat{H}_{\text{Dirac}} = c\,\hat{\boldsymbol{\alpha}}\,\hat{\mathbf{p}} + \hat{\rho}_3 m_0 c^2 \tag{3.35}$$

and at this point the information presented in the operator $\hat{\boldsymbol{\pi}}_{\uparrow(\downarrow)}$ passes on to the $\hat{\boldsymbol{\alpha}}$-matrices.

Although such an idea has some sense, because it discloses a possible inner origin of the $\hat{\boldsymbol{\alpha}}$-matrices and spinors of the Dirac equation, the additional term written in the Hamiltonian (3.34) has not been observed in spectra of particles.

Nevertheless, based on his experiments Oudet [143–145] showed that all the spinor components ψ_j $(j = \overline{1,\ 4})$ in the Dirac equation could be regarded as being an exchange by "grains" between the electron and its field, where he understood a "grain" to be an element of the total electron mass. So Oudet's studies insist on involving some intrinsic processes inside the electron, which are responsible for its spin.

An important investigation was performed by Lévy-Leblond [146] (see also Ref. [147]): he was able to linearize the Schrödinger equation. The Schrödinger equation initially is presented in the operator form:

$$\hat{S}\psi = 0,\ \hat{S} \equiv i\hbar\partial/(\partial t) + \hbar^2\Delta/(2m) = \hat{E} - \hat{\mathbf{p}}^2/(2m) \tag{3.36}$$

The equation is symmetric with respect to time $(\partial/\partial t)$ and space $(\partial/\partial \mathbf{r})$ derivatives, but is quadratic in $\hat{\mathbf{p}}$. To reach the symmetry, Lévy-Leblond constructed a wave equation as below

$$\widehat{\Theta}\psi = (\widehat{A}\widehat{E} + \widehat{\boldsymbol{B}} \cdot \hat{\mathbf{p}} + \widehat{C})\psi = 0 \tag{3.37}$$

where $\widehat{A}, \widehat{\boldsymbol{B}}$ and $\widehat{C}$ are linear operators, rather than matrices. Finally, the result, which splits the Schrödinger equation to four linear equations, is as follows

$$\left[-i\begin{pmatrix} 0 & 0 \\ \mathbf{1} & 0 \end{pmatrix}\widehat{E} + \begin{pmatrix} \widehat{\sigma} & 0 \\ 0 & \widehat{\sigma} \end{pmatrix} \cdot \hat{\mathbf{p}} + 2mi\begin{pmatrix} 0 & \mathbf{1} \\ 0 & 0 \end{pmatrix}\right]\begin{pmatrix} \phi \\ \eta \end{pmatrix} = 0 \tag{3.38}$$

Here, the wave ψ-function becomes a 4-component matrix, in which we obtain

$$\phi = \begin{pmatrix} \phi_1 \\ \phi_2 \end{pmatrix}, \ \eta = \begin{pmatrix} \eta_1 \\ \eta_2 \end{pmatrix};$$

$\widehat{\sigma}$ is the vector whose components are the three Pauli matrices, $\widehat{\sigma} = \{\widehat{\sigma}_1, \ \widehat{\sigma}_2, \widehat{\sigma}_3\}$ and

$$\mathbf{1} = \begin{pmatrix} 1 & 0 \\ 0 & 1 \end{pmatrix}$$

is the unit matrix. Writing the matrix equations (3.38) results in the coupled system of equations for two-component spinors $\phi = (\phi_1, \ \phi_2)$ and $\eta = (\eta_1, \ \eta_2)$:

$$\widehat{\sigma} \cdot \hat{\mathbf{p}}\,\phi + i2m\eta = 0, \ \ \widehat{\sigma} \cdot \hat{\mathbf{p}}\,\eta - i\widehat{E}\phi = 0 \tag{3.39}$$

In the presence of an external electromagnetic field, the gauge invariance of the Schrödinger equation requires the substitution

$$i\hbar\partial/\partial t \ \rightarrow \ i\hbar\partial/\partial t - eV(\mathbf{r}, \ t), \ -i\hbar\nabla \ \rightarrow \ -i\hbar\nabla - e\mathbf{A}(\mathbf{r}, \ t)$$

and then the linear equations of motion (3.39) become

$$\widehat{\sigma} \cdot (\hat{\mathbf{p}} - e\mathbf{A})\phi + i2m\eta = 0, \widehat{\sigma} \cdot (\hat{\mathbf{p}} - e\mathbf{A})\eta \ - i(\widehat{E} - eV)\phi = 0 \tag{3.40}$$

After some transformations equations (3.40) are finally transferred to the Pauli equation

$$\left[\hat{E} - eV - \frac{1}{2m}\left(\hat{\mathbf{p}} - e\mathbf{A}\right)^2 + \frac{e\hbar}{2m}\hat{\sigma}\cdot\mathbf{B}\right]\phi = 0 \tag{3.41}$$

where $\hat{\mathbf{p}} = -i\hbar\nabla$ and the magnetic induction $\mathbf{B} = \nabla \times \mathbf{A}$.

The last term in equation (3.41) describes the interaction energy of the intrinsic magnetic moment of the electron with the external magnetic field.

Then Greiner [147] points out that the intrinsic magnetic moment $\hat{\mu} = e\hbar\hat{\sigma}/(2m)$ can be presented via the spin operator $\hat{\mathbf{S}} = \frac{1}{2}\hat{\sigma}$ of the particle studied:

$$\hat{\mu} = \frac{e\hbar}{m}\hat{\mathbf{S}} = g_{\text{spin}}\,\mu_{\text{B}}\,\hat{\mathbf{S}} = 2\mu_{\text{B}}\hat{\mathbf{S}} \tag{3.42}$$

where the spin-Landé factor, or the gyromagnetic ratio g_s, is equal to 2, and μ_{B} is the Bohr magneton. So the linearized theory establishes the correct intrinsic magnetic moment of a spin-1/2 particle. Greiner emphasized that the existence of spin is a consequence of the linearization of the wave equations; furthermore, Lord initially wrote the field equations in linearized form, i.e., a system of two coupled differential equations of first order, and then coupled the electromagnetic field. These linear equations coupled with the electromagnetic field arrive at the Pauli equation (3.41).

Thus it seems that the linearized form of the equations of motion appear to be the most fundamental, as it leads to the phenomenon of spin and is the basis for both the Pauli and the Schrödinger equations. Besides, the Dirac equation, which is obtained by linearization of the Hamiltonian operator $\hat{H} = \sqrt{\hat{\mathbf{p}}^2 c^2 + m_0^2 c^2}$, also discloses the presence of spin in the particle.

Recent experimental studies of spin properties of electrons and neutrons have displayed the importance of the structure of the space in their vicinity. The mobility of electrons in the graphene honeycomb lattice indicates that the electron's half-integer spin originates from the nearest space around the particles, rather than from the particles themselves [6]. In an experiment with neutrons that were passed through a perfect silicon crystal interferometer, Denkmayr et al. [148] performed weak measurements of the particle's location and its magnetic moment; the experimental results allowed them to suggest that the neutrons were passing through one beam path, while their magnetic moment travelled along another.

Nevertheless, two other research teams [149, 150] criticized the interpretation of "the neutron and its spin are spatially separated."

Stuckey et al. [149] note that those weak values were measured with both a quadratic and a linear magnetic field B_z interaction. Those two opposite effects were of the same value. Therefore weak values can only be interpreted as belonging to the particle. In their opinion, the separation can really be possible only in case the quadratic contribution being negligible.

Corrêa et al. [150] note that the measurements [148] can be interpreted in another way, not via the detachment of spin properties from the particle. They emphasize that neutrons from the second beam path having the appropriate phase in their wave functions were able to influence the neutrons from the first beam path in the area of the second splitter. So the claimed result can be interpreted as simple quantum interference, with no separation between the quantum particle and its internal degree of freedom.

The above studies [148–150] are interesting because they point to the fact that the spin of neutrons is not local and has an extension in space; the spin wave function is quite extended, such that it is able to influence other neutrons at a distance (even if the spin wave function was only attached to the particle for a short time).

The discovery and characterization of gravitationally bound quantum states of neutrons has been reviewed by Baeßler [151] where the lowest neutron quantum states in a gravitational potential were distinguished and characterized by a measurement of their spatial extent. In particular, Baeßler reported an effect of a spin-dependent extra interaction of the ultra cold neutron with a gravitational potential. One neutron spin-component would be attracted, the other repelled at short distances and even unpolarized neutrons are sensitive to such spin-dependent interactions.

Thus, the experiments point to the fact that all the information about spin lies in components of the wave function that is the product of a quantum particle. So basically, from the theoretical point of view, the situation with spin appears as follows. A particle has a dense kernel that moves with a velocity $\boldsymbol{v}$ and its motion is featured by four components of its wave function, or four subwave functions. The named characteristics when coupled with an external electromagnetic field create a new characteristic of the particle called spin, which is manifested as an intrinsic form of the particle's angular momentum.

Any further theoretical study of the phenomenon of spin is impossible without understanding the nature of the wave function and the reasons for splitting it into four components. The statistical interpretation of the wave ψ-function stated by Born [152, 153] is shaky. De Broglie [11] constantly emphasized the need for searching for the physical meaning of the wave ψ-function. Now Hofer [142] claims that the wave ψ-function represents the density of a particle (3.32) and in the case of the electron, its wave function additionally receives one more term, a Poynting-like vector of the electromagnetic energy flux that also oscillates by the same rule as the density. However, why does the density oscillate with this specific frequency? Hofer does not consider any idea that could lead to the splitting of ψ.

What are the reasons behind the necessity for splitting the ψ-function? In Chapter 2 we have discussed the classical relativistic Lagrangian (2.45) and its modification – the radical has been rewritten in the form that reveals the inner motion of a particle, (2.47) and (2.70), which has been split to two sub motions: the particle and its inerton cloud. This is the first step in understanding the nature of splitting of ψ. However, we cannot consider the separation of the particle from the system {particle + its inerton cloud} as the first step to the description of spin, because this system as the whole is the primary object whose projection to a phase space is an abstract wave ψ-function that satisfies the Schrödinger equation.

A hint to the reasons of the decomposition of ψ is in the inner oscillations of the particle mass in the framework of its deformation coat (3.27) and de Broglie's [154] work that considers a dynamics of a particle with a variable mass. So, does mass admit variability? If yes, we must recognize that mass, as a volumetric fractal deformation of a cell of the tessellattice, is a variable characteristic: the volumetric fractal deformation is periodically transferred to a tense state of the cell.

Hence the particulate cell is 'breathing' and as we know, there is no life without breathing.

Thereby we admit the variability of the mass: the particle's mass undergoes periodical transformations to another physical state, a state of tension (meaning the physical condition of being stretched or strained). Accordingly in any quantum mechanical Lagrangian/Hamiltonian we shall represent the classical parameter of "mass" as a variable parameter that is periodically substituted by this condition of tensity. The frequency of the oscillations of the particle is the same as that of its inerton cloud, which has been discussed in Chapter 2.

Thus, one pair $\phi = (\phi_1, \phi_2)$ of a set of four sub wave functions shall be related to the particle's kinetic mass m and corresponding tension Ξ. These two sub wave functions have to be presented by two antagonistic components: ϕ_1 is for mass (which is responsible for attraction) and ϕ_2 is for the tension (which is responsible for repulsion). In this case we cover Hofer's idea expressed in relation (3.33), namely, that spin is in fact related to the density of the particle. The second pair $\eta = (\eta_1, \eta_2)$ of the four sub functions should have a structure similar to that of the $\phi = (\phi_1, \phi_2)$. All the sub wave functions spread out to a distance covered by the amplitude Λ of the inerton cloud, which usually far exceeds the value of λ_{Com} (3.31).

The initial conditions for a fermion are formed when it acquires a new momentum at the point of scattering. Therefore for the $\phi = (\phi_1, \phi_2)$-pair of four sub wave functions we have two opposite tendencies: (i) the fermion acquiring the vector velocity **v** occupies the mass state (the sub function ϕ_1); (ii) the fermion acquiring the velocity **v** is in the tensity state (the sub function ϕ_2). This is the real physical sense of antisymmetric wave functions of fermions. Along the particle path in odd sections λ the particle emits inertons and the tension gradually grows in them; in even sections λ the inertons come back to the particle and their tension gradually drops down to the state of mass. So the particle's inerton cloud is a carrier of the deformation potential of the particle. The mass appears as a local deformation, which is responsible for attraction. The tensity component has to have an opposite property – it will induce a local repulsive potential.

We immediately find the confirmation regarding the structure of the said four sub wave functions in Baeßler's [151] review paper on properties on neutrons: "one neutron spin-component would be attracted, the other repelled at short distances."

In this manner, we naturally come to the Pauli exclusion principle declaring that two identical fermions cannot occupy the same quantum state simultaneously. In terms of the submicroscopic concept this means that if two fermions are found in the same place they will be attracted if they are characterized by the opposite sub wave functions, ϕ_1 and ϕ_2; if they are both featured by the same sub wave functions (i.e., both by ϕ_1 and ϕ_1 or ϕ_2 and ϕ_2) the particles will be repelled. We will further consider the pair (ϕ_1, ϕ_2) in Section 3.6 devoted to the interpretation of the wave function.

However, the Pauli equation (3.40) and (3.41) impose an additional condition on spinors: they must react to an applied magnetic field. We will come back to this issue at the end of Chapter 4.

The Pauli exclusion principle only works in the case when two fermions are close enough, i.e., when their inerton clouds overlap or at least touch each other.

3.5 THE MOTION OF INERTONS

Let us consider the motion of inertons emitted by the moving particle. Their migration is similar to the movement of a standing wave that forms the profile of a string (Figure 2.5). The particle deforms an oncoming cell in the direction perpendicular to the particle's path. However, owing to the elasticity, each *n*th cell regains its original state by passing the *n*th inerton by a relay mechanism into the depth of the tessellattice. At the maximum distance the *n*th inerton stops and returns to the particle. The return of the *n*th inerton is due to rebound by the tessellattice. The motion of such quasi-particle is described by the equation

$$\mu_n d^2\chi_n / dt^2 = -\gamma\chi_n \tag{3.43}$$

where γ is an elasticity constant of the tessellattice. Figure 3.2 depicts two states of the *n*th inerton: near the particle when it possesses the *n*th fractal fragment (i.e., some mass) and the state at the maximum distance when the fractal fragment is completely converted to the tension state. At this distant location of the *n*th inerton the appropriate cell should be shifted from the equilibrium position in the tessellattice, which allows the tessellattice to push the excitation backward to the particle. Such a motion shows that the kinetic energy $\mu_n \dot{\chi}_n^2 / 2$ of the inerton passes to the potential energy $\gamma\omega_n^2\Lambda_n^2 / 2$ at the point $\chi_n|_{\max} = \Lambda_n$ and is then returned to $\mu_n \dot{\chi}_n^2 / 2$, and so on. The cyclic frequency is given by the elasticity of the tessellattice that leads and directs the *n*th inerton with the mass μ_n: $\omega_n = \sqrt{\gamma / \mu_n}$.

Such motion of an inerton from the whole inerton cloud of the particle in question is included in the appropriate Lagrangian studied in Chapter 2.

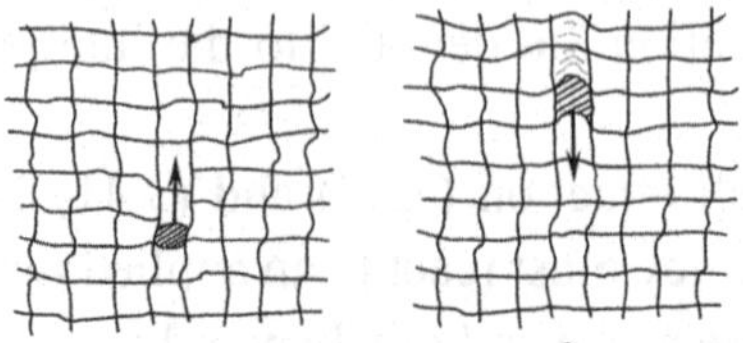

FIGURE 3.2 Two states of the inerton excitation: *a* – the cell with the inerton is deformed (near the particle when the inerton has the volumetric fractal deformation, i.e., mass); *b* – the cell at the maximum distance from the particle where the fractal deformation is transferred to the local tension of the cell (which can also be called a local rugosity of the tessellattice).

How many inertons are there in the particle's inerton cloud? The answer is simple: this is a ratio of the particle's de Broglie wavelength to the Planck length, $N_{\text{inertons}} = \lambda_{\text{dB}} / \ell_{\text{P}}$, because the particle emits as many inertons as meet oncoming cells of the tessellattice. For example, if $\lambda_{\text{dB}} = 5 \cdot 10^{-10}$ m, then $N_{\text{inertons}} \sim 10^{25}$.

From relation $m_0 \upsilon_0^2 = \mu_{0(n)} c^2$ (2.14) we can estimate an average mass of the electron's inerton, which for example moves with the speed $\upsilon = 10^6$ m·s^{-1}, and mass $\langle \mu \rangle = m_{\text{electron}} \upsilon^2 / c^2 \cdot N_{\text{inertons}}^{-1} \sim 10^{-60}$ kg. When the electron mass drops to 10^{-40} kg and the speed to 10^{-5} m·s^{-1}, the mass of the appropriate *n*th inerton will be $\mu_n = m_{\text{electron}}^{(n)} \upsilon_{(n)}^2 / c^2 \sim 10^{-67}$ kg. Since the mass of inertons decreases significantly when the particle approaches the edge of the odd section λ of the particle's path (the same occurs with the particle too), the mass of the first emitted inertons has to be large in comparison with the average value above. How much more? Theoretically, the relation (2.14) shows that the maximum possible mass of a single inerton at the given velocity of the electron is about 10^{-35} kg.

Note that inertons in the inerton cloud of the electron are electrically polarized and that is why they carry also fragments of electromagnetic field (Figure 1.13*c*). Probably this polarization plays a role in limiting the speed of inertons belonging to the particle's inerton cloud. Therefore we assume their speed equal to the speed of light *c*.

When a particle is scattered by other particles or accelerated by an electromagnetic field, some inertons of course have to release from the particle's inerton cloud. In such a case, the velocity of a free inerton is able to exceed the speed of light *c*. Let us designate the speed of a free inerton as c_{in}.

What is the equation of motion of a free inerton? Before writing the Lagrangian of a free inerton, let us determine the notions that we will use. It is advisable to quote Clifford's [33] words when he discussed the origin of matter. Clifford related the appearance of matter to the curvature of local pieces of space as follows: (i) small curved portions of space could be treated as matter because "the ordinary laws of geometry are not valid in them"; (ii) a curved/distorted property could travel from one portion of space to another like a wave; (iii) this variation of curvature happens in the motion of matter; and (iv) in the physical world only this variation occurs.

So, following Clifford let us now consider how a tiny piece of space moves in the tessellattice.

The inerton has a mass μ and a characteristic size, i.e., the size of a cell (which can be equal to the Planck length ℓ_{P}). The mass μ is a volumetric fractal deformation of a cell of the tessellattice; the inerton fractal deformation is much smaller than the fractal deformation of a particulate cell. The inerton migrates in the tessellattice hopping from cell to cell with a constant speed c_{in} and at each hop its inner state changes. After running the distance of Λ during the time T, the inerton changes its volumetric fractal state to a state of tension state that can be noted by a vector $\xi = \xi\,(\mathbf{r}, t)$. Then coming to the next section Λ, the inerton's state is restored to the initial state, and so on. The volumetric fractals forming the inerton can be treated as a local scalar field (the mass scalar field); the same will apply to the local condition of tension: it has to be a vector field (because any typical tension, as a force, has direction in space).

Now we can write the Lagrangian of a free moving inerton

$$L_{\text{inerton}} = \tfrac{1}{2}\ell_{\mathrm{P}}^2\,\dot{\mu}^2 + \tfrac{1}{2}\mu_0^2\,\dot{\xi}^2 - \tfrac{1}{4}c_{\mathrm{in}}^2\,\mu_0^2\left(\nabla\cdot\xi\right)^2 + \tfrac{1}{\sqrt{2}}c_{\mathrm{in}}\,\mu_0\,\ell_{\mathrm{P}}\,\dot{\mu}\cdot\nabla\xi \qquad (3.44)$$

Since the mass and the tension are field parameters, they should be considered in terms of local derivatives in line with field theories used in hydrodynamics. The Lagrangian is constructed in such a way that each term on the right hand side is the square form by mass, length and time (its dimension is energy multiplied by mass, i.e., $\mathrm{kg^2 \cdot m^2 \cdot s^{-2}}$). These square forms are scalar for both the scalar and vector fields. In the Lagrangian

(3.44) μ_0 is the initial mass of the inerton and $c_{\rm in}$ is the inerton speed that is constant. In expression (3.44) the first term is the kinetic energy of the mass component, the second term is the kinetic energy of the component of tension, the third and fourth terms describe the interaction between the mass and the tension inside the cell in which the inerton is located at the moment t (recall, in mathematical terms the mass here is a volumetric fractal fragment of the cell!).

Proceeding to the Euler-Lagrange equations for variables μ and ξ we have to use them due to the term $\nabla\xi$ in the form (see, e.g., Ref. [114], p. 173)

$$\partial / \partial t\,(\partial L / \partial \dot{Q}) - \delta L / \delta Q = 0 \tag{3.45}$$

where the functional derivative is

$$\frac{\delta L}{\delta Q} = \frac{\partial L}{\partial Q} - \frac{\partial}{\partial x}\frac{\partial L}{\partial(\partial Q/\partial x)} - \frac{\partial}{\partial y}\frac{\partial L}{\partial(\partial Q/\partial y)} - \frac{\partial}{\partial z}\frac{\partial L}{\partial(\partial Q/\partial z)} \tag{3.46}$$

The equations for μ and ξ obtained from Eqs. (3.45) and (3.46) in one dimensional approximation (for instance, along the X-axis) are as follows

$$\ddot{\xi} - \tfrac{1}{2}c_{\rm in}^2 \frac{\partial^2 \xi}{\partial x^2} + \tfrac{1}{\sqrt{2}}\frac{c_{\rm in}\,\ell_{\rm P}}{\mu_0}\frac{\partial \dot{\mu}}{\partial x} = 0 \tag{3.47}$$

$$\ddot{\mu} + \tfrac{1}{\sqrt{2}}\frac{c_{\rm in}\,\mu_0}{\ell_{\rm P}}\frac{\partial \dot{\xi}}{\partial x} = 0 \tag{3.48}$$

Integrating Eq. (3.48) and then substituting $\dot{\mu}$ to Eq. (3.47) we arrive at two differential equations below

$$\ddot{\xi} - c_{\rm in}^2 \frac{\partial^2 \xi}{\partial x^2} = 0 \tag{3.49}$$

$$\dot{\mu} = -\tfrac{1}{\sqrt{2}}\frac{c_{\rm in}\,\mu_0}{\ell_{\rm P}}\frac{\partial \xi}{\partial x} + C \tag{3.50}$$

where C is a constant of integration. The initial conditions are $\mu(0, 0) = \mu_0$ and $\xi(0, 0) = 0$. For the wave equation (3.49) the boundary conditions have zero values of the tension ξ at the wave nodes: $\xi(q\Lambda,\ qT) = 0$ where $q = 0, 1, 2, \ldots$ is the number of the cycle of the migrating inerton whose spatial and time period are respectively Λ and T. Such conditions satisfy the solution in the form of a plane wave

$$\xi(x,\ t) = \xi_{\max}\,[1 - \cos(kx - \omega t)] \tag{3.51}$$

and from Eq. (3.50) we obtain the solution for the mass of the free inerton, also in the form of a plane wave:

$$\mu(x,\ t) = \tfrac{1}{2}\mu_0 \cdot [1 + \cos(kx - \omega t)] \tag{3.52}$$

where $k = 2\pi / \Lambda$, $\omega = 2\pi / T$, $\Lambda / (T/2) = c_{\text{in}}$, $\xi_{\max} = \ell_{\text{P}} / \sqrt{2}$ and the constant of integration $C = \mu_0 / 2$.

Thus a free travelling inerton periodically changes its mass m to the tension ξ, namely, the volumetric fractals in contact with the ambient space periodically disappear, i.e., are destroyed, and then space returns the fractal state back to the inerton. Such conditions indeed correspond to the solutions in the form of a plane wave (3.51) and (3.52). These inner properties uninterruptedly oscillate changing each into the other. Figure 3.3 depicts the travelling inerton.

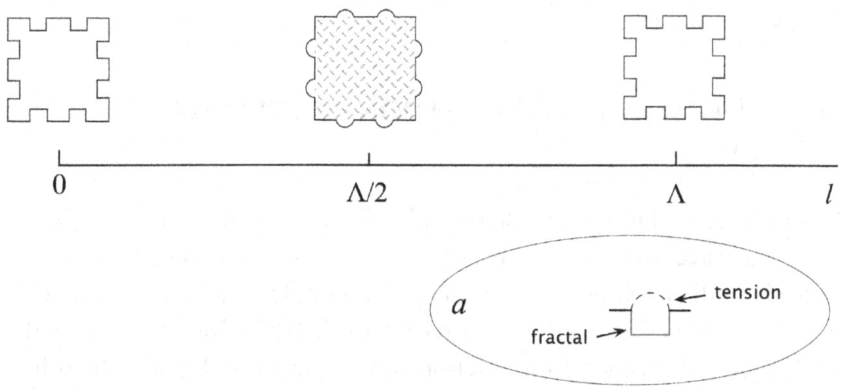

FIGURE 3.3 Free inerton moving in the tessellattice. The inerton periodically loses its fractal state (i.e., mass) passing to the tension state.

The mother-particle, which creates inertons, moves in a jerky manner caused by its continuous periodic stopping (Figure 2.6), and its inerton cloud behaves similarly. In contrast, a free inerton travels by hopping from cell to cell with the constant velocity c_{in} and its value could exceed the speed of light c. This is because the mechanisms of the motion of a bound inerton and a free inertons are different: the mobility of the bound inerton is given by the whole tessellattice, though the mobility of a free inerton is set by local processes occurring in cells through which the excitation is hopping.

Recently Chang and Lee [155] have examined the possibility to treat particles as excitation waves in a vacuum that behaves like a physical medium. They note that using such a model, the phenomenon of wave-particle duality can be explained naturally and the key question is only to find out what kind of physical properties this vacuum medium may have. In particular, they show that the vacuum medium should possess two fundamental (sound) velocities. One is for the speed of light c and the other one – for what? Of course this second fundamental velocity is the speed of inertons c_{in}.

The value of the speed c_{in} was preliminary estimated [156] in the framework of a submicroscopic model that described some thermodynamic transitions occurring with an exchange of mass. In that model all the interactions between masses were examined for the situation when free inertons executed the mass exchange (bound inertons were not taken into account). The calculations showed that c_{in} is about two orders higher than the speed of light c.

3.6 THE PHYSICAL INTERPRETATION OF THE WAVE ψ-FUNCTION

Although the majority of physicists still adhere to the statistical interpretation of the wave ψ-function, some experimentalists studying quantum effects write about direct observations of the ψ-function, for example for electrons [157] and for neutrons [151] and recent work [158]. Moreover, recently a possibility of fission of the electron wave function in liquid helium has been reported [159]. Hofer [134–136, 142] directly states that the ψ-function describes the density of a quantum particle, particularly an electron.

This brings to mind the familiar expression of Louis de Broglie, which he said basing on his experience of radio-engineering work at the Eiffel tower in the time of the First world war: "When one dirtied one's hands, pushing day and night to operate the large generators that were used for radio transmission, it was difficult to believe that the wave could only be probabilistic."

The submicroscopic concept allows us to easily understand a real physical meaning of the wave ψ-function [99]. Indeed, the tessellattice is a continuous substrate. That is why a moving particle together with its cloud of inertons can be treated as a fluid element in this continuum. Since the inerton cloud carries the particle's mass, the fluid element in question can be specified with both mass and volume. Hence, it possesses a density ρ. In its motion the fluid element induces a tension in its immediate surrounding, which is characterized by a displacement vector ξ.

For the description of such a moving fluid element, we may employ the known results of field theories used in hydrodynamics. Indeed, we may begin with the Lagrangian density (see, e.g., Ref. [114], p. 214)

$$\mathcal{L} = \tfrac{1}{2}(\dot{\Xi}\cdot\dot{\Xi}) - \tfrac{1}{2}\bar{\upsilon}^2(\nabla\cdot\dot{\Xi})^2 \tag{3.53}$$

where $\bar{\upsilon}$ is the velocity of the fluid element, which is a constant. Matter with the density ρ is available only in a volume $\mathcal{V}$ of the space occupied by the fluid element; let ρ_0 be the initial density, or the equilibrium value. Now we may write the continuity equation for our fluid element

$$\dot{\rho} + \rho_0(\nabla\cdot\dot{\Xi}) = 0 \tag{3.54}$$

The Euler-Lagrange equations constructed on the basis of the Lagrangian density (3.53) and Eq. (3.54) culminate in equations

$$\ddot{\Xi} - \bar{\upsilon}^2\nabla\cdot(\nabla\cdot\Xi) = 0 \tag{3.55}$$

$$\Delta\rho - \frac{1}{\bar{\upsilon}^2}\ddot{\rho} = 0 \tag{3.56}$$

The most interesting of these is Eq. (3.56) which describes the propagation of the density of the system {particle + inerton cloud}; it takes the form of the wave equation for sound waves. The solution to Eq. (3.56) can be found to be proportional to $\cos(2\pi\nu t)$, which results in

$$\Delta\rho + \frac{(2\pi\nu)^2}{\bar{\upsilon}^2}\rho = 0 \qquad (3.57)$$

Using the correlation $\nu = \upsilon_0 / (2\lambda_{\text{dB}})$ (2.64) and setting $\bar{\upsilon} = \upsilon_0 / 2$, Eq. (3.57) changes to

$$\Delta\rho + \left(\frac{2\pi}{\lambda_{\text{dB}}}\right)^2 \rho = 0 \qquad (3.58)$$

which exactly coincides with equation (2.67). So finally we gain the Schrödinger equation

$$\Delta\rho + \frac{2m[E - U]}{\hbar^2}\rho = 0 \qquad (3.59)$$

derived for the case of the density ρ of a moving fluid element. Since the ψ-function is considered to be dimensionless in quantum mechanics, the normalized function ρ / ρ_0 should exactly correspond to it: $\psi = \rho / \rho_0$. A complex part of ψ means that the appropriate part of the particle's mass and its inerton cloud is transferred to a state of tension. The wave ψ-function is shown in Figure 3.4.

Note in Eq. (3.59) the function ρ represents the mass field that is exposed to variation rather than the volume $\mathcal{V}$, however, in the second term the value of the parameter m must remain fixed (m represents the initial value of mass, i.e., the particle's inert mass $m = m_0 / \sqrt{1 - \upsilon^2 / c^2}$). Therefore, we may proceed from a more sophisticated Lagrangian than (3.53). In fact, we may adopt the Lagrangian written for a free inerton (3.45) and then no additional equation, such as the continuity equation (3.54), is required.

Now let us consider a {particle + its inerton cloud} in motion. The system can be described by the Lagrangian

$$\mathcal{L}_{\{\text{part.+in.}\}} = \tfrac{1}{2}\lambda_{\text{dB}}^2\,\dot{m}^2 + \tfrac{1}{2}m_0^2\,\dot{\Xi}^2 - \tfrac{1}{4}\upsilon_0^2\,m_0^2\left(\nabla\cdot\Xi\right)^2 + \tfrac{1}{\sqrt{2}}\upsilon_0 m_0 \lambda_{\text{dB}}\,\dot{m}\nabla\Xi \qquad (3.60)$$

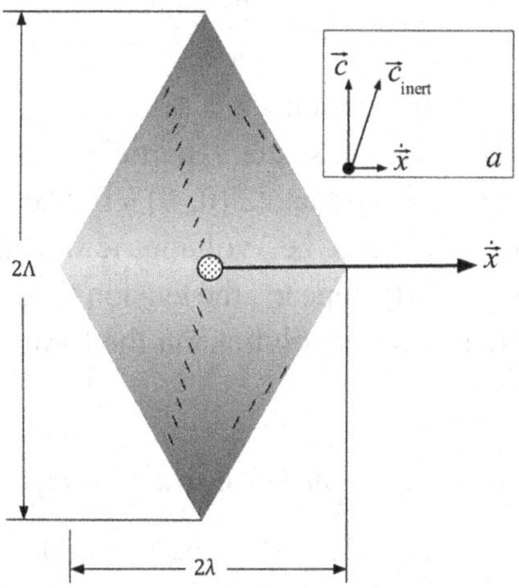

FIGURE 3.4 This is the wave ψ-function: a particle moving together with its inerton cloud in the real space.

The dimension of the function (3.60), which gathers square forms of the variables, is energy multiplied by mass, i.e., kg²·m²·s⁻². Here, m_0 is the mass of the system {particle + its inerton cloud}, which is equal to the mass of the naked particle; λ_{dB} is the particle's de Broglie wavelength, i.e., the size of the system {particle + its inerton cloud} along particle's path; υ_0 is the system's constant velocity. Variables m and Ξ describe the current mass and the tension of the system {particle + its inerton cloud}, respectively, in the point x of the particle's path in the moment of time t.

The equations of motion are

$$\ddot{\Xi} - \tfrac{1}{2}\upsilon_0^2 \nabla^2 \Xi + \tfrac{1}{\sqrt{2}} \frac{\upsilon_0}{m_0} \nabla \dot{m} = 0 \tag{3.61}$$

$$\ddot{m} + \tfrac{1}{\sqrt{2}} \frac{\upsilon_0 m_0}{\lambda_{dB}} \nabla \dot{\Xi} = 0 \tag{3.62}$$

The system of equations (3.61) and (3.62) is transformed to two equations

$$\ddot{\Xi} - \upsilon_0^2 \nabla^2 \Xi = 0 \tag{3.63}$$

$$\dot{m} = -\tfrac{1}{\sqrt{2}} \frac{\upsilon_0 m_0}{\lambda_{\text{dB}}} \nabla\Xi + C \tag{3.64}$$

where C is a constant of integration.

The initial conditions for the system {particle + its inerton cloud} can be of two types: (i) $m(0,\ 0) = m_0$ and $\Xi(0,\ 0) = 0$, and (ii) $m(0,\ 0) = 0$ and $\Xi(0,\ 0) = \Xi_{\text{max}}$. Hence, the mass, i.e., volumetric fractals, are periodically destroyed and a new property appears, the tension (Figure 3.5). These conditions result in the two pairs of solutions in the form of a plane wave (in one dimension for simplicity):

$$m(x,\ t) = \tfrac{1}{2} m_0 \,[1 + \cos(kx - \omega t)],$$
$$\Xi(x,\ t) = \Xi_{\text{max}} \,[1 - \cos(kx - \omega t)] \tag{3.65}$$

and

$$m(x,\ t) = \tfrac{1}{2} m_0 \,[1 - \cos(kx - \omega t)],$$
$$\Xi(x,\ t) = \Xi_{\text{max}} \,[1 + \cos(kx - \omega t)] \tag{3.66}$$

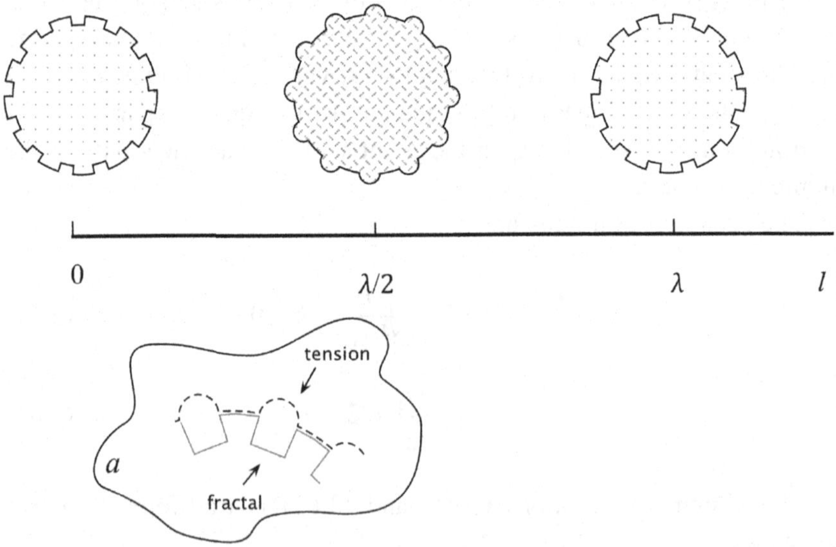

FIGURE 3.5 Moving particle in the tessellattice. The particle periodically loses its fractal state (i.e., mass) passing to the tension state.

where $k = 2\pi / \lambda_{dB}$, $\omega = 2\pi / T$, $\lambda_{dB} = \upsilon_0 / (T/2)$ (see Section 2.3) and $\Xi_{max} = \lambda_{dB} / \sqrt{2}$.

Thus we have derived not only the wave ψ-function, which is the ratio $\psi = m(x, t) / m_0$. We also have obtained the pair of sub functions $\phi = (\phi_1, \phi_2)$, which are responsible for the manifestation of spin, as has been discussed in Section 3.4:

$$\phi_1(x, t) = \tfrac{1}{2}\,[1 + \cos(kx - \omega t)] \tag{3.67}$$

$$\phi_2(x, t) = \tfrac{1}{2}\,[1 - \cos(kx - \omega t)] \tag{3.68}$$

The spinor ϕ_1 describes the state of the particle when in the initial moment it emits its inertons, as its mass decreases (3.65). The spinor ϕ_2 describes the state of the particle when in the initial moment it absorbs its inertons, as its mass increases (3.66). These two situations are depicted in Figure 3.6.

Let us now look at the Lagrangian (3.60) and the equations (3.63) and (3.64) from the macroscopic point of view. It means that the minimum size of a fluid element is restricted by the size of the particle's inerton cloud, because in hydrodynamics any quantum fluctuation should be suppressed. Thus hydrodynamics works at a scale larger than λ_{dB} along the particle path and larger than Λ in transverse directions (see also Section 2.5); hence the criterion for the use of hydrodynamics is $r >> \Lambda$ and $t >> T$. Especially interesting is Eq. (3.64): dividing both hand sides by a volume $\mathcal{V}$ of the fluid element, the equation can be rewritten in terms of the element density

$$\dot{\rho} = -\tfrac{1}{\sqrt{2}}\,\frac{\upsilon_0 \rho_0}{\lambda_{dB}}\,\nabla \Xi + C \tag{3.69}$$

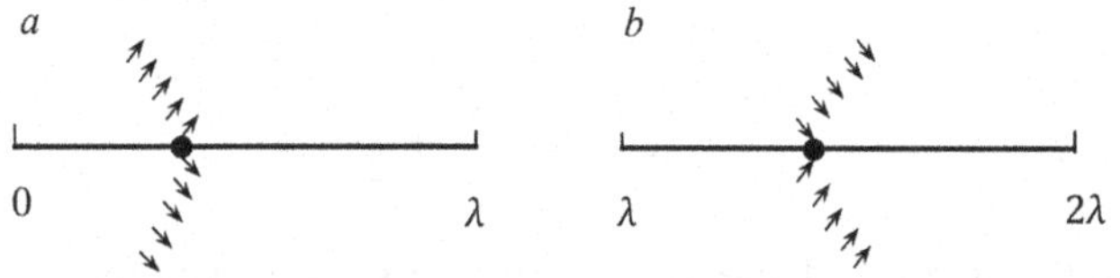

FIGURE 3.6 Particle and its inertons: a – the particle emits inertons and its mass disintegrates, which accounts for the solution (3.67); b – the particle absorbs inertons, or inertons guide the particle, and the particle mass is restored, which elucidates the solution (3.68).

but using $\lambda_{\rm dB} / \upsilon_0 = T / 2$, i.e., a minimum time interval for the fluid element and hence it must be the time derivative. Besides, in the hydrodynamic approximation we put the constant of integration $C = 0$, because in the case of an ideal liquid the difference between the density of a fluid element and the density of the equilibrium liquid is $(\rho - \rho_0)\,|_{t=0} \approx 0$. However, in consideration of the total system {particle + its inerton cloud} the difference between the density of the system and the density of the environment is $(\rho - 0)\,|_{t=0} = \rho$. Then, since the factor $2^{-1/2}$ can be inserted in the short time interval T (as hydrodynamics works only at $t >> T$), Eq. (3.69) in the three dimensional case is transferred to Eq. (3.54), i.e., the continuity equation, which is very important in hydrodynamics.

Thus, the submicroscopic concept is able to propose the origin of hydrodynamic equations, which underlies the possibilities of the formalism of conventional quantum physics.

It is necessary to mention here important studies of the wave function of the Schrödinger wave equation, which have recently been carried out by Shpenkov [98]. He emphasizes that all three variables r, θ, and ϕ (in the spherical polar frame of reference) have to be taken into account simultaneously, not only its major argument, the distance r. In particular, he notes that since the s-state of the wave function is characterized by spherical symmetry, quantum mechanics attributes a similar symmetry to the hydrogen atom. Namely, his analysis shows that atoms are constructed on the basis of the symmetries available in the ψ-function.

3.7 STATISTICS OF FERMI-DIRAC AND BOSE-EINSTEIN

For a system of identical fermions the average number of fermions in a single-particle state i obeys the Fermi–Dirac distribution

$$\overline{n}_i = \frac{1}{e^{(\varepsilon_i - \mu)/(k_{\rm B}\Theta)} + 1} \tag{3.70}$$

where ε_i is the energy of the single-particle state i, μ is the total chemical potential, which is always positive, $k_{\rm B}$ is Boltzmann's constant, and Θ is the absolute temperature. The distribution (3.70) is based on using the Pauli exclusion principle, because it is this principle that separates particles from the same position in space. The particles, owing to their inerton clouds must be separated to a distance $2\lambda_{\text{dB}}$ along the total vector velocity of two particles and to a distance 2Λ in transverse directions. The above said results in inequalities $0 < \overline{n}_i < 1$, which allows the many-particle system to be described in terms of single-particle energy states.

The Fermi-Dirac statistics work for half-integer spins. However, what do the spin values equal to 3/2, 5/2, … really mean? The submicroscopic concept can only suggest spin equal to ½. Whether other half-integer spins can exist, the theory can't say, because it is not clear what kinds of sub wave functions are permissible in such cases (whether they exist in principle?).

The Bose-Einstein distribution describes the statistical behavior of boson particles that are believed to have integer spin. At low temperatures, practically an unlimited number of bosons can be collected into the same energy state. The Bose-Einstein distribution reads

$$\overline{n}_i = \frac{1}{e^{(\varepsilon_i - \mu)/(k_{\text{B}}\Theta)} - 1} \tag{3.71}$$

where $\varepsilon_i > \mu$ and $\overline{n}_i$ is the number of particles in state i. Once again, the value of $\overline{n}_i$ is practically unlimited.

Of course, as seen in the case of fermions, bosons also have to be separated to a distance $2\lambda_{\text{dB}}$ along the total vector velocity of paired particles and to a distance 2Λ in transverse directions. But under these conditions there will not be any interaction between two bosons that are in proximity to each other. So, what is the difference between fermions and bosons? From the viewpoint of submicroscopic mechanics, bosons are combined particles, i.e., each elementary boson consists of two fermions with two opposite projections of their spin. A couple of such fermions revolve around each other forming a vortex. This problem is well known in quantum mechanics as the rigid rotor (see, e.g., Ref. [160]).

In the problem of the rigid rotor, the potential depends only on a distance $\hat{r}$ between the fermions. The center of mass is determined by the relation

$$m_1 r_1 + m_2 r_2 = 0, \; \hat{r} = r_1 + r_2 \tag{3.72}$$

Each of the rotating partners has an angular momentum $J_i = I_i \omega$ where ω is the angular velocity, which is the same for each partner, and $I_i = m_i r_i^2$ is the moment of inertia for the particle.

The Hamiltonian of the rigid rotor in the two-dimension problem is as below

$$H = \frac{J_1^2}{2I_1} + \frac{J_2^2}{2I_2} = \tfrac{1}{2}(mr_1^2 + mr_2^2)\,\omega = \tfrac{1}{2}\widehat{m}\widehat{r}^2 \tag{3.73}$$

where the reduced mass $\widehat{m}$ of the two rotating particles is defined as

$$1/\widehat{m} = 1/m_1 + 1/m_2 \tag{3.74}$$

So the system of two fermion particles behaves as a single system with a reduced mass $\widehat{m}$ rotating at the distance $\widehat{r}$ from the origin. The de Broglie relation $p = h/\lambda_{\text{dB}}$ allows a condition for quantization of angular motion $J^{(z)} = h\widehat{r}/\lambda_{\text{dB}}$.

The quantum mechanical Hamiltonian $\widehat{H} = \widehat{\mathbf{J}}^2/(2I)$ allows one to write the Schrodinger equation in the spherical coordinates

$$-\frac{\hbar^2}{2I}\left[\frac{1}{\sin\vartheta}\frac{\partial}{\partial\vartheta}\left(\sin\vartheta\frac{\partial}{\partial\vartheta}\right) + \frac{1}{\sin^2\vartheta}\frac{\partial^2}{\partial\phi^2}\right]Y_J^m(\vartheta,\phi) = E_J Y_J^m(\vartheta,\phi) \tag{3.75}$$

The result is that the energy spectrum of the rigid rotor is

$$E_J = \frac{\hbar^2}{2I}J(J+1),\; J = 0,\,1,\,2,\,\ldots \tag{3.76}$$

Here, the letter J is chosen for rotation of a diatomic molecule, while for an electron moving around the nucleus the chosen letter is l.

So if a boson is a particle that combines two rotating fermions, it must be a typical rigid rotor and the spectrum (3.76) can also be applied to it. In the case of spin the letter s is chosen and then the eigenvalues of the square of the spin moment $\widehat{\mathbf{S}}^2$ for bosons are

$$S^2 = \hbar^2 s(s+1) \tag{3.77}$$

Relation (3.77), which most often is written in the linear form for the spin momentum $S = \hbar\sqrt{s(s+1)}$, is used in applications for bosons all the time (simply as an analogue to the real orbital momentum). However, as we have seen above, bosons are indeed rigid rotors formed by a couple of fermions.

Thus bosons are not elementary particles, they are combined particles. A couple of fermions with opposite spin projections form a rigid rotor and such a vortex supports the stability of the created boson to support its stability for some time. Bosons do not have an inner moment that is called spin. The boson's spin is the real angular momentum of the boson, which is a combined particle of a couple of elementary fermions whose spin projections are ± ½.

In particular, Buckholtz [161] shows in a recent book that some effects in high-energy physics can be explained only if we are considering bosons with the spin quantum number $s = 2$. From the submicroscopic viewpoint this means that in this particular case the investigated particles have been typical rigid rotors with the energy spectrum (3.76) in which one has to put $J = s = 2$.

The situation with the so-called Bose-Einstein condensation means that non-interacting vortexes are gathered in some place (or types of non-rotating vortexes with $s = 0$). In the case of diluted gases cooled to a temperature near absolute zero, the system of atoms may experience a form of condensation that resembles the abstract theoretical pattern of the Bose-Einstein condensation – an ensemble of non-interacting particles. Because the atoms are confined to movement in strict synchrony and phase, it is this peculiar movement that holds the atoms together (A detailed submicroscopic consideration of the phenomenon is given in Section 5.3.).

Regarding photons, as carriers of electromagnetic interaction, we can say they travel in space like elementary vortexes. The size of a photon is practically a point (more exactly, $\ell_{\mathrm{P}} \sim 10^{-35}$ m, which together with local perturbations of the tessellattice can spread to $\sim 10^{-31}$ m, see Section 1.8.8). Hence the concentration of photons in a local place may be enormous, which satisfies the conditions of the Bose-Einstein statistics. The photon spin $s = 1$ means that something rotates in this quasi-particle. That is indeed very close to the truth and we will get acquainted with the appropriate mechanism of inner "rotation" of the photon in Chapter 4.

KEYWORDS

- **inertons**
- **Pauli exclusion principle**
- **spin**
- **tunneling**
- **uncertainty relation**
- **wave ψ-function**

CHAPTER 4

ELECTROMAGNETIC PHENOMENA IN THE TESSELLATTICE

CONTENTS

In quantum electrodynamics (see, e.g., Feynman [5]) an electron emits or absorbs a photon at a certain place and time and this is a way for all other electrically charged particles to interact, namely, exchange by photons. The working mechanism of quantum electrodynamics is a quantum field theory that, however, can be only an approximation to the description of nature, just as the Fourier approximation is applied to a concrete continuous function in mathematics.

What is the photon? What is the charge? The Modern Standard Model of Particles does not know. Nevertheless, some desperate researchers try to disclose the nature of the photon [162, 163].

Below in this Chapter we will see how methods of conventional quantum field theory lead to a logical conclusion about the discreteness of space. Then we will see how such notions as the electric and magnetic fields naturally arise from a dynamic of the surface fractals of the particulate ball.

4.1 PHOTON: FROM THE QUANTUM FIELD THEORY TO A DISCRETE LATTICE OF SPACE

The methodology of the second quantization is widely used for the description of free photons and their interaction with particles. However, the description of the electromagnetic field in terms of creation and annihilation operators is justified in the wave vector presentation. Quantum electrodynamics does not study the problem of spatial pattern of the photon. Researchers working in the area of optics, usually talk about a photon as something that occupies a volume $\sim \lambda^3$ where λ is the photon wavelength (because quantum electrodynamics normalizes photons just in this way: one photon to the volume λ^3). On the other hand, photons obey the Bose-Einstein statistics and therefore a large quantity of photons could fill a volume significantly less than λ^3.

Theorists who work on problems related to the motion of single photons through a fiber optic cable, quantum teleportation of light, entangled photon states, and so on sometime become confused when reporting to an audience that are actively asking questions. They talk about the wave function of a single photon, rules of its interaction with other photons, a resonator, switcher, and the optic cable. Listening to the flow of questions (what is the volume occupied by the photon?, what does the photon's wave function look like?, how far does this wave function spread?, etc.), which stop the speaker in his tracks, he himself begins to ask the audience: "But there is no clear definition of the photon! Can you give me the definition?"

The definition of the photon given in Subsection 1.1.8 of course is very short and cannot be directly applied to the study of specific issues of quantum optics. So let us dive into the heart of the matter.

Bussey [164] noted that the atomic displacements u_p which create the phonon state in the crystal lattice, resemble the field elements $\phi(x)$ of a particle, since quantum field theory describes elementary particles as excitations of fields whose ground state is the vacuum. An existing parallel between excitations of vacuum fields and phonons of the crystal lattice has long been recognized [165]. Let us get into the details [166].

Phonons determined in the **k**-space appear due to spontaneous vibrations of atoms in the crystal lattice. Actually, owing to the atom-atom interaction, one can write the Lagrangian

$$L_{\text{crystal}} = \tfrac{1}{2}\sum_{\mathbf{n}} m\dot{\mathbf{r}}_{\mathbf{n}}^{2} - \tfrac{1}{2}\sum_{\substack{\mathbf{n},\,\mathbf{m}\\(\mathbf{n}\neq\mathbf{m})}} \gamma_{\mathbf{nm}}\,\mathbf{r}_{\mathbf{n}}\mathbf{r}_{\mathbf{m}} \tag{4.1}$$

where m is the mass of the **n**th atom, $\mathbf{r}_{\mathbf{n}}$ and $\dot{\mathbf{r}}_{\mathbf{n}}$ are the displacement of the atom from the equilibrium position and the velocity of the atom respectively, $\gamma_{\mathbf{nm}}$ is the tensor of elasticity interaction of atoms, the dot over the vector $\dot{\mathbf{r}}_{\mathbf{n}}$ means the differentiation with respect to the proper time of the crystal. As is well known expression (4.1) can be rewritten via the generalized, or canonical, coordinates $\mathcal{A}_{\mathbf{k}s}$ and $\dot{\mathcal{A}}_{\mathbf{k}s} \equiv \mathcal{P}_{\mathbf{k}s}$

$$L_{\text{crystal}} = \tfrac{1}{2}\sum_{\mathbf{k},\,s}\left(\dot{\mathcal{A}}_{\mathbf{k}s}\dot{\mathcal{A}}_{-\mathbf{k}s} - \varpi_{s}^{2}(\mathbf{k})\mathcal{A}_{\mathbf{k}s}\mathcal{A}_{-\mathbf{k}s}\right) \tag{4.2}$$

where $\varpi_s^2(\mathbf{k})$ is the frequency of the sth branch of acoustic vibrations of atoms. $\dot{\mathcal{A}}_{-\mathbf{k}s}$ denotes the generalized momentum $\mathcal{P}_{\mathbf{k}s}$. Coordinates $\mathcal{A}_{\mathbf{k}s}$ and $\mathcal{P}_{\mathbf{k}s}$ are substituted for the corresponding operators

$$\mathcal{A}_{\mathbf{k}s} \to \widehat{\mathcal{A}}_{\mathbf{k}s} = \sqrt{\hbar / 2\varpi_{s}(\mathbf{k})}\,(\hat{b}_{\mathbf{k}s} + \hat{b}^{+}_{-\mathbf{k}s}) \tag{4.3}$$

$$\mathcal{P}_{\mathbf{k}s} \to \widehat{\mathcal{P}}_{\mathbf{k}s} = \sqrt{\hbar\varpi_{s}(\mathbf{k}) / 2}\,(\hat{b}^{+}_{-\mathbf{k}s} - \hat{b}_{\mathbf{k}s}) \tag{4.4}$$

Here, $\hat{b}^{+}_{\mathbf{k}s}(\hat{b}_{\mathbf{k}s})$ is the Bose operator of creation (annihilation) of a phonon. In terms of the second quantization operators, the energy operator of the lattice vibrations takes the form

$$\widehat{H}_{\text{crystal}} = \sum_{\mathbf{k},\,s}\hbar\varpi_{s}(\mathbf{k})\left(\hat{b}^{+}_{\mathbf{k}s}\hat{b}_{\mathbf{k}s} + \tfrac{1}{2}\right) \tag{4.5}$$

The phonon having the wave vector **k** envelops a great number of the lattice sites, namely, the phonon occupies a volume $\sim k^{-3}$ where $k = 2\pi/\lambda$

is the phonon's wave number and λ is the wavelength of the appropriate acoustic excitation.

Now let us proceed to the inspection of the energy operator of a free electromagnetic field, which has the same form

$$\widehat{H}_{\text{photon}} = \sum_{\mathbf{k},s} \hbar\omega_s(\mathbf{k}) \left(\hat{a}^{+}_{\mathbf{k}s} \hat{a}_{\mathbf{k}s} + \tfrac{1}{2} \right) \tag{4.6}$$

where $\hat{a}^{+}_{\vec{k}s}$ $(\hat{a}_{\vec{k}s})$ is the Bose operator of creation (annihilation) of a photon, an elementary excitation of the electromagnetic field; $\omega_s(\mathbf{k})$ is the cyclic frequency of the photon with the wave vector $\mathbf{k}$ and the s polarization. In spite of the similarity of expressions (4.5) and (4.6), their original classical Lagrangians are absolutely different. In the case of phonons we start from the discrete function (4.1), but in the case of photons we emanate from the continuous Lagrangian density

$$L_{\text{photon}} = -\frac{1}{4\mu_0} F_{ij} F^{ij} \tag{4.7}$$

where μ_0 is the magnetic constant and F_{ij} and F^{ij} are covariant and contravariant field strength tensors, respectively, determined as

$$F_{ij} = \partial_i A_j - \partial_j A_i \tag{4.8}$$

The Lagrangian (4.7) can be rewritten as below

$$L_{\text{photon}} = -\frac{1}{2\mu_0} \left[\frac{(-\vec{\nabla}\varphi - \dot{\vec{A}})^2}{c^2} - (\vec{\nabla} \times \vec{A})^2 \right] \tag{4.9}$$

where $\vec{E} = -\vec{\nabla}\varphi - \partial\vec{A}/\partial t$ and $\vec{B} = \vec{\nabla} \times \vec{A}$. When we introduce the Weyl gauge $\varphi = 0$ (also known as the Hamiltonian, or temporal gauge), the Lagrangian density (4.9) then reduces to the form

$$L_{\text{photon}} = -\frac{1}{2\mu_0} \left[\frac{1}{c^2} \left(\frac{\partial \vec{A}}{\partial t} \right)^2 - (\vec{\nabla} \times \vec{A})^2 \right] \tag{4.10}$$

If we formally carry out an analysis trying to advance from the energy operator (4.6) to the classical Lagrangian keeping the phonon scheme above, we find an interesting outcome.

First, the operators of canonical variables expressed in terms of $\hat{a}^{+}_{\mathbf{k}s}$ and $\hat{a}_{\mathbf{k}s}$ are

$$\hat{A}_{\mathbf{k}s} = \sqrt{2\pi\hbar c^2 / \omega_{\mathbf{k}}}\ [\hat{a}_{\mathbf{k}s}(t) + \hat{a}^{+}_{-\mathbf{k}s}(t)] \quad (4.11)$$

$$\hat{P}_{\mathbf{k}s} = i\sqrt{\hbar\omega_{\mathbf{k}} / 8\pi c^2}\ [\hat{a}^{+}_{\mathbf{k}s}(t) - \hat{a}_{-\mathbf{k}s}(t)] \quad (4.12)$$

Then the corresponding canonical variables are

$$\vec{A}(\mathbf{r}, t) = \frac{1}{\sqrt{\mathcal{V}}} \sum_{\mathbf{k},s} \vec{e}_{\mathbf{k}s}\, A_{\mathbf{k}s}(t)\, e^{i\mathbf{k}\mathbf{r}} \quad (4.13)$$

$$\vec{P}(\mathbf{r}, t) = \frac{1}{\sqrt{\mathcal{V}}} \sum_{\mathbf{k},s} \vec{e}_{\mathbf{k}s}\, P_{\mathbf{k}s}(t)\, e^{-i\mathbf{k}\mathbf{r}} \quad (4.14)$$

where $\vec{e}_{\mathbf{k}s}$ is the unit vector. Formulas (4.13) and (4.14) present the Fourier-series expansion of the classical variables $\vec{A}(\mathbf{r}, t)$ and $\vec{P}(\mathbf{r}, t) \equiv (1/4\pi c^2)\partial\vec{A}/\partial t$ in the volume $\mathcal{V}$. Indeed, the expansion adequately depicts the actual discrete structure of the electromagnetic field, i.e., the photonic nature of the field. Thus one would rewrite the Lagrangian density (4.10) as

$$L_{\text{photon}} = -\frac{1}{2\mu_0 \mathcal{V}} \sum_{\mathbf{k},s} \left[\frac{1}{c^2} \frac{\partial A_{\mathbf{k}s}}{\partial t} \frac{\partial A_{-\mathbf{k}s}}{\partial t} - \left(\nabla \times \vec{e}_{\mathbf{k}s} A_{\mathbf{k}s}\right)\left(\nabla \times \vec{e}_{\mathbf{k}s} A_{-\mathbf{k}s}\right) \right] \quad (4.15)$$

The canonical variables (4.13) can be written also for a discrete lattice

$$\vec{A}_{\mathbf{n}s}(\mathbf{r}_{\mathbf{n}}, t) = \frac{1}{\sqrt{\mathcal{V}}} \sum_{\mathbf{k}} \vec{e}_{\mathbf{k}s} A_{\mathbf{k}s}(t) e^{i\mathbf{k}\mathbf{n}},\ A_{\mathbf{k}s} = \frac{1}{\sqrt{\mathcal{V}}} \sum_{\mathbf{k}} \vec{e}_{\mathbf{k}s} A_{\mathbf{n}s}(t) e^{-i\mathbf{k}\mathbf{n}} \quad (4.16)$$

and similarly for the variable $\vec{P}(\mathbf{r}, t)$ (4.14). Then the right-hand side of the function (4.15) written in the **k**-wave vector presentation can be inverted to that written in terms of discrete spatial variables

$$L_{\text{photon}} = -\frac{1}{2\mu_0 \mathcal{V}} \sum_{\mathbf{n},s} \left[\frac{1}{c^2} \left(\frac{\partial \vec{A}_{\mathbf{n}s}}{\partial t} \right)^2 - \left(\nabla_{\mathbf{n}} \times \vec{A}_{\mathbf{n}s} \right)^2 \right] \tag{4.17}$$

In this new presentation (4.17) of the Lagrangian density for photons, the vector **n** plays the role of the radius vector that defines the **n**th knot of a lattice of space, or the appropriate cell of the space net.

The equation of motion derived on the basis of the Lagrangian density (4.17) appears very similar to a typical wave equation that specifies the behavior of the polarization $\vec{A}_{\mathbf{n}s}$ localized in the **n**th cell of space in the moment *t*.

Therefore the photon, in fact, could be regarded as an elementary excitation, or quasi-particle that travels by hopping from cell to cell in the space net rather than a particle-wave of an undetermined nature moving in an empty space, or vacuum. The motion of the photon 'core' features the equation

$$\dot{\mathbf{r}}_n = \frac{\mathbf{n}}{|\mathbf{n}|} c \tag{4.18}$$

The vector potential $\vec{A}_{\mathbf{n}s}$ induced in the **n**th cell can be interpreted as some kind of surface polarization, or deformation (to the right or the left, due to the index *s*). We will see below that both the scalar and vector potentials of a photon obey the wave equation and to prove this we shall deviate from the Weyl gauge

$$\varphi = 0 \tag{4.19}$$

The wavelength λ of the photon will then mean the spatial period at which the cell's polarization restores its initial state. In other words, the photon wavelength λ is the distance that the photon runs by hopping from cell to cell in order to reach the initial phase of its polarization. The period of the polarization oscillations $T = 1/\nu$ is related to λ by the relation $c = \lambda / T$.

Thus the photon should be excluded from the list of canonical elementary particles; it is not a particle but an elementary excitation of the tessellattice.

4.2 THE ELECTRIC CHARGE

In quantum field theory and its branches, such as string theory, the electric charge is treated as a generator of a $U(1)$ symmetry that then is promoted to a gauge symmetry (i.e., local symmetry). The $U(1)$ symmetry and the gauge field are components of a low-energy effective action. So the problem of the charge is reduced to some manifestations of its presence through local symmetries.

Different approaches related to the electric charge have been reviewed in a study [167] that also developed the theory of the charge and electromagnetic field being in the tessellattice.

In Sections 1.8.5–1.8.8 we have seen how the notions of charge and photon appear in the tessellattice. Physics has to deal with the motion of these objects.

4.2.1 THE GEOMETRY OF THE SURFACE OF A PARTICULATE BALL

It has been shown that the tessellattice is composed of topological balls. In the body of the universe these balls are slightly contracted in size in comparison with (conventionally free) balls outside the universe. Our universe can be treated as a mega-huge cluster and the possibility arises that the total universal tessellattice may include a number of such mega-clusters. Between these universes topological balls (free balls) may have a large diameter, which form an insurmountable energy barrier between the universes.

Let the volume of a free ball be $\mathcal{V}^{(o)}$ and an average volume of the universe's degenerate ball is $\mathcal{V}^{\deg}$. Then the inequality holds $\mathcal{V}^{(o)} > \mathcal{V}^{\deg}$ and all real elementary particles and field particles emerge just from such degenerate balls at their fractal transformations.

In the Section 1.8.5 we have associated surface fractals with the appearance of the electric charge in the particulate ball. Now we can further examine the state of the particulate ball's surface. Below we will use the following axiom: *Whatever changes will occur with a topological ball, the area of its surface will be preserved, or in other words, the area of the surface of a ball is invariant.* Indeed, if we take a sheet of paper of format A4, we can crumple it by hand into a large ball and then squeeze it into a smaller one. The volume changes dramatically, but the total surface of the ball, which is determined by the 4A size, remains unchanged.

Let $\mathcal{R}_{\text{o}}$ and $\mathcal{R}_{\text{deg}}$ be typical radii of a free ball and the universe's ball, respectively. Then we can write the relation

$$4\pi\mathcal{R}_{\text{o}}^{2} = 4\pi\mathcal{R}_{\text{deg}}^{2} + \Delta S \tag{4.20}$$

where the surface defect ΔS of the ball is in the state of a collective motion of the surface fragments. The surface defect ΔS consists of a composition of surface elements, or amplitudes of the surface vibrations that are directed inward and outward from the surface formed by the radius $\mathcal{R}_{\text{deg}}$ (Figure 4.1), such that $\Delta S = \sum_{n=1}^{\mathcal{N}} \sigma_n$ where $\mathcal{N} = (R_{\text{universe}} / \ell_{\text{P}})^3 \sim 10^{183}$ is the number of balls that form our universe ($R_{\text{universe}} \sim 10^{26}$ m is the visible radius of the universe and $\ell_{\text{P}} \sim 10^{-35}$ m is the Planck length). Homogeneous transformations of the surface elements of the sphere from inside $\sigma_n^{(\text{in})}$ to outside $\sigma_n^{(\text{out})}$ do not change the local neutrality and on average $\sum_n \sigma_n^{(\text{in})} = \sum_n \sigma_n^{(\text{out})}$.

However, in the case of fractal transfigurations of the ball's surface, the surface defect ΔS will consist of exclusively unidirectional surface fractals $\sigma_n^{(\text{in})}$ or $\sigma_n^{(\text{out})}$. The transformation of a ball to the positive or negative charged particle can be presented as follows

$$4\pi\mathcal{R}_{\text{o}}^{2} = 4\pi\mathcal{R}_{\text{particle}}^{2} + \sum_{n=1}^{\mathcal{N}} \sigma_n^{(\text{out})} \quad \text{or} \quad 4\pi\mathcal{R}_{\text{o}}^{2} = 4\pi\mathcal{R}_{\text{particle}}^{2} + \sum_{n=1}^{\mathcal{N}} \sigma_n^{(\text{in})} \tag{4.21}$$

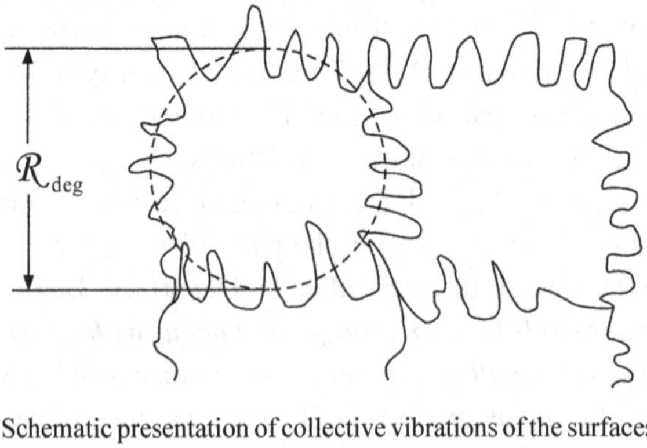

FIGURE 4.1 Schematic presentation of collective vibrations of the surfaces of topological balls in the degenerate space (these are homeomorphic transformations of cell surfaces).

The only difference of the charge of a lepton from that of a quark is the length of the surface amplitudes that in the case of the lepton can be longer than that of the quark (i.e., the surface of a lepton could be covered by longer spikes than the surface of a quark owing to the inequality $\mathcal{R}_{\text{lepton}} < \mathcal{R}_{\text{deg}} < \mathcal{R}_{\text{quark}}$, see Figure 1.9).

The "chestnut" model of the electric charge (Figs. 17 and 1.8) in which $\mathcal{N} \sim 10^{183}$ spikes protrude inside or outside the sphere surface should be sustainable; this is the quantum of the surface fractal. For this purpose we shall set a shape of a spike and investigate conditions of its stability.

Let the shape of the surface spike be defined as a surface generated by rotating a two-dimensional curve

$$y = \hbar \cdot (x / r - 1)^2 \tag{4.22}$$

about the Y-axis. The surface of revolution is the typical spike (Figure 4.2). Here h is the height of the spike and r is the radius of the base of the spike. Then the surface area of the spike is

$$\sigma_{\text{spike}} = 2\pi \int_0^{\hbar} x(y)\, dl = 2\pi \int_0^{\hbar} r(1 + y/\hbar)\sqrt{1 + r^2/(4\hbar y)}\, dy$$
$$= \frac{\pi r^4}{6\hbar^2}\left[\left(1 + \frac{4\hbar^2}{r^2}\right)^{3/2} - 1\right] + \frac{\pi r^4}{2\hbar}\left[\frac{2\hbar}{r}\sqrt{1 + \frac{4\hbar^2}{r^2}} + \ln\left(\sqrt{1 + \frac{4\hbar^2}{r^2}}\ \frac{2\hbar}{r}\right)\right] \tag{4.23}$$

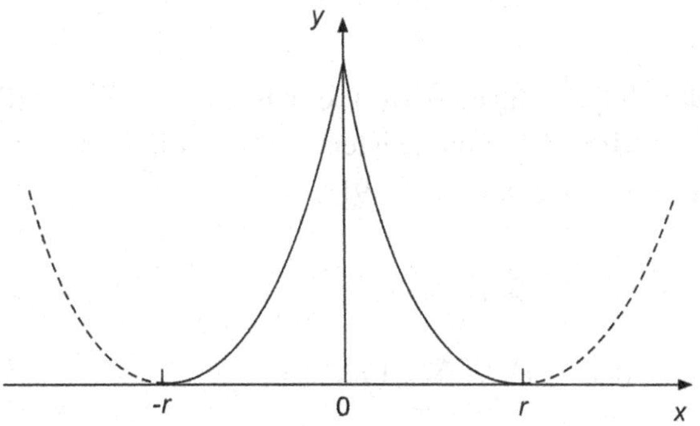

FIGURE 4.2 Graphic display of the function (4.22).

The total area of the particle's surface is

$$S_{\text{particle}} = 4\pi \mathcal{R}^{2}_{\text{particle}} - \mathcal{N}\sigma(r) + \mathcal{N}\sigma_{\text{spike}} \tag{4.24}$$

where

$$\sigma(r) = 2\pi \mathcal{R}^{2}_{\text{particle}} \left[1 - \cos(r / \mathcal{R}_{\text{particle}})\right] \tag{4.25}$$

is the area of the spike base (i.e., circle) on the sphere with the radius $\mathcal{R}_{\text{particle}}$ (see, e.g., Ref. [109], p. 18).

We exclude the area $\mathcal{N}\sigma(r)$ from the surface defect ΔS, because the area of the spike's base is not part of the particle's surface. Since $r \lll \mathcal{R}_{\text{partcile}}$, we can reduce expression (4.25) to $\sigma(r) \cong \pi r^2$. Then the area of the particle surface reads

$$S_{\text{particle}} = 4\pi \mathcal{R}^{2}_{\text{particle}} + \Delta S \tag{4.26}$$

where

$$\Delta S = \mathcal{N}\sigma_{\text{spike}} \tag{4.27}$$

The surface defect described by expressions (4.27) and (4.23) is a function of two parameters, $\hbar$ and r. If ΔS has minimum values, $\hbar = \hbar_0$ and $r = r_0$, which are the solutions to equations

$$\Delta S'_{\hbar} = 0 \tag{4.28}$$

$$\Delta S'_{r} = 0 \tag{4.29}$$

then the spike-shaped surface of the particle in fact will be stable. The necessity conditions for the existence of the minimum are the inequalities below (see, e.g., Refs. [168, 169])

$$\text{Det} = \Delta S''_{\hbar\hbar}(\hbar_0, r_0)\Delta S''_{rr}(\hbar_0, r_0) - \Delta S''_{\hbar r}(\hbar_0, r_0)\Delta S''_{r\hbar}(\hbar_0, r_0) > 0$$

$$\Delta S''_{\hbar\hbar}(\hbar_0, r_0) > 0, \; \Delta S''_{rr}(\hbar_0, r_0) > 0 \tag{4.30}$$

Equations (4.28) and (4.29) in the explicit form are respectively reduced to

$$\frac{1}{\kappa}\ln\left(\sqrt{1+\kappa^2}+\kappa\right)=\tfrac{1}{3}\left\{2+\sqrt{1+\kappa^2}-\tfrac{8}{3}\frac{1}{\kappa^2}\left[(1+\kappa^2)^{3/2}-1\right]\right\} \qquad (4.31)$$

$$\frac{1}{\kappa}\ln\left(\sqrt{1+\kappa^2}+\kappa\right)=3\sqrt{1+\kappa^2}-\tfrac{4}{3}\frac{1}{\kappa^2}\left[(1+\kappa^2)^{3/2}-1\right] \qquad (4.32)$$

where the notation

$$\kappa = 2\hbar / r \qquad (4.33)$$

is introduced. Compatibility of Eqs. (4.31) and (4.32) is provided by equating their right-hand sides. It gives the following equation for κ

$$6\kappa^2+5=(\kappa^2-5)\sqrt{1+\kappa^2} \qquad (4.34)$$

The solution of Eq. (4.34) is

$$\kappa_0 \cong 0.8889025353 \qquad (4.35)$$

Now we shall calculate the second derivatives of ΔS:

$$\Delta S''_{\hbar\hbar}=\mathcal{N}\pi\left\{\frac{16}{\kappa^4}\left[(1+\kappa^2)^{3/2}-1\right]+\frac{8}{\kappa^3}\ln\left(\sqrt{1+\kappa^2}+\kappa\right)-16\frac{2+\kappa^2}{\kappa^2}\sqrt{1+\kappa^2}\right\} \qquad (4.36)$$

$$\Delta S''_{rr}=\mathcal{N}\pi\left\{\frac{8}{\kappa^2}\left[(1+\kappa^2)^{3/2}-1\right]+\frac{6}{\kappa}\ln\left(\sqrt{1+\kappa^2}+\kappa\right)-4\frac{3+2\kappa^2}{\sqrt{1+\kappa^2}}-2\right\} \qquad (4.37)$$

$$\Delta S''_{\hbar r}=\mathcal{N}\pi\left\{-\frac{32}{3\kappa^3}\left[(1+\kappa^2)^{3/2}-1\right]-\frac{6}{\kappa^2}\ln\left(\sqrt{1+\kappa^2}+\kappa\right)+2\frac{11+7\kappa^2}{\kappa\sqrt{1+\kappa^2}}\right\} \qquad (4.38)$$

$$\Delta S''_{r\hbar}=\mathcal{N}\pi\left\{-\frac{32}{3\kappa^3}\left[(1+\kappa^2)^{3/2}-1\right]-\frac{6}{\kappa^2}\ln\left(\sqrt{1+\kappa^2}+\kappa\right)+2\frac{11+10\kappa^2}{\kappa\sqrt{1+\kappa^2}}\right\} \qquad (4.39)$$

Substituting the four derivatives (4.36)–(4.39) into inequalities (4.30) we obtain at $\kappa = \kappa_0$ (4.35):

$$\begin{aligned} \mathrm{Det}\big|_{\kappa_0} &= (\mathcal{N}\pi)^2 \{2.645513 \times 3.835575 - (-0.531893) \times 4.518111\} \\ &= (\mathcal{N}\pi)^2 \times 12.550216 > 0 \end{aligned} \tag{4.40}$$

$$\Delta S''_{\hbar\hbar}(\kappa_0) = \mathcal{N}\pi \times 2.645513 > 0,\ \Delta S''_{rr}(\kappa_0) = \mathcal{N}\pi \times 3.835575 > 0 \tag{4.41}$$

Expressions (4.40) and (4.41) satisfy the necessary conditions (4.30), which means that the function ΔS, which is defined by expressions (4.27) and (4.23), reaches the minimum at $\hbar = \hbar_0$ and $r = r_0$. Using the calculated value of κ_0 (4.35) and the definition (4.33), we can rewrite expression (4.26) for the surface defect in the form

$$\Delta S = 4\pi \mathcal{R}^2_{\text{particle}} + \mathcal{N}\pi r_0^2 \times 2.4157446 \tag{4.42}$$

where r_0 determines the most optimal radius of the spike base on the particle.

Let us calculate the volume of the spike, which can easily be done in view of the rotation of the curve (4.21) around the Y-axis:

$$v_{\text{spike}} = \pi \int_0^{\hbar} x^2(y)\, dy = \pi \int_0^{\hbar} \{r(1 + \sqrt{y/\hbar})\}^2\, dy = \tfrac{11}{12}\pi r^2 \hbar \tag{4.43}$$

In the point $(\hbar_0,\ r_0)$:

$$v_{\text{spike}}(\hbar_0,\ r_0) = 0.81482729 \times \pi r_0^3 \tag{4.44}$$

Clearly, the total volume of $\mathcal{N}$ spikes on the particle should satisfy the inequality $\mathcal{N} v_{\text{spike}} << 4\pi \mathcal{R}^3_{\text{particle}}/3$, which sets the restriction to the value of the spike radius r_0:

$$r_0 < \mathcal{R}_{\text{particle}} / \mathcal{N}^{1/3} \tag{4.45}$$

Inequality (4.30) with respect to the value of r_0 is the distinctive adiabatic condition: each of the two states of the particle's surface polarization (outward or inward) is stable when the surface amplitudes are as small as the condition (4.45) requires. In other words, homeomorphic transformations of the particle surface stop when the amplitudes reach the size given by the values of r_0 and $\hbar_0$, which launches the fractal mechanism of the surface organization.

The effective curvature of a spike is considerably greater than that of the basic surface of the particle. That is why it is reasonable that the impact of spikes with very small parameters r_0 and $\hbar_0$ upon surrounding cells is much stronger than that caused by the particle's sphere with the radius $\mathcal{R}_{\text{particle}}$.

One more surface of revolution was tested [167], namely the surface of a parabola-like curve

$$y = \hbar \cdot (1 - x^2 / r^2) \tag{4.46}$$

The curve is depicted in Figure 4.3. The analysis has shown that such surface of revolution does not satisfy the conditions of minimum (4.30), which means that surface amplitudes of this type cannot generate a stable fractal configuration on the particle surface.

4.2.2 THE SURFACE FRACTALS AND THE ELECTRODYNAMIC POTENTIALS

Each nth spike on a particulate ball can be regarded as the normal vector to the spherical surface. If we designate the normal dimensionless unit vector

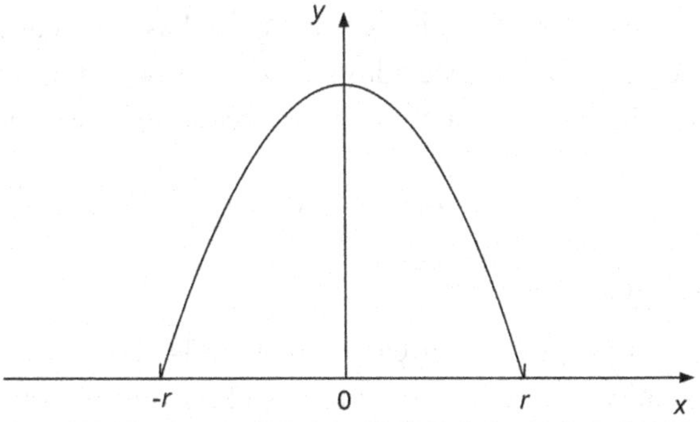

FIGURE 4.3 Parabola-like curve (4.46).

as $\vec{u}_n^{\perp}$, the combination $\vec{u}_n^{\perp}\hbar_0$ can be interpreted as an elementary vector of the electric field, i.e.,

$$\vec{\mathcal{E}}_n = \vec{u}_n^{\perp}\,\hbar_0\,C^{\text{dimension}} \tag{4.47}$$

where $C^{\text{dimension}}$ is the dimension constant. The flux U_n of the vector $\vec{\mathcal{E}}_n$ through the surface $\delta s_n = \pi\, r_0^2$ of the nth spike base can be written in the form of scalar product

$$U_n = \vec{\mathcal{E}}_n \cdot \delta\vec{s}_n \tag{4.48}$$

Then the electric charge can be interpreted as the total sum of all fluxes U_n through the particle surface (see Figure 1.7)

$$\tilde{e} = \sum_{n=1}^{\mathcal{N}} U_n = \mathcal{N}\left(\vec{\mathcal{E}}_n \cdot \delta\vec{s}_n\right) \tag{4.49}$$

The constancy of the charge $\tilde{e}$, for instance, for the lepton series, when the particle's radius decreases $\mathcal{R}_e > \mathcal{R}_\mu > \mathcal{R}_\tau$, should be provided by appropriate changes in the parameters $\hbar_0$ and r_0 with holding the relation (4.33) and (4.35). The charge $\tilde{e}$ induces the same polarization in the surfaces of ambient cells of the tessellattice, which spreads gradually decreasing (Figure 4.3) up to a distance determined by the classical radius of the electron r_e (1.90).

Since particles are constantly in a state of motion, their spikes collide with oncoming cells. Let us assume that the height of the spike can vary between values $\hbar_0$ and $\hbar_0 - \delta\hbar$, for example owing to splitting (which can change over time by merging). The surface of each spike is stretched on the base πr_0^2 that is motionless. This type of motion of the spike has a potential and hence all states of the nth spike surface admit the description by a scalar function $\varphi_n(\hbar) \propto \hbar$. Then the field vector $\vec{\mathcal{E}}_n$ can be related to this function

$$\vec{\mathcal{E}}_n(\hbar) = -\nabla_{\hbar}\,\varphi_n(\hbar) \tag{4.50}$$

so $\vec{\mathcal{E}}_n$ is a co-vector.

On the other hand, the tip of each nth spike is able to deviate from its equilibrium position, i.e., a bending of the spike from its axis of symmetry cannot be ruled out. It is obvious that the value of displacement decreases from the tip to the base that is fixed. Therefore this kind of motion can be related to a vector field rather than to a vector, because a vector describes only a displacement of a separate point. Let us denote this vector field as $\vec{A}_n$.

We do not know motionless particles. Particles exist only in the state of motion. So let us treat the rectilinear uniform motion of the charge $\tilde{e}$. The most important characteristic of the motion of a particle is its velocity or, as we have seen in the preceding sections, the initial velocity $\vec{\upsilon}_0$. In turn, υ_0 is a combination of the particle's spatial period λ and time T of collisions of the particle with its inerton cloud, $\upsilon_0 = 2\lambda / T$. Due to these imminent collisions, the velocity υ_0 also becomes an important feature of the inner motion of the particle's spikes. Since the environment of the charge $\tilde{e}$ is polarized (Figure 4.4), we must assume that the inerton cloud, which is created at collisions of the particle with oncoming cells, consists of polarized inertons (Figure 1.13). Besides, Figure 4.4 clearly demonstrates that the charge and the ambient cells are typical gears, which automatically means that the rectilinear motion of the charge $\tilde{e}$ generates a rotary motion in its spikes (owing to the tangential component of the motion).

So, James Clerk Maxwell was right when he involved imaginary cogwheels constructing the equations of motion of the electromagnetic field!

Thus when a charged particle is moving, it periodically emits and absorbs its inerton-photon cloud. During such motion the charge's spikes experience two kinds of inner movements: (i) oscillations of the spike's height $\hbar_0$ (it is gradually split at collisions and then restored), i.e., oscillations of the potential $\varphi_n(\hbar)$, and (ii) oscillations of the spike relative to the axis of symmetry (the Y-axis in Figure 4.2), i.e., oscillations of the vector potential $\vec{A}_n$.

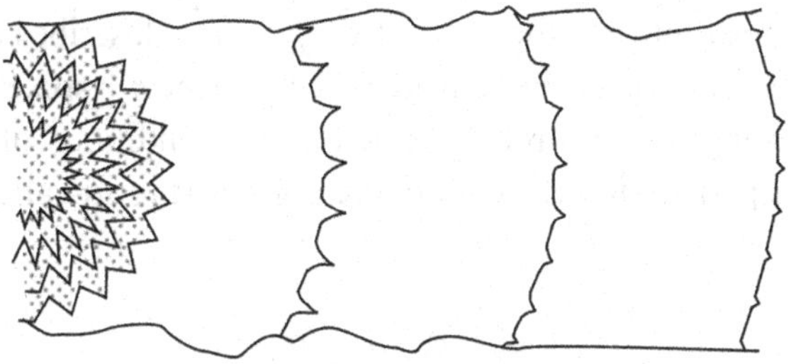

FIGURE 4.4 Electric charge polarizes the ambient cells of the tessellattice.

Consider the charge $\tilde{e}$ (Figure 1.7) that moves with the velocity υ_0 and let the *n*th spike be described by the scalar $\varphi_n(\hbar)$ and vector $\vec{A}_n$ fields. Dimensionality of these fields corresponds to length. Therefore, their velocities are $\dot{\varphi}_n(\hbar)$ and $\dot{\vec{A}}_n$. For the sake of simplicity we will describe the inerton-photon cloud as a single quasi-particle that has the same quantity of spikes as the charge $\tilde{e}$ itself. Let ϕ_n and $\vec{\alpha}_n$ be the scalar and vector fields, respectively, of the *n*th effective spike of the cloud. Then the velocities of the given fields are $\dot{\phi}_n$ and $\dot{\vec{\alpha}}_n$. Now we may construct the Lagrangian density for the motion of the *n*th particle's spike and the *n*th cloud's spike taking into account their mutual interaction:

$$L_n = C\left\{\begin{array}{l}\frac{1}{2}\dot{\varphi}_n^2 + \frac{1}{2}\dot{\vec{A}}_n^2 + \frac{1}{2}\dot{\phi}_n^2 + \frac{1}{2}\dot{\vec{\alpha}}_n^2 + \upsilon_0(\dot{\vec{A}}_n \cdot \vec{\nabla}\phi_n + \dot{\vec{\alpha}} \cdot \vec{\nabla}\varphi_n) - \\ \frac{1}{2}\upsilon_0^2(\vec{\nabla}\times\vec{A}_n)(\vec{\nabla}\times\vec{\alpha}_n)\end{array}\right\} \tag{4.51}$$

where C is a constant with dimensionality of the density, kg×m^{-3}. Here, the first four quadratic forms relate to the kinetic energy of the four fields – the particle's φ_n and $\vec{A}_n$, and the cloud's ϕ_n and $\vec{\alpha}_n$. The last two terms describe the interaction between the fields of the particle and the cloud. In expression (4.51), the fields are differentiated by the time t set as the proper time of the particle, i.e., $t = l / \upsilon_0$ is a natural parameter where l is the length of the particle's path. It is important to emphasize that owing to the introduction to the Lagrangian density (4.51) such operators as the divergence and the curl, the derivatives $\dot{\varphi}_n$, $\dot{\phi}_n$ and $\dot{\vec{A}}_n$, $\dot{\vec{\alpha}}_n$ should be treated as the partial derivatives: $\partial\varphi_n / \partial t$, $\partial\phi_n / \partial t$ and $\partial\vec{A}_n / \partial t$, $\partial\vec{\alpha}_n / \partial t$. These time derivatives describe changes in time of the corresponding fields, which occurs in the point $\vec{r}$ of the particle's path l. Vortexes $\vec{\nabla}\times\vec{A}_n$ and $\vec{\nabla}\times\vec{\alpha}_n$ are taken in the same point $\vec{r}$. More exactly, the operator $\vec{\nabla}$ is constructed on the spatial vector $\vec{r}_n = \vec{r} + \overrightarrow{(\mathcal{R}_{\text{particle}} + \hbar_n)}$ where $\vec{r}$ is the radius vector of the center of mass of the particle and the vector $\overrightarrow{(\mathcal{R}_{\text{particle}} + \hbar_n)}$ describes the tip of the *n*th spike. In the same point we consider the tip of the *n*th spike of the cloud.

The following obvious relations determine conditions of passing the field from the particle to the cloud in the point $\vec{r}_n$ (the so-called continuity of the fields):

$$\varphi_n = \phi_n, \ \vec{A}_n = \vec{\alpha}_n \tag{4.52}$$

The conditions (4.52) make it possible to split the Lagrangian density (4.51) to two sub parts

$$L_n = L_n^{\text{particle}} + L_n^{\text{cloud}} \tag{4.53}$$

where

$$L_n^{\text{part}} = C\left\{\tfrac{1}{2}\dot{\varphi}_n^2 + \tfrac{1}{2}\dot{\vec{A}}_n^2 + \upsilon_0 \dot{\vec{A}}_n \cdot \vec{\nabla}\varphi_n - \tfrac{1}{2}\upsilon_0^2(\vec{\nabla}\times\vec{A}_n)^2\right\} \tag{4.54}$$

$$L_n^{\text{cloud}} = C\left\{\tfrac{1}{2}\dot{\phi}_n^2 + \tfrac{1}{2}\dot{\vec{\alpha}}_n^2 + \upsilon_0 \dot{\vec{\alpha}}_n \cdot \vec{\nabla}\phi_n - \tfrac{1}{2}\upsilon_0^2(\vec{\nabla}\times\vec{\alpha}_n)^2\right\} \tag{4.55}$$

Since the Lagrangian densities (4.54) and (4.55) have the same form, we can treat only the behavior of the potentials of the particle's spike. The Euler-Lagrange equations for the Lagrangian density (4.54) will be constituted in line with Eqs. (3.45) and (3.46); this is correct for the variables $\dot{\varphi}_n$ and $\vec{\nabla}\varphi_n$. However, in the case of variables $\dot{\vec{A}}_n$ and $\vec{\nabla}\times\vec{A}_n$ we shall use Eq. (3.45) and the equation, which substitutes Eq. (3.46), namely

$$\frac{\delta L_n}{\delta Q} = \vec{\nabla}\times\frac{\partial L_n}{\partial(\vec{\nabla}\times\vec{A}_n)} \tag{4.56}$$

where in the present case $Q \equiv \vec{\nabla}\times\vec{A}_n$. As the result we obtain a system of two equations:

$$\ddot{\varphi}_n + \upsilon_0 \vec{\nabla}\dot{\vec{A}}_n = 0 \tag{4.57}$$

$$\ddot{\vec{A}}_n + \upsilon_0 \vec{\nabla}\dot{\varphi}_n + \upsilon_0^2 \vec{\nabla}\times(\vec{\nabla}\times\vec{A}_n) = 0 \tag{4.58}$$

From Eq. (4.58) we get

$$\dot{\varphi}_n = -\upsilon_0 \vec{\nabla}\vec{A}_n + \text{const} \tag{4.59}$$

Substituting $\dot{\varphi}_n$ from Eq. (4.59) into Eq. (4.58) we get

$$\ddot{\vec{A}}_n - \upsilon_0^2\left[\vec{\nabla}(\vec{\nabla}\vec{A}_n) - \vec{\nabla}\times(\vec{\nabla}\times\vec{A}_n)\right] = 0 \tag{4.60}$$

With the known formula

$$\vec{\nabla}(\vec{\nabla}\vec{b})-\vec{\nabla}\times(\vec{\nabla}\times\vec{b})=\vec{\nabla}^2\vec{b}$$

Equation (4.60) can be reduced to the form of a typical wave equation

$$\ddot{\vec{A}}_n-\upsilon_0^2\,\vec{\nabla}^2\vec{A}_n=0 \tag{4.61}$$

Integrating Eq. (4.58) over t we can solve it relative to the variable $\dot{\vec{A}}_n$:

$$\dot{\vec{A}}_n=-\upsilon_0\vec{\nabla}\varphi_n-\upsilon_0^2\int\vec{\nabla}\times(\vec{\nabla}\times\vec{A}_n)\,dt \tag{4.62}$$

Substituting the obtained expression for $\dot{\vec{A}}_n$ (4.62) into Eq. (4.58) we arrive at the equation

$$\ddot{\varphi}_n-\upsilon_0^2\,\vec{\nabla}^2\varphi_n-\upsilon_0^3\,\vec{\nabla}\left[\int\vec{\nabla}\times(\vec{\nabla}\times\vec{A}_n)\,dt\right] \tag{4.63}$$

Since the formula

$$\vec{\nabla}(\vec{\nabla}\times\vec{b})=0$$

is valid for any vector field $\vec{b}$, the last term in Eq. (4.63) equals zero and we finally arrive at a wave equation for the potential φ_n

$$\ddot{\varphi}_n-\upsilon_0^2\,\vec{\nabla}^2\varphi_n=0 \tag{4.64}$$

Wave equations for the fields $\vec{\alpha}_n$ and ϕ_n that are similar to equations (4.61) and (4.64) can be derived in the same way from the Lagrangian density (4.55) of the cloud.

The total fields of the particle and the cloud are respectively:

$$\varphi=\sum_{n=1}^{\mathcal{N}}\varphi_n,\ \vec{A}=\sum_{n=1}^{\mathcal{N}}\vec{A}_n \tag{4.65}$$

$$\phi=\sum_{n=1}^{\mathcal{N}}\phi_n,\ \vec{\alpha}=\sum_{n=1}^{\mathcal{N}}\vec{\alpha}_n \tag{4.66}$$

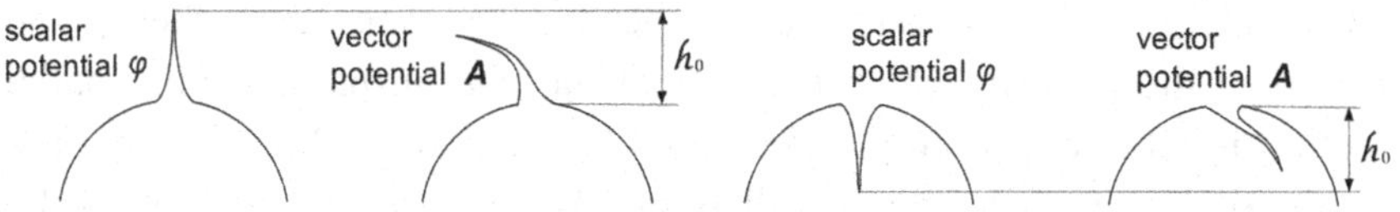

FIGURE 4.5 Scalar φ and vector $\vec{A}$ potential as the normal surface spike and the twisted spike, respectively, which manifest themselves in the wave equations (4.64) and (4.61). a – surface fractals are oriented outward; b – surface fractals are oriented inward.

We can see from the wave equations (4.61) and (4.64) that the initial velocity υ_0 of the motion of the particle determines the rate of change for each of the fields. Figure 4.5 illustrates the spike behavior on the surface of the appropriate ball. The initial conditions and the solution to the equations of motion for the fields φ and $\vec{A}$ (which have here the dimensionality of length) are analyzed below in Section 4.3.

4.3 THE MAXWELL EQUATIONS AS THE MANIFESTATION OF HIDDEN DYNAMICS OF SURFACE FRACTALS

Now we can write the Lagrangian for a free photon. We can do it in a system of units in which the scalar potential φ is dimensionless (i.e., it is determined by the relative length of spikes on the surface of the photon-excited cell) and the vector potential $\vec{A}$, which is a field that twists the spikes, has a dimension of length. The Lagrangian reads

$$L_{\text{photon}} = \tfrac{1}{2}c^2T^2\dot{\varphi}^2 + \tfrac{1}{2}\dot{\vec{A}}^2 + c\lambda\,\dot{\vec{A}}\cdot\vec{\nabla}\varphi - \tfrac{1}{2}c^2(\vec{\nabla}\times\vec{A})^2 \quad (4.67)$$

and the equations of motions for the potentials are

$$\ddot{\varphi} - c^2\nabla^2\varphi = 0 \quad (4.68)$$

$$\ddot{\vec{A}} - c^2\nabla^2\vec{A} = 0 \quad (4.69)$$

where the relationship $\lambda/T = c$ has been used. Here, λ is the spatial period of the photon, which means that the photon passing this spatial section is returned to the photon's initial state and T is the time period. The period T describes the oscillations of spikes on the surface of the photon. Namely: oscillations of the height along the vector normal to the surface of the cell (the scalar potential φ); torsional oscillations of spikes (the vector potential $\vec{A}$).

The Lagrangian (4.67) can be written in conventional physical terms of the SI units:

$$L_{\text{photon}} = \frac{\varepsilon_0}{2c^2}\dot{\varphi}^2 + \frac{\varepsilon_0}{2}\dot{\vec{A}}^2 + \varepsilon_0 \dot{\vec{A}} \cdot \vec{\nabla}\varphi - \frac{\varepsilon_0 c^2}{2}(\vec{\nabla} \times \vec{A})^2 \quad (4.70)$$

In the SI units the charge is measured in C (Coulomb); the scalar potential φ in kg·m^2/(C·s^2); the vector potential $|\vec{A}|$ in kg×m/(C·s); the electric constant ε_0 in C^2·s^2/(kg·m^3). The equations of motion of the appropriate potentials φ and $\vec{A}$ are exactly the same as written above, (4.68) and (4.69), respectively.

Now we can write the Lagrangian density of the electromagnetic field, i.e., a flux of free photons, which interacts with a charged particle. In standard symbols

$$L = \frac{\varepsilon_0}{2c^2}\dot{\varphi}^2 + \frac{\varepsilon_0}{2}\dot{\vec{A}}^2 + \varepsilon_0 \dot{\vec{A}} \cdot \vec{\nabla}\varphi - \frac{\varepsilon_0 c^2}{2}(\vec{\nabla} \times \vec{A})^2 - \rho \cdot (\varphi_0 - \varphi) + \rho \vec{\upsilon}_0 \vec{A} \quad (4.71)$$

Here, ρ is the charge density; φ_0 is the reference point of the potential φ, because as we know in physics in reality it is the difference of the potentials between two points, which in the present case is $\varphi_0 - \varphi$; $\vec{\upsilon}_0$ is the vector velocity of the particle studied (i.e., $\vec{\upsilon}_0$ is the particle's initial velocity, which oscillates along the whole particle path, Figure 2.6).

The Euler-Lagrange equations for the Lagrangian density (4.71) read

$$\ddot{\varphi} / c^2 - \rho / \varepsilon_0 + \vec{\nabla}\dot{\vec{A}} = 0 \quad (4.72)$$

$$\ddot{\vec{A}} + \vec{\nabla}\dot{\varphi} + c^2 \vec{\nabla} \times (\vec{\nabla} \times \vec{A}) - \vec{\upsilon}_0 \rho / \varepsilon_0 = 0 \quad (4.73)$$

From Eq. (4.73) we find

$$\dot{\vec{A}} = -\vec{\nabla}\varphi - \int c^2\vec{\nabla}\times(\vec{\nabla}\times\vec{A})\,dt + \vec{\upsilon}_0 t\rho/\varepsilon_0 \tag{4.74}$$

Setting $\dot{\vec{A}}$ from expression (4.74) into Eq. (4.72), we gain

$$\ddot{\varphi}/c^2 - \nabla^2\varphi = \rho/\varepsilon_0 \tag{4.75}$$

From Eq. (4.72) we get

$$\dot{\varphi} = \rho t c^2/\varepsilon_0 - c^2\vec{\nabla}\vec{A} + \text{const} \tag{4.76}$$

Setting $\dot{\varphi}$ (4.76) into equation (4.73) we obtain

$$\ddot{\vec{A}} - c^2\nabla^2\vec{A} = \vec{\upsilon}_0\rho/\varepsilon_0 \tag{4.77}$$

The obtained Maxwell equations (4.75) and (4.77) can now be rewritten in the conventional d'Alembert's form:

$$\frac{1}{c^2}\frac{\partial^2\varphi}{\partial t^2} - \nabla^2\varphi = \frac{\rho}{\varepsilon_0} \tag{4.78}$$

$$\frac{1}{c^2}\frac{\partial^2\vec{A}}{\partial t^2} - \nabla^2\vec{A} = \mu_0\vec{j} \tag{4.79}$$

where μ_0 is the permeability of free space (magnetic constant), which is related to ε_0, the absolute dielectric permittivity of a classical vacuum (electric constant) through the relationship $\varepsilon_0\mu_0 = c^{-2}$; $\vec{j} = \rho\vec{\upsilon}_0$ is the current density.

If the charge is absent, then putting $\rho = 0$ in the equation of motion (4.72) we immediately obtain the so-called FitzGerald-Lorenz gauge condition:

$$\vec{\nabla}\cdot\vec{A} + \frac{1}{c^2}\frac{\partial\varphi}{\partial t} = 0 \tag{4.80}$$

So there is no need to invent an additional gauge condition, the condition is a part of the system of Euler-Lagrange equations of motion of potentials φ and $\vec{A}$ that describe the behavior of surface fractals of the appropriate excited cell in the tessellattice.

Note that in contrast to the proposed Lagrangian density (4.71), the standard Lagrangian of the electromagnetic field (see, e.g., expression (4.9)) does not contain the variable $\dot{\varphi}$ because so far there has not been an understanding of either the nature of the charge or the inner physical processes that lead to the Maxwell equations. In classical electrodynamics the problem has been compensated by the supplemented FitzGerald-Lorenz condition (4.80), which includes the missing derivative $\dot{\varphi}$. The origin of charge, photon, potentials φ and $\vec{A}$ were unknown and so far these important parameters have remained undetermined.

Now let us look at the conventional Maxwell equations:

$$\begin{aligned}
&\text{(i)}\ \nabla\cdot\boldsymbol{\mathcal{E}} = \rho/\varepsilon_0\\
&\text{(ii)}\ \nabla\cdot\mathbf{B} = 0\\
&\text{(iii)}\ \nabla\times\boldsymbol{\mathcal{E}} = -\partial\mathbf{B}/\partial t\\
&\text{(iv)}\ \nabla\times\mathbf{B} = \mu_0\mathbf{j} + \varepsilon_0\mu_0\partial\boldsymbol{\mathcal{E}}/\partial t
\end{aligned}\tag{4.81}$$

What are the fields $\mathcal{E}(\mathbf{r}, t)$ and $\mathbf{B}(\mathbf{r}, t)$? Below Griffiths [170] answers in the following way. He notes that it is not an easy problem, and it pays to begin by representing the fields in terms of potentials. Then since in electrostatics $\nabla\times\mathcal{E} = 0$, the field $\mathcal{E}$ can be written as the potential: $\mathcal{E} = -\nabla\varphi$. The field **B**, which is non-divergent, can be written in the form

$$\mathbf{B} = \nabla\times\mathbf{A}\tag{4.82}$$

Putting this into the equation (iii) (4.81) yields

$$\nabla\times\left(\mathcal{E} + \partial\mathbf{A}/\partial t\right) = 0\tag{4.83}$$

Here **A** is a quantity, unlike $\mathcal{E}$ alone, whose curl does not vanish. So it can be written as a gradient of a scalar, such that

$$\mathcal{E} = -\nabla\varphi - \partial\mathbf{A}/\partial t\tag{4.84}$$

The potential representations (4.83) and (4.84) automatically fulfil the two homogeneous Maxwell equations, (ii) and (iii). Putting (4.84) into equation (i) (4.81), we find that

$$\nabla^2\varphi + \partial(\nabla\cdot\mathbf{A})/\partial t = -\rho/\varepsilon_0 \tag{4.85}$$

Putting (4.82) and (4.84) into equation (iv) (4.81) yields

$$\nabla\times(\nabla\times\mathbf{A}) = \mu_0\mathbf{j} - \varepsilon_0\mu_0\nabla(\partial\varphi/\partial t) - \varepsilon_0\mu_0\partial^2\mathbf{A}/\partial t^2 \tag{4.86}$$

which using the vector identity $\nabla\times(\nabla\times\mathbf{A}) = \nabla(\nabla\mathbf{A}) - \nabla^2\mathbf{A}$ is transformed to

$$\left(\nabla^2 - \varepsilon_0\mu_0\partial^2\mathbf{A}/\partial t^2\right) - \nabla\left(\nabla\cdot\mathbf{A} + \varepsilon_0\mu_0\partial\varphi/\partial t\right) = -\mu_0\mathbf{j} \tag{4.87}$$

Equations (4.85) and (4.87) contain all the information in Maxwell's equations. This is the end of the classical interpretation expressed by Griffiths [170]. Exactly the same conversion from **E** and **B** to φ and **A** was demonstrated by Tamm [171].

Now applying the FitzGerald-Lorenz gauge condition (4.80) to Eq. (4.85) and to the second term in the left hand side of Eq. (4.87), we immediately gain equations (4.78) and (4.79), respectively.

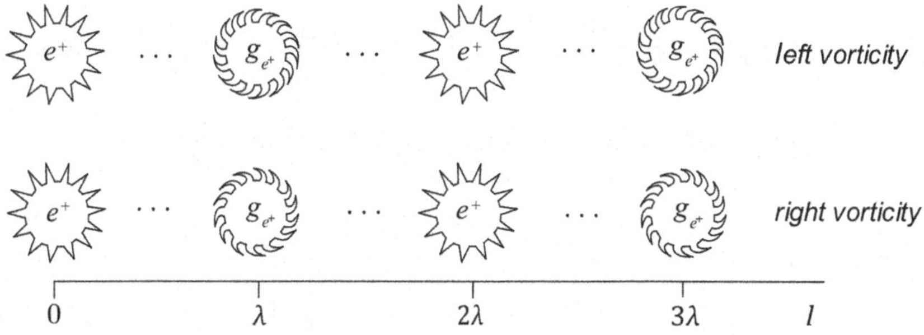

FIGURE 4.6 Positive charged particle in motion. Due to the interaction with the particle's inerton-photon cloud, it changes its pure electric state "e" to the pure magnetic monopole state "g" each section λ of its path.

Thus, the scalar and vector potentials are responsible for the electromagnetic phenomena. In turn these two parameters are directly related to the surface fractals and their appropriate dynamics on the surface of a particulate ball (or a cell that carries the photon excitation).

Figure 4.6 discloses the hidden dynamics of a charged particle behind the Maxwell equations. The particle is always created as a charged ball that has a radial symmetry (Figures 1.7–1.9). However, when the charged particle is moving, the symmetry of its surface changes from the radial to axial; the surface fractals behave as a pseudovector (or axial vector), which is transformed like a vector under a proper rotation. Periodic changes from the typical chestnut state to the monopole state occur in every section λ (or the time interval $t = T/2$). In the case of a particle, this is the de Broglie wavelength λ_{dB}, in the case of a photon, this is half the photon wavelength $\lambda/2$ (T is the time interval between collisions of the particle with its inerton-photon cloud, where for a free photon $T = 1/\nu$ is its period).

Uncoupling the potentials φ and $\vec{A}$, (4.72) – (4.76), results in the two independent equations of motion (4.78) and (4.79). This means that φ and $\vec{A}$ do not affect each other on the surface of the same ball. Hence, the following canonical Lagrangian density can be suggested for a charged particle moving in an external electromagnetic field:

$$L_{\text{canon}} = L_{\varphi} + L_{\vec{A}} \tag{4.88}$$

$$L_{\varphi} = \frac{\varepsilon_0}{2c^2}\dot{\varphi}^2 - \varepsilon_0(\vec{\nabla}\varphi)^2 - \rho\cdot(\varphi_0 - \varphi) \tag{4.89}$$

$$L_{\vec{A}} = \frac{\varepsilon_0}{2}\dot{\vec{A}}^2 - \frac{\varepsilon_0 c^2}{2}\left[(\vec{\nabla}\cdot\vec{A})^2 - (\vec{\nabla}\times\vec{A})^2\right] + g\vec{\upsilon}_0\vec{A} \tag{4.90}$$

The Euler-Lagrange equations for φ and $\vec{A}$ have the form

$$\varepsilon_0 / c^2 \cdot \ddot{\varphi} - \varepsilon_0 \nabla^2\varphi - \rho = 0 \tag{4.91}$$

$$\varepsilon_0 \ddot{\vec{A}} + \varepsilon_0 c^2\left[\vec{\nabla}(\vec{\nabla}\vec{A}) - \vec{\nabla}\times(\vec{\nabla}\times\vec{A})^2\right] - g\vec{\upsilon}_0 = 0 \tag{4.92}$$

where for the vector potential we introduced the magnetic monopole state g, because it is exactly this state of the charged particle, which corresponds to the formation of its proper axial field $\vec{A}$. It seems that the total sum (4.65) of all *n*th fragments of the surface fractals for both the φ and $\vec{A}$ form the electric "*e*" charge and the magnetic "*g*" monopole.

Of course, equations (4.91) and (4.92) are respectively transferred to the equations of motion (4.78) and (4.79), as the expression $g\vec{\upsilon}_0$ is the density of the current: $\vec{j} = g\vec{\upsilon}_0$. It should be noted that Jackson [172] discussed a possibility of introducing a magnetic charge into the second Maxwell's equation (ii) (4.81). He insisted that it should be new species.

However, we have found that the electric charge periodically changes its state of shared existence (or incarnation, hypostasis) to another state. Namely, the pure electric state (electric charge) is periodically transferred to the pure magnetic state (magnetic monopole). Each canonical particle is a dynamic dyon, as its electric property periodically changes to the magnetic one (Figure 4.6). This finding allows us to reveal a hidden symmetry of the Maxwell equations (4.81): the incarnation of the electric charge manifests itself in the static Gauss's law, Eq. (i), and the incarnation of the magnetic monopole manifests itself in the dynamic Ampère's law, Eq. (iv).

The canonical Lagrangian density (4.88)–(4.90) makes it possible to obtain canonical momenta $p_\varphi = \partial L_{\text{canon}} / \partial \dot{\varphi}$ and $p_{\vec{A}} = \partial L_{\text{canon}} / \partial \dot{\vec{A}}$ and introduces the appropriate canonical Hamiltonian, however, in the present case we cannot introduce mass, which is defined as a volumetric fractal contraction (1.50) of a particulate ball. The mass of the surface spikes is rather a convolution of regular defects of the surface and its dimension varies between 2D and 3D. Nevertheless, the Hamilton-Jacobi theory in which the transformation from variables *p* and *q* to *J* and *w* (the increment of the action per period and the angle, respectively) can be used to calculate frequencies of various motions without completely solving the

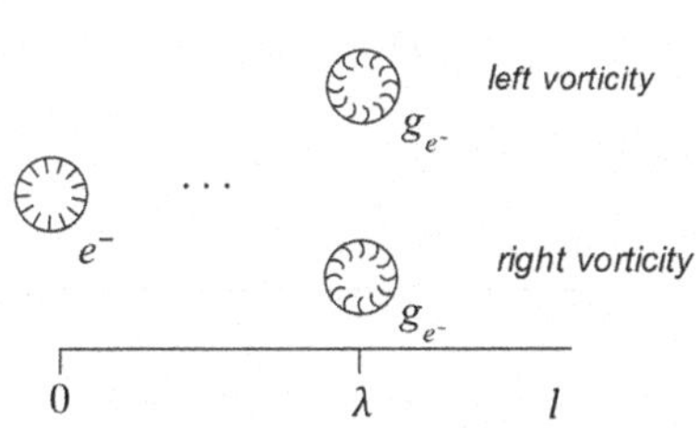

FIGURE 4.7 Schematic sketch of the two states of a negatively charged particle: the electric state and the monopole state.

problem if the motion of the system is both separable and periodic. For periodic motions the Hamiltonian of the system becomes a function of only one action J per period, $E(J)$, which is also true particularly in the case of libration (see, e.g., Ref. [108], § 6.2) when spikes on the surface of a charged particulate ball are twisted (Fig 4.6, Figure 4.7). This also depicts the surface of a photon (Figure 4.7).

The characteristic J we find from the canonical equations of motion:

$$J = \text{constant and } \dot{w} = \partial E / \partial J = \nu \tag{4.93}$$

Therefore $\nu = \partial E / \partial J$ is the frequency of motion.

Ascribing to J the smallest value of increment action, namely identifying it with the Planck constant h, we get for both a charged particle and a free photon

$$E = h\nu \tag{4.94}$$

Thus the fundamental relation (4.94) of quantum theory can easily be derived for a subquantum charged/polarized system in the framework of the Hamilton-Jacobi formalism.

When we ascribe the relationship (4.94) to a particle, we automatically ascribe it a frequency ν, which of course is related to the appropriate period T of oscillation (or an inner libration in the case of the charged state of the surface). Since the particle is in motion, the period T will be connected with a distance λ along which this periodic process takes place in the particle and the particle velocity υ_0: $\upsilon_0 = 2\lambda / T$. Since any real particle possesses a mass, it also has the kinetic energy $\frac{1}{2} m \upsilon_0^2$ and then the

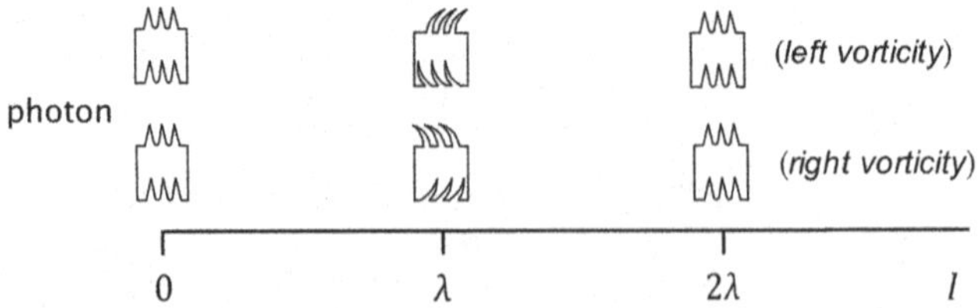

FIGURE 4.8 Photon in two extreme states: the electric state when spikes are normal to the surface (in points λn where $n = 0, 1, 2, \ldots$) and the magnetic state when spikes are twisted to the tangential direction (in points $\lambda/2 + \lambda n$ where $n = 0, 1, 2, \ldots$).

increment of the action per period (2.59) results in the de Broglie relationship $\lambda = h/(m\upsilon_0)$. In such a manner we can see that a periodic process in the motion of a field particle (photon, inerton) or a material particle results in an analogous wave behavior of these particles.

4.3.1 VORTICITY AND SPIN

In Section 3.4 we have considered the phenomenon of spin, associating the particle's spin with two possible initial states of the particle: the mass state (m) and the tension state (Ξ), which are described by the two sub wave functions ϕ_1 and ϕ_2, respectively (3.65)–(3.68).

The Pauli equation (3.40) and (3.41) imposes an additional condition – spinors should provide the interaction with an applied magnetic field.

The submicroscopic consideration of the Maxwell equations allows us to conclude that the spinor sub wave functions have to relate to two states of vorticity of the particle: left and right, which are clearly illustrated in Figure 4.6, because each charged moving particle together with its inerton-photon cloud could be treated as a vortex in which the surface fractals are subject to libration, left or right. Let the sub wave function ϕ_1 describe for the left libration of the surface spikes of the particle (the unit polarization vector $\mathbf{e}_{\text{left}}$) and ϕ_2 is responsible for the right libration (the unit polarization vector $\mathbf{e}_{\text{right}}$). Then the spinor components ϕ_1 (3.67) and ϕ_2 (3.68) have to be rewritten as below:

$$\phi_1(x, t) = \tfrac{1}{2}\mathbf{e}_{\text{left}}\,[1 + \cos(kx - \omega t)] \tag{4.95}$$

$$\phi_2(x, t) = \tfrac{1}{2}\mathbf{e}_{\text{right}}\,[1 - \cos(kx - \omega t)] \tag{4.96}$$

In such a manner we may completely clarify the hidden mechanism of the Pauli exclusion principle. Two particles, whose separation is close to their de Broglie wavelengths along the line of the sum of their vector velocities or is closer than the amplitudes of their inerton clouds, will interact through their inertons. In other words, the inerton clouds of these particles must overlap. The particles will be attracted if their sub wave functions are in counter phase, namely, if one particle is characterized by the mass state and the left vorticity and the other one by the tension state and the right vorticity. If the two characteristics in two different particles are the same, they will be repelled. For example, two particles shall be repelled if each of them is in the mass state (or in the tension state) and has the same vorticity.

Usually in particle physics researchers use the term 'helicity' – a combination of the spin and the linear motion of a subatomic particle. Besides, in electrodynamics, particularly in optical physics, researchers use the term 'polarization' (left, right, circular, etc.) in application to photons. In this section we use the term 'vorticity' just to demonstrate the origin of the phenomenon. Nevertheless, Figs. 1.7 and 1.8 clearly indicate that the vorticity of spikes will involve a 3D-dynamics, i.e., all three parameters of the spikes' spherical geometry (height h, azimuthal angle φ, and polar angle θ) must be activated. This exhibits Ampère's law (the fourth Maxwell equation (iv) (4.81)): a current induces a helical magnetic field: $\mathbf{j} \propto \nabla \times \mathbf{B} = \nabla \times (\nabla \times \mathbf{A})$. Hence all three characteristics – vorticity, helicity and polarization – work together (though the term 'polarization' is generally the most universal).

4.3.2 STATIONARY MAGNETIC FIELD AND THE PARTICLE SPIN

Formally a stationary magnetic field is generated by a current, as the fourth Maxwell equation prescribes, $\nabla \times \mathbf{B} = \mu_0 \mathbf{j}$. From the submicroscopic view

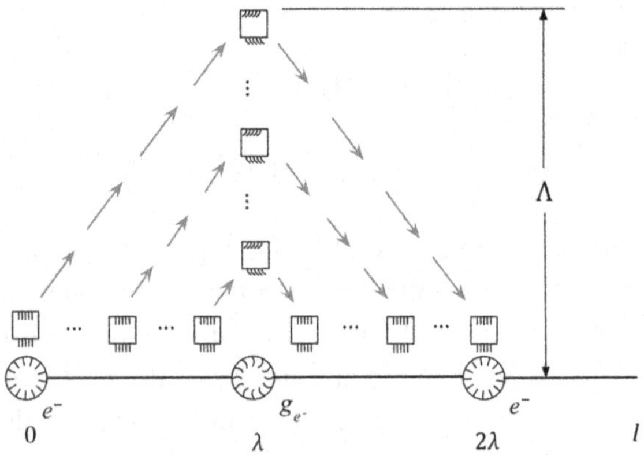

FIGURE 4.9 Motion of the charged particle that creates inerton-photons. Tangential state of spikes of the surface fractals, which is related to the creation of the magnetic field, appears on inerton-photons at the distance Λ from the electron in transverse directions to the line of the electron path. Λ is the amplitude of the electron's inerton-photon cloud.

point the situation looks as follows. In a flow of charged particles each of the particles creates its proper inerton-photon cloud that spreads up to a distance provided by the amplitude of the inerton cloud (in the present case, the inerton-photon cloud) $\Lambda = \pi \lambda_{\rm dB} c / \upsilon_0$ where $\lambda_{\rm dB}$ is the particle's de Broglie wavelength and υ_0 is the particle's velocity. In the cloud, the state of inerton-photons gradually changes from pure electrical polarization (near the particle) to pure magnetic polarization (at the distance L in transverse directions to the particle path). Figure 4.9 accounts for the mechanism of the formation of magnetic field. The vector potential is the origin of the magnetic field, which is evident from the fourth Maxwell equation that we can write in terms of the vector potential **A** and the magnetic monopole g:

$$\nabla \times (\nabla \times \mathbf{A}) = g \mathbf{v}_0 \text{, or } \nabla (\nabla \cdot \mathbf{A}) - \nabla^2 \mathbf{A} = g \mathbf{v}_0 \quad (4.97)$$

Hence the magnetic monopole g is the source for the vector potential **A** and, therefore, for the magnetic field **B**. Recall the magnetic monopole state of a charged particle is schematically shown in Figures 4.6 and 4.7.

In Section 3.4 starting from the quantum mechanical formalism, we have considered how the spin interacts with an electromagnetic field. Now analyzing electrodynamics at the submicroscopic level, we may ask: How does a stationary magnetic field interact with the particle spin? For a charged particle like the electron, it is obvious: inerton-photons that bring the stationary magnetic field to the electron (these inerton-photons are shown in Figure 4.9 as cells with bending spikes) touch the electron that is in the monopole state g, which turns the electron to one of the two possible trajectories. Namely, the energy of the electron changes by the value of $E_{\uparrow(\downarrow)} = \hat{\mu}_z B_z$, where $\hat{\mu}_z$ is the z-projection of the electron's intrinsic magnetic moment and B_z is the stationary magnetic field directed along the Z-axis. Explicitly, form the change in energy is

$$E_{\uparrow(\downarrow)} = \pm \tfrac{1}{2} e\hbar B_z / m \quad (4.98)$$

where m and e are the electron mass and charge, respectively. If the orientation of B_z corresponds to the vorticity of g, then in expression (4.98) we shall choose the sign "+"; if the orientation of B_z is opposite to the vorticity of g, we choose the sign "-."

When a test particle is the proton, the consideration is exactly the same. The situation with the neutron spin is also similar. We will come back to these baryons and their spin in Chapter 6. But looking ahead, we can emphasize the following primary characteristics of a canonical particle, which relate to its notion of spin-1/2:

- it is the initial state of the particle's wave ψ-function (mass or tension), as has been discussed in Section 3.4;
- it is the direction of vorticity (or polarization) of the monopole state of the particle's charge: left or right;
- it is the mandatory presence of the charged state on the particle (electron, muon, positron) or the presence of a quasi-free charged particle inside a combined particle (one non-compensated charged quark is available in both the proton and neutron).

If particles are either combined (bosons) or a quasi-particle (photon), then their spin is integral, since it is associated with a real vorticity of these particles.

The Pauli exclusion principle is due to the vorticity vector (left or right) of the surface fractals of particles. The particle's inerton-photon cloud spreads this vorticity in the ambient space around the particle. However, if we deal with a system of purely neutral particles (like diluted cool quantum gases, for example, a gas of cesium atoms), which do not interact with a stationary magnetic field, each particle will be sensitive to the oscillations of its neighbor, such that the initial states of all the wave functions (mass and tension) in the ensemble become synchronized. In other words, the Pauli principle can also be applied to such system in which it will put every atom in its place,

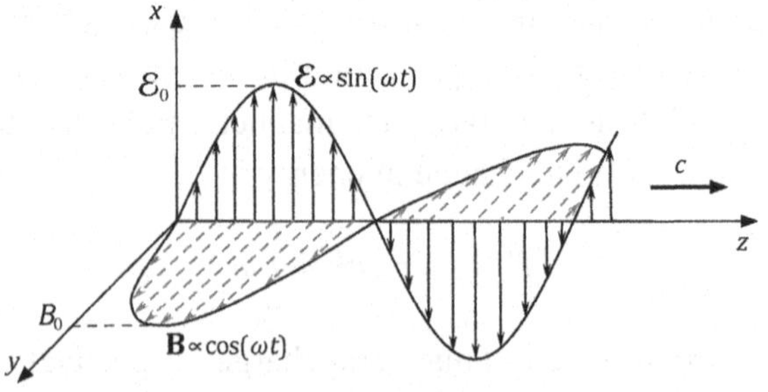

FIGURE 4.10 Distribution of the electric field ε and the magnetic field **B** in a monochromatic plane wave. The direction of ε sets the polarization of an electromagnetic wave.

such that the whole system of weakly interacting particles will exhibit the phenomenon known as the Bose-Einstein condensation (see Chapter 5).

4.3.3 STANDING WAVES VERSUS PLANE WAVES

In classical electrodynamics the solutions to the Maxwell equations for a free electromagnetic field are presented as monochromatic plane waves (see, e.g., Ref. [170], p. 376; Ref. [171], p. 472; Ref. [172], p. 296):

$$\mathcal{E} = \mathcal{E}_0 \exp(i\mathbf{k}\cdot\mathbf{x} - i\omega t) \tag{4.99}$$

$$\mathbf{B} = \mathbf{B}_0 \exp(i\mathbf{k}\cdot\mathbf{x} - i\omega t) \tag{4.100}$$

These solutions mean that vectors $\mathcal{E}$ and $\mathbf{B}$ are in phase and change synchronously, though they develop in perpendicular directions. To be consistent with Maxwell's equations, the absolute values of the amplitudes are related as follows: $\mathcal{E}_0 / B_0 = c$. Figure 4.10 illustrates this widely accepted opinion regarding the distribution of electric and magnetic fields in a travelling electromagnetic wave.

The energy density of the wave is written as (see, e.g., Ref. [170], p. 380; Ref. [171], p. 419, 422)

$$W \propto \mathcal{E}^2 + \mathbf{B}^2 \tag{4.101}$$

The time-averaged energy density is correspondingly $\langle W \rangle \propto \mathcal{E}\mathcal{E}^* + \mathbf{B}\mathbf{B}^* = |\mathcal{E}_0|^2/2$ (see, e.g., Ref. [172], p. 298) and the energy density of the wave as a function of time is

$$W \propto \mathcal{E}_0^2 \cos^2(\omega t - \mathbf{k}\mathbf{x} + \delta) \tag{4.102}$$

which is of course in agreement with the solutions (4.99) and (4.100) and Figure 4.9.

What does expression (4.102) really mean for the reader of this book? In the reference frame associated with your eyes consider the coordinate $\mathbf{x} = 0$ and also the phase $\delta = 0$. In this case photons will come to your eyes having the energy $\mathcal{E}_0^2 \cos^2(\omega t)$. This means that those photons, which bring you the information from the book's pages, periodically disappear and hence you see this book only from time to time... In other words, the book together with the photons play hide and seek with you! But if you do not have gaps in your vision and the book is constantly there, then you have met an actual paradox in classical electrodynamics, on level ground.

Besides, a stationary magnetic field also contravenes expressions (4.99)–(4.102). As we know a stationary magnetic field exists by itself without presence of a stationary or time-dependent electric field, whereas a changing magnetic field is generated by a moving charged particle(s) (see Figure 4.9). This statement guarantees: (i) Zeeman's [173] experiment in which a static magnetic field splits a spectral line into several components; (ii) Stern-Gerlach's [174] experiment in which a heterogeneous magnetic field splits an electron beam; (iii) a current in the solenoid coil, which creates a stationary longitudinal magnetic field inside the solenoid, etc.

The paradox is due to the fact that the search for the solution of wave equations (4.68) and (4.69), or homogeneous equations (4.78) and (4.79), have been done in terms of a plane wave. In a plane wave, the compression and the stretching periodically disappear and substitute each other. In particular, this situation is realized in both the case of a free inerton (see the solutions (3.51) and (3.52)) and the case of the particle's wave ψ-function (see the solutions (3.65) and (3.66)). In the event of the surface fractals, i.e., spikes, they do not disappear during the period, but only change the orientation from normal to tangential. This is a typical behavior of an oscillating string; at transverse oscillations it periodically runs a set of permissible positions. These are standing vibrations and a corresponding solution should consist of the multiplication of spatial and temporal trigonometric functions.

Figure 4.5 clarifies that at the moment $t = 0$ the scalar potential $\varphi(0) = \varphi_0$, i.e., the charge state of the particle surface is entirely characterized by the central symmetry – $\mathcal{N}$ elementary vectors protrude outside of the surface and the nth spike generates the maximum electric field $\mathcal{E}_{0n} = -\nabla\varphi_n\big|_{\hbar=\hbar_0}$; at the moment $t = T/2$ (or the spatial half-period $\lambda/2$), the field φ reaches the minimum, which is zero; at the moment $t = T$ (or the spatial period λ), again $\varphi(T) = \varphi_0$, and so on. For the vector potential: $\mathbf{A}(0) = \mathbf{e}_{\text{left (right)}}\upsilon_0$ where $\mathbf{e}_{\text{left (right)}} = \pm 1$ is the polarization vector; this condition implies that at the moment $t = 0$ all the spikes on the sphere take the same tangential velocity (to the left for $\mathbf{e}_{\text{left}} = +1$ or to the right for $\mathbf{e}_{\text{right}} = -1$ relative to the particle's path l) and they move synchronously. At the moment $t = T/2$ the tangential

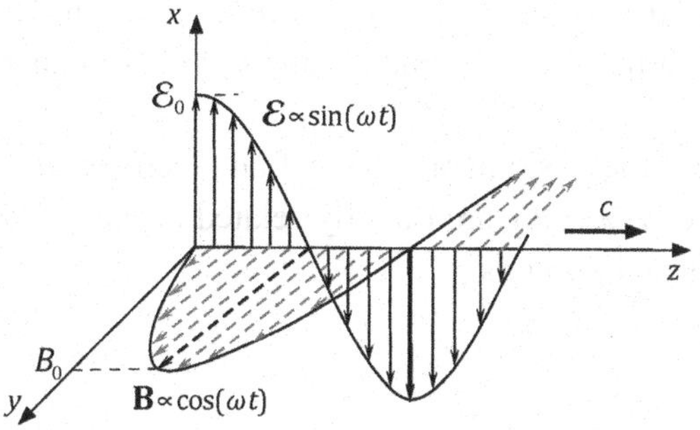

FIGURE 4.11 Correct distribution of the electric field ε and the magnetic field **B** in a travelling electromagnetic wave.

velocity of the spikes decreases to zero but at this moment they have the maximum bending and the charge state of the particle's surface is entirely characterized by the axial symmetry. Then at $t = T$ the velocity of the vector field $\mathbf{A}(T) = \mathbf{e}_{\text{left (right)}}\, \upsilon_0$ and so on. Hence the state of the charged particle at moments $t = T/2 + \text{n}\, T$, where n = 0, 1, 2, …, configures a typical monopole.

Thus the solution to equations (4.68) and (4.69) (or homogeneous equations (4.78) and (4.79)) should be chosen in the form

$$\varphi(\mathbf{x}, t) = \tfrac{1}{2}\varphi_0\left[1 + \cos(\mathbf{k}\cdot\mathbf{x})\cos(\omega t)\right] \tag{4.103}$$

$$\mathbf{A}(\mathbf{x}, t) = \tfrac{1}{2}\mathbf{A}_0\left[1 - \cos(\mathbf{k}\cdot\mathbf{x})\cos(\omega t)\right] \tag{4.104}$$

These expressions indeed satisfy equations (4.68) and (4.69), respectively. Then the electric and magnetic fields become:

$$\mathcal{E}(\mathbf{x}, t) = -\nabla\varphi - \partial\mathbf{A}/\partial t = \tilde{\mathcal{E}}(\mathbf{x})\sin(\omega t) \tag{4.105}$$

$$\mathbf{B}(\mathbf{x}, t) = \nabla\times\mathbf{A} = \tilde{\mathbf{B}}(\mathbf{x})\cos(\omega t) \tag{4.106}$$

where $\tilde{\mathcal{E}}(\mathbf{x}) = \mathcal{E}_0\cdot(-\omega\varphi_0 - \mathbf{A}_0\cdot\mathbf{k})\cos(\mathbf{k}\cdot\mathbf{x})$ and $\tilde{\mathbf{B}}(\mathbf{x}) = \text{rot}\,\mathbf{A}_0\cos(\mathbf{k}\cdot\mathbf{x})$. Then the energy density reads

$$W \propto \tilde{\mathcal{E}}^2(\mathbf{x})\sin^2(\omega t) + \tilde{\mathbf{B}}^2(\mathbf{x})\cos^2(\omega t) \tag{4.107}$$

and all the time this value W is larger then zero. Figure 4.11 exhibits the correct distribution of electric and magnetic fields in an electromagnetic wave.

Nevertheless, the basic parameters of an electromagnetic field are the potentials φ and **A**, as they are directly related to the subtle surface structure of particles and photons.

KEYWORDS

- **electric charge**
- **electric field**
- **magnetic field**
- **photon**
- **surface fractal**
- **tessellattice**

CHAPTER 5

INERTONS IN CONDENSED MEDIA

CONTENTS

The previous chapters have clearly demonstrated that the submicroscopic concept is a powerful approach to the study of the microcosm. In this chapter we will see how the submicroscopic deterministic mechanics will allow us to investigate various problems, clarify difficulties and disclose unusual paradoxes presented in physics of condensed media.

Let us recall the major aspects of submicroscopic mechanics. The momentum p of a particle is decomposed to the mass m and the velocity υ and each of these parameters is characterized by its own behavior in the line of a particle's path. The whole particle path is subdivided by the particle's de Broglie wavelength λ: along the section λ the particle's velocity changes from υ to zero and in the next section λ is reinstated to υ; owing to the emission and re-absorption of inertons by the particle, its mass also varies, $m \rightarrow 0 \rightarrow m \rightarrow 0 \ldots$ because inertons transfer fragments of the total mass of the particle: due to the particle's motion a local deformation of space (which is a volumetric fractal deformation of a cell, i.e., mass m) periodically changes to a local tension of space (which is a displacement of the volumetric fractal deformation from its equilibrium state, Ξ). Furthermore, it has been revealed that periodic transformations of physical parameters of quantum systems from their original state to a tension state are universal and such transformations only occur in the section 2λ of a particle's path. The periodic transformation of the surface state of the particle kernel from the chestnut shape to the inclined spikes-shape corresponds to the electric charge and the magnetic monopole, respectively, which is described by the conventional Maxwell equations (4.89)–(4.90). In the case of a free photon, which moves hopping from cell to cell, the section λ is equal to its wavelength, the photon's electric polarization (which stands for fractal spikes normal to the surface of an oncoming cell) periodically transforms to the magnetic polarization (in this state fractal needles are inclined, such that they are tangential to the surface).

In condensed media the major entity is an atom or molecule and through a paired potential the entities interact one with another. If the interaction is weak, the appropriate entities will be in a gaseous phase, but if the interaction is strong enough, the entities will arrange a network that we call a liquid or even the crystal lattice that is typical for a solid.

5.1 INERTONS IN THE CRYSTAL LATTICE

In the crystal lattice atoms/molecules are separated by a distance a, which is called the lattice constant. Atoms continuously vibrate near their equilibrium positions. Even at the zero temperature atoms undergo the zero point fluctuations. Why? The only correct answer is: Because this is the quantum world.

However, let us try to look at this quantum world through a submicroscope eyepiece. As we know from the previous chapters, a moving particle is surrounded with its inerton cloud. Hence in the crystal lattice any movement of an atom shall be accompanied with an inerton cloud that emerges owing to the interaction of the atom with oncoming cells of the tessellattice. In the crystal lattice the behavior of nodes is defined by the Lagrangian

$$L_{\text{vibr.}} = \tfrac{1}{2}\sum_{\mathbf{n}}\left(m\,\delta\dot{r}_{\mathbf{n}}^2 - \gamma\,\delta r_{\mathbf{n}}^2\right) \qquad (5.1)$$

where m_{n} is the mass of the $\mathbf{n}th$ node, $\delta r_{\mathbf{n}}$ is the deviation of this node from its equilibrium position and γ is the force constant. It is well known that based on the Lagrangian (5.1) the Euler-Lagrange equations can be constructed, which completely determine the dynamics of the nodes.

Since massive nodes vibrate near their equilibrium positions, i.e., are in motion in the tessellattice, they emit and re-absorb clouds of inertons. Therefore inertons periodically remove a part of the mass from vibrating nodes and bring it back. Such behavior can be described in terms of the Lagrangian below (for simplicity of consideration we consider the one-dimensional lattice), which is similar in form to the expression (5.1) but is different in dimension, as its variables are masses

$$L_{\text{mass}} = \sum_{\mathbf{n}}\left\{\tfrac{1}{2}\dot{m}_{\mathbf{n}}^2 - \frac{\pi}{2T}\left(\dot{m}_{\mathbf{n}}\,\mu_{\mathbf{n}} + \dot{m}_{\mathbf{n+g}}\,\mu_{\mathbf{n}}\right) + \tfrac{1}{2}\dot{\mu}_{\mathbf{n}}^2\right\} \qquad (5.2)$$

where m_{n} and μ_{n} are variations of mass of the **n**th node and its cloud of inertons, respectively, which occur due to the overlapping of inerton clouds of neighbor nodes; **g** is the lattice vector; T is the period of collision of the mass located in the **n**th node with its inerton cloud. The dot over mass means the derivative in respect to the time t treated as a natural parameter.

Instead of variables m_n and μ_n we may pass on to collective variables $\Phi_{\mathbf{k}}$ and $\phi_{\mathbf{k}}$ by rules

$$m_{\mathbf{n}} = \frac{1}{\sqrt{N}} \sum\nolimits_{\mathbf{k}} \Phi_{\mathbf{k}} e^{i\mathbf{kg}} \tag{5.3}$$

$$\mu_{\mathbf{n}} = \frac{1}{\sqrt{N}} \sum\nolimits_{\mathbf{k}} \phi_{\mathbf{k}} e^{i\mathbf{kg}} \tag{5.4}$$

Substituting expressions (5.3) and (5.4) into the Lagrangian (5.2), we obtain

$$L_{\text{mass}} = \frac{1}{N} \sum_{\mathbf{n}} \left\{ \frac{1}{2} \dot{\Phi}_{\mathbf{k}} \dot{\Phi}_{-\mathbf{k}} - \frac{\pi}{2T} \left(\dot{\Phi}_{\mathbf{k}} \phi_{-\mathbf{k}} + \dot{\Phi}_{\mathbf{k}} \phi_{-\mathbf{k}} e^{i\mathbf{kg}} \right) + \frac{1}{2} \dot{\phi}_{\mathbf{k}} \dot{\phi}_{-\mathbf{k}} \right\} \tag{5.5}$$

In the Lagrangian (5.5) each summand must be real, in particular, the term $\dot{\Phi}_{\mathbf{k}} \phi_{-\mathbf{k}}$. Then in the term $\dot{\Phi}_{\mathbf{k}} \phi_{-\mathbf{k}} e^{i\mathbf{kg}}$ the exponent $e^{i\mathbf{kg}}$ must also be real. This means that the Lagrangian (5.5) is real and reduces to

$$L_{\text{mass}} = \frac{1}{N} \sum_{\mathbf{k}} \left\{ \frac{1}{2} \dot{\Phi}_{\mathbf{k}} \dot{\Phi}_{-\mathbf{k}} - \frac{\pi}{2T} (1 + \cos \mathbf{kg}) \; \dot{\Phi}_{\mathbf{k}} \phi_{-\mathbf{k}} + \frac{1}{2} \dot{\phi}_{\mathbf{k}} \dot{\phi}_{-\mathbf{k}} \right\} \tag{5.6}$$

The Euler-Lagrange equation

$$[\partial L_{\text{mass}} / \partial \dot{Q}] / dt - \partial L_{\text{mass}} / \partial Q = 0 \tag{5.7}$$

where the variable Q stands for $\Phi_{\mathbf{k}}$ and $\phi_{\mathbf{k}}$, becomes

$$\ddot{\Phi}_{\mathbf{k}} - \omega(\mathbf{k}) \dot{\phi}_{-\mathbf{k}} = 0 \tag{5.8}$$

$$\ddot{\phi}_{\mathbf{k}} + \omega(\mathbf{k}) \dot{\Phi}_{\mathbf{k}} = 0 \tag{5.9}$$

where we designate

$$\omega(\mathbf{k}) = \frac{\pi}{2T} (1 + \cos(\mathbf{kg})) \tag{5.10}$$

Periodical solutions to equations (5.8) and (5.9), which satisfy the physical characteristics of the system of varying masses, can be chosen as follows

$$\Phi_{\mathbf{k}} = \Phi_0 + \Phi_1 \cos(\omega(\mathbf{k})t) \tag{5.11}$$

$$\phi_{\mathbf{k}} = -\Phi_1 \cos(\omega(\mathbf{k})t) \tag{5.12}$$

where, as follows from relations (5.3) and (5.4), parameters Φ_0 and Φ_1 are proportional to the rest mass of the system's particles and the mass of their inerton clouds, respectively, and inversely proportional to the square root of the total number of particles $N^{-1/2}$.

The above mentioned arguments point out that the variables $\Phi_{\mathbf{k}}$ and $\phi_{\mathbf{k}}$ represent collective mass excitations in the lattice: $\Phi_{\mathbf{k}}$ describes collective mass excitations of the nodes of the lattice and $\phi_{\mathbf{k}}$ characterizes the mass field of inertons that like dust fill the entire space between the nodes in the lattice. The spectrum of these collective mass excitations, is the dispersion law (5.10) that in the long-wave approximation becomes

$$\omega(\mathbf{k}) \cong \frac{\pi}{2T}\left(1 + \tfrac{1}{2}(\mathbf{k}\mathbf{g})^2\right) \tag{5.13}$$

It should be emphasized that these mass excitations are completely independent from the phonons of the lattice, because phonons are associated with collective changes of positions of nodes (atoms). Mass excitations described by the variable $\Phi_{\mathbf{k}}$ represent the collective mass state of nodes at the moment t and the variable $\phi_{\mathbf{k}}$ depicts the collective mass state of the total inerton cloud of the lattice. In other words $\Phi_{\mathbf{k}}(t)$ shows how the mass of nodes with the wave vector **k** changes with time; $\phi_{\mathbf{k}}(t)$ describes the mass of the inerton cloud with the wave vector **k** at a function of t.

This means that there should be an amplitude δm of oscillations of the mass of the **n**th node. The amplitude can crudely be estimated as a ratio of the dispersion of the **n**th node's inerton cloud at the maximal distant object ($r = \Lambda$) and the nearest node ($r = g$)

$$\delta m \approx m_{\mathbf{n}} \frac{g}{\Lambda} \sim 10^{-4} m_{\mathbf{n}} \tag{5.14}$$

where Λ is the amplitude of inerton cloud of the **n**th node. In accordance with correlation (2.69), the mentioned amplitude is related to the node's de Broglie wavelength, $\lambda_{\mathbf{n}} \equiv \delta r_{\mathbf{n}}$, the node's velocity (sound velocity) υ, and the inerton velocity in the lattice can be equal to the velocity of light, c; then

$$\Lambda = \delta r_{\mathrm{n}} \frac{c}{\upsilon} \geq \delta r_{\mathrm{n}} \cdot 10^5 \sim 10^4 g \approx (3-4) \times 10^{-6} \text{ m} \qquad (5.15)$$

which is about a few micrometers. In relation (5.15) we put approximately that the amplitude of oscillations δr of a node in the lattice is approximately 0.1 of the lattice constant g (in the case of heavy metals δr may be much less). The evaluation (5.15) results in the estimate (5.14).

Thus, we can see that in a solid there exists an additional physical field, which is the inerton field that so far has not been practically taken into account. In fact, vibrations of atoms result in a series of acoustic waves. But the space between atoms is filled with inertons, which appear owing to the interaction of vibrating atoms with the tessellattice. Overlapping of inerton clouds and the mobility of atoms, which is the source of these inertons, brings about the formation of inerton waves (with their own harmonics) in the solid as well.

5.2 CLUSTER FORMATION IN CONDENSED MATTER

When particles in the system studied are characterized by different types of interactions, they may form clusters [175]. Let us consider the mechanism of such clustering, which is provided by different paired potentials, which act in an ensemble of particles (molecules or atoms). Then we will supplement the potentials with an inerton component that is also present in the same ensemble. Could the inerton component change the balance in the ensemble of particles? This will be seen in the next sections.

So let us start from an approach that makes it possible to describe systems of interacting particles by statistical methods taking into account their heterogeneous spatial distribution, i.e., cluster formation. This approach is very suitable for an examination of nanosystems.

We may begin with the construction of the Hamiltonian for a system of two kinds of interacting particles, for example: particles of kind α and particles of kind β. Let these particles form a 3D lattice and let $n = \{0, 1\}$ be the filling number of the sth lattice knot. The energy for such a system can be written in the form [176]

$$H = \sum_{\mathbf{r},\,\mathbf{r}'} V_{\alpha\alpha}(\mathbf{r},\,\mathbf{r}')\, c_\alpha(\mathbf{r}) c_\alpha(\mathbf{r}') + 2\sum_{\mathbf{r},\,\mathbf{r}'} V_{\alpha\beta}(\mathbf{r},\,\mathbf{r}') c_\alpha(\mathbf{r}) c_\beta(\mathbf{r}')$$
$$+ \sum_{\mathbf{r},\,\mathbf{r}'} V_{\beta\beta}(\mathbf{r},\,\mathbf{r}')\, c_\beta(\mathbf{r}) c_\beta(\mathbf{r}') \tag{5.16}$$

where V_{ij} is the interaction potential of particles of two kinds: $i,\ j = \alpha,\ \beta$. They occupy knots in Ising's lattice described by the radius vectors $\mathbf{r}$ and $\mathbf{r}'$ and $c_i(\mathbf{r}) = \{0,\ 1\}$ are the random functions, which satisfy condition

$$c_\alpha(\mathbf{r}) + c_\beta(\mathbf{r}) = 1 \tag{5.17}$$

The Hamiltonian (5.16) can be written as follows

$$H = H_0 + \tfrac{1}{2}\sum_{\mathbf{r},\mathbf{r}'} V(\mathbf{r},\,\mathbf{r}') c_\beta(\mathbf{r}) c_\beta(\mathbf{r}') \tag{5.18}$$

where

$$H_0 = \tfrac{1}{2}\sum_{\mathbf{r}} [1 - 2c_\beta(\mathbf{r})] \sum_{\mathbf{r}'} V_{\alpha\alpha}(\mathbf{r},\,\mathbf{r}') + \sum_{\mathbf{r}} c_\beta(\mathbf{r}) \sum_{\mathbf{r}'} V_{\alpha\beta}(\mathbf{r},\,\mathbf{r}') \tag{5.19}$$

$$V(\mathbf{r},\,\mathbf{r}') = V_{\alpha\alpha}(\mathbf{r},\,\mathbf{r}') + V_{\beta\beta}(\mathbf{r},\,\mathbf{r}') - 2V_{\alpha\beta}(\mathbf{r},\,\mathbf{r}') \tag{5.20}$$

Let us rewrite the Hamiltonian (5.18) in the form

$$H = H_0 - \tfrac{1}{2}\sum_{\mathbf{r},\mathbf{r}'} V^{\text{att}}_{\mathbf{r}\mathbf{r}'}\, c(\mathbf{r}) c(\mathbf{r}') + \tfrac{1}{2}\sum_{\mathbf{r},\mathbf{r}'} V^{\text{rep}}_{\mathbf{r}\mathbf{r}'}\, c(\mathbf{r}) c(\mathbf{r}') \tag{5.21}$$

where the index α is omitted at the function $c(\mathbf{r})$ and the following designations are introduced:

$$V^{\text{att}}_{\mathbf{r}\mathbf{r}'} = V_{\alpha\beta}(\mathbf{r},\mathbf{r}') \tag{5.22}$$

$$V^{\text{rep}}_{\mathbf{r}\mathbf{r}'} = \tfrac{1}{2}[V_{\beta\beta}(\mathbf{r},\mathbf{r}') + V_{\alpha\alpha}(\mathbf{r},\mathbf{r}')] \tag{5.23}$$

If the potentials $V^{\text{att}}_{\mathbf{r}\mathbf{r}'},\ V^{\text{rep}}_{\mathbf{r}\mathbf{r}'} > 0$ the second term in the right-hand side of the Hamiltonian (5.19) corresponds to the effective attraction (5.22) and the third term conforms to the effective repulsion (5.23). This allows one to represent the Hamiltonian (5.19) in the form that is typical for the model of ordered particles, which is characterized by a particular nonzero order parameter,

$$H(n) = \sum_s E_s n_s - \tfrac{1}{2}\sum_{s,s'} V_{ss'}^{\text{att}} n_s n_{s'} + \tfrac{1}{2}\sum_{s,s'} V_{ss'}^{\text{rep}} n_s n_{s'} \tag{5.24}$$

Here, E_s is the additive part of the particle's energy (the kinetic energy) in the sth state. The main point of our approach is the initial separation of the total nucleon potential into two terms: the repulsion and attraction components. So, in the Hamiltonian (5.24) the potential $V_{ss'}^{\text{att}}$ represents the paired energy of attraction and the potential $V_{ss'}^{\text{rep}}$ is the paired energy of repulsion. The potentials take into account the effective paired interaction between nucleons located in states s and s'. The filling numbers n_s can have only two meanings: 1 (the sth knot is occupied in the model lattice studied) or 0 (the sth knot is not occupied in the model lattice studied). The signs before positive functions $V_{ss'}^{\text{att}}$ and $V_{ss'}^{\text{rep}}$ in the Hamiltonian (5.24) directly specify the proper signs of attraction (minus) and repulsion (plus).

5.2.1 THE FORMALISM OF CLUSTER FORMATION FOR QUANTUM PARTICLES

The statistical sum of a system of interacting particles

$$Z = \sum_{\{n\}} \exp\left[-H(n)/(k_{\text{B}}\Theta)\right] \tag{5.25}$$

can be presented in the field form [173]

$$Z = \int_{-\infty}^{\infty} D\varphi \int_{-\infty}^{\infty} D\psi \sum_{\{n\}} \exp\left[\begin{array}{l} -\sum_s \tilde{E}_s n_s + \sum_s (\psi_s + i\varphi_s) n_s \\ +\tfrac{1}{2}\sum_{s,s'}\left(\tilde{V}_{ss'}^{\text{att}\,-1}\psi_s\psi_{s'} - \tilde{V}_{ss'}^{\text{rep}\,-1}\varphi_s\varphi_{s'}\right) \end{array}\right] \tag{5.26}$$

due to the following representation known from the theory of Gauss integrals

$$\exp\left(\tfrac{1}{2}\rho\sum_{s,s'} W_{ss'} n_s n_{s'}\right) = \text{Re}\int_{-\infty}^{\infty} D\chi \exp\left(\rho\sum_s n_s\chi_s - \tfrac{1}{2}\sum_{s,s'} W_{ss'}^{-1}\chi_s\chi_{s'}\right) \tag{5.27}$$

$$\sum_{s''} W^{-1}_{ss''} W_{s''s'} = \delta_{ss'} \tag{5.28}$$

where $D\chi \equiv \prod_s \sqrt{\det \| W_{ss'} \|}\, 2\pi\, d\chi_s$ implies the functional integration with respect to the field χ, $\rho^2 = \pm 1$ in relation to the sign of interaction (+1 for attraction and −1 for repulsion). The dimensionless energy parameters $\tilde{V}^{\text{att}}_{ss'} = V^{\text{att}}_{ss'} / (k_{\text{B}}\Theta)$, $\tilde{V}^{\text{rep}}_{ss'} = V^{\text{rep}}_{ss'} / (k_{\text{B}}\Theta)$ and $\tilde{E}_s = E_s / (k_{\text{B}}\Theta)$ where Θ is the absolute temperature, are introduced into expression (5.26). Further, we will use the known formula

$$\frac{1}{2\pi i} \oint dz\, z^{N-1-\sum_s n_s} = 1 \tag{5.29}$$

which makes it possible to settle the quantity of particles in the system, $\sum_s n_s = N$, and, consequently, we can pass to the consideration of the canonical ensemble of N particles. Thus the statistical sum (5.26) is replaced with

$$Z = \text{Re} \frac{1}{2\pi i} \oint \int D\varphi \int D\psi \exp \begin{bmatrix} -\frac{1}{2} \sum\limits_{ss'} \left(\tilde{V}^{\text{att}\,-1}_{ss'} \psi_s \psi_{s'} + \tilde{V}^{\text{rep}\,-1}_{ss'} \varphi_s \varphi_{s'} \right) \\ + (N-1) \ln z \end{bmatrix}$$
$$\times \sum_{\{n_s\}=0}^{1} \exp\left\{ \sum_s n_s (\psi_s + i\varphi_s - \tilde{E}_s) - \ln z \right\} \tag{5.30}$$

Summing over n_s we obtain

$$Z = \text{Re} \frac{1}{2\pi i} \int D\varphi \int \psi \oint dz\, e^{S(\varphi,\, \psi,\, z)} \tag{5.31}$$

where

$$S = \sum_s \left\{ -\frac{1}{2} \sum_{s'} \left(\tilde{V}^{\text{att}\,-1}_{ss'} \psi_s \psi_{s'} + \tilde{V}^{\text{rep}\,-1}_{ss'} \varphi_s \varphi_{s'} \right) + \eta \ln \left| 1 + \frac{\eta}{z} e^{-\tilde{E}_s} e^{\psi_s} \cos\varphi_s \right| \right\}$$
$$+ (N-1) \ln z \tag{5.32}$$

Here, the symbol η describes the type of quantum statistics: Fermi-Dirac ($\eta = +1$) or Bose-Einstein ($\eta = -1$).

Let us put $z = \xi + i\zeta$ and consider the action S on a transit path that passes through the saddle-point at a fixed imaginable variable $\mathrm{Im}\, z = \zeta = \zeta_0$. In this case S can be regarded as the functional that depends on the two field variables φ_s and ψ_s, and the fugacity $\xi = e^{-\mu/(k_{\mathrm{B}}\Theta)}$ where μ is the chemical potential. The extremum of functional (5.32) is realized at the solutions of the equations $\delta S / \delta\varphi_s = 0$, $\delta S / \delta\psi_s = 0$, and $\delta S / \delta\xi = 0$, or explicitly

$$\sum_{s'} \tilde{V}_{ss'}^{\mathrm{rep}\,-1} \varphi_{s'} = -\frac{2\exp\left(-\tilde{E}_s + \psi_s\right)\sin\varphi_s}{\xi + \eta\exp(-\tilde{E}_s + \psi_s)\cos\varphi_s} \tag{5.33}$$

$$\sum_{s'} \tilde{V}_{ss'}^{\mathrm{att}\,-1} \psi_{s'} = \frac{2\exp\left(-\tilde{E}_s + \psi_s\right)\cos\varphi_s}{\xi + \eta\exp(-\tilde{E}_s + \psi_s)\cos\varphi_s} \tag{5.34}$$

$$\sum_{s} \frac{\exp\left(-\tilde{E}_s + \psi_s\right)\cos\varphi_s}{\xi + \eta\exp(-\tilde{E}_s + \psi_s)\cos\varphi_s} = N - 1 \tag{5.35}$$

Equations from (5.33) to (5.35) completely solve the problem of the statistical description of systems with any type of interaction. Any state of the system is realized in accordance with the solution of the nonlinear equations among which there are solutions that correspond to the spatial nonhomogeneous distribution of particles.

If we introduce designation

$$\mathcal{N}_s = \frac{\exp\left(-\tilde{E}_s + \psi_s\right)\cos\varphi_s}{\xi + \eta\exp(-\tilde{E}_s + \psi_s)\cos\varphi_s} \tag{5.36}$$

then inserting $\mathcal{N}_s$ into the left hand side of Eq. (5.35) we will see that the total sum $\sum_s \mathcal{N}_s$ is equal to the number of particles in the system studied

$$\sum_s \mathcal{N}_s = N - 1 \tag{5.37}$$

So it follows from Eqs. (5.33)–(5.35) that the parameter $\mathcal{N}_s$, which combines variables φ_s, ψ_s and ξ, is a typical variable that characterizes the number of particles contained in the sth cluster.

Multiplying the two sides of Eq. (5.34) by the function $\tilde{V}_{ss'}^{\text{att}}$ and then sum Eq. (5.34) over s, we obtain

$$\psi_{s'} = 2\sum_{s} \tilde{V}_{ss'}^{\text{att}} \mathcal{N}_s \tag{5.38}$$

Now let us multiply the same equation (5.34) by ψ_s; then summing it over s we acquire

$$\sum_{ss'} \tilde{V}_{ss'}^{\text{att}\,-1} \psi_{s'} \psi_s = 4\sum_{ss'} \tilde{V}_{ss'}^{\text{att}} \mathcal{N}_{s'} \mathcal{N}_s \tag{5.39}$$

Multiplying Eq. (5.33) by $\tilde{V}_{s''s}^{\text{rep}}$ and summing it over s we get

$$\sum_{s'} \tilde{V}_{ss'}^{\text{rep}\,-1} \tilde{V}_{s''s}^{\text{rep}} \varphi_{s''} = -\sum_{s} \tilde{V}_{ss''}^{\text{rep}} \frac{2\exp\left(-\tilde{E}_s + \psi_s\right)\sin\varphi_s}{\xi + \eta \exp(-\tilde{E}_s + \psi_s)\cos\varphi_s} \tag{5.40}$$

Using condition (5.28) for the left-hand side of Eq. (5.40) we transform the equation to the form

$$\varphi_s = -2\sum_{s'} \tilde{V}_{ss'}^{\text{rep}} \mathcal{N}_{s'} \tan\varphi_{s'} \tag{5.41}$$

Multiplying the two sides of Eq. (5.33) by the function φ_s and then summing the equation over s, we obtain

$$\sum_{ss'} \tilde{V}_{ss'}^{\text{rep}\,-1} \varphi_{s'} \varphi_s = -\sum_{s} \mathcal{N}_s \varphi_s \tan\varphi_s \tag{5.42}$$

Substituting in Eq. (5.42) the variable φ_s for its presentation in the right hand side of Eq. (5.41) we gain

$$\sum_{ss'} \tilde{V}_{ss'}^{\text{rep}\,-1} \varphi_{s'} \varphi_s = 2\sum_{ss'} \tilde{V}_{ss'}^{\text{rep}} \mathcal{N}_{s'} \mathcal{N}_s \tan\varphi_{s'} \tan\varphi_s \tag{5.43}$$

Both tangents in the right hand side of Eq. (5.43) can be presented via the combined variable and the fugacity. Indeed from Eq. (5.36) we get

$$\cos\varphi_s = \eta \frac{\xi \mathcal{N}_s}{1-\mathcal{N}_s} \exp\left(\tilde{E}_s - \psi_s\right) = \eta \frac{\xi \mathcal{N}_s}{1-\mathcal{N}_s} \exp\left(\tilde{E}_s - 2\sum_{s'} \tilde{V}_{s's}^{\text{att}} \mathcal{N}_{s'}\right) \tag{5.44}$$

in which we have used the relation (5.38) to eliminate the variable $\psi_{s'}$. Then, since $\tan\varphi_s = \sqrt{1-\cos^2\varphi_s} / \cos\varphi_s$, we are able to transform the expression (5.43) to the form

$$\sum_{ss'} \tilde{V}_{ss'}^{\text{rep}\,-1} \varphi_{s'}\varphi_s = 2\sum_{ss'} \tilde{V}_{ss'}^{\text{rep}} \mathcal{N}_{s'}\mathcal{N}_s \left\{ \frac{(1-\mathcal{N}_s)^2}{\xi^2 \mathcal{N}_s^2} \exp\left(2\tilde{E}_s - 4\sum_{s''} \tilde{V}_{s''s}^{\text{att}} \mathcal{N}_{s''} \right) - 1 \right\}^{1/2}$$
$$\times \left\{ \frac{(1-\mathcal{N}_{s'})^2}{\xi^2 \mathcal{N}_{s'}^2} \exp\left(2\tilde{E}_{s'} - 4\sum_{s''} \tilde{V}_{s''s'}^{\text{att}} \mathcal{N}_{s''} \right) - 1 \right\}^{1/2} \tag{5.45}$$

One more term in the action (5.32), namely, $\eta \ln[1+\eta \exp(-\tilde{E}_s + \psi_s) \cos\varphi_s / \xi]$ in the extremum point can be found using expressions for $\cos\varphi_s$ (5.44) and for ψ_s (5.38); it becomes equal to $\eta \ln|1-\mathcal{N}_s|$.

Now using relations (5.39) and (5.45) we can rewrite the action (5.32) in the point of extremum as follows:

$$S = -2\sum_{s,s'} \tilde{V}_{ss'}^{\text{att}} \mathcal{N}_{s'}\mathcal{N}_s + \eta\sum_s \ln|\mathcal{N}_s - 1| + (\mathcal{N}-1)\ln\xi$$
$$-\sum_{ss'} \tilde{V}_{ss'}^{\text{rep}} \mathcal{N}_{s'}\mathcal{N}_s \left\{ \frac{(1-\mathcal{N}_s)^2}{\xi^2 \mathcal{N}_s^2} \exp\left(2\tilde{E}_s - 4\sum_{s''} \tilde{V}_{s''s}^{\text{att}} \mathcal{N}_{s''} \right) - 1 \right\}^{1/2}$$
$$\times \left\{ \frac{(1-\mathcal{N}_{s'})^2}{\xi^2 \mathcal{N}_{s'}^2} \exp\left(2\tilde{E}_{s'} - 4\sum_{s''} \tilde{V}_{s''s'}^{\text{att}} \mathcal{N}_{s''} \right) - 1 \right\}^{1/2} \tag{5.46}$$

Expression (5.46), as a function of only one variable $\mathcal{N}_s$ and the fugacity ξ, can be applicable to any kinds of interactions, even though the inverse operator of the paired interaction is unknown. The great advantage of expression (5.46) lies in the fact that it provides a way of finding such characteristics of the system in question according to its size, the number of particles in a cluster and the temperature of phase transition.

It is now appropriate to conduct further analysis of the continuous variables. Inasmuch as we are interested in the nonhomogeneous distribution of particles in an indeterminate volume, let a radius R be the fitting parameter of the system studied. Assume that the density of particles is distinguished from zero only in the cluster's volume. Then Eq. (5.37) can be written as

$$\mathcal{N} K = N - 1 \tag{5.47}$$

where K is the number of clusters in the system and is the variable defined in expressions (5.35) and (5.36), which can be called a combined variable that can be interpreted as the mean quantity of particles in a cluster.

Since in statistical mechanics clusters should involve a large quantity of particles, it is advisable to pass to the continual presentation of the action (5.46). The passage to the continual presentation is realized by the substitution

$$\sum_s f_s = K \frac{1}{\mathcal{V}} \int_{\text{cluster}} f(\vec{r})\, d\vec{r} \tag{5.48}$$

here the integration extends for the volume of a cluster and $\mathcal{V} = 4\pi g^2/3$ is an effective volume occupied by one particle where g is the distance between particles (the lattice constant in a solid).

Thus in the continuous representation the action (5.46) takes the form

$$\begin{aligned} S = &-\frac{2}{\mathcal{V}^2} \int d\vec{r} \int d\vec{r}\,' \tilde{V}^{\text{att}}(\vec{r} - \vec{r}\,') \mathcal{N}(\vec{r}) \mathcal{N}(\vec{r}\,') \\ &- \frac{1}{\mathcal{V}^2} \int d\vec{r} \int d\vec{r}\,' \tilde{V}^{\text{rep}}(\vec{r} - \vec{r}\,') \mathcal{N}(\vec{r}) \mathcal{N}(\vec{r}\,') \\ &\times \left\{ \frac{ \frac{1}{\xi^2 \mathcal{N}(\vec{r})} (1 - \mathcal{N}(\vec{r}))^2 }{ \exp\left(-2\tilde{E}(\vec{r}) + 4 \int \tilde{V}^{\text{rep}}(\vec{r} - \vec{r}\,'') \mathcal{N}(\vec{r}\,'') d\vec{r}\,'' \right) - 1 } \right\}^{1/2} \\ &\times \left\{ \frac{ \frac{1}{\xi^2 \mathcal{N}(\vec{r}\,')} (1 - \mathcal{N}(\vec{r}\,'))^2 }{ \exp\left(-2\tilde{E}(\vec{r}\,') + 4 \int \tilde{V}^{\text{rep}}(\vec{r}\,' - \vec{r}\,'') \mathcal{N}(\vec{r}\,'') d\vec{r}\,'' \right) - 1 } \right\}^{1/2} \\ &+ \eta \frac{1}{\mathcal{V}} \int \ln| \mathcal{N}(\vec{r}\,') - 1 | d\vec{r}\,' + (N-1) \ln \xi \end{aligned} \tag{5.49}$$

The integration is effected by the rule

$$\frac{1}{\mathcal{V}}\int_{\text{cluster}} f(\vec{r})d\vec{r} = \frac{1}{\frac{4\pi}{3}g^3}\int_0^{2\pi} d\phi \int_0^{\pi} \sin\theta d\theta \int_g^R f(r)r^2 dr$$

$$= \frac{1}{\frac{4\pi}{3}g^3} 4\pi \int_g^R f(r)r^2 dr \qquad (5.50)$$

We shall normalize the integrals to the number of particles $\mathcal{N}$ in a cluster:

$$\frac{1}{\mathcal{V}}\int_{\text{cluster}} d\vec{r} = \frac{1}{\frac{4\pi}{3}g^3} 4\pi \int_g^R f(r)r^2 dr = \frac{R^3 - g^3}{g^3} = \mathcal{N} - 1 \cong \mathcal{N} \qquad (5.51)$$

Relation (5.51) allows us to introduce the dimensionless variable $x = r/g$ in integral (5.50), such that

$$\frac{1}{K}\sum_s f_s = 3\int_1^{\mathcal{N}^{1/3}} f(gx)x^2 dx \qquad (5.52)$$

Having integrated the action (5.49), we shall exploit the following relationships:

$$\frac{1}{\mathcal{V}^2}\int d\vec{r}\int d\vec{r}'\,\tilde{V}^{\text{att}}(\vec{r}-\vec{r}')\mathcal{N}(\vec{r})\mathcal{N}(\vec{r}')$$

$$= \frac{1}{\mathcal{V}^2}\int d\vec{r}\,\tilde{V}^{\text{att}}(\vec{r})\mathcal{N}(\vec{r})\int d\vec{r}'\,\mathcal{N}(r')$$

$$= \frac{1}{\mathcal{V}^2}\int d\vec{r}\,\tilde{V}^{\text{att}}(\vec{r})\mathcal{N}(\vec{r})\cdot\mathcal{N}\cdot\int d\vec{r}'$$

$$= \frac{\mathcal{N}^2}{\mathcal{V}}\int d\vec{r}\,\tilde{V}^{\text{att}}(\vec{r}) = \mathcal{N}^2\cdot 3\int_1^{\Gamma^{1/3}} dx x^2\tilde{V}^{\text{att}}(gx)$$

$$\frac{1}{\mathcal{V}^2}\int d\vec{r}\int d\vec{r}'\,\tilde{V}^{\text{rep}}(\vec{r}-\vec{r}')\mathcal{N}(\vec{r})\mathcal{N}(\vec{r}')$$

$$= \mathcal{N}^2\cdot 3\int_1^{\Gamma^{1/3}} dx x^2\tilde{V}^{\text{rep}}(gx) \qquad (5.53)$$

With the transformations, we have used the Heaviside step function: $\mathcal{N}(r) = \mathcal{N} \cdot \vartheta(\mathcal{N} - r^3 / g^3)$ where $\vartheta(\mathcal{N} - r^3 / g^3) = 0$ if $r^3 / g^3 \geq \mathcal{N}$ and $\vartheta(\mathcal{N} - r^3 / g^3) = 1$ if $r^3 / g^3 < \mathcal{N}$. Besides,

$$\begin{aligned} &\frac{1}{\mathcal{V}^2} \int d\vec{r} \int d\vec{r}' \tilde{V}^{\text{rep}}(\vec{r}, \vec{r}') f(\vec{r}) f(\vec{r}') \\ &= \frac{1}{\mathcal{V}^2} \int d\vec{r}\, \tilde{V}^{\text{rep}}(\vec{r}) f(\mathcal{N}) \cdot \int d\vec{r}' f(\mathcal{N}) \\ &= f^2 \cdot 3 \int_1^{\mathrm{N}^{1/3}} dx x^2 \tilde{V}^{\text{rep}}(gx) \end{aligned} \tag{5.54}$$

where $f(\vec{r}) = \{\exp[-2\tilde{E} + 4\mathcal{N} \cdot \int d\vec{r}\, V^{\text{att}}(\vec{r})]\}^{1/2}$.

These rules allow us to transform the action (5.49) to the presentation

$$\begin{aligned} S = K \cdot \Bigg\{ &-2 \cdot 3 \int_1^{\mathcal{N}^{1/3}} dx x^2 \tilde{V}^{\text{att}}(gx) \mathcal{N}^2 - 3 \int_1^{\mathcal{N}^{1/3}} dx x^2 \tilde{V}^{\text{rep}}(gx) \mathcal{N}^2 \\ &\times \left[\frac{(1-\mathcal{N})^2}{\xi^2 \mathcal{N}^2} \exp\left(-2\tilde{E} + 4 \cdot 3 \int_1^{\mathcal{N}^{1/3}} dx x^2 \tilde{V}^{\text{att}}(gx) \Gamma \right) - 1 \right] \\ &+ \eta \cdot 3 \int_1^{\mathcal{N}^{1/3}} dx x^2 \ln|\mathcal{N} - 1| + (N-1) \ln \xi \Bigg\} \end{aligned} \tag{5.55}$$

which can be rewritten in the compact form as

$$\begin{aligned} S = K &\left\{ \begin{aligned} &[a(\mathcal{N}) - 2b(\mathcal{N})]\, \mathcal{N}^2 \\ &- a(\mathcal{N})(\mathcal{N} - 1)^2 e^{4b\mathcal{N}} \exp\left(\frac{-2E + \mu}{k_{\mathrm{B}} \Theta} \right) + \eta \ln|\mathcal{N} - 1| \end{aligned} \right\} \\ &+ (N-1) \ln \xi \end{aligned} \tag{5.56}$$

where the following dimensionless functions are introduced

$$a(\mathcal{N}) = 3 \int_1^{\mathcal{N}^{1/3}} \tilde{V}^{\text{rep}}(gx) x^2\, dx \tag{5.57}$$

$$b(\mathcal{N}) = 3 \int_{1}^{\mathcal{N}^{1/3}} \tilde{V}^{\text{att}}(gx) x^2 \, dx \tag{5.58}$$

The expression (5.56) can further be simplified. In fact, for a system of quantum particles that obey the Fermi-Dirac statistics, the second term in the curly brackets can be negligible, because at a low temperature the chemical potential μ being positive approaches the Fermi energy and hence the exponent $\exp[(-2E + \mu)/(k_{\mathrm{B}}\Theta)]$ is reduced zero.

In the case of particles that obey the Bose-Dirac statistics the chemical potential $\mu < 0$ and at a low temperature it tends to go to zero. This means that with decreasing temperature, the exponent $\exp[(-2E + \mu)/(k_{\mathrm{B}}\Theta)]$ also approaches zero.

Therefore, for both statistics at a low temperature the factor $\exp[-2E/(k_{\mathrm{B}}\Theta)] << 1$ becomes dominant, such that it cuts out the second term on the right hand side of the expression (5.56). In such a manner we arrive at the following action for a cluster

$$S \cong K \cdot \left\{ \left[a(\mathcal{N}) - 2b(\mathcal{N}) \right] \mathcal{N}^2 + \eta \ln \mathcal{N} \right\} + N \ln \xi \tag{5.59}$$

If we minimize the action (5.59) by the variable $\mathcal{N}$, we will be able to define the number of particles in a cluster and the cluster size. It must be emphasized that the definition of the parameters of nonhomogeneous formations requires only explicit forms of potentials of interparticle interactions. Then the system itself selects the realization that will provide the minimization of the free energy. Nonetheless, by first assuming the availability of clusters the conditions and parameters of their existence are then defined.

5.2.2 THE FORMALISM OF CLUSTER FORMATION FOR CLASSICAL PARTICLES

In a classical system the mean occupation number of the sth energy level obeys the inequality

$$n_s = e^{-\tilde{E}_s} / \xi = e^{(\mu - E_s)/(k_{\mathrm{B}}\Theta)} << 1 \tag{5.60}$$

as in the classical system the chemical potential $\mu < 0$ and $|\mu|$ a few times larger then $k_B\Theta$. By this means in the action (5.32) we may expand the logarithm $\ln|1+\eta e^{(\mu-E_s)/(k_B\Theta)}e^{\psi_s}\cos\varphi_s|$ into a Taylor series in respect to the small member (5.60) and preserve only the first term. As a result, we get the action that describes the ensemble of interacting particles, which are subjected to the Boltzmann statistics:

$$S \cong \sum_s \left\{ -\tfrac{1}{2}\sum_{s'}\left(\tilde{V}_{ss'}^{\text{att}\,-1}\psi_s\psi_{s'} + \tilde{V}_{ss'}^{\text{rep}\,-1}\varphi_s\varphi_{s'}\right) + \frac{1}{\xi}e^{-\tilde{E}_s}e^{\psi_s}\cos\varphi_s \right\} + (N-1)\ln\xi \tag{5.61}$$

The extremum of the functional (5.61) will be realized in the solution of equations $\delta S/\delta\varphi_s = 0$, $\delta S/\delta\psi_s = 0$ and $\delta S/\delta\xi = 0$. The corresponding equations are

$$\sum_{s'}\tilde{V}_{ss'}^{\text{rep}\,-1}\varphi_{s'} = -\frac{2}{\xi}e^{-\tilde{E}_s}e^{\psi_s}\sin\varphi_s \tag{5.62}$$

$$\sum_{s'}\tilde{V}_{ss'}^{\text{att}\,-1}\psi_{s'} = \frac{2}{\xi}e^{-\tilde{E}_s}e^{\psi_s}\cos\varphi_s \tag{5.63}$$

$$\frac{1}{\xi}\sum_{s'}e^{-\tilde{E}_{s'}}e^{\psi_{s'}}\cos\varphi_{s'} = N-1 \tag{5.64}$$

We may introduce the combined variable

$$\mathcal{N}_s = \frac{1}{\xi}e^{-\tilde{E}_s}e^{\psi_s}\cos\varphi_s \tag{5.65}$$

inserting $\mathcal{N}_s$ in the left hand side of equation (5.64) we obtain

$$\sum_s \mathcal{N}_s = N-1 \tag{5.66}$$

So the combined variable (5.65) describes the number of particles in the sth cluster.

Utilizing the approach used in the previous section, we can find:

$$\psi_{s'} = 2\sum_{s} V_{ss'}^{\text{att}} \mathcal{N}_{s'} \tag{5.67}$$

$$\sum_{ss'} V_{ss'}^{\text{att}\,-1} V_{s's}^{\text{att}} \psi_s \psi_{s'} = 4\sum_{s} V_{ss'}^{\text{att}} \mathcal{N}_{s'} \mathcal{N}_s \tag{5.68}$$

$$\varphi_{s'} = -2\sum_{s} V_{ss'}^{\text{rep}} \mathcal{N}_s \tan\varphi_s \tag{5.69}$$

$$\sum_{ss'} V_{ss'}^{\text{rep}\,-1} V_{s's}^{\text{rep}} \varphi_s \varphi_{s'} = 2\sum_{s} V_{ss'}^{\text{rep}} \mathcal{N}_{s'} \mathcal{N}_s \tan\varphi_s \tan\varphi_{s'} \tag{5.70}$$

$$\cos\varphi_s = \mathcal{N}_s \xi \exp\left(\tilde{E}_s - \psi_s\right) = \mathcal{N}_s \xi \exp\left(\tilde{E}_s - 2\sum_{s} V_{ss'}^{\text{att}} \mathcal{N}_{s'}\right) \tag{5.71}$$

$$\tan\varphi_s = \left\{\frac{1}{\mathcal{N}_s^2 \xi^2} \exp\left(-2\tilde{E}_s + 2\sum_{s} V_{ss'}^{\text{att}} \mathcal{N}_{s'}\right) - 1\right\}^{1/2} \tag{5.72}$$

Relations (5.67)-(5.72) allow us to transfer the action (5.62), which will depend only on one variable $\mathcal{N}_s$ and the fugacity ξ

$$\begin{aligned} S = & -2\sum_{s,s'} \tilde{V}_{ss'}^{\text{att}} \mathcal{N}_{s'} \mathcal{N}_s - \tilde{V}_{ss'}^{\text{rep}} \mathcal{N}_{s'} \mathcal{N}_s \left\{\frac{1}{\mathcal{N}_s^2 \xi^2} \exp\left(-2\tilde{E}_s + 2\sum_{s} V_{ss'}^{\text{att}} \mathcal{N}_{s'}\right) - 1\right\}^{1/2} \\ & \times \left\{\frac{1}{\mathcal{N}_s^2 \xi^2} \exp\left(-2\tilde{E}_s + 2\sum_{s} V_{ss'}^{\text{att}} \mathcal{N}_{s'}\right) - 1\right\}^{1/2} \\ & + \sum_{s} \mathcal{N}_s (1 + \ln\xi) \end{aligned} \tag{5.73}$$

Continuing the development of the scheme described in the previous subsection, we shall come to the action

$$S = K \cdot \left\{ \begin{array}{l} \left[a(\mathcal{N}) - 2b(\mathcal{N})\right]\mathcal{N} - 2\dfrac{a(\mathcal{N})}{\mathcal{N}^2 \xi^2} \\ \exp\left[-2\tilde{E} + 4b(\mathcal{N})\mathcal{N}\right] + \mathcal{N}\ln\xi \end{array} \right\} \tag{5.74}$$

where the dimensionless functions a and b are determined in expressions (5.58) and (5.59), respectively. The second term includes three factors that direct it towards zero. In fact the fugacity $\xi = \exp(\mu / k_B \Theta)$ and the exponent $\exp[(-2E+\mu)/(k_B \Theta)]$ are small, as for classical particles $\mu < 0$ and at least a few times $k_B \Theta$. Besides, the factor $\mathcal{N}^{-2}$ dramatically suppresses the whole second term. Hence neglecting the second term in the expression (5.74) we finally acquire a simple action

$$S = K \cdot \left\{ \left[a(\mathcal{N}) - 2b(\mathcal{N}) \right] \mathcal{N} + \mathcal{N} \ln \xi \right\} \tag{5.75}$$

The extremum (minimum) of the action (5.75) is reached at the value of $\mathcal{N}$ that is found from the equation $\delta S / \delta \mathcal{N} = 0$ and satisfies the inequality $\delta^2 S / \delta \mathcal{N}^2 > 0$. The value of $\mathcal{N}$ obtained in such a way will consistent with the number of particles that form the cluster.

5.3 BOSE-EINSTEIN CONDENSATION: SUBTLE NUANCES

Laser cooling, trapping states and the mechanism of Bose-Einstein condensation in dilute gases with concentrations of 10^{12} to 10^{15} cm^{-3} of atoms have been impressively described by Pethick and Smith [177]. In general, the discovered phenomena only represented collective effects in a quantum gas but since then the cooling mechanism and the Bose-Einstein condensation have been studied in greater detail. Nevertheless, the submicroscopic approach allows us to shed new light upon those phenomena. Indeed, we may apply the statistical mechanical approach that has been developed and presented in the previous sections to account for collective effects of cooling gases [178]. This approach makes it possible to investigate how the quantum mechanical field (a substructure of the matter waves, i.e., inertons, carriers of particles' field of inertia) is able to affect the collective behavior of atoms.

Today the phenomenon of Bose-Einstein condensation in diluted cold gases is explained as follows. When atoms are cooled to the temperature where the thermal de Broglie wavelength $\lambda_{\text{th}} \cong h / (3m\, k_B \Theta)^{1/2}$ is comparable to the mean distance g between atoms, the atomic wave functions overlap and according to Ketterle [179] the atoms "become a 'quantum soup' of indistinguishable particles" and in such a "soup" the atoms are in the coherent state. However, real gases are non-uniform. That is why in experiments they are contained in a trap, which typically provides a harmonic-oscillator potential. The trap potential is characterized by a cyclic frequency $\omega_{\text{trap}} \sim 10^3\ \text{s}^{-1}$ and a radius of the atom cloud $R \sim 10^{-3} - 10^{-4}$ m. The role of the trap potential is to set a small oscillating bias field that prevents loss of particles from the volume of the trap (Ref. [177], p. 6).

On the other hand, the phenomenon of the whole coherent state in which the motion of all the atoms is synchronized appears when the de Broglie wavelength λ_{th} exactly becomes equal to the distance g between atoms. In this case the (**n** – **g**)th atom emits its inerton cloud that then is fully absorbed by the **n**th atom; the **n**th atom emits its own cloud of inertons, which then is fully absorbed by the (**n** + **g**)th atom, etc. In other words, the coherent exchange of inerton clouds between atoms when an inerton cloud emitted by one atom hops to the neighbor atom and is absorbed by it, relates to the phenomenon of Bose-Einstein condensation.

The effect of synchronous transferring of inerton clouds from atom to atom occurs when the distance between atoms is equal to the thermal de Broglie wavelength. The applied trap potential allows the whole ensemble of atoms in the trap to oscillate in phase, which supports the coherent exchange of inerton clouds between them.

Let us investigate whether the occurrence of clusters can be possible in a Bose-Einstein condensate. The action S for an ensemble of N interacting boson particles, which tend to form clusters with $\mathcal{N}$ particles in a cluster, has been derived above (5.59):

$$S \cong K \cdot \{[a(\mathcal{N}) - 2b(\mathcal{N})]\mathcal{N}^2 - \ln \mathcal{N}\} + N \ln \xi \qquad (5.76)$$

where the functions a and b are determined in expressions (5.57) and (5.58), respectively. Note that in the case of Bose-Einstein condensation the chemical potential of atoms $\mu \to 0$, the fugacity $\xi = e^{\mu/(k_B\Theta)} \to 1$ and hence we may neglect the term $N\ \ln\ \xi$ in expression (5.76).

The potential of repulsion can be chosen in any form, because it does not play a principal role in a solution describing a collective state (due to the fast-falling character of the potential). Let us consider it in Lennard-Jones' form

$$V^{\text{rep}}(\rho) = \frac{C_{12} / r^{12}}{\rho^{12}} \qquad (5.77)$$

where $\rho = r / g$ here and below is the dimensionless distance. The attraction potential should include at least three terms: (i) the dispersion potential of interatomic interaction, which is usually written as $-C_6 / r^6$, (ii) a potential formed by a trap, which can be modeled by a harmonic potential $V_{\text{trap}} = \frac{1}{2} m \omega_{\text{trap}}^2 r^2$, and (iii) the harmonic potential caused by small spatial oscillations of atoms near their equilibrium positions $V_{\text{atom}} = \frac{1}{2} m \omega^2 \delta r^2$, i.e., stipulated by the inerton elastic interaction – because even the smallest alteration of the atom position leads to the creation of an inerton cloud bounded with the atom. Since the trap harmonic potential V_{trap} will possess the same frequency as the frequency of emitted inerton clouds, we may combine them in one potential. So, the whole attraction potential reads

$$V^{\text{att}}(\rho) = C_6 / (r\rho)^6 - \tfrac{1}{2} m \omega^2 (\delta r)^2 \rho^2 \tag{5.78}$$

where m is the mass of an atom, r is the distance between atoms, ω is the cyclic frequency of proper oscillations of an atom, and δr is the appropriate amplitude. Note the correct sign of V^{att} is opposite to that in expression (5.78), which is included in the processing of the action S.

Let us substitute the potentials (5.77) and (5.78) into the functions a (5.57) and b (5.58), and the action (5.76):

$$a = \frac{C_{12} / r^{12}}{3 k_{\text{B}} \Theta} \tag{5.79}$$

$$b = \frac{C_6 / r^6}{k_{\text{B}} \Theta} - \frac{3 m \omega^2 \delta r^2}{10 k_{\text{B}} \Theta} \mathcal{N}^{5/3} \tag{5.80}$$

$$S = K \cdot \left\{ \frac{C_{12} / r^{12}}{3 k_{\text{B}} \Theta} \mathcal{N}^2 - \frac{2 C_6 / r^6}{k_{\text{B}} \Theta} \mathcal{N}^2 + \frac{3 m \omega^2 \delta r^2}{5 k_{\text{B}} \Theta} \mathcal{N}^{11/3} - \ln \mathcal{N} \right\} \tag{5.81}$$

where we have preserved only the terms that are larger in comparison with unit, as we anticipate that $\mathcal{N} >> 1$.

Proper oscillations of atoms, which are characterized by the frequency ω, are produced by the movement of the atoms. In other words, the origin of the frequency ω is due to the collision of atoms with their inerton clouds: $\omega = 2\pi / T$ where T is the time period of collisions of an atom with its inerton cloud. However, in the present case of cooling diluted gases we have to talk about collisions of the inerton cloud of an atom with its neighboring atom: one emits the cloud and the other absorbs the cloud. Then the parameter T becomes the time of the free path of the atom and it is related to the amplitude δr of oscillations of an atom and its velocity υ, $1/T = \upsilon / \delta r$ because the amplitude δr of oscillations of the atom is nothing but the atom's de Broglie wavelength λ.

The equation for the number $\mathcal{N}$ of atoms, which are gathered in a cluster, comes from the equation $\partial S / \partial \mathcal{N} = 0$ if the inequality $\partial^2 S / \partial \mathcal{N}^2 |_{\mathcal{N} = \mathcal{N}_{\text{in cluster}}} > 0$ holds. The solution is

$$\mathcal{N} = \left\{ \frac{10}{33} \frac{6C_6 / r^6 - C_{12} / r^{12}}{m\omega^2 \delta r^2} \right\}^{3/5} \tag{5.82}$$

Let us assign numerical values to the parameters δr and ω for the case of cesium atoms whose mass $m_{\text{Cs}} = 2.207 \times 10^{-25}$ kg. Since the amplitude δr of oscillations of an atom is associated with its de Broglie wavelength λ, we may write $\delta r = \lambda = h / (m_{\text{Cs}} \upsilon)$. However, in dilute gases a lattice of atoms is not formed because the overlapping of their inerton clouds is not strong enough. That is why the oscillation of atoms is caused only by their thermal motion: $\upsilon \approx \upsilon_{\text{th}} = \sqrt{3 k_{\text{B}} \Theta / m_{\text{Cs}}} \cong 3 \cdot 10^3$ m·s^{-1} where we put a typical temperature of Bose-Einstein condensate $\Theta = 50$ nK. So, $\delta r = \lambda_{\text{th}} = h / (m_{\text{Cs}} \upsilon_{\text{th}}) = 10^{-6}$ m and then the cycle frequency of atom oscillations become $\omega = 2\pi\upsilon / (2\delta r) \cong 9.43 \times 10^3$ s^{-1}. Here, in expression (5.82) a fitting parameter is the numerator, i.e., the difference between the six-fold attraction and the repulsion energies at an equilibrium distance r of interacting atoms, which can be put $6C_6 / r^6 - C_{12} / r^{12} = 0.43 \times 10^{-19}$ J, or 1.81 kJ·mol.

The above-mentioned numerical values of the parameters allow the evaluation of the number of atoms that assemble in a Bose-Einstein cluster: $\mathcal{N} \approx 1.95 \times 10^5$.

A variation of the parameters in expression (5.82) allows us to construct the dependence of the number of atoms in a cluster $\mathcal{N}$ versus the thermal de Broglie wavelength, or amplitude δr, Figure 5.1. In fact these estimates are in line with experimental observations that accommodate up to around 10^6 atoms being in the state of Bose-Einstein condensation [179].

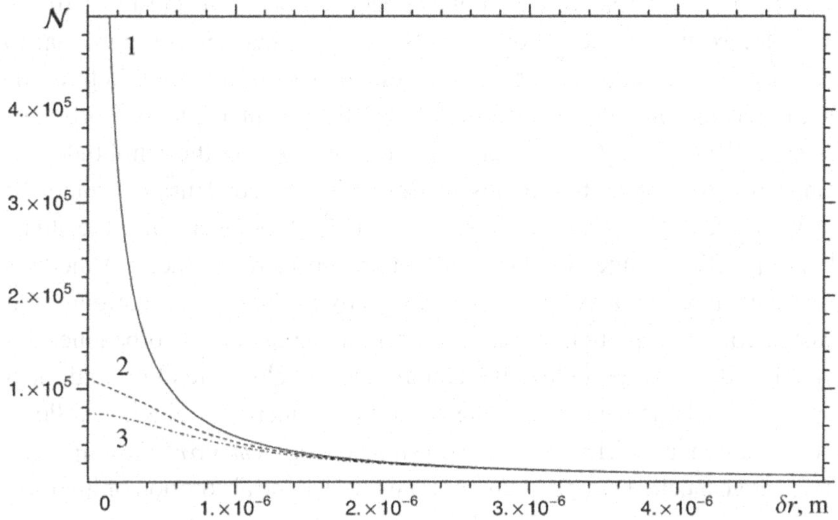

FIGURE 5.1 Number of atoms $\mathcal{N}$ in a Bose-Einstein condensate cluster versus the amplitude δr of thermal motion of atoms, i.e., the solution of Eq. (5.82) as the function of the de Broglie thermal wavelength $\delta r = \lambda_{\text{th}} = h/(m_{\text{Cs}} \upsilon_{\text{th}}) = h/\sqrt{3 k_{\text{B}} \Theta\, m_{\text{Cs}}}$. Curve 1: two potentials, V_{trap} and V_{atom}, merged in one $\frac{1}{2} m_{\text{Cs}} \omega^2 \delta r^2 \approx 10^{-29}$ J. Curve 2: the potential $V_{\text{trap}} = 2.5\times10^{-29}$J is not synchronized with V_{atom} and it is additionally introduced to the denominator of expression (5.77). Curve 3: the potential $V_{\text{trap}} = 2.5\times10^{-28}$ J is added to the denominator of expression (5.82).

Thus, from this analysis, it is now possible to account for the quantity of matter that is found in the condensed state: the interatomic interactions may subdivide the system of atoms to clusters but the cluster state is only realized when the absolute value of an attraction potential starts to exceeds the thermal energy, $V^{\text{att}} \geq k_{\text{B}}\Theta$. This inequality holds for the case calculated above: the attraction energy $\frac{1}{2} m_{\text{Cs}} \omega^2 \delta r^2 \approx 10^{-29}$ J exceeds the thermal energy $k_{\text{B}} \Theta \approx 7\times10^{-31}$ J. However, is the whole system of diluted gas atoms gathered in one cluster or is the system subdivided to either a few or many clusters? The question is open.

When the drift/diffusive rate of ultracold atoms increases (a typical observed speed is around 1 cm·s^{-1}), the thermal de Broglie wavelength λ_{th} decreases, which brings about a rearrangement of atoms, because at such conditions the increasing density of the gas initiates the collapse of the Bose-Einstein clusters.

Would an external inerton field affect Bose-Einstein clusters? In the present case the frequency of oscillations ω is inversely proportional to the particle amplitude $\omega = 2\pi \upsilon_{\text{th}} / \delta r$ (the same as is the case of a free particle). Because of that in the solution (5.82) the potential energy $\frac{1}{2} m\omega^2 \delta r^2$ is reduced to $3\pi^2 k_{\text{B}} \Theta$ and hence the dependence on the amplitude δr vanishes. This means that atoms in Bose-Einstein condensates cannot be further frozen by exposure with an inerton field. Moreover, since the atoms do not practically interact, they will not absorb inertons. Incident inertons can be captured by a system of particles only in the case of a strong or at least intermediate particle-particle interaction; in this case absorbed inertons are aligned so that particles' overlapping inerton clouds tend to hold them together. In a lattice of particles the generation of inerton excitations follows the dispersion law (5.13). So in the potential energy $\frac{1}{2} m\omega^2 \delta r^2$ the amplitude δr is not cancelled but plays an important role in the absorption of inertons.

5.4 INERTONS VIOLATE THE STABILITY OF HOMOGENEOUS MEDIA

5.4.1 CLUSTER FORMATION OF WATER MOLECULES

The infrared spectrum of water shows strong activity in the range of the hydrogen bond around the frequency of 200 cm^{-1}, which means that water molecules are characterized by a strong ionic polarizability and then the study of the statistical behavior of an aqueous system might be treated in the framework of a typical ionic lattice. In such a case the pair potential of interacting water molecules may consist of an ionic crystal potential (which includes the dimensionless Madelung constant α that falls within the range from unity to two, see e.g., Kittel [180]), the potential of dipole-dipole interaction and a fluctuation quantum potential:

$$V^{\text{att}}(\rho) = \frac{\alpha e^2}{4\pi\varepsilon_0 \varepsilon(g, v) g \rho} + \frac{\sqrt{2/3}}{4\pi\varepsilon_0 \varepsilon(g, v)} \frac{d^2}{g^3 \rho^3} - \tfrac{1}{2}\gamma(\delta r)^2 \rho^2 \quad (5.83)$$

(note the correct sign of V^{att} is opposite to that in expression (5.83) and also below in (5.89), which is included in processing of the action S).

Here the second term in the right hand side represents for the average dipole-dipole interaction when all reciprocal orientations of dipoles are equiprobable.

The last term in the right hand side of expression (5.83) is the direct consequence of the availability of inerton clouds around moving/vibrating material entities, as discussed above, because, as it follows from expressions (5.15), the cloud of inertons spreads over about four orders of the lattice constant's size. This last term is crucial for the derivation of the collective states of molecules. In expression (5.83) d is the dipole momentum of a node in the water network/lattice; $\varepsilon(g, v)$ is the permittivity as a function of the lattice constant g and frequency v; γ is the force constant. ρ is the dimensionless distance.

The potential of repulsion can be chosen in any form, because it does not play a principal role in a solution describing a collective state (due to the fast-falling character of the potential). Let us consider it in Lennard-Jones' form

$$V^{\text{rep}}(\rho) = V_0 / (g\rho / g)^{12} = V_0 / \rho^{12} \tag{5.84}$$

The ensemble of N interacting particles, which obey the Boltzmann statistics and tend to form clusters, is specified by the action (5.75). Using the potentials (5.83) and (5.84) we obtain the functions a (5.57), b (5.58), and the action S in the explicit form:

$$a = V_0 / (3k_{\text{B}}\Theta) \tag{5.85}$$

$$b = \frac{3\alpha e^2}{8\pi\varepsilon_0\varepsilon(g, v) g\, k_{\text{B}}\Theta} \mathcal{N}^{2/3} + \frac{\sqrt{2/3}\, d^2}{4\pi\varepsilon_0\varepsilon(g, v) g^3\, k_{\text{B}}\Theta} \ln \mathcal{N} - \frac{3\gamma(\delta r)^2}{10 k_{\text{B}}\Theta} \mathcal{N}^{5/3} \tag{5.86}$$

$$S / K = \frac{V_0}{3k_{\text{B}}\Theta} \mathcal{N} - \frac{3\alpha e^2}{4\pi\varepsilon_0\varepsilon(g, v) g\, k_{\text{B}}\Theta} \mathcal{N}^{5/3} - \frac{\sqrt{2/3}\, d^2}{2\pi\varepsilon_0\varepsilon(g, v) g^3\, k_{\text{B}}\Theta} \mathcal{N} \ln \mathcal{N} + \frac{3\gamma(\delta r)^2}{5 k_{\text{B}}\Theta} \mathcal{N}^{8/3} + \mathcal{N} \ln \xi \tag{5.87}$$

The equation for the number of particles combined in a cluster comes from the equation $\partial S / \partial \mathcal{N} = 0$. Explicitly,

$$\mathcal{N}^{5/3} - \frac{25}{32\pi} \frac{\alpha e^2}{\varepsilon_0 \varepsilon(g,v) g \gamma \delta r^2} \mathcal{N}^{2/3}$$
$$- \frac{5\sqrt{2/3}}{16\pi} \frac{d^2}{\varepsilon_0 \varepsilon(g,v) g^3 \gamma \delta r^2} (\ln \mathcal{N} + 1)$$
$$+ \frac{5}{24} \frac{V_0}{\gamma \delta r^2} + \frac{5}{8} \frac{k_B \Theta}{\gamma \delta r^2} \ln \xi = 0 \tag{5.88}$$

This equation can be solved numerically. For water, the numerical values of the corresponding parameters are: $d = 1.84$ Debye $= 6.17 \times 10^{-30}$ C·m; Madelung's constant $\alpha = 1.7$; permittivity $\varepsilon(g, v) = 5$ (it was shown [181] that the dielectric constant ε changes from 5 to 10 when the depth of the water layer varies from 1 to 12 nm that could be the size of a cluster); the lattice constant, which also determines the effective size of a molecule in the lattice, $g \cong 2.8 \times 10^{-10}$ m; δr, as the amplitude of vibration of a molecule, can also be considered as an amplitude of fluctuation of the size of a molecule in the lattice and, as the rule $\delta r \sim 10^{-11}$ m.

In Eq. (5.88) the fugacity $\xi = \exp[-|\mu|/(k_B\Theta)]$; the chemical potential μ in the first approximation can be chosen as is the case for the gas phase $\mu \approx 3k_B\Theta \ln(\langle\lambda_{th}\rangle n^{1/3})$ where $n = 3\times 10^{28}$ m^{-3} is the concentration of water molecules and the thermal de Broglie's wavelength $\langle\lambda_{th}\rangle \approx h/(3mk_B\Theta)^{1/2}$. The mass of a water molecule $m = 2.99\times 10^{-26}$ kg and hence putting $\Theta = 300$ K we get $\langle\lambda_{th}\rangle \cong 3.4\times 10^{-11}$ m, and thus $\langle\lambda_{th}\rangle n^{-1} \cong 0.105648$. The initial potential of repulsion V_0 can be chosen as an average value of the energy of a hydrogen bond in water and at room temperature $V_0 \approx 10k_B\Theta_{300} \cong 4.25\times 10^{-20}$ J where $\Theta_{300} = 300$ K. The force constant γ of the water network is a fitting parameter, the value of which is anticipated in the order of 1 N/m (in solids γ varies approximately from 1 N/m to tens of N/m); below we put $\gamma = 1$ N/m.

Figure 5.2 depicts the solution to Eq. (5.88) as a function of amplitude δr. We can see the smaller the amplitude δr, the larger the number of particles involved in a cluster. The other solution for $\mathcal{N}$ falls within the interval from 0 and 1, which signifies that the system of N water molecules under consideration represents a kind of a mixture of two phases: clusters and individual molecules.

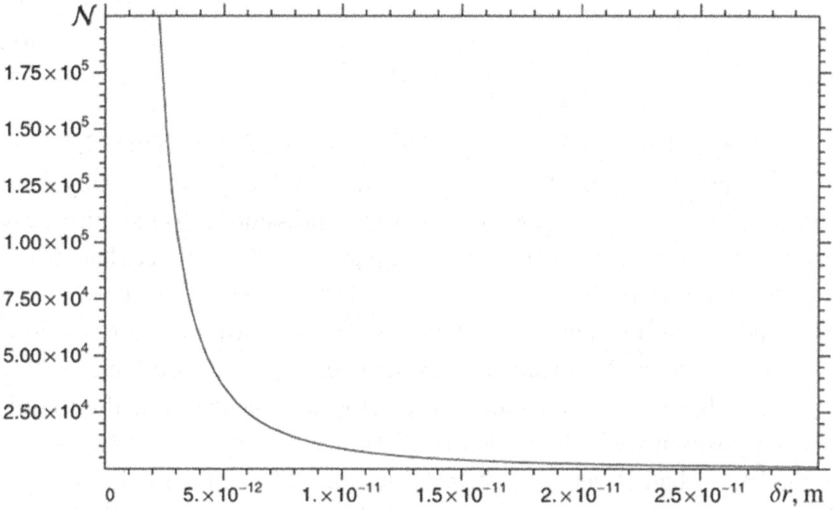

FIGURE 5.2 Numerical solution $\mathcal{N}$ to Eq. (5.88) as a function of amplitude δr. Values of the parameters are $V_0 \cong 4.25 \times 10^{-20}$ J and $\gamma = 1$ N/m.

5.4.2 CLUSTER FORMATION IN MODEL SYSTEMS

Let us examine a model system of N molecules interacting through the Lennard-Jones pair potential. In this case, taking into account the inerton interaction $\frac{1}{2}\gamma\delta r^2$ between molecules, the attraction and repulsion parts of the pair potential respectively are

$$V^{\text{att}} = \frac{V_0}{\rho^6} - \tfrac{1}{2}\gamma(\delta r)^2\rho^2 \tag{5.89}$$

and expression (5.84) for V^{rep}. For these potentials the action (5.75) is in the explicit form

$$S/K = -\frac{V_0}{3k_{\text{B}}\Theta}\mathcal{N} + \frac{3\gamma(\delta r)^2}{5k_{\text{B}}\Theta}\mathcal{N}^{8/3} + \mathcal{N}\ln\xi \tag{5.90}$$

and the equation for the number of molecules in a cluster $\partial S/\partial\mathcal{N} = 0$ results in the solution (if we neglect the small contribution on the side of $\ln\xi$)

$$\mathcal{N} \approx \left(\frac{5}{24} \frac{V_0}{\gamma \delta r^2} \right)^{3/5} \tag{5.91}$$

The dependence of $\mathcal{N}$ as a function of δr is shown in Figure 5.3

The appearance of clusters with a fractional degree, given by both the major term in Eq. (5.88) and also in expression (5.91) significantly changes the behavior of the system. In particular, we can see that reducing δr automatically increases $\mathcal{N}$. However, this is possible only when the amplitude δr of oscillations of particles near their equilibrium positions begins to decrease, i.e., becomes less than approximately 10^{-10} m.

As we have seen the oscillation of atoms/molecules near their equilibrium positions, which are described by the term $\frac{1}{2}\gamma \delta r^2$, or $\frac{1}{2} m \omega^2 \delta r^2$, is the basic characteristic of cluster formation of many particle systems. The absorption of incident inerton radiation by this system will result in an increase of the mass of the system's entities. Consequently, because in condensed matter the mass of an atom oscillates, as has been discussed in Section 5.1, this will then automatically lead to a reduction of

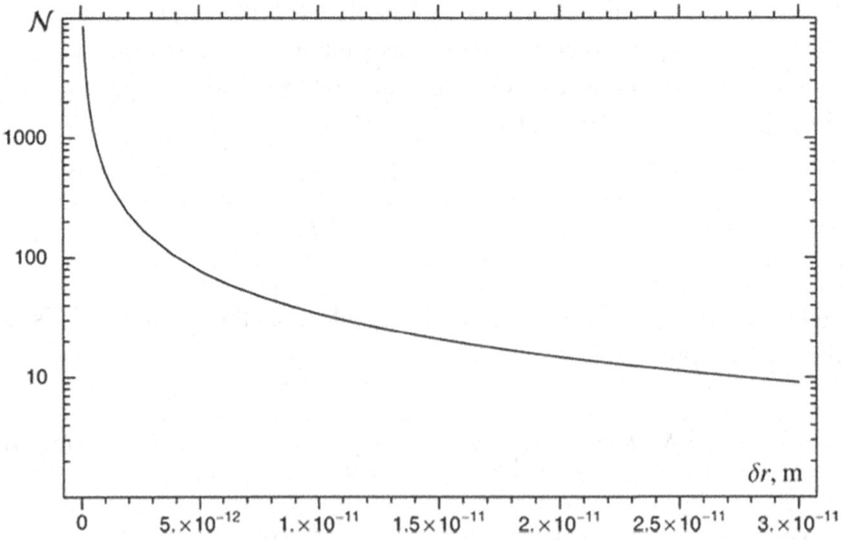

FIGURE 5.3 Numerical solution $\mathcal{N}$ (5.91) versus δr; values of the parameters are: $V_0 = 5 \times 10^{-20}$ J and $\gamma = 0.3$ N·m.

the amplitude δr, which promotes clustering, since increasing the mass reduces the amplitude of oscillation.

For example, an experimental study [182] of the spin-lattice relaxation rates for protons in water samples showed that water by its nature is a metastable liquid. The degree of binding of its molecules depends essentially on the past history of the water system. The pulsed nuclear magnetic resonance technique allowed the direct determination of a peculiar critical temperature Θ_c of a structural transition and this value of Θ_c changed from day to day. The recorded activation energy E_a of water molecules, which is responsible for diffusion of individual molecules, also dramatically changed every day. In general a water system requires a modification of thermodynamics [183]. Moreover, such modification should be applied to any physical-chemical system, inasmuch as thermodynamics call for an additional thermodynamic function describing a mass exchange [156, 184, 185].

5.4.3 EXPERIMENTALLY REVEALED INERTON EFFECTS IN AQUEOUS SOLUTIONS AND BIOSYSTEMS

Our first experimental studies [186–189] of inerton field effects that were generated in aqueous solutions by a Teslar watch© exhibited sizable changes in the samples under consideration. A Teslar watch is one of the applications of inerton fields; the watch includes a special chip, a Teslar chip – a narrow ferromagnetic film folded and sealed in the form of a typical Möbius strip. Before inserting into the watch, the Teslar chip (TC) was induced in a special way with a magnetic signal of about 8 Hz. When placed in the watch the TC, which was charged as described, is affected by the electromagnetic field generated by the watch's usual mechanism and the Möbius strip starts to react to the electromagnetic signal compensating the north and south magnetic components. So the strip generated only a scalar field.

Our experiments [186, 187] in fact proved the existence of a scalar field generated by the TC at the frequency of about 8 Hz. We observed that the scalar field of the TC was able to locally influence chemical physical processes. In the experiments, we investigated how the capacity of aqueous solution varies with time both under the influence of the TC, and after the influence of the TC. Analysis of the measurements showed an unusual behavior of the dielectric properties of a 50%-aqueous solution of alcohol.

The key results of the experiments can briefly be stated as follows. Upon exposure to radiation from the TC the alcohol component evaporated more intensively from the solution; this was evident from studying the solution density [186, 187]. At the same time the state of the water component is specified by a 'frozen' state. These results are associated with the behavior of the permittivity of the aqueous solution, because the capacity C is proportional to ε and the permittivity of the solution components are $\varepsilon_{\mathrm{w}} = 81$ (water) and $\varepsilon_{\mathrm{a}} = 26$ (alcohol). Figure 5.4 shows the usual behavior of the aqueous solution studied without the application of the TC but in the presence of a conventional quartz watch used as an imitator. Figure 5.5 depicts the behavior of the sample affected by the TC; we can see that initially the real part of the permittivity ε of the solution increases remarkably with time but then the following decrease of ε is provoked by the evaporation of the remaining water.

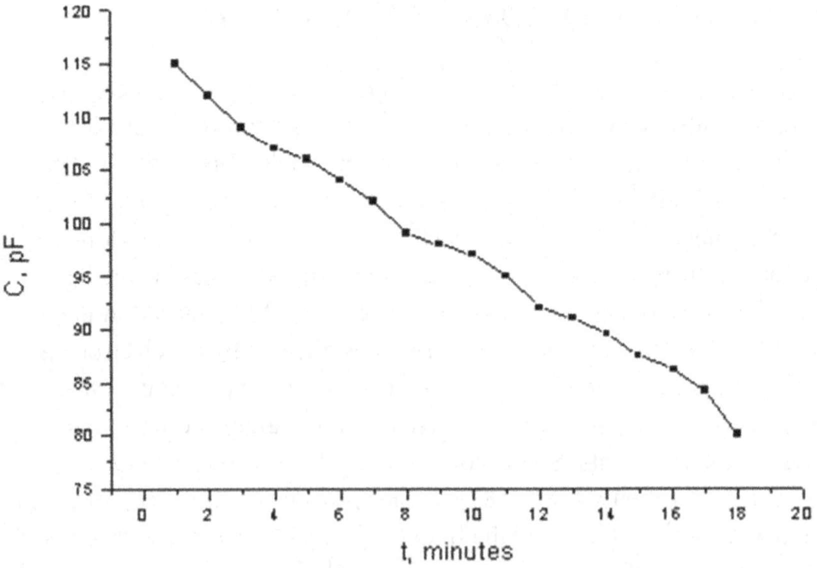

FIGURE 5.4 Capacitance of the 50%-aqueous solution of alcohol as a function of time in an electric field of frequency $f_{\mathrm{meas}} = 1$ kHz. A conventional quartz watch (an imitator) was put under the cuvette (with permission from Ref. [186]).

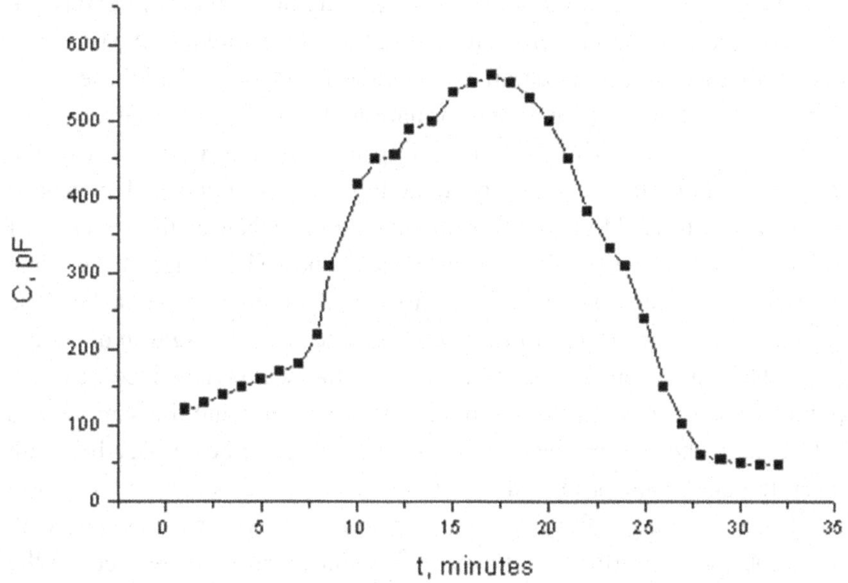

FIGURE 5.5 Capacitance of the 50%-aqueous solution of alcohol affected by the Teslar watch, which was put under the cuvette, as a function of time in an electric field of frequency f_{meas} = 1 kHz (With permission from Ref. [186]).

The observed anomalies of the permittivity of the aqueous solution of alcohol $\varepsilon_{\mathrm{w+a}}$ cannot be explained in the framework of classical electrodynamics in principle. The only explanation is that in a liquid with two or more components the molecules absorbed inertons from an external source, which decreased their amplitudes of oscillation δr and altered the initial equilibrium distance r_0 between the different types of molecules, $r_0 \rightarrow \tilde{r}_0$. This rearrangement of molecules would have resulted in their clustering, which is specified with the critical temperature T_c. Molecules involved in the formation of such clusters are 'frozen' of course and they evaporate more slowly, which was experimentally demonstrated (Figure 5.5).

One more peculiarity was observed [187]. A water molecule is a double dipole, namely, due to its configuration the negative charge is shifted to the oxygen atom and the positive charge is shifted to two hydrogen atoms, which results in a stationary large dipole moment d = 1.84 Debye. When an inerton field affects H_2O molecules, they will be aligned along the field. Since O is the more massive component of the molecule and H_2 is the less massive component, all the molecules will tend orient their more massive aspect to the source of the mass field, i.e., the inerton field. In the presence of a stationary (or quasi-stationary) electric or magnetic field these water molecules will be unstable to the formation of long-lived clusters.

Holographic experiments [188, 189] have been carried out with the use of the holographic interferometer IGD-3, developed and produced in the Institute of Physics of Semiconductors of National Academy of Sciences of Ukraine, Kyiv. The laser beam passed through the cuvette with the object under study and then arrived at the finely dispersed diffuse scatterer which according to Lambert's law scatters the light in all directions. The light from the entire surface of the scatterer arrived at every point of the light sensitive thermoplastic in which plane the sample was selected. In the thermoplastic's plane, the objective beam together with the reference beam produced by the mirror and the objective lens form a holographic image of the object under study, in this case the cuvette with a aqueous solution of biomolecules. The hologram was observed on the monitor screen with a camera connected to a computer, which was used for recording and processing the holograms. If an external factor causes an alteration of the refraction index n even at the value of 10^{-6}, the fringe pattern within the limits of the object's profile will change.

The Teslar watch was put onto the top of a quartz cuvette filled with the solution being studied. As the model of primary reception able to react to an external inerton field we chose the following systems: (i) double-distilled water, (ii) a saturated aqueous solution of amino acids (tyrosine, tryptophan, and alanine), and (iii) a diluted aqueous solution of human blood plasma (in the case of blood plasma biomolecules, the system in question was non-equilibrium and able to respond to very small stimuli.) In systems (i) and (ii) the results showed typical slight changes of the fringe pattern within 400 seconds or longer time interval. These changes should be associated with the inner drift of liquid parameters. The curve of long-time dynamics did not show any influence of the TC on the cuvette.

A different pattern of behavior was observed in system (iii) of the blood plasma solution. Blood plasma solutions had been diluted in the ratio 1:50 and 1:100 with distilled water. The time between blood extraction and the holographic measurement was 4 hours. Without the TC action, the solution showed stable and reproducible characteristics for more than

4 hours. In Figure 5.6 one can see the image of the cuvette with plasma blood solution affected by the TC. Black dots indicate the center position of one of the interference bands on the image plane. The value of the shift relating to zero line characterizes the degree of influence of the TC. Black rectangles (Figure 5.6, right) show the positions of two Teslar chips relating to cuvette; the fringe pattern of the solution relating to the chip is deformed in different ways in different zones (short, mid, and far-distance). A physical mechanism of the change is associated with the increase of the refraction index *n* in the short-distance zone; *n* remains unchanged in the mid-distance zone and decreases in the far-distance zone. Moreover, it seems that slow laminar flows have been induced by the TC near the front wall and directed to it.

If the refraction index of the solution changes in one place under the influence of an external factor, the length of the optical path will also change. For the purpose registering these changes, a tool was designed in such a way that the 'initial interferogram' constitutes a group of horizontal bands of equal thickness. Depending on the character of changes of the optical density in the cuvette volume, the bands can be distorted (local changes of *n*), even though gaps between bands can expand without deformations (volumetric decreases of *n*). Thus, arbitrary deformations of the fringe pattern are caused by a combination of local and global changes of the optical density. Changes of *n* are produced by changes in the structure

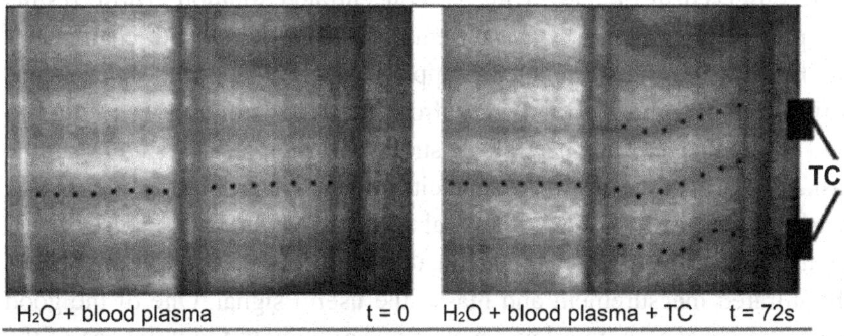

FIGURE 5.6 Dynamics of the fringe pattern of the aqueous solution of plasma of human blood at the insertion of 2 TC. The strong disturbance of the optical density of the solution had emerged already even within 72 s, right figure. The two TC are located at 4 mm from the right wall of the cuvette (From Ref. [188]).

of the network of hydrogen bonds of water, which being under the influence of oxygen, biomolecules, and the inerton field generated by the TC, form long-lived structures. In the mentioned network, those new structures try to minimize the total energy relative to the volume occupied by the water system. These changes (restructuring the aqueous solution) occur sufficiently slowly to be optically recorded.

The influence of the TC on water leads to only minor changes of the fringe pattern ($\Delta n = 2\times10^{-5}$) whereas the behavior of proteins is markedly influenced by the TC. The effects associated with the TC are opposite to those associated with heating.

The maximum change of n of the protein solution affected by the TC reached the value of $\Delta n = 2\times10^{-4}$, which was an order of magnitude larger than the temperature changes of n. Thus, the numerical estimates and the experimental data demonstrated that changes of n caused by the influence of the TC were conditioned by nonthermal changes of the dielectric constant of the aqueous solution being studied. Changes in the aqueous solution affected by the TC covered the entire macroscopic volume of the sample. This behavior of the sample was associated with both inner convective flows (similar to the case of Benard cells) and structural changes of the water. The latter brought about changes in the reflective index n of the solution and the fringe pattern.

In a series of tests with the TC, we particularly studied samples of different concentrations of H_2O_2 in distilled water. The most drastic changes were observed in the case with the concentrated solution. Those results are presented in Figure 5.7. The temperature in the sample chamber of the Bruker IFS66 instrument was approximately 30°C. Under these conditions the evaporation of oxygen from the H_2O_2 solution was found to be very intensive. The spectral registration in such a case is not a simple task. The evaporation of oxygen excites the aqueous solution stimulating a diffusion reflection, etc., which is also supplemented with a noise caused by the non-plate surface and chemical reactions. This strongly influences the infrared measurement and masks the useful signal. One of the good H_2O_2 spectra (i.e., with a low level of noise) without the influence of the TC is shown in Figure 5.7, curve 1. The spectra have changed with time. As is seen from Figure 5.7, in the region 1600–1700 cm^{-1} two bands of deformation vibrations due to the different hydrogen bonds are available

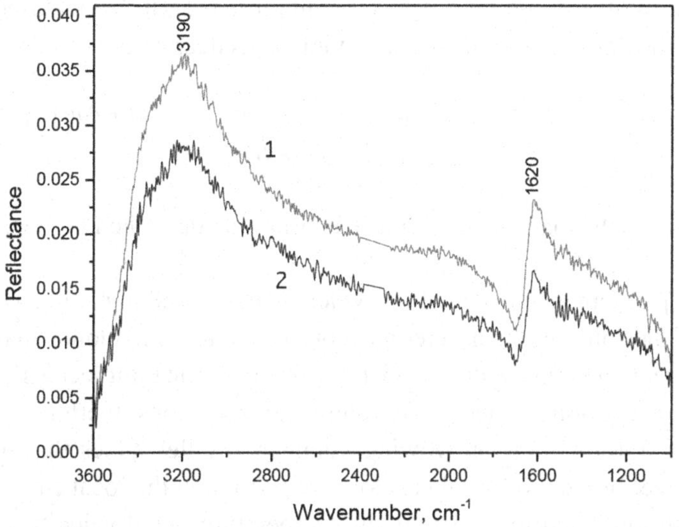

FIGURE 5.7 IR-RAS spectra of H_2O_2 aqueous solution 15 minutes after pouring the solution into the cuvette. 1 – without the TC; 2 – under the influence of the TC.

at the start of the experiment. 15 minutes later only one band of the deformation vibration of the hydrogen bond is registered.

When the TC was inserted, it became evident that the chip's inerton radiation suppressed the vibration of molecules in the aqueous solution and, in particular, strongly 'froze' vibrations of O–H groups, which is especially seen in the vicinity of the maximum 3200 cm^{-1} (Figure 5.7, curve 2). This behavior of curve 2 with an intensity around 20% less than that of the curve 1 is very unusual and can only be understood with the involvement of an inerton field.

The TC strongly affects a significantly non-equilibrium system, and the hydrogen bond is a good marker for the observation of changes in the system in question. In the case of the spectroscopic study of aqueous solutions of hydrogen peroxide, the micro dynamics were associated with the liberation of oxygen in the solution.

Thus, the inerton field transferred mass changing potential properties of the environment. Consequently, the mass defect Δm becomes an inherent property not only of atomic nuclei but also of any physical and physical chemical systems (including biophysical ones). Physical entities are able to absorb inertons, which results in the entity's increased mass. Hence, the amplitude of the entity's vibrations decreases

$$m \to m + \Delta m, \quad \delta r = h/(m\upsilon) \to \delta r' = h/[(m + \Delta m)\upsilon],$$
$$\text{that is,} \quad \delta r' < \delta r$$

which starts the mechanism of cluster formation described in the previous sections.

The phenomenon discussed is widely encountered in nature. The presence of inertons and their heterogeneous distribution in clouds may result in different snowflakes that condensed from identical molecules of water within locally homogeneous environmental conditions. Inertons influence the cluster formation for growing younger, smaller flakes that begin to form lower down the (weather) cloud, even though the local environmental conditions (humidity, dew point, temperature, wind velocity, etc.) are identical. A variety of shapes of snowflakes can be obtained mathematically in the framework of the formalism of cluster formation described in this chapter; the only requirement is to represent the term $\frac{1}{2}\gamma(\delta r)^2$ in the 3D format, which allows asymmetry: $\frac{1}{2}\gamma(\delta x)^2$, $\frac{1}{2}\gamma(\delta y)^2$ and $\frac{1}{2}\gamma(\delta z)^2$.

Thus, when the atmospheric conditions become unstable, inertons may influence the gases of the air. In this case, as shown, a new distribution of clusters in gases will be established inducing new dynamic properties of the air. Therefore, we may conjecture that during unstable weather conditions, any living bodies coming into contact with such perturbed air will be more strongly influenced by those inertons, which may explain why some people are predisposed or more sensitive to changes in weather conditions (since the blood of sensitive people could become more dense, or rather viscous, in the periods of so-called magnetic storms when a compass needle is dancing, which is observed at the appearance of black spots on the sun).

An important medical development involving inerton induced cluster formation may be considered to underlie the main principle of homeopathy and of its modern more powerful branch of medicine called information medicine (where electro-acupuncture points having bad conductivity can be corrected through injection of a mass field, or a field of spatial fractals, i.e., an inerton field, in a special frequency range, which corresponds to the specific oscillation frequency of the organ in question. Pioneering research that used electro-acupunction was conducted by Reinhold Voll, a German doctor of medicine, in 1950s, and clinically used by Zenovii Skrypniuk, a Ukrainian doctor of biology, in 1990–2000s.

The effects of inerton induced cluster formation may provide the basis for understanding the different crystallites of frozen water obtained and investigated by Emoto's team [190], which were formed under different environmental conditions (including the effect of human consciousness on the molecular structure of water). Whatever may have stimulated the diverse formations of those crystallites, the results could only be associated with the presence of incident inerton radiation; this radiation would be generated in parallel with acoustic waves, as has been argued above, namely, the space between oscillating atoms is filled with inertons, which can be treated as a gas, an inerton gas. These inertons being adsorbed by water molecules lead to the formation of clusters when drops of water are cooling below 0°C and additional elastic inerton interaction members $\frac{1}{2}\gamma(\delta x)^2$, $\frac{1}{2}\gamma(\delta y)^2$ and $\frac{1}{2}\gamma(\delta z)^2$ will result in various shapes of frozen drops.

A brief excursion: A further example is worthy of mention from my own personal experience. One evening in 2007 I suddenly experienced acute pain in the lower right part of the abdomen – that was a clear inflammation of the appendix. In a couple of hours the temperature of my body increased up to 41°C, and I was shivering with fever. So, what should I do? The situation was critical, I had to call an ambulance – the appendix should be removed surgically. Nevertheless, a thought came to my mind: The local inflammation required cooling from inside. Accordingly, I put my Teslar watch with two Teslar chips on the relevant area of the abdomen where the pain was felt. Within half an hour the feverish shivering ceased, the temperature fell down to 36.7°C and the sharp local pain had subsided. I attached the watch by adhesive tape to the appendix area and went to sleep. Two days later I removed the Teslar watch from the appendix area since there was no longer any pain. Thanks to inertons I have avoided the surgeon.

5.5 INERTONS IN SOME PRACTICAL APPLICATIONS

Liquid substances such as gels, hand creams, lotions and ointments, etc., after processing with an inerton field become less viscous, demonstrating a kind of a superfluidity. In particular, the rate of penetration of hand creams through the skin increases 20–30%, as measured by a Dermal

Penetration Test used in cosmetology. Figures 5.8 and 5.9 demonstrate how viscosity of these substances changes after the treatment.

Figures 5.8 and 5.9 show that the substances affected by the inerton field becomes less viscous and most easily absorbed 40 hours after the processing. This means that the dynamics of these substances have been modified, which last for many hours. After three days the viscosity started to increase reaching the initial value after about two weeks. In some substances a few days after the processing a kind of a polymerization was observed. It was found that the substances affected did not break down into their separate components; i.e., the components had not formed homogeneous clusters.

Typical changes in the viscosity of sorbent clay samples irradiated by an inerton field are shown in Figures 5.10 and 5.11. The dynamic viscosity and the shear viscosity of the sample under consideration significantly depend on the exposure time in an inerton field, which is shown in Figure 5.12 for measurements conducted approximately half an hour after the inerton irradiation.

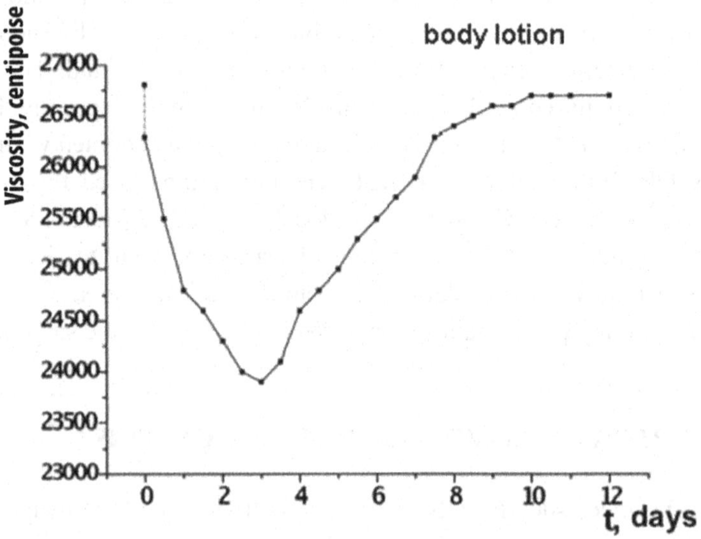

FIGURE 5.8 Viscosity of one brand of typical body lotions/gels, processed with an inerton field, versus time.

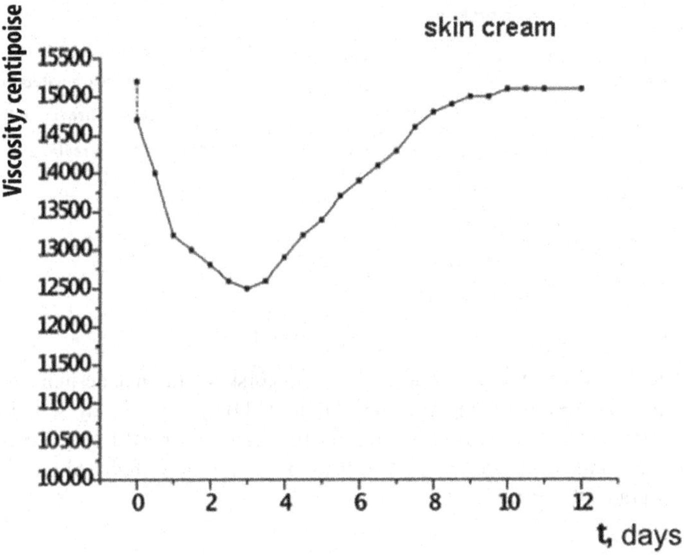

FIGURE 5.9 Viscosity of one brand of typical skin creams, processed with an inerton field, versus time.

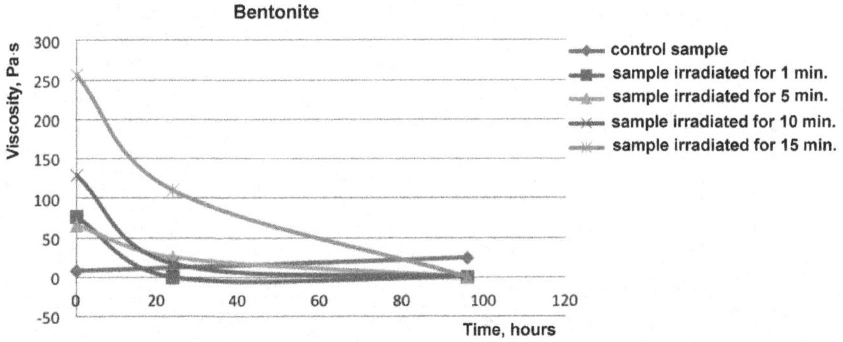

FIGURE 5.10 Viscosity of a bentonite clay $((Na, Ca)_{0.33}(Al, Mg)_2(Si_4O_{10})(OH)_2 \cdot nH_2O)$ affected by the inerton field at different expositions (viscosity vs. hours). Note 15 minutes of irradiation increased the viscosity of the bentonite up to 250 times.

In contrast, samples of water irradiated by an intensive inerton field demonstrated an increase in the water density up to 1.2 kg/liter, as registered by a hydrometer. This indicates a change in both the density and

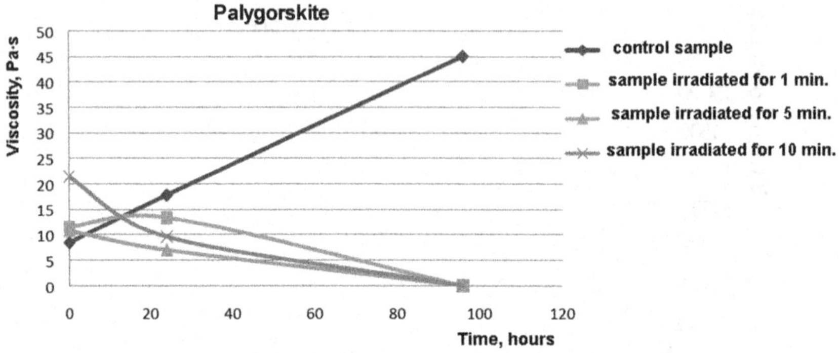

FIGURE 5.11 Viscosity of samples of palygorskite (a magnesium aluminum phyllosilicate with formula (Mg, Al)$_2$Si$_4$O$_{10}$(OH)·4(H$_2$O), a type of clay soil) irradiated with an inerton field. The viscosity of the control sample increased under atmospheric conditions with time, contrasted by the viscosity of the samples irradiated by the inerton field having gradually decreased.

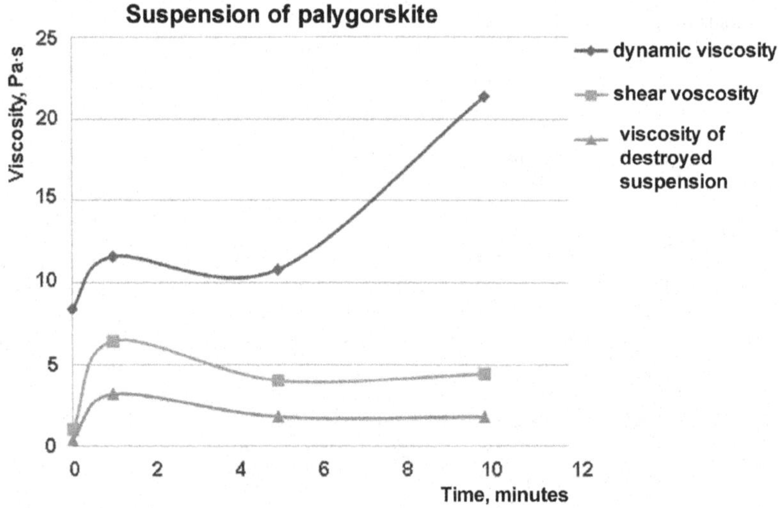

FIGURE 5.12 Viscosity versus time of an inerton irradiated aqueous suspension of palygorskite.

the viscosity of water processed with an inerton field. Plants given such water show a significant decrease of transpiration, which is important for agriculture in hot dry regions.

Wine or any other alcohol drink acquires a noticeably pleasant mild flavor after processing in the bottle in an Lx unit (see Figure 5.13) for only 15–30 seconds.

In the chemical industry an inerton field is able to play the role of a field catalyst or, in other words, an inerton field can serve to control the rate of chemical reactions. In the reactive chamber of the Lx unit we are able to generate an inerton field by using magnetostriction agents: owing to the striction the agents contract non-adiabatically, which culminates in the irradiation of sub matter, i.e., inertons, from the agents. In the reactive chamber the inerton field 'freezes' the molecules of the reacting substances. If in the reactive chamber substances move rapidly, the motion will hasten reactions of breaking of long molecules 'frozen' by the inerton field into shorter molecules (in other words, long molecules become massive and inactive, which makes it possible to cut them by fast metal agents). This procedure does not require high temperature and high pressure or long term mechanical mixing (over hours), which are typical for modern chemical technologies. Under the influence of inerton radiation, the formation of a new substance takes place within a few seconds, though usually these chemical reactions last hours. One such application of inerton fields is in the manufacture of biodiesel (the methyl transesterification) from a medley of vegetable oil and methanol (with catalyst KOH), which has been described in our patent application [191]. Figures 5.13 and 5.14 demonstrate the laboratory unit and the industrial set-up, respectively, which generate a powerful inerton field in the frequency range of about 10 Hz to 20 kHz.

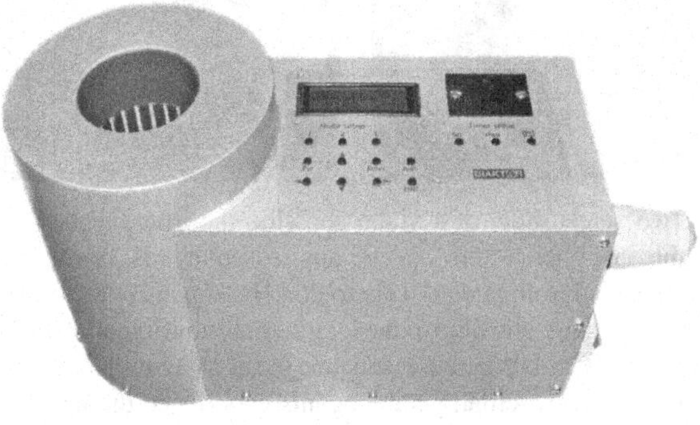

FIGURE 5.13 Laboratory batch unit Lx (http://biaktor.com). The device is intended for intensification of physicochemical reactions in liquid and gel substances.

Figure 5.15 shows the interior of the reaction chamber at two different moments. Metal agents oscillate in the chamber generating an inerton field.

FIGURE 5.14 Industrial unit Bx that produces biodiesel, 1 ton/hour (http://biaktor.com).

FIGURE 5.15 Two screenshots of a video showing the iron agents inside the reactor chamber of the laboratory batch unit Lx. The applied voltage is 380 V; the frequency of the current can be changed from about 10 Hz to 20 kHz. Agents cling to the surface of the cylindrical chamber by a magnetic field of the longitudinal configuration that is induced due to a special winding of the coil. When in resonance the agents oscillate near fixed positions in the chamber and at these oscillations the agents experience the magnetostriction that causes the inertons to radiate from the agents. Each agent (a Ferrum cylinder) has a length 16 mm and a diameter 0.9 mm.

The generation of an inerton field in the reaction chamber also affects the current in the coil; preliminary measurements show that the current increases 3 to 7 times under different conditions reaching 10–14 A. This can be explained by absorption of inertons by copper atoms in the wire, which effectively 'freezes' them: the copper atoms become more massive and the amplitudes of their equilibrium vibrations decrease, which in turn reduces the electron-atom scattering and hence the resistance of the wire is decreased.

5.6 AN ELECTRON DROPLET

The first electron droplets, or clusters, were generated in experiments on the electric discharge at a room temperature by Shoulders [192, 193] in the 1990s. Shoulders [192, 193] and then others [194–197] showed that electron clusters, consisting of around 10^{10} electrons and ranging from several to tens of micrometers, could easily move at a low speed in either a vacuum or air over macroscopic distances. In particular, Shoulders named his clusters "EV" and demonstrated micrographs of foil and examples of crystals pierced with holes by those EV, which indicated that it was possible to get more energy out of electron clusters than was put in.

Kukhtarev et al. [198, 199] have studied the self-organized spatiotemporal pattern formation produced by the scattering of laser light in photorefractive ferroelectric crystals of $LiNbO_3$. Some of these patterns clearly demonstrated charged formations. Laser-induced nonlinear holographic scattering in an electro-optic crystal revealed space-charge waves formed within the crystal volume by recording a set of holographic gratings. These holographic gratings were recorded by interference patterns arranged between the pump beam and the scattered waves, which produced moving space-charge waves inside the crystal. Due to electro-optic effects those space-charge waves modulated the refractive index (forming holographic gratings) and the pump beam diffraction forming these holographic gratings provided a visualization of the space-charge waves. These results were relatively well understood for the volume space-charge waves.

However, in addition to the known effects it was also revealed an unusual behavior of the specular reflection (i.e., the reflection of the laser beam from the frontal crystal surface) that resembled the visualization of the volume space-charge waves. Those moving scattering spots were recognized as the manifestation of charged plasma clusters (electron droplets) formed near the crystal surface. The appearance of those droplets was accompanied by strong radio-frequency signals (with the maximum in MHz frequency range) that were picked-up by the needle-shaped and plain electrodes with capacitive coupling. The results were finally published with a detailed theory according for the discovered phenomenon [200]. The electron droplets, or plasma clusters, had been generated by the illumination from a focused laser beam (CW green laser, $\lambda = 532$ nm, $P = 100$ mW) of the $LiNbO_3$ crystal whose surface was covered with a thin metal film. The droplets with the size about 100 μm were stable and slowly moved (with the velocity about 0.5 $m \cdot s^{-1}$) a few centimeters across the charged exterior surface of the crystal before annihilation.

Thus the specular reflection of the focused laser beam had periodically generated bright droplets that moved along the crystal surface (Figure 5.16).

The creation and annihilation of plasma clusters, or droplets, generated nanosecond radio field pulses and (when placed in a modest vacuum) bursts of X-rays. The intensity of X-ray emission from ferroelectric crystals was strong enough to produce a shadow image on commercial dental X-ray films.

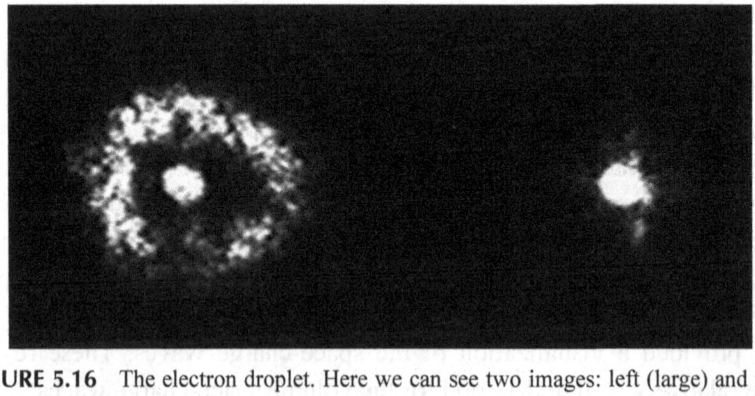

FIGURE 5.16 The electron droplet. Here we can see two images: left (large) and right (small). The right picture is the specular reflection from the surface exhibiting the droplet. This picture changes in time, as the droplet moves and slowly evolves (which is seen on our recorded video). The left picture depicts nonlinear scattering that originated mainly in the crystal volume. This left picture also pulsates in time and shows the dynamic of the volume space-charge waves.

Electron clusters named "EV" by Shoulders [192, 193] and "ectons" by Mesyats [196, 197] were clusters produced in plasma discharges. The important difference between their results and ours [200] is that in our case there was no externally applied electric field; the electric field was generated inside the crystal due to photogalvanic and pyroelectric effects induced by the laser illumination. So our clusters would be "photogalvanic ectons" (or "pectons").

Since in our experiments [200] the velocity of a relatively stable droplet was about 0.5 $cm \cdot s^{-1}$, which is very small, it seems that the droplet cannot be composed solely of electrons. At the same time the droplet must be charged, because it created the strong electric field (several kV/cm) needed to modulate the optical reflection from the crystal surface. The electron droplet was created in a triple point, i.e., the contact of crystal and metal with the air bubbles near the crystal surface. In our case the metal contact is due to the antireflection coating, and the electron clouds are formed due to both the photo ionization of the crystal surface and the electric field ionization of the air near the surface. Because of that, it is reasonable to assume that electron clouds may also include positive ions from the air, though their quantity seems rather unimportant.

Since the velocity of droplets revealed in the experiment is extremely low, we must conjecture that electrons in a droplet are found under very peculiar conditions, because as a matter of fact free electrons have never been observed at a velocity lower than 10^5 $m \cdot s^{-1}$. However, those charged droplets in fact were observed in our experiments. They moved longitudinally to the crystal surface, which in the present case of lithium niobate is characterized by a strong longitudinal electric field (several $kV \times cm^{-1}$) perpendicular to the axis of crystal polarization.

As known the Wigner crystallization [201] of electrons predicted in the 1930s received a remarkable support in the 1970s when electron crystals/clusters were experimentally observed on the surface of liquid helium bubbles (see, e.g., Ref. [202]). This allowed further theoretical studies, for instance, recent studies on electron solvation in molecular clusters [203]. How do electrons survive in a cluster? Electrons should be governed by a competition between the Coulomb repulsion and a kind of a confinement field; Wigner [201] associated this confinement field with the wave nature of electrons that obeys the Schrödinger equation. In the case of a crystal, such a field would be associated with phonons. But what is the origin of a confinement field in an electron droplet?

First of all we may assume that the laser beam along with the generation of photoelectric electrons also knocks a batch of inertons from the crystal, because the crystal is a strictor. Inertons driven out of the crystal envelope the released electrons, which result in binding the electrons together. The Hamiltonian of the system of binding electrons has to be specified by the repulsive and attractive pair potentials given by the expression (5.24). The repulsion is due to the Coulomb force and the attraction is due to the induced viscosity in the space filled with inertons between electrons. The statistical sum of the system of electrons interacting via these two pair potentials is presented in expression (5.25). Then we finally come to the action (5.59) that in the case of one cluster takes the form

$$S_1 \cong \left[a(\mathcal{N}) - 2b(\mathcal{N})\right] \mathcal{N}^2 - \ln \mathcal{N} + \mathcal{N} \ln \xi \qquad (5.92)$$

where the functions a and b are determined in relations (5.57) and (5.58), respectively.

It is obvious that the repulsive paired potential for electrons is

$$V^{\text{rep}}(\rho) = \frac{1}{4\pi\varepsilon_0} \frac{e^2}{r\rho} \qquad (5.93)$$

An elastic interaction of electrons through the inerton field, which plays here the role of a viscous substrate, is presented in the form of a typical harmonic potential

$$V^{\text{att}}(\rho) = \tfrac{1}{2} m^* \omega^2 (\delta r \rho)^2 \qquad (5.94)$$

Here, ρ is the dimensional distance; $m^* \omega^2$ is the elastic constant of the cluster substance where m^* is the effective mass of an electron in the cluster and ω is the cyclic frequency of its oscillations. Substituting calculated a and b into the action (5.92) we get the action in the explicit form

$$S_1 = \frac{3e^2/r}{8\pi\varepsilon_0 k_B \Theta} \mathcal{N}^{8/3} - \frac{3m\omega^2 \delta r^2}{5k_B \Theta} \mathcal{N}^{11/3} - \ln \mathcal{N} + \mathcal{N} \ln \xi \qquad (5.95)$$

The equation for the number of electrons in the cluster is $\partial S_1 / \partial \mathcal{N} = 0$, or explicitly

$$\frac{11m\omega\delta r^2}{5k_{\mathrm{B}}\Theta}\mathcal{N}^{8/3} - \frac{e^2/r}{\pi\varepsilon_0 k_{\mathrm{B}}\Theta}\mathcal{N}^{5/3} + \frac{1}{\mathcal{N}} - \frac{\mu}{k_{\mathrm{B}}\Theta} = 0 \tag{5.96}$$

where we put for the fugacity $\xi = \exp[\mu / k_{\mathrm{B}}\Theta]$.

Now we shall determine the value of the chemical potential. For a non-interacting Fermi gas the Fermi energy is (see, e.g., Kittel [180], Ch. 6, pp. 142–144)

$$E_{\mathrm{F}} = \frac{\hbar^2}{2m}\left(\frac{3\pi^2\mathcal{N}}{4\pi R^3/3}\right)^{2/3} \tag{5.97}$$

Since in our case $\mathcal{N} = 10^{10}$ and $R = 5\times10^{-5}$ m, $E_{\mathrm{F}} = 3.4\times10^{-23}$ J. In common metals at room temperature $k_{\mathrm{B}}\Theta = 0.01E_{\mathrm{F}}$ and μ is closely equal to E_{F}. However, in the electron cluster, which is at room temperature, we have an opposite relation $k_{\mathrm{B}}\Theta \approx 125\,E_{\mathrm{F}}$ (since $k_{\mathrm{B}}\Theta\,|_{300\,\mathrm{K}} = 4.25\times10^{-21}$ J). This means that electrons in the cluster are classical, though strongly interacting one with another. When electrons obey classical statistics, the value of $|\mu|$ is limited by the value of a few $k_{\mathrm{B}}\Theta$.

Thus neglecting the last term in the Eq. (5.96) and also the penultimate term $1/\mathcal{N}$ that is very small, we gain the solution to Eq. (5.96):

$$\mathcal{N} = \frac{10}{5}\,\frac{e^2/(4\pi\varepsilon_0 r)}{m^*\omega^2\,\delta r^2/2} \tag{5.98}$$

Since in the cluster electrons behave rather as classical charged particles, we could probably use the appropriate action (5.75) suitable for the classical statistics. Then we arrive at the solution below

$$\mathcal{N}_{\mathrm{class}} = \frac{10}{11}\,\frac{e^2/(4\pi\varepsilon_0 r)}{m^*\omega^2\,\delta r^2/2} \tag{5.99}$$

which has the same order as the solution for quantum statistics (5.98).

The lattice constant for the electron cluster is $\overline{r} = n^{-1/3} = 3.7\times10^{-8}$ m. Then the repulsion energy between two electrons separated by $\overline{r}$ is 7×10^{-20} J. The amplitude of oscillations of an electron near its equilibrium position should satisfy the inequality $\delta r \ll \overline{r}$ and hence can lie in the range 10^{-10} to 10^{-9} m. The parameter of viscosity, $\gamma = m^{*}\omega^{2}$, is given first of all by a defect mass Δm that an electron acquires from absorbing inertons. An order of the parameter ω is known, because in the experiment [194] radio-frequency signals, which were measured, fell exactly within the MHz region. Therefore we may put $\omega \sim 10^{7}$ s^{-1}. Then the effective mass of an electron is $m^{*} \approx (10^{6}$ to $10^{7})\, m_0$, which exceeds the rest mass of an electron millions of times.

Thus it is the heaviness of electrons in the droplet, which provides for the droplet stability. Due to the increased mass the kinetic energy of electrons in the droplet, which is proportional to $\omega \sim 10^{7}$ s^{-1}, is very small in comparison with the kinetic energy of free electrons and electrons in conductors and semiconductors where this energy is proportional to $\omega \geq 10^{15}$ s^{-1}.

Since the droplet can be formed only in the course of non-adiabatic processes, the time of formation of a droplet should be much less than the relaxation time of electrons in a metal ($t << \tau_{\text{relax.}} \leq 10^{-12}$ s) and the inverse Debye frequency ($t << \nu_{\text{D}}^{-1} \approx 10^{-13}$ to 10^{-12} s). The binding energy of an electron in the droplet can be evaluated as a composition

$$E_{\text{bond}} = \mathcal{N}\hbar\omega \sim 10^{-17}\ \text{J} \tag{5.100}$$

This energy is injected in the crystal through a surface area comparable with the size of the droplet by the laser beam with power $P = 100$ mW during a time of about $t \sim 10^{-16}$ s. This time satisfies the above inequalities.

An electron can be ejected from the droplet by an energetic ultraviolet photon with the frequency $E_{\text{bond}} / h = \nu_{\text{ph}} \approx 10^{16}$ s, or $\lambda_{\text{ph}} = 30$ nm. When knocked out, the electron loses its heavy mass. The mass defect $\Delta m = m^{*} - m_0$ turns into inerton(s). The rest mass m_0 of the electron is restored and the inertons emitted.

Since electron droplets consist of a massive substance, they are able to behave as conventional matter droplets that are studied in the framework of fluid, or hydrodynamic mechanics. That is, moving electron droplets change their shape with time and the hydrodynamic laws indeed prescribe such behavior of droplets (see, e.g., Refs. [204–206]).

Thus, the theory of an electron droplet as presented enables a natural description of the electron's confinement in terms of the submicroscopic concept in which the basic role is played by inertons – carriers of inertia, mass and the fractal properties of the tessellattice. These inertons accumulate a quite high energy in a droplet.

5.6.1 INERTONS AS A BINDER OF BOTH CHEMICAL BONDS AND SUPERCONDUCTING ELECTRONS

Putz [207] has studied the chemical bond between elements and in particular between an electron and an atom starting from the de Broglie-Bohmian quantum formalism with the $U(1)$ and $SU(2)$ gauge transformations of the wave-function and the relativistic spinor, within the Schrödinger and Dirac quantum pictures of the motion of an electron. By using such an approach, he showed that the chemical bonding could be described in terms of chemical bonding quanta $\hbar c/e$. These bondon quanta represent a chemical bonding field and are characterized by a mass (m_{B}), velocity (υ_{B}), charge (e_{B}), and life-time (t_{B}). In this approach the wave ψ-function is subdivided to a core and a tail, which definitely means an implicit involvement of inertons in the consideration of bound particles. Hence the description of a chemical bond enables a substructure, i.e., bondons, whose origin is inertons of interacting particles and chemical elements. Putz's hypothesis then was developed in more detail especially in applications to nanosystems [208–210].

The submicroscopic concept allows us to look deeply inside the quantum phenomenon known as superconductivity. In a superconductor there is no resistance because current electrons are in a specific bound coherent state in which they are not scattered by the crystal lattice and its defects. The coherent state of electrons is associated with the so-called Cooper pairs. The mechanism by which negatively charged electrons are bound together is still not well understood for modern superconducting systems such as the copper oxides or alkali metal fullerides, though it is considered to be quite clear in low-temperature metal superconductors to which the Bardeen-Cooper-Schrieffer (BCS) theory is applied.

In the superconductor state the binding energy of a pair of electrons features a gap in the energy spectrum at the Fermi energy, which separates the paired states from the normal state of single electrons. The size of a Cooper pair is given by the coherence length $l \sim 100$ nm for metal, which is about 3 nm in copper oxides. In the BCS theory, the electrons are bound together by their interaction with the vibrations of the crystal lattice.

However, the space between nodes of the crystal lattice is filled with the crystal's inertons. That is why from the submicroscopic point of view electrons can gather together when they involve inertons, which has been demonstrated in several cluster models above and, particularly, an electron droplet. So, behind the BCS theory the clustering of electrons with the participation of the crystal's inertons must take place. This means that the coherent length l of the BCS pair becomes the diameter of a cluster of electrons and then realistically we shall talk not about a coherent pair of electrons, but about an electron droplet of many electrons collected in a cluster. Then the many-electron BCS wave function becomes a function that describes the cluster of $\mathcal{N}$ electrons and the number $\mathcal{N}$ becomes the solution of the appropriate equation as, for example, the Eq. (5.96) for the electron droplet that has been described above.

We have only expressed here a possible submicroscopic mechanism of superconductivity. A detailed development will require separate studies both theoretical and experimental.

5.7 THE PHENOMENON OF DIFFRACTIONLESS LIGHT

The problem of diffraction is not as simple as it is presented to students at universities and described in the appropriate textbooks. A quantum theory of diffraction of light was raised by Epstein and Ehrenfest [211, 212] in 1920s and the next paper on the theory only appeared in 2010 [213]. Ehrenfest and Epstein [212] finished the second paper with the confidence that some kinds of diffraction, e.g., the Fresnel ones, could not be explained by purely corpuscular considerations and that essential features of the wave theory in a form suitable for the quantum theory would be needed. They believed that quanta of light should feature phase and coherence similar to the waves of the classical theory. Furthermore, they assumed that the papers on modern quantum mechanics by de Broglie and

Schrödinger, which had just appeared at that time, would bring researchers much nearer to solving the problem.

As we know the problem was resolved by introducing an undetermined notion of "wave-particle," though Louis de Broglie, the 'father' of quantum relationships $E = h\nu$ and $\lambda_{dB} = h / (m\upsilon)$ for a particle was against such unification. Nevertheless, by using this strange mutant called the wave-particle duality, physicists were able to describe quantitatively many phenomena.

However, there is one more, a very natural way of solving the problem, which implies an intervention of the environment on passing photons. Perhaps many in childhood studied various optical phenomena and, in particular, a change of focus for a light beam passing through the gap created by the thumb and forefinger of both hands. If the photon is a corpuscle, then surrounding walls of the gap could influence the photon via an induction of some field. For instance, such a type of field every child (i.e., also a physical corpuscle) feels when swinging on a swing or a carousel. Hence, vibrations in the surrounding medium could be able to affect a beam of photons. In other words, periodic processes that occur in the surroundings would superimpose on the rectilinearly travelling photons. Then these photons could be perceived as wave-particles.

Let us look now at rigorous experiments that tested the phenomenon of diffraction of light. Panarella [214] wrote a wonderful review paper dedicated to the experimental testing of the wave-particle duality notion for photons. He reviewed the results of many researchers and also presented his own data and analysis. In particular, he emphasized that his experimental results brought new evidence that a diffraction pattern on a photographic plate was not formed when the intensity of light was extremely low, even when the total number of photons reaching the film was larger than that which was needed to form a clear diffraction pattern. Hence it was established that a diffraction pattern did not follow the linear principle with decreasing light intensity, as the wave-particle duality required. He obtained the same results by using photoelectric detection and oscilloscope recording of the diffraction pattern.

In particular, Panarella noted that with a flux (generated by an optical laser) of around 10^{10} statistically independent photons per second in the interferometer, a clear diffraction pattern was recorded on the oscilloscope. At a photon flux of around 10^8 photons per second, no clear diffraction pattern appeared. Further decrease of the intensity showed an increase of nonlinearity in the behavior of photons. Moreover, a flux in the interferometer of 10^4 photons per second clearly demonstrated a single particle phenomenon – no diffraction at all. Analyzing the experiments of previous researchers who dealt with fluxes of only tens of photons per second, Panarella rightly intimated that they were unable unambiguously to determine whether their sources of light produced individual/single photons or the sources produced packets of photons.

Panarella [214] concluded: "The series of experiments reported here on the detection of diffraction patterns from a laser source at different low light intensities instead of confirming the wave nature of collections of photons tends to dispute it, or provides no clear proof of it, for single photons."

Further on, Panarella tried to account for the absence of the diffraction at the low intensity of single photons by suggesting a "photon clump" model in which he hypothesized a possible interaction between single photons in a low intensity photon flux, which gathered photons in clumps, such that they did not show wave properties at the diffraction. However, such a hypothesis raises a question about the inner nature of such interaction (sub-electromagnetic interaction between photons?).

Dontsov and Baz [215] were probably the first who observed and reported the absence of the diffraction phenomenon at an extremely low intensity of statistically independent photons, namely, 100–200 photons per second, using a monochromatic source of light. They reported that at such a low intensity of statistically independent photons, which passed through a Fabry-Perot interferometer, photons did not form the interference pattern. However, when the intensity of photons was so large that photons emitted by a lamp were in correlated states, the interference pattern was quite distinguishable.

So, the phenomenon of diffraction is absent in the case of low intensity single photons. For single photons the "wave-particle" does not work. Does this mean that the mutant wave-particle is not real? Let us consider this phenomenon in detail [216].

A photon is an excitation of the tessellattice that migrates by hopping from cell to cell. A photon travelling in an interferometer must interact with the interferometer's substance (lenses, mirrors, etc. and a foil with a pinhole or slit). In the interferometer the density of the substance influences the photon through its refractive index n, such that the wavelength of the photon $\lambda \rightarrow \lambda n$. Besides, the photon scatters by some nodes. The scattering momentum vector $\hbar \mathbf{q}$, which acquires the crystal lattice, is defined as the difference between incident and scattered propagation momentum vectors of the photon

$$\hbar \mathbf{q} = \hbar \mathbf{k}_{\text{incid}} - \hbar \mathbf{k}_{\text{scat}} \tag{5.101}$$

where $\mathbf{k} = \mathbf{e} n 2\pi/\lambda$ and $\mathbf{e}$ is the unit vector showing the direction of motion, λ is the photon's spatial amplitude (widely known as the photon wavelength) and n is the refractive index. The angle between $\mathbf{k}_{\text{incid}}$ and $\mathbf{k}_{\text{scat}}$ is called the scattering angle θ. From relation (5.101) we obtain

$$\begin{aligned} q^2 &= k_{\text{incid}}^2 - k_{\text{scat}}^2 - 2\mathbf{k}_{\text{incid}} \cdot \mathbf{k}_{\text{scat}} \cong 2k_{\text{incid}}^2 (1 - \cos\theta) \\ &= 4k_{\text{incid}}^2 \sin^2(\theta/2) \end{aligned} \tag{5.102}$$

This acoustic excitation is a typical non-equilibrium acoustic phonon with a cyclic frequency Ω much less than $10^7\,\text{s}^{-1}$ (the so-called basic acoustic scattering of light), because excitations with higher frequencies fall within a range of the Brillouin light scattering. As Wolfe [217] noted: "There are myriad ways to generate non-equilibrium phonons in crystals." So due to the dispersion equation $\varpi(q) = \upsilon_{\text{s}} q$, where υ_{s} is the sound velocity, and owing to a very small value of the angle θ, we can write expressions for the wave number and cycle frequency of acoustic phonons:

$$q = 2\pi n\theta/\lambda, \; \varpi = 2\pi \upsilon_s n\theta/\lambda \tag{5.103}$$

These non-equilibrium phonons are the major subject of our study. Entities in condensed media vibrating near equilibrium positions create inerton clouds that accompany the entities. Hence non-equilibrium acoustic excitations (which are phonons in the k-space) are also accompanied with their inerton clouds. In the time τ the acoustic excitations decay and hence all their inertons shall be irradiated, as free inertons, roughly in a steric angle 4π.

These non-equilibrium acoustic phonons can be formed in optical parts of the interferometer (lenses, mirrors, etc.) or/and in a foil with a pinhole or slit. The mentioned relaxation time τ of generated acoustic excitations is about 10^{-11} s in a metal [218] and 10^{-10} to 10^{-8} s in semiconductors and dielectrics [219–221].

In the photon flux the forward photons, which generate free inertons into the interferometer, are able to affect subsequent photons by means of the emitted inertons. The pictures below clearly demonstrate this mechanism.

A similar situation takes place in a foil near the pinhole. Photons bombard the foil and generate non-equilibrium phonons that then relax in time τ generating free inertons. Some of these inertons intersect the photon flux

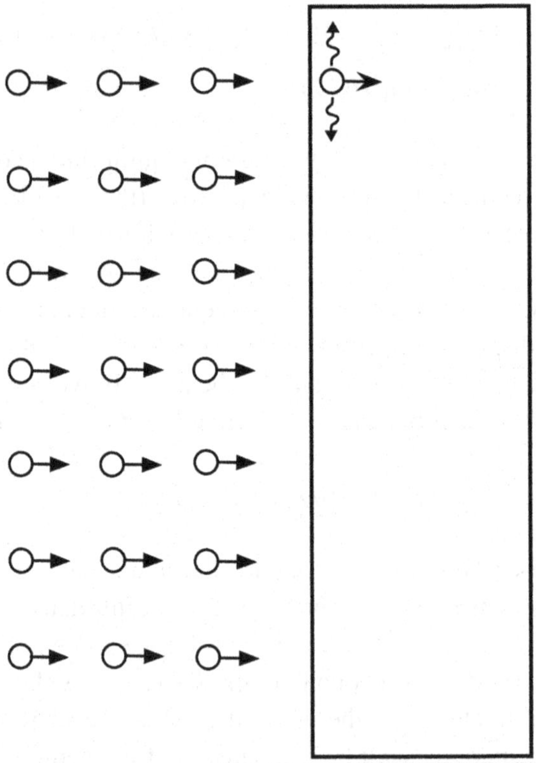

FIGURE 5.17 The first photon enters the interferometer. The photon creates the acoustic excitation that in turn generates its inerton cloud (non-relevant photons are shown before the interferometer).

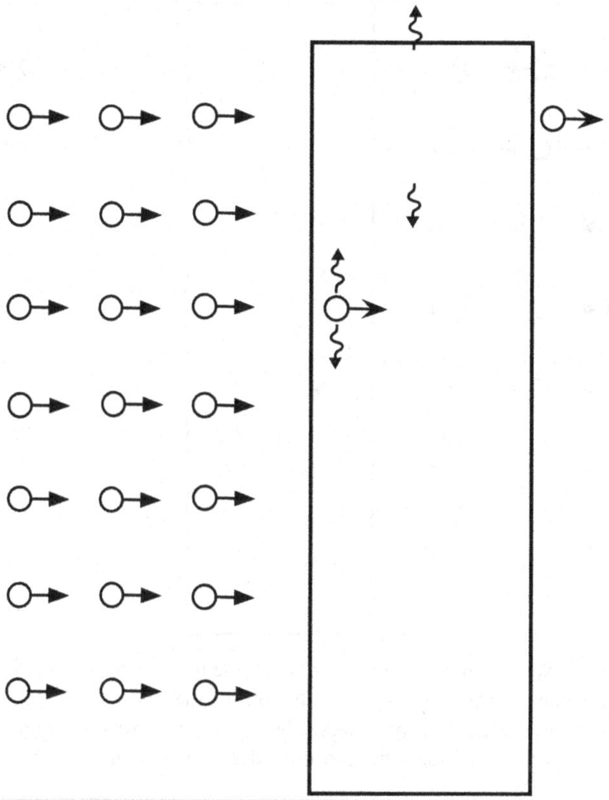

FIGURE 5.18 The first photon leaves the interferometer. Its non-equilibrium phonon has decayed and released free inertons that run around the approaching the path of the second photon. The second photon just enters the interferometer; the photon creates the appropriate acoustic excitation, which generates its own inerton cloud.

in the pinhole, which deflects photons from their direct line motion to the target. (We can mention in passing that a couple of incident photons, among which one is polarized in the horizontal mode and another in the vertical mode, i.e., the so-called "entangled photons", could generate a weaker phonon excitation in the interferometer, which finally would result in a clear diffraction patter.)

The values of N_i used by Panarella [214] were $N_1=10^{10}$, $N_2=10^8$, and $N_3=10^4$ photons per second. In subsequent work [216] the concentration of photons in the focal volume was evaluated for further analysis. However,

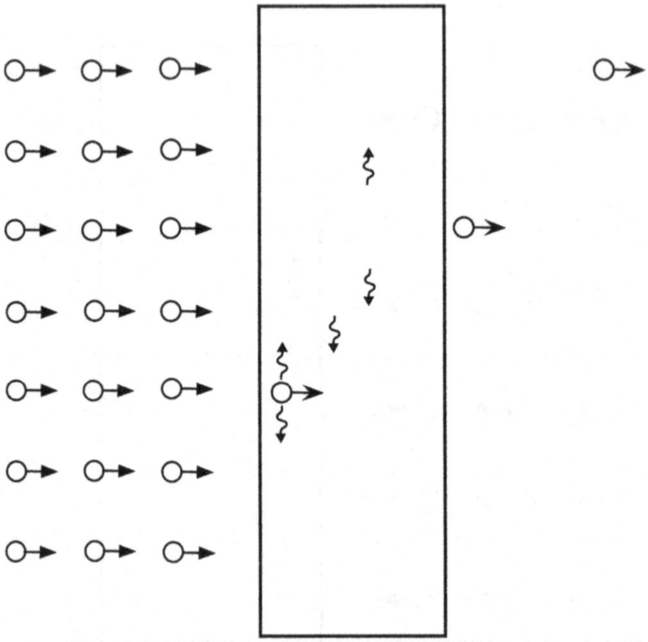

FIGURE 5.19 The two photons have already left the interferometer: the second one has experienced a sideways action through the inertons of the first photon. The third photon just enters the interferometer; it experiences sideways action through inertons generated by the two previous photons (through the respective decayed phonons).

we may directly roughly estimate the appropriate time interval t_i between traveling photons:

$$t_1=10^{-10}\text{ s},\ t_2=10^{-8}\text{ s, and } t_3=10^{-4}\text{ s} \tag{5.104}$$

Now let us compare the calculated time intervals (5.104) with the relaxation time τ of non-equilibrium acoustic phonons. Why is it interesting to compare t and τ? Because of the change of the inequality $t<<\tau$ to the opposite one is able to dramatically change the physical pattern of the phenomenon of diffraction.

The lifetime τ of non-equilibrium phonons for dielectrics, as mentioned above, varies from 10^{-10} to 10^{-8} s [219–221] and the inequality $t<\tau$ or $t\lesssim\tau$ is held. However, in the case of a very low intensity ($N_3 = 10^4$) the inequality becomes opposite

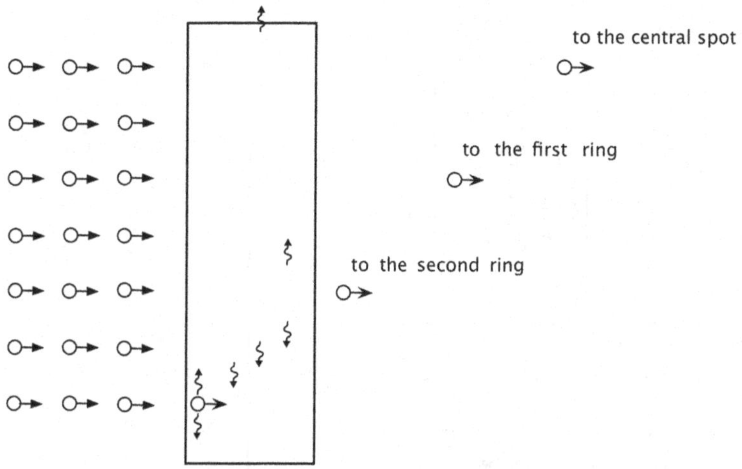

FIGURE 5.20 The three photons have already left the interferometer and the fourth photon that has entered the interferometer undergoes sideways action from three flows of inertons generated by the previous photons. The first three photons follow their own trajectories: (i) the first one, which has not been affected by inertons, moves to the center of the target; (ii) the second photon, which was influenced by the first photon (through inertons of the appropriate phonon), proceeds to form the first ring of the Airy diffraction pattern; (iii) the third photon, which underwent the influence of the double flow of inertons (from the two first photons), is deflected to form the third ring of the Airy pattern, and so on...

$$t_3 >> \tau \tag{5.105}$$

The inequality (5.105) means that each following photon will arrive at the interferometer at the moment when inertons generated by the previous photon will no longer be there. Therefore, the next photon does not experience a transverse action and will continue to follow its path to the central peak on the target. The inequality (5.105) holds for the case of the lowest intensity of photons, $N_3 \approx 10^4$. Hence the mechanism described is capable to account for Panarella's experiments in which the diffraction fringe was absent.

The distribution of photons into the rings of the diffraction pattern is described in classical optics [222]: the first subsidiary maximum should have an amplitude 0.0175 times the amplitude of the central peak; the second subsidiary maximum has an amplitude 0.0042 times the central amplitude. These results point out that the intensity of transverse inerton

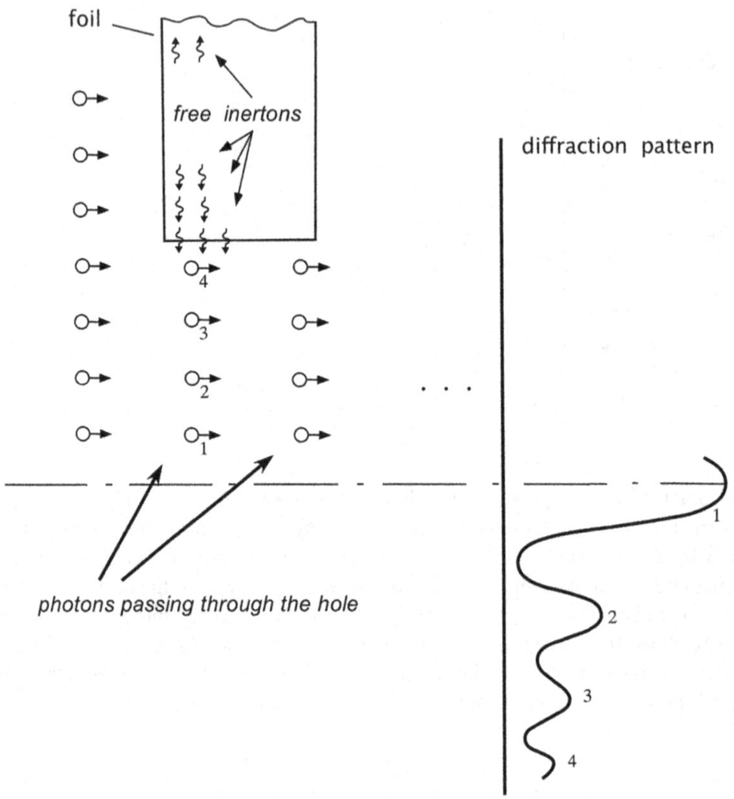

FIGURE 5.21 Free inertons, which have been released after decay of non-equilibrium acoustic phonons, leave the foil intersecting the pinhole where they affect the flux of photons. This lateral inerton wind blows the photons away resulting in the formation of the diffraction pattern on the target: photons 1, 2, 3, and 4 form respectively maxima 1, 2, 3, and 4 (2, 3, and 4 are affected by inertons).

flows in the interferometer, which deflects photons from their direct line to the central peak, is not negligible in the case of a comparative high intensity N of the photon flux. What is the reason for such perceptible intensity of inertons?

If the energy of an incident photon is $h\nu$, then the energy of the acoustic excitation produced by the photon is $\hbar\varpi \approx h\nu \cdot \upsilon_{\text{sound}} / c \approx 10^{-5}\, h\nu$. The energy $\hbar\varpi$ is quenched during the time τ and inertons emitted at the phonon decay carry away an energy no more than $10^{-5}\, h\nu$. This value of energy is not enough to deflect a subsequent photon from the direct line;

this would simply spread the width of the central spot from the diameter d_0 to $(1+10^{-5})\cdot d_0$.

However, the coefficient of transparency for glass may vary from 45 to 95%, which means that the lost intensity goes to the creation of a number of non-equilibrium acoustic phonons. The number of these phonons may reach thousands, which is quite enough to intensify flows of free inertons in the interferometer and the foil with the pinhole or slit.

We also have to discuss experiments of other researchers who were searching for the reduction of optical interference effects at low light levels, but did not reveal any changes in the interference pattern.

Jeffers and collaborators [223, 224] tried to repeat the Panarella's experiments [214]. They conducted a similar series of Panarella's low-intensity diffraction experiments using two different optoelectronic detectors. The lowest intensity of a photon flux reached by Jeffers et al. was the same as Panarella's, i.e., 10^4 photons/second. However, in their experiments the diffraction pattern did not disappear on decreasing the intensity of a photon flux down to the value of 10^4. The same was reported by other researchers (see, e.g., Ref. [225]): they did not reveal changes in a classical diffraction pattern at a low intensity of photons.

So, why the results are so different and which is really true? The answer is simple, but requires the use of submicroscopic concept. Panarella's [214] pinhole from which photons proceed to the target had the diameter 50 μm. Dontsov and Baz [215] did not mention the size of their exit slit used in their Fabry-Pérot interferometer, but it seems it was not too narrow. The width of a typical averaged exit slit or aperture in such an interferometer is 30 μm.

Jeffers [226] started his experiments with an intensity of a photon flux around $N = 10^{10}$ photons per second and with the width of an exit slit about 100 μm. However, below $N = 10^7$ he narrowed the exit slit, such that it became exactly equal to the width of the diffraction central maximum at its base. This value of the slit width was 4.8 μm [224].

The mentioned scales of holes completely clarify the situation. Let us look at the relation (5.15). It indicates the distance to which an inerton cloud of an atom under consideration spreads. This distance is equal 3 to 4 μm. Therefore a metal plate in which the hole was cut is able to induce a fairly powerful inerton field inside the hole. The thickness of this inerton interface is 3–4 μm. In the Panarella's experiment the area of the hole covered by the metal's inerton interface is small or even negligible. The same has to be correct for the experiment of Dontsov and Baz [215].

However, in the Jeffers' experiment the width of the slit was completely covered by the inerton interface induced by the surrounding material(s). The inerton interface is a continuation of the solid, as inerton clouds of atoms are directed beyond the material surface. Therefore, the interface was following oscillations of the body's atoms. When photons, which move to the target, pass through the oscillating interface, they are scattered by the inerton clouds and as a result the momentum of photons acquires an additional component:

$$\mathbf{k}_{\text{incid}} + \mathbf{Q} \rightarrow \mathbf{k}_{\text{scat}} + \mathbf{Q}' \tag{5.106}$$

As can be seen from the Eq. (5.106), the scattered photon is specified by the new momentum. Due to the geometry of the hole, the inerton interface oscillates in the plane of the material and hence the photon's new momentum possesses a transverse component that will decline it to the right or to left from the direct line of motion. The structure of the $\mathbf{k}_{\text{scat}}$ is as follows:

$$\mathbf{k}^2_{\text{scat}} = k_{\|}^2 + k_{\perp}^2, k_{\perp}^2 = \Delta Q^2 \tag{5.107}$$

where ΔQ is the momentum passed to the photon by the inerton.

In conclusion it should be emphasized that collisions of photons with a surface (initially the surface of materials made of glass in the interferometer) within a time interval can be considered as a quasi-periodical process, or modulation, which is especially true in the case of a laser beam. The relaxation time τ for non-equilibrium acoustic phonons of glass materials is a stable parameter. Until $t < \tau$, or for the corresponding rates $N > \tau^{-1}$, the process of relaxation plays a dominant role in the phenomenon of diffraction. However, when the arrival rate of photons becomes less than the relaxation rate, i.e., when $N < \tau^{-1}$, the mechanism of diffraction changes (see the relation (5.105)), namely, the diffraction disappears from a large hole/slit and continues to be present from a small hole/slit caused by vibrating atoms inside the material. Thus there are three characteristic parameters: the rate of photons N, the relaxation time τ of phonons in the interferometer's materials, and the width l of an exit hole. The superposition of two quasi-stationary processes (which are characterized by parameters N and τ) is able to produce some phenomena, such as the diffraction, which masquerade as a purely natural wave appearance. In the case of a low intensity photon flux and a small hole or narrow slit, the diffraction phenomenon is provided by an inerton interface that is present inside the hole.

5.8 DOUBLE-SLIT EXPERIMENT: SOLVING THE PROBLEM

In 1803 Thomas Young performed his famous double-slit experiment (strictly speaking, a double hole experiment), about which he reported as "the effect of the interference of double lights" in his lectures of 1807 [227]. Since then many researchers all around the world have repeated the experiment many times providing it with newer and newer increasing modifications. An important modification is an increase of the number of slits from initial two to 7, 10 and more.

In modern double-slit experiments with light the slit width equals a few tens of micrometers and the width of a column between the nearest slits has the same order of magnitude. Though the exit slits are quite wide, the diffraction fringes are still preserved even at a low intensity of light, 10^5 photons per second (see e.g., Ref. [228]).

To explain the diffraction fringes in the double-slit experiment, the researchers [229] considered the interference pattern as a result of the internal operation of the detector because they believed that nothing else could cause the interference pattern to appear. In their corpuscular approach, the interference appears as a result of processing individual events, but not of "wave-like" ingredients. In their model, each particle carries its own local oscillator. This oscillator only serves to mimic the frequency of the individual particle (photon). The particle hits the detector, the detector "observes" the state of the oscillator that is attached to this particular particle and determines its time-of-flight, which is not reduced to the interference by summing wave functions like $a_k e^{-iwt_k}$ where t_k is the transit time of the particle.

Kolenderski et al. [230] experimentally studied the time-resolved double-slit interference pattern generated by entangled photons. The two sets of fringes were extracted from the same measurement data, which means that the corpuscular model of the double-slit experiment by Jin et al. [229] gave an alternative description for the buildup of the interference pattern (and hence the entanglement resided outside of the scope of the corpuscular model).

Shafiee et al. [231] considered a particle associated with a field partaking of the energy of the particle, which are both described by deterministic causal equations of motion. After the passage through one of the slits, the particle shares some of its energy with its surrounding field and a particle-field system is again formed and its motion is governed by a deterministic dynamics during the motion of the particle towards the screen. The interference pattern is explained by showing how the final location of each particle-field system is distributed according to an angular distribution at the screen.

We have mentioned works [229, 231] as their authors approached closely to the solution of the double-slit diffraction phenomenon. In this phenomenon the interface inertons, which have been discussed in the previous section, play rather a minor role. The major role belong to inertons

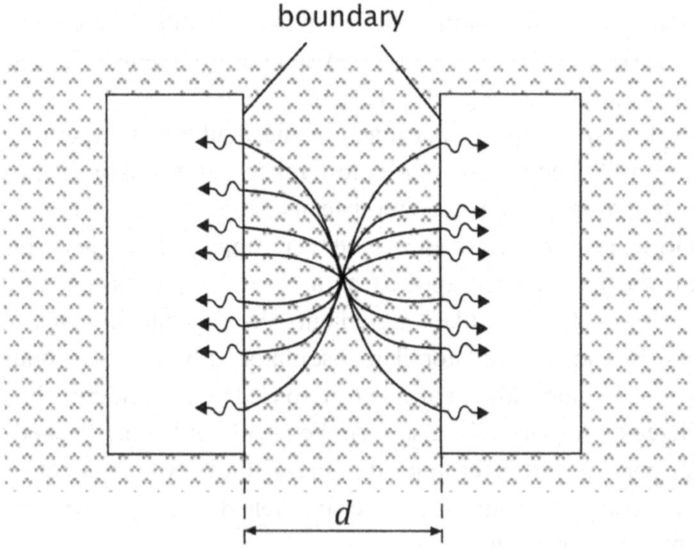

FIGURE 5.22 Diffraction grating. The two nearest slits are separated by a column with the width *d*. The standing wave and its harmonics present in the column are shown. Flows of inertons that leave the column and move in the plane of the slits are depicted with wavy arrows.

generated by the material from which the diffraction grating is made, which deflect photons from the direct line to the target. A mechanism for this is as follows. Let us look at a diffraction grating in which the column width between two slits is equal to d (Figure 5.22). The edges of the column are boundaries between the solid material and the very tenuous air. These boundaries are not fixed; they are free. Therefore the column shall vibrate with a resonance frequency defined by the column width d and the grating material. The wavelength of the main standing acoustic wave is $\lambda/4 = d/2$. The node of this wave is located inside the column and the antinode is located on the edges of the column. The temperature supports these oscillations that also have a set of harmonics. At the edge of the column the energy of this standing acoustic wave is released into the air. However, the acoustic wave, i.e., phonons with the wave number $k = 2\pi / (2d)$, is accompanied with its cloud of inertons. At the boundary the wave number k is transferred to k_{air}, but for the inerton cloud the conditions are rather different. The acoustic wave impacts the boundary and the wave's inerton cloud is stripped from the column. So the stripped inerton cloud continues to move along the initial vector **Q** inside the slit and then is absorbed by the diffraction grating or goes away from the system under consideration.

Thus in the case of the double-slit experiment each slit is almost completely filled with inertons released from emitted standing acoustic waves. These inertons travel in the plane of the slit and hence photons passing through the slit will be scattered by inertons. In this situation we have now found the basic reason of the problem: Why do photons passing through the double-slit diffract? Within the slit the photons acquire a transverse component to their momentum due to being scattered by inertons that travel in the transverse direction across the plane of the slit.

Let us estimate the parameters of inertons moving across a slit. If the column width $d = 20\ \mu$m, then the standing acoustic wave has a wavelength $\lambda = 40\ \mu$m, the energy of the wave is $\hbar\varpi = \hbar\upsilon_{\text{sound}} / (\lambda / 2\pi)$ and the number of the appropriate generated acoustic phonons is in the order of $k_{\text{B}}\Theta / (\hbar\varpi) \sim 10^5$ if we put $\upsilon_{\text{sound}} \sim 10^3$ m·s^{-1}. Hence in the slit there are constantly no less than 10^5 inertons with the momentum $\propto \lambda_{\text{ac}}^{-1} \sim [(4 \cdot 10^{-5})]^{-1}$ m. In principle the number of generated phonons may be up to ten times larger owing to a series of harmonics. The momentum of optical photons $\propto \lambda_{\text{phot}}^{-1} \sim [(4-7) \cdot 10^{-7}]^{-1}$. Hence $\lambda_{\text{ac}}^{-1} / \lambda_{\text{phot}}^{-1} \sim 0.01 - 0.1$, which means that over a long period of time the inertons will be able to deflect a sufficient quantity of photons at any intensity of the photon flux, which is required to create an interference pattern on the target.

Thus we can now explain the underlying mechanism of Huygens' Principle. In fact in 1678 Christian Huygens proposed that every point of a wave front of light acts like a new source of a spherical light wave. He did not explain the deviations from rectilinear propagation when light encounters edges, apertures and other obstacles, which we call the phenomenon of diffraction. Modern physics has also not gone deeply into the causes of the diffraction phenomenon; it is enough to use the Huygens's Principle: at the edge of a slit a straight beam of light suddenly begins to move as a spherical wave. In 1927 Ehrenfest [212] suggested that diffraction could be explained by a unification of the wave and particle properties of photons. However, why does the wave-particle suddenly change its trajectory inside a slit? Until now no idea has been expressed. So far the reason, or the mechanism of interaction of the particle and the slit, still has remained a puzzle.

The angle of deviation from normal incidence of light is defined by the diffraction law derived by W. H. and W. L. Bragg, and independently by Yu. V. Wulff in 1913.

So, the nature of the appearance of an angle of deviation is the key point of the diffraction phenomenon. Figure 5.23 sheds light on the emerging the angle of deviation θ in a slit as the result of a direct collision of passing photons with a transverse flow of inertons.

If the scattered photons interfere on the target effectively, they should remain in phase since the path length of each wave is equal to an integer multiple of the wavelength. In the Bragg's law the path difference between two waves that have undergone interference is given by $2d \sin \theta$, i.e., $2d \sin \theta = n\lambda$ where $n = 0, \pm 1, \pm 2, \ldots$ Further formalism involves the distribution of intensities on the target,

$$I = I_0 \frac{\sin^2(\bar{N}\pi d \sin \theta)}{\sin^2(\pi d \sin \theta)} \tag{5.108}$$

which in turn is complicated by a modulating multiplier $\sin^2(\pi d \sin \theta) / (\pi d \sin \theta)^2$ and then the total intensity expression is used

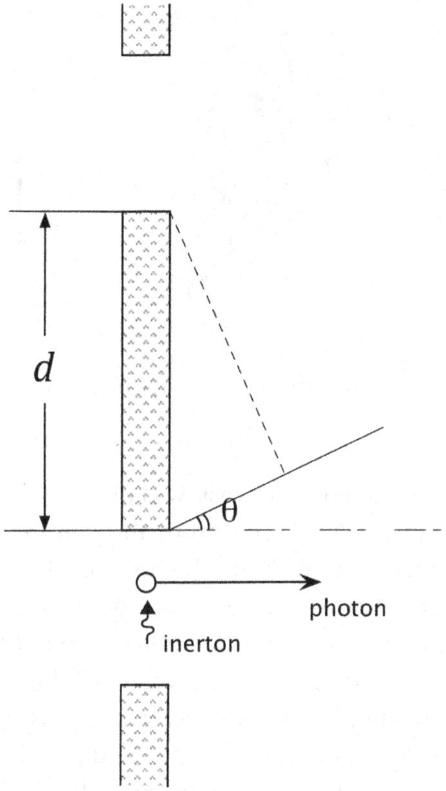

FIGURE 5.23 Appearance of the angle of deviation θ of light (the scattering angle) from the normal incidence. Within the slit inertons scatter photons by passing them a transverse component of the momentum.

for the comparison with an observed pattern (here, $\bar{N}$ is the number of slits in the diffraction grating).

The double-slit experiment was also conducted with electrons [232, 233] (Figure 5.24): the authors demonstrated diffraction from single, double, and multiple micro-slits. The parameters were as follows: the slit width was 0.6 μm, the grating constant 2 μm, and the electron's de Broglie wavelength $\lambda_{\mathrm{dB}} = 6.1\times10^{-12}$ m [232]; the slit width was 0.3 μm, the grating constant was 1 μm, and the electron's de Broglie wavelength $\lambda_{\mathrm{dB}} = 5.6\times10^{-12}$ m [233].

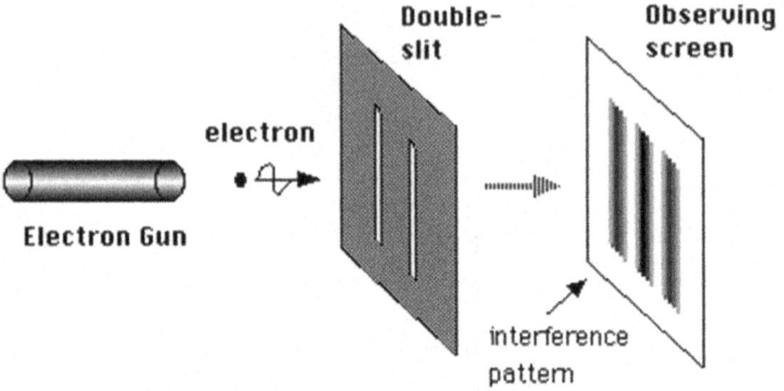

FIGURE 5.24 Double-slit experiment with electrons (from the web).

In recent work [234] a movable mask was placed in front of a double-slit to control the transmission of electrons through the individual slits. Bach et al. [234] observed probability distributions for single- and double-slit arrangements. They were able to record single electron detection events diffracting through a double-slit and built up a diffraction pattern from individual events. The parameters were as follows: the slit width 62 nm, the column width 210 nm; the electron's de Broglie wavelength $\lambda_{\rm dB} = 5.2\times10^{-11}$ m, velocity $\upsilon = 1.4\times10^{7}$ m·s^{-1} and hence the amplitude of the inerton cloud $\Lambda = \lambda_{\rm dB}\, c/\upsilon \approx 1.12$ nm (in papers [232, 233] $\Lambda \le 10^{-10}$ m). Hence the size of an electron together with its inerton cloud was over 50 times the size of the slit.

So in the experiments [232–234] electrons were rather tiny objects inside the slit and hence they were not able to interact with the slit because the electron's inerton cloud having a short amplitude Λ was unable to touch the walls of the slit. But only feedback on the side of the wall is able to turn the electron out of the strait line of its initial path. Accordingly it was those inertons emitted from the columns that separated the slits, which deflected the electrons from their rectilinear motion, as has been accounted above for the double-slit experiment with light.

The double-slit experiments with neutrons were reviewed by Zeilinger et al. [235] and they also reported their own results on the diffraction of cold neutrons at single- and double-slits that had widths 20 to 100 μm. The authors emphasized that their double-slit diffraction experiment is its most precise realization hitherto for matter waves, because the experimental results show an ideal coincidence with the calculated theoretical diffraction patterns. The theoretical predictions used a mathematical formalism to describe the emergence of the diffraction pattern. The formalism utilizes the diffraction angle θ as a given parameter, which has been clarified in the paragraphs above.

What was the real mechanism that resulted in the deviation by an angle θ of neutrons inside the slit? The wavelength of neutrons studied was about 20 nm. Hence their speed was 2000 m·s^{-1} and the amplitude of the neutron's inerton cloud was $\Lambda = \lambda_{\rm dB} c / \upsilon \cong 3$ mm. Such macroscopic value of Λ, which about two orders of magnitude exceeds the slit width, allows the neutron to directly interact with the walls of the slit. Since the inerton cloud of the moving neutron touches the walls it follows that if the neutron does not move exactly along the central axis of the slit, the walls will elastically reflect it. The extended inerton cloud behaves like a kind of a tentacle of every moving particle. In such a way neutrons automatically acquire their major contribution to a scattering angle, θ_1.

In the experiments [235] an approximate width of the column that separated the slits could roughly be estimated as 50 μm and hence the generated standing acoustic wave should have the momentum $\propto \lambda_{\rm ac}^{-1} \sim [(5 \cdot 10^{-5})]^{-1}$ (this value can be large owing to a few shorter harmonics). The momentum of a neutron is $\propto \lambda_{\rm dB}^{-1} \sim [(2 \cdot 10^{-9})]^{-1}$. Hence a ratio $\lambda_{\rm ac}^{-1} / \lambda_{\rm phot}^{-1} \sim 4 \cdot 10^{-5}$, which is very small. However, the number of phonons that emitted these inertons is $k_{\rm B}\Theta / (\hbar\varpi) \sim 10^5$ and hence the intensity of inertons is also about 10^5. This means that inertons emitted by the column in the plane of the slit by acoustic phonons are able to pass a diffraction angle θ_2 to photons that go through the slit.

Thus two contributions discussed above form the total diffraction angle $\theta = \theta_1 + \theta_2$ that further leads to the classical pattern of quantum diffraction that was demonstrated in work [235].

Zeilinger et al. [235] suggested that any proposals for alternative theories should be checked in great detail against their experimental evidence. Now the mentioned researchers and their followers have the opportunity to express an opinion regarding the presented submicroscopic concept of physics, which at last has disclosed the underlying mechanism for the phenomenon of diffraction that they have been observing for decades.

The double-slit experiment has also been tested with different atoms and more recently with molecules [236–240]. Among molecules, which were used for the double-slit experiments, there were C_{60} and C_{70} fullerenes, tetraphenylporphyrins, PFNS10, carbon nanospheres with ten perfluroalkyl chains, structural isomers TPPF152, and a TPP derivative with 152 fluorine atoms. The size of the molecules studied varied from 1 to 10 nanometers, and the molecular configuration varied from chains and cruciform to spherical. All the entities studied created an interference pattern that sometimes looked quite complicated.

The researchers [236–240] noted that all theoretical curves, which were calculated, included an averaged measured velocity distribution in the molecular beam studied. In particular, in their calculations they took into account the interaction of particles in the beam, namely: a quantum wave model including the attractive van der Waals interaction between the polarizable C_{70} molecule and the gold grating wall and the van der Waals attraction by the asymptotic form of the Casimir-Polder interaction.

Typical widths of exit slits used in the diffracting gratings varied from hundreds to tens of nanometers. Since the slit width is small, every molecule that passes through the slit has to be strongly affected by the inertons from columns that separate the slits. Besides, since the velocity of large molecules in a molecular beam is low, the molecules will have extended inertons clouds as in the case of cold neutrons described in previous paragraphs. For example, in the case of fullerene molecules C_{60} [236]: $\lambda_{\rm dB} = 2.5 \times 10^{-12}$ m, the maximum velocity of molecules $\upsilon = 220$ m·s^{-1} and hence the amplitude of inerton cloud of a molecule C_{60} is $\Lambda = \lambda_{\rm dB} c / \upsilon \approx 3.4$ μm; the appropriate slit width is a few tens of nanometers. In the case of the phthalocyanine molecule PcH_2 and its derivative $F_{24}PcH_2$ [237]: $\lambda_{\rm dB} = 5.2 \times 10^{-12}$ m, the velocity of molecules $\upsilon = 150$ m·s^{-1} and therefore the amplitude of inerton cloud of such a molecule is $\Lambda = \lambda_{\rm dB} c / \upsilon \approx 10$ μm (the slid width is also a few tens of nanometers). So we can see that in double-slit experiments with large molecules the value of Λ significantly exceeds the slit width and thus molecules inside the slit move in contact with the walls, which automatically generates a diffraction angle θ_1 in the molecule's trajectories. Besides, inerton flows from columns that separate slits also add their contribution to the diffraction angle, θ_2, such that the total diffraction angle becomes $\theta = \theta_1 + \theta_2$.

The researchers of diffraction fringes continually repeat that the passage of single molecules through a grating present an unambiguous demonstration of the wave–particle duality of quantum physics: "It is only explicable in quantum terms, independent of the absolute value of the interference contrast. In contrast to photons and electrons, which are irretrievably lost in the detection process, fluorescent molecules stay in place to provide clear and tangible evidence of the quantum behavior of large molecules" [237].

It seems that if the researchers could somehow push an ordinary brick through a diffraction grating, they would also start to talk about its quantum behavior and wave-particle duality… Reading contemporary articles devoted to the diffraction of particles, a natural question arises: Why is it that researchers having so remarkable ultramodern precision instruments have never sought for the reasons of deviation of particles inside slits of the diffraction grating?

It should be emphasized that a large heavy particle could not be considered as a wave in the diffraction experiments mentioned above. For example, the de Broglie wavelength of large molecules is equal to a few picometers, but the size of a slit is tens of nanometers and hence the ratio of the scales is roughly 1:1000. Hence following the standard logic of quantum mechanics, the large molecule in question cannot overtake the obstacle in principle. Besides, a large empty hole cannot also be an obstacle for such a molecule by definition. Nevertheless, the diffraction fringe is present. The reason is that a periodic process is hidden in the slow motion of the particle and the periodical emission of the particle's inerton cloud plays a role of a tentacle that guides the particle.

The rectilinear motion is not uniform: the particle has a spatial amplitude $\lambda \equiv \lambda_{\text{dB}}$ and the particle velocity oscillates in the section 2λ from the initial value υ to zero, back to υ, and in the next spatial period 2λ again to zero, and so on, because the motion occurs in the tessellattice with which the particle uninterruptedly interacts. It is known from analytical mechanics that for periodic motions the Hamiltonian of the system studied becomes a function of only an action J per period. For quantum systems J is the Planck constant, which results (see the end of Section 4.3) in the relation $E = h\nu$.

Thus in quantum systems the energy of a particle can be characterized by two expressions $E = \frac{1}{2}m\upsilon^2$ and $E = h\nu$. The period T of oscillations of the particle together with the particle's spatial amplitude λ relate to the particle's initial velocity: $\upsilon = \lambda / (T/2)$. These three relationships give rise to the de Broglie claim $\lambda_{\text{dB}} = \lambda = h/(m\upsilon)$ (see Section 2.4) and give the impression of a travelling material wave. The situation is not so simple, as the construct of particle-wave duality; we are dealing instead with a complicated submicroscopic mechanics of the {particle + its inerton cloud}-system, which is developed in the tessellattice. We can see that such concept is in line with de Broglie's views: only this double solution theory will be able to shed light on the correct interpretation of wave mechanics.

5.9 ANOMALOUS MULTIPHOTON PHOTOELECTRIC EFFECT

The first reports on the experimental demonstration of laser-induced gas ionization occurrence at a frequency below the threshold at which the photoelectric effect is manifested appeared in the mid-1960s. In this effect the photon energy of the incident light is essentially smaller than the ionization potential of atoms of rarefied noble gases and the work function of the metal. Agostini and Petite [241] reviewed studies exploiting the prevailing multiphoton theory. The multiphoton concept is based on the typical interaction Hamiltonian

$$H_{\text{int}} = -e\vec{z}\vec{\mathcal{E}}_0 \cos(\omega t) \tag{5.109}$$

that characterizes the interaction between the dipole moment $e\vec{z}$ of an atom and the incident electromagnetic field $\vec{\mathcal{E}} = \vec{\mathcal{E}}_0 \cos(\omega t)$. The concept starts from a standard time dependent perturbation theory describing the probability of transition of an atom from the bound state to a state $\langle \mathrm{c}|$ in the continuum per unit time. On the next stage the concept modifies the simple photoelectric effect to a nonlinear one [242, 243] in which the atom is ionized by a gradual absorption of several photons. As the result, the absorption of each next photon occurs step by step and the *N*th order time dependent perturbation theory changes the usual Fermi golden rule to the *N*-photon absorption that produces a complicated probability

$$w_N = \frac{2\pi}{\hbar}\left(\frac{2e^2}{\varepsilon_0 c}\right)^N \sum_{\mathrm{c}} \left| \sum_{i,j,\dots,k} \frac{\langle g|z|i\rangle\langle i|z|j\rangle \dots \langle k|z|\mathrm{c}\rangle}{(E_g + \hbar\omega - E_i)(E_g + 2\hbar\omega - E_j)\dots} \right|^2 \tag{5.110}$$

where $|i\rangle, |j\rangle, \ldots |k\rangle$ are the atomic states, $|c\rangle$ is the continuum states with the energy $E_g + \hbar\omega$ and E_g is the energy of the ground state $|g\rangle$. The probability of these multiphoton processes results in the so-called generalized cross section [241]

$$s_N = 2\pi(8\pi\alpha)^N r^{2N}\omega^{-N+1} \tag{5.111}$$

where $r \sim 0.1$ nm is the effective atom radius and $\alpha = 1/137$ is the fine structure constant. The Einstein law $E = \hbar\omega$ characterizing the simple photoelectric effect changes to the relation specifying the nonlinear photoelectric effect

$$E_c = N\hbar\omega - E_i \tag{5.112}$$

The N-photon ionization rate (5.110) is proportional to I^N where I is the intensity of the laser beam. This prediction was verified experimentally up to $N = 22$ photons and at a laser intensity of about 10^{15} W·cm^{-1}.

On the other hand, a series of experiments [244] doubted the dependence I^N: Xe, Kr, and Ar simultaneously absorbed N photons; the linear slope held up to 2×10^{13} W·cm^{-1} and the maximum value was $N = 14$.

In review papers Panarella [245, 246] analyzed many other experiments devoted to laser-induced gas ionization and laser-irradiated metal and convincingly demonstrated the inconsistency of generally accepted multiphoton methodology. In particular, Panarella studied the following series of experiments: (i) variation of the total number N_i of ionized gas atoms as a function of the laser intensity I_p; (ii) variation of the total number N_i as a function of time t of the increase in intensity of laser pulse; (iii) variation of the breakdown intensity threshold against the gas density; (iv) focal volume dependence of the breakdown threshold intensity; and others.

Experimental data obtained with the use of nanosecond and picosecond light/laser pulses already in the 1960s and 1970s also failed to support the multiphoton approach. Modern studies [247] show that this number of absorbed photons N_i is not a constant depending only on the characteristics of the metal and light, but varies with the interaction duration in ultrashort time scales. The phenomenon occurs when electromagnetic energy is transferred, via ultrafast excitation of electron collective modes, to conduction electrons in a duration less than the electron energy damping time, which has been observed through a significant increase of electron production.

So, Panarella [245, 246] noted that new physics should be present in the phenomena described above and he proposed an effective photon theory. He postulated that the photon energy expression $\varepsilon = h\nu$ had to be modified "ad hoc" into the novel one:

$$\varepsilon = \frac{h\nu}{1 - \beta_\nu f(I)} \tag{5.113}$$

where $f(I)$ is the function of the light intensity and β_ν is a coefficient. In this manner Panarella's theory holds that, at the extremely high intensities of light, the photon-photon interaction begins to play a significant role in the light beam such that the photon energy becomes a function of the photon flux intensity. To develop an effective photon concept it was pointed out that the number density of photons in the focal volume is much larger than λ^{-3} where λ is the wavelength of laser's irradiated light. In this respect he came up with the proposal to reduce the photon wavelength in the focal volume. He assumed that it unquestionably followed from quantum electrodynamics that photons could not approach each other closer than λ.

The effective photon concept satisfied all available experimental facts and was successfully applied by Panarella to his own experiments on electron emission from a laser irradiated metal surface. The remarkable success of his formula (5.113) gave rise to the confidence that some hidden reasons might lead to understanding the principles of effective photons formation.

Nevertheless, the phenomenon of anomalous photoelectric effects under an intensive laser pulse of low energy photons can be solved [248] quite naturally in the framework of the submicroscopic deterministic concept. Having described ionization of atoms of gas and photoemission from a metal in terms of the submicroscopic approach, an effort can be made to try to develop a theory of the anomalous photoelectric effect in which the electron's wide spread inerton-photon cloud (for simplicity, the inerton cloud) simultaneously absorbs a number of coherent photons from the intensive laser pulse (Figure 5.25). Thus the theory combines Panarella's

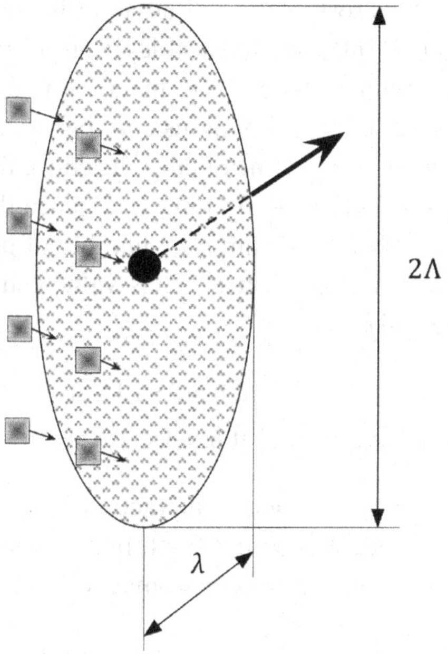

FIGURE 5.25 Electron surrounded by its cloud of inertons is attacked by N photons.

idea on the anomalous photoelectric effect and the idea of the multiphoton concept on simultaneous absorption of N photons. In this approach we do not need any hypothesis regarding the unexplainable reduction of the photon wavelength in the focal volume.

In the first approximation atoms of gas and the metal may be considered as systems of quasi-free electrons. The Fermi velocity of s and p electrons in an atom is equal to $(1-2)\times 10^6$ m·s^{-1}. Setting $\upsilon_F = \upsilon_0 \cong 2 \times 10^8$ m·s^{-1} one obtains $\lambda_{dB} = h/(m\upsilon_0) \cong 0.36$ nm (m is the electron mass) and then in accordance with relation (2.69) the amplitude of oscillations of the inerton cloud equals $\Lambda = \lambda_{dB}\, c / \upsilon_0 \cong 5.5$ nm. The cloud has anisotropic properties: it spreads over the distance λ along the electron path, i.e., along the velocity vector $\vec{\upsilon}_0$, and over Λ in transverse directions. This means that the cross section σ of the electron together with its inerton cloud satisfies inequalities:

$$\lambda^2 < \sigma < \Lambda^2, \text{ or } 10^{-19}\ \text{m}^2 < \sigma < 3\times 10^{-16}\ \text{m}^2 \tag{5.114}$$

At the same time the cross-section of an atom is only $\sim 10^{-20}$ m^2. The intensity of light in (10–100)-psec focused laser pulses used for the study of gas ionization and photoemission from metals was within the order of 10^{12} to 10^{15} W·cm^{-2}, i.e., 10^{30} to10^{33} photons·cm^{-2} per second. Dividing this intensity into the velocity of light one obtains the concentration of photons in the focal volume $n \approx 3\times(10^{25} - 10^{28})$ m^{-3} and hence the mean distance between photons is $n^{-1/3} \approx (30-3)$ nm. The number of photons bombarding the electron's inerton cloud is $\sigma n^{2/3}$; this value can be estimated, in view of inequality (5.114), as

$$1 < \sigma n^{2/3} < 10^3 \text{ at } n \approx 3\times 10^{23} \text{ m}^{-3},$$
$$1 < \sigma n^{2/3} < 10^5 \text{ at } n \approx 3\times 10^{26} \text{ m}^{-3} \quad (5.115)$$

Now we have to write the model interaction between the electron's inerton cloud and an incident coherent light. In an ordinary classical representation an electron in an applied electromagnetic field is characterized by the energy

$$E = \frac{1}{2m}(\vec{p} - e\vec{A})^2 \quad (5.116)$$

where $\vec{A}$ is the vector potential of the electromagnetic field. This usually implies that the vector potential $\vec{A}$ in Ampére's formula (5.116) relates to the field of one photon. In the formalism of quantum field theory this means that both the wave function of the electron and the wave function of the photon are normalized to one particle in the same volume V, Berestetskii et al. [249].

However, the electron together with its inerton cloud is an extended object. Therefore it can interact with many photons simultaneously and the density of an incident photon flux should define the coupling function between the electron and the applied coherent electromagnetic field. Hence in addition to the existing weak electron-photon interaction we can introduce the approximation of a strong electron-photon coupling

$$E = \frac{1}{2m}(\vec{p} - e\vec{A}_{\text{eff}})^2 \quad (5.117)$$

which should be correct in the case of simultaneous absorption/scattering of N photons by the electron. Here, in expression (5.117)

$$\vec{A}_{\text{eff}} = e\vec{A}N,\ N = \sigma n^{2/3} \tag{5.118}$$

In the experiments involving noble gases discussed by Panarella [245, 246] the laser pulse intensity had the triangular shape. So let the intensity be changed over the duration Δt of the laser pulse which intensity runs along the two equal sides of an isosceles triangle, namely, from $I = 0$ at $t = 0$ to the peak intensity $I = I_{\text{p}}$ at $t = t_{\text{p}} = \Delta t / 2$ and then to $I = 0$ at $t = \Delta t$. Then the $\vec{A}_{\text{eff}}$ becomes time dependent. Let us present it in the form

$$\vec{A}_{\text{eff}}(\vec{r},\, t) = \vec{A}_{\text{p}} e^{i\vec{k}\vec{r} - i\omega t} N(t) \tag{5.119}$$

where $\vec{A}_{\text{p}}$ is the vector potential of the electromagnetic field at the peak intensity of the pulse and

$$N(t) = \sigma_{\kappa} n_{\kappa}^{2/3} \frac{t}{t_{\text{p}}} \tag{5.120}$$

is the number of photons absorbed by the electron, where $n_{\kappa}^{2/3}$ is the effective photon density in the unit area at the threshold intensity of the laser pulse when the energy of $\sigma_{\kappa} n_{\kappa}^{2/3}$ photons reaches the absolute value of the ionization potential of atoms or the work function of the metal, i.e., $h\nu\, \sigma_{\kappa} n_{\kappa}^{2/3} = W$. As relation (5.120) indicates the cross section σ_k of the electron inerton cloud also depends upon the threshold intensity; very probably σ_k is not constant and depends on the velocity of the electron, the frequency of incident light and the light intensity. The presentation (5.119) is correct within the time interval $\Delta t/2$, i.e., $t \in [0,\ t_{\text{p}}]$.

The Hamiltonian operator of an electron in an intensive electromagnetic field reads

$$\hat{H} = \frac{\hat{\vec{p}}^{\,2}}{2m} - \frac{e}{m}\vec{A}_{\text{eff}}(\vec{r},\, t)\,\hat{\vec{p}} \tag{5.121}$$

where we limit our consideration by the linear field effect, much as is the case in the theory of the ordinary photoelectric effect (see, e.g., Ref. [249], p. 231). The problem of the anomalous photoelectric effect occurring due to the operator (5.121) can be solved using the perturbation theory. Namely, we will be able to calculate the probability of transition of an electron from an atom or a metal irradiated by an intensive laser pulse to a free state.

In the absence of an external field the Schrödinger equation

$$i\hbar\frac{\partial\psi_0}{\partial t} = \hat{H}_0\psi_0 \tag{5.122}$$

is characterized by the eigenfunction $\psi_0 = e^{-i\frac{\hat{H}_0}{\hbar}t}$. In the presence of an electromagnetic field Eq. (5.122) is transferred to

$$i\hbar\frac{\partial\psi}{\partial t} = [\hat{H}_0 + \hat{W}(t)]\psi \tag{5.123}$$

The function ψ can be presented in the form

$$\psi(\vec{r}, t) = e^{-i\frac{\hat{H}_0}{\hbar}t}\sum_l a_l(t)\ \psi_l(\vec{r}) \tag{5.124}$$

where $a_l(t)$ is a coefficient at the eigen function $\psi_l(\vec{r})$. Substituting function (5.124) into Eq. (123) and multiplying the new equation by $\psi_f^*(\vec{r})$ to left and then integrating over $\vec{r}$ we obtain

$$i\hbar\frac{\partial a_f(t)}{\partial t} = \sum_l a_l(t)\langle f|\hat{W}|l\rangle e^{i\omega_{fl}t} \tag{5.125}$$

where $\hbar\omega_{fl} = E_f - E_l$ is the eigenvalue of Eq. (5.122) and the matrix element

$$\langle f|\hat{W}|l\rangle = -\frac{e}{m}\int\psi_f^*\vec{A}_{\text{eff}}(\vec{r}, t)\hat{\vec{p}}\,\psi_l d\vec{r} \tag{5.126}$$

In the first approximation the coefficient is

$$a_f^{(1)} \cong -\frac{i}{\hbar}\int_0^t\langle f|\hat{W}|l\rangle e^{i\omega_{fl}\tau}d\tau \tag{5.127}$$

The possibility of transition of an electron from the atom state E_l to the state of ionized atom E_f (or the possibility of ejection of an electron out of the metal) is given by the expression

$$P(t) \equiv P_f(t) = \left| a_f^{(1)}(t) \right|^2 \tag{5.128}$$

or explicitly

$$P(t) = \left| -\frac{i}{\hbar}\frac{e}{m}\langle f \,|\, \vec{A}_{\mathrm{p}}\hat{\vec{p}} \,|\, l\rangle \right|^2 \left| \int_0^t N(\tau) e^{i(\omega_{fl} - \omega)\tau} d\,\tau \right|^2 \tag{5.129}$$

The first multiplier in expression (5.129) is well known in the conventional photoelectric effect, because it defines the probability of the electron's transition from the atomic state $|l\rangle$ to a free state $\langle f|$. Let us designate this factor by $|M|^2$ and extract from expression (5.129) in the explicit form (see, e.g., Blokhintsev [75], p. 407)

$$|M|^2 \equiv 16\pi \frac{e^2\hbar^2}{m^2 \mathcal{V}} \vec{A}_{\mathrm{p}}^2 \left(\frac{Z}{a_0} \right)^5 \left(\frac{\hbar}{\vec{p}_{\mathrm{free}}} \right)^6 \frac{\sin^2\theta \cos^2\phi}{(1 - \frac{\upsilon}{c}\cos\theta)^4} \tag{5.130}$$

where V is a normalized volume of the system under consideration, a_0 is Bohr's radius, Z is the number change, and $\vec{p}_{\mathrm{free}}$ is the momentum of the stripped electron.

The last multiplier in expression (5.129) shows that the momentum $\vec{p}_{\mathrm{free}}$ falls within the solid angle $d\Omega$ (υ is the velocity of the free electron and $|\vec{p}_{\mathrm{free}}| = m\upsilon$).

The vector potential $\vec{A}$ is connected with the intensity I of the electromagnetic field through relations

$$I = \varepsilon_0 c^2 \,|\vec{\mathcal{E}}\,|^2, \ \vec{\mathcal{E}} = -\frac{\partial \vec{A}}{\partial t} = i\omega \vec{A}_{\mathrm{p}} e^{i(\omega t - \vec{k}\vec{r})} \tag{5.131}$$

which makes it possible to derive

$$\vec{A}_{\mathrm{p}}^2 = \frac{1}{\varepsilon_0 c^2 \omega^2} I_{\mathrm{p}} \tag{5.132}$$

The relation (5.132) allows us to substitute $\vec{A}_{\mathrm{p}}^2$ for I_{p}, which is measurable, in expression (5.130):

$$|M|^2 = |\mathcal{M}|^2 \, I_{\mathrm{p}}, \; |\mathcal{M}|^2 \equiv 16\pi \frac{e^2 \hbar^2}{\varepsilon_0 c^2 \mathcal{V} m^2} \left(\frac{Z}{a_0}\right)^5 \left(\frac{\hbar}{\vec{p}_{\mathrm{free}}}\right)^6 \frac{\sin^2\theta \cos^2\phi}{(1 - \frac{\upsilon}{c}\cos\theta)^4} \tag{5.133}$$

The second multiplier in expression (5.129) can be subdivided to two multipliers:

$$|\mathcal{I}(t)|^2 = \int_0^t N^*(\tau) e^{-i(\omega_{fl} - \omega)\tau} d\tau \int_0^t N(\tau) e^{i(\omega_{fl} - \omega)\tau} d\tau \tag{5.134}$$

Let us calculate the integral $\mathcal{I}(t)$:

$$\mathcal{I}(t) = \int_0^t N(\tau) e^{i(\omega_{fl} - \omega)\tau} d\tau = \frac{\sigma_\kappa n_\kappa^{2/3}}{t_{\mathrm{p}}} \int_0^t \tau e^{i(\omega_{fl} - \omega)\tau} d\tau$$

$$= \frac{\sigma_\kappa n_\kappa^{2/3}}{t_{\mathrm{p}}} \left[\frac{t}{i(\omega_{fl} - \omega)} e^{i(\omega_{fl} - \omega)t} + \frac{t}{(\omega_{fl} - \omega)^2} \left(e^{i(\omega_{fl} - \omega)t} - 1 \right) \right] \tag{5.135}$$

Substituting $\mathcal{I}(t)$ and $\mathcal{I}^*(t)$ into expression (5.134) we obtain

$$|\mathcal{I}(t)|^2 = \frac{(\sigma_\kappa n_\kappa^{2/3})^2}{t_{\mathrm{p}}^2 (\omega_{fl} - \omega)^2} \left\{ \begin{aligned} & t^2 - \frac{2t}{(\omega_{fl} - \omega)} \sin[(\omega_{fl} - \omega)t] \\ & + \frac{2}{(\omega_{fl} - \omega)^2} \left(1 - \cos[(\omega_{fl} - \omega)t]\right) \end{aligned} \right\} \tag{5.136}$$

In our case $\omega_{fl} - \omega = (E_f - E_l)/\hbar - \omega$ where $\omega = 2\pi v$ and v is the frequency of incident light. As $\omega_{fl} - \omega >> \omega$, we can put $\omega_{fl} - \omega \cong \omega_{fl}$. Besides we only consider the approximation when $t << t_{\mathrm{p}} = \Delta t / 2 \approx 10^{-8} - 10^{-7}$ s. Hence for the wide range of time ($\omega^{-1}{}_{fl} << t << t_{\mathrm{p}}$) the inequality $\omega_{fl} t >> 1$ holds and the expression (5.136) can be replaced by

$$\mathcal{I}(t)|^2 = \left(\frac{\sigma_\kappa n_\kappa^{2/3}}{t_\mathrm{p}\,\omega_{fl}}\right)^2 t^2 \tag{5.137}$$

The matrix element ω_{fl} in expression (5.137) can be eliminated by substituting the absolute value of the ionization potential of atoms or the work function of the metal, W, i.e., $\omega_{fl} \to W/\hbar$. Now substituting expression (5.137) into the probability (5.129) we finally get

$$P(t) = |\mathcal{M}|^2 \left(\frac{\sigma_\kappa n_\kappa^{2/3}}{t_\mathrm{p}\,\omega_{fl}}\right)^2 I_\mathrm{p} t^2 = |\mathcal{M}|^2 \left(\frac{\hbar N}{W t_\mathrm{p}}\right)^2 I_\mathrm{p} t^2 \tag{5.138}$$

If an incident laser pulse can be considered as a perturbation that is not time dependent, the interaction operator $\hat{W}$ of electron-photon interaction can be regarded as a constant value

$$\hat{W} = -\frac{e}{m}\vec{A}_\mathrm{eff}(\vec{r})\hat{p} \equiv -\frac{e}{m} N \vec{A}_0 e^{i\vec{k}\vec{r}}\,\hat{\vec{p}} \tag{5.139}$$

between the moments of cut-in and cut-off and $\hat{W} = 0$ behind the time interval Δt corresponding to the duration of the laser pulse. When we have the interaction operator (5.139), we may apply the Fermi golden rule and obtain the probability of the anomalous photoelectric effect (compare with the theory of the simple photoelectric effect, e.g., Ref. [75], p. 407)

$$P_0 = \frac{2\pi}{\hbar}|\mathcal{M}|^2 N^2 I \mathcal{V} m |\vec{p}_\mathrm{free}| \Delta t\, d\Omega \tag{5.140}$$

Here, $|\mathcal{M}|^2$ is the matrix element defined above (5.133), $N = \sigma_\kappa n_\kappa^{2/3}$ is the number of photons adsorbed by an atom or a metal simultaneously, I is a typical intensity of the laser pule, and $\mathcal{V} m |\vec{p}_\mathrm{free}|$ is the density of states ($\mathcal{V}$ is the normalizing volume, m is the electron mass and $\vec{p}_\mathrm{free}$ is the momentum of the stripped electron). Then for the energy absorbed by the electron we obtain a simple formula $\varepsilon = h\nu N$ instead of Panarella's "ad hoc" formula (5.113).

Thus, we can see that the interaction between an intensive laser pulse and gas atoms or electrons in a metal is not nonlinear. The electron's inerton cloud is characterized by the large absorption cross section (5.114), which in a dense photon beam becomes the receptor (receiver) of many photons. Expression (5.140) was successfully verified [248] by applying to many experiments devoted to laser-induced gas ionization and the electron emission from a laser-irradiated metal in which the electron absorbed 10 or more low-energy photons simultaneously.

5.10 SONOLUMINESCENCE AS AN INERTON-PHOTON PHENOMENON

Single bubble sonoluminescence is the phenomenon in which light is produced from an acoustically driven collapsing gas bubble trapped in a liquid. In other words, a standing acoustic wave is able to influence a gas bubble in the center of the liquid system studied in such a way that the collapsing bubble emits a monochrome flash of light. A description of detailed experiments, which demonstrate the unique properties of the system studied, can be read in review papers [250, 251]. The authors discuss the classical theory of bubble dynamics, the gas dynamics inside the bubble, the hydrodynamic and chemical stability of the bubble, and the conditions that bring about stable single-bubble sonoluminescence and so on. The Rayleigh-Plesset equation is usually used for the analysis of bubble wall dynamics [251, 252] and in addition the equation is supplemented by the bubble-wall Mach number [253], which includes effects caused by the compressibility of the liquid. The general opinion has prevailed that the phenomenon of sonoluminescence pushes fluid mechanics beyond its known limit and despite the great efforts of many researches the phenomenon has remained completely unexplained.

Though the emission of light from bubbles in many aspects is similar to the behavior of a black body, many researchers have tried to involve plasma processes to account for the generation of light at the collapse of the bubble. Nevertheless, they note that a state of matter that would admit photon–matter equilibrium under such conditions is a mystery [254]. The inability to reconcile the long photon mean free path with the smallness of the hot spot suggests new physics for the modeling of sonoluminescence. There have been attempts to consider electrical and/or thermal processes that attract formulas for thermal emission such as electron-neutral Bremsstrahlung. Plasma can be created at the ionization potential around 15 eV but recent studies [255] have shown that dilute plasma models of sonoluminescence are not valid. The researchers [256] concluded that collective processes caused by acoustic oscillations could uncouple the charge needed to create plasma in the bubble; however, by their most optimistic estimation the acoustic energy could only reduce the ionization potential by 75%, although another very recent study [257] reduces this value to 60–70%.

The phenomenon of sonoluminescence occurs under the following conditions. The nature of liquid in which the phenomenon is studied influences the intensity of the flash of light, nevertheless most results have been obtained in experiments with water. The optimal acoustic pressure is 1.2 to 1.5 times the atmospheric pressure p_0 and each water molecule is vibrated by ultrasound waves with an energy of about 1.5×10^{-30} J. The concentration of gas in water is close to the standard dissolved value, though the glow is brighter when a noble gas is added in the concentration of only a few percent. The maximum radius of an expanded bubble is around 50 μm, which collapses to the value of around 0.5 μm in time 15 μs. This means that the speed of collapse is at least 4 times the velocity of sound in the liquid studied; under these conditions the interfacial Mach number approaches unity. As the radius of the bubble is started to approach the minimal critical value R_c (the van der Waal's hard core), the pre-collapsed bubble emits a broad spectral flash of light. The maximum of the luminescence spectrum is in the ultraviolet region. Each photon originating from a single atom of the bubble region (i.e., an atom of gas within the bubble) has an energy about 10^{-20} J (i.e., 6 eV) or higher. Hence the ratio of the emitted energy density to the absorbed acoustic energy is a factor over 10^{10}. The duration of emission is circa 50 to 100 picoseconds. The intensity of emitted light is much higher in the case of such liquids as sulfuric and phosphoric acids. Researchers have noted about a strong pressure in the collapsing bubble, which can reach a few thousand bar. Nevertheless, this pressure is estimated by the rate of an ejected gaseous-liquid jet created after the collapse of the bubble and it can be associated with the relaxation of the compressed liquid rather than with the inner state of the bubble.

The mechanism of the light emission remained uncertain until investigations were made into the role played by inertons [258]. Moreover, Brennan and Fralick [259] obtained a new and very interesting experimental result: they measured the timing of sonoluminescence by observing laser light scattered from a single sonoluminescing bubble and found that the flash typically occurs 100 nanoseconds before the minimum radius. This outcome now allows us to completely clarify the phenomenon of sonoluminescence.

Let us consider the situation in water. In the equilibrium state the surface layer that separates the bubble from the bulk of water is composed of water molecules with a typical mean distance d between the molecules of 0.31 nm. With increasing exposure, the ultrasound will further compress the water and hence the radius of the bubble will grow (since the volume of the whole water-bubble system should be preserved). The mean distance d between water molecules in the boundary layer will also expand to a critical value d_c. When the distance between molecules reaches the value of d_c, the boundary layer becomes unstable to rupture and through the torn layer the squeezed water quickly expands (four times of the sound velocity) into the cavity causing the bubble to collapse.

The energy and other characteristics of the mass transfer carried out by inertons can be evaluated taking into account a set of parameters of the layer's molecules of the collapsing bubble. The total pressure inside the water under consideration is the sum of the atmospheric pressure $p_0 = 101.325$ kPa at room temperature and the ultrasound pressure $p_u = (1.2 \text{ to } 1.5)\, p_0$. Let $p_u = 1.3 p_0$, then the total pressure in the water is around $p = 2.3 p_0$. A typical frequency of ultrasound is 20 to 30 kHz, which means that during the time of collapse (15 μs) the pressure in the bubble practically does not change. Hence the work produced by the expanded water, which results in the bubble collapsing, can be estimated as

$$W = V_1 p_1 - V_2 p_2 \tag{5.141}$$

Here, the volumes are $V_1 = 4\pi R_1^3 / 3$, $R_1 = 50 \times 10^{-6}$ m, and $V_2 = 4\pi R_2^3 / 3$, $R_2 = 0.5 \times 10^{-6}$ m; the pressures are $p_1 = p + \sigma / R_1$ and $p_2 = p + \sigma / R_2$ where σ is the coefficient of surface tension of water; in the case of the water-air interface $\sigma \cong 73 \times 10^{-3}$ N·m^{-1} at room temperature. Calculating the work (5.141), we obtain

$$W \cong \frac{4\pi}{3} R_1^3 p = \frac{4\pi}{3} R_1^3 \times 2.3 p_0 \approx 1.22 \times 10^{-7}\,\text{J} \tag{5.142}$$

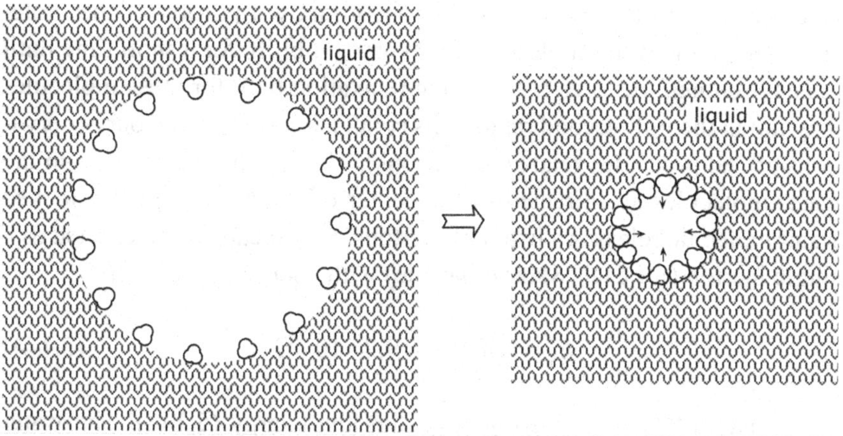

FIGURE 5.26 Collapse of the bubble.

Let us now calculate the number of water molecules in the boundary layer that separates the bulk of water from the bubble of gas. The boundary layer is shown in Figure 5.26. We assume that the number N of water molecules is preserved in the boundary during the collapse of the bubble (these molecules squeeze to the point when they are 'touching' each other as in the initial equilibrium state).

The value of N can be evaluated by using a simple expression

$$N = 4\pi R_1^2 / d_c^2 \tag{5.143}$$

where $4\pi R_1^2$ is the area of the boundary layer at the moment of time before the bubble collapse; d_c may exceed the equilibrium value 0.31 nm two times, so let us put $d_c = 0.63$ nm. This allows us to estimate the number of molecules in the boundary layer by using expression (5.143): $N = 1.55 \times 10^{11}$.

The boundary layer, which is quickly moving to the center of the bubble, experiences an abrupt resistance at the radius $R_2 \approx 0.5\ \mu$m. The gas in the bubble is compressed by the collapsing boundary layer up to the point of becoming a liquid state.

In being reduced from the large bubble radius R_1 to the small radius R_2 the boundary layer experiences a sudden shock and a portion of inertons is stripped from the inerton clouds of the N water molecules of the boundary layer. In loosing those inertons, and accordingly a portion of the molecule's mass, the water molecules lose some mobility for a short time until the environment restores it again; the relaxation time for these molecules may fall in the millisecond range (this conclusion stems from our observations of fast processes related to inertons). The inertons released from the water molecules are directed to the center of the bubble. Each shocked water molecule releases a batch of inertons, which can be represented as a single inerton (for each water molecule) with energy

$$E_{\text{inert}} = W / N \cong 9.7 \times 10^{-19}\,\text{J} \tag{5.144}$$

So each of these N inertons transfers the energy (5.144). The water molecules in the boundary layer will reabsorb some of these inertons, but others go to the center of the bubble. Those inertons that go straight to the center of the bubble carry additional mass to the gas atoms. The mass μ of a single inerton escaping from the boundary layer can be evaluated by an expression $E_{\text{inert}} = \mu c^2$, which gives $\mu \cong 1.08 \times 10^{-35}$ kg.

In the equilibrium state the radius of the bubble is around 50 μm. Such a volume of water under these conditions can contain 1.75×10^{13} molecules of water whereas the number of gaseous atoms that fill this bubble is about 4–5 orders of magnitude smaller, i.e., the bubble contains 10^8 to 10^9 gaseous atoms (a mixture of air and a noble gas). Typically the flash of light emitted from the collapsed bubble contains around 10^6 photons (in the case of the ultrasound not being too powerful). This means that we may identify the number 10^6 of emitted photons with the number N of inertons, which came from the boundary layer into the bubble where they have been absorbed by N gaseous atoms.

The electrons of the outer shell of a gas atom rotate surrounded by the atom's inerton-photon cloud, which we have discussed in Chapter 4 (see Sections 4.2.2, 4.3 and Figure 4.6) and Chapter 1 (see Section 1.8.8 and Figure 1.13*c*). This inerton-photon cloud is able to absorb an inerton coming from the boundary layer with the effect of increasing the rest mass of an electron: $m \to m + \mu$. This act of inerton absorption alters the known conditions of the initial equilibrium state of the electron in the atom:

$$\frac{e^2}{4\pi\varepsilon_0 r^2} = \frac{m\upsilon^2}{r} \tag{5.145}$$

$$m\upsilon r = n\hbar \qquad (5.146)$$

which result in the Bohr radius

$$r_0 = \frac{4\pi\varepsilon_0 \hbar^2 n^2}{e^2 m} \qquad (5.147)$$

Due to the absorption of the additional mass μ, the electron must occupy a lower intermediate orbit with a radius

$$r_0^* = \frac{4\pi\varepsilon_0 \hbar^2 n^2}{e^2 \cdot (m+\mu)} \qquad (5.148)$$

However the lower orbit with the radius r_0^* is unstable. Since the inerton impacts the electron's inerton-photon cloud creating a mass excitation in it – the cloud becomes locally smaller in size. In turn the cloud has to respond to the impact, which should result in an oscillation process. Eventually the inerton excitation should gradually spread along the whole inerton-photon cloud of the electron. When the inerton excitation is uniformly absorbed in the cloud, this will stop the oscillation process creating the moment of the decay of the inerton excitation. As a result the compressed cloud emits a photon, which immediately restores the inerton-photon cloud and once more stabilizes the electron in a stationary orbit with the initial radius (5.147).

The problem under discussion can be described by a differential equation of damped oscillations of the radius r, i.e., r oscillates between values of $(r_0 \pm \delta r)$, Figure 5.27:

$$\delta\ddot{r} + 2\beta\,\delta\dot{r} + (2\pi\nu)^2 \delta r = 0 \qquad (5.149)$$

where 2β is the damping coefficient and $2\pi\nu$ is the cyclic frequency of the electron in the orbit of the radius r. It is reasonable to assume that the damping coefficient relates to the ratio of two masses multiplied by the rotation period ν^{-1} of the electron; namely, $2\beta \equiv 2\nu^{-1}\mu / m$. Then the solution to Eq. (5.149) is

$$\delta r(t) \cong \delta r_0 \exp(-\nu^{-1} t \mu / m) \cos(2\pi\nu\, t) \qquad (5.150)$$

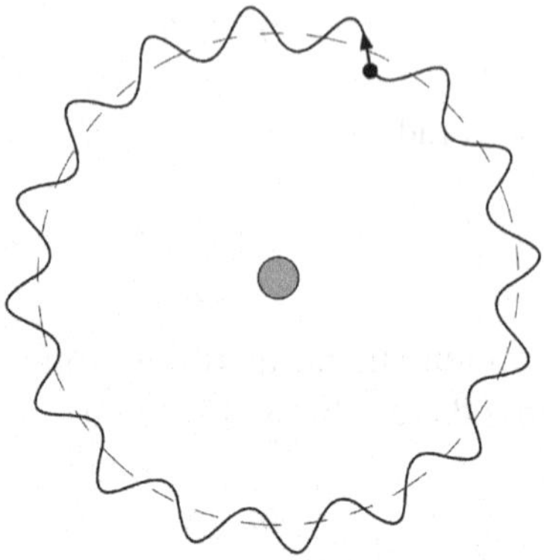

FIGURE 5.27 Trajectory of the electron in the atom, which absorbed an inerton excitation.

In the time $t = (\nu^{-1}\mu / m)^{-1}$ the oscillations practically decay, which means that this value determines the lifetime of the inerton excitation. In our problem, $m \approx 9 \times 10^{-31}$ kg, $\mu \approx 10^{-35}$ kg and $\nu^{-1} \sim 10^{-15}$ s. Therefore the lifetime $t_{\text{life}} = (\nu^{-1}\mu / m)^{-1} \sim 10^{-10}$ s. After that a photon is emitted from the electron's inerton-photon cloud. The energy of the emitted photon is exactly the same as the energy of the absorbed inerton (5.144), $h\nu_{\text{ph}} = E_{\text{inert}} = 9.85 \times 10^{-19}$ J, or 6.16 eV, which is between the middle and far region of the ultraviolet spectrum; this result is in agreement with the experiments on sonoluminescence. The structure of the outer electron shell in noble gases, in particular argon, permits the gases to emit monochromatic light.

The estimated lifetime of $t_{\text{life}} = 100$ ps of the inerton excitation in the system {electron and its inerton-photon cloud} corresponds to the time delay between the emission of photons and the bubble collapse measured in the experiments on sonoluminescence [259].

Finally we have to emphasize that the transfer of a group of inertons from water molecules to gaseous atoms in the bubble cannot be considered as the emission of free inertons with their subsequent absorption. The process of transfer has rather been a nonradiative process: groups of inertons (or single inertons) have been passed from one entity to another in the framework of a closed system.

5.11 PYRAMID POWER

As we have already seen above, inertons contribute to vibrations of the crystal atoms. This means that an external inerton field is able to intensify atom vibrations in particular in a metal, which is impossible for an electromagnetic field due to a strong screening effect on behalf of free electrons.

Let us consider how such intensification occurs, initially theoretically and then experimentally [260].

The total Lagrangian of the crystal lattice is

$$L = \frac{m}{2}\sum_{\vec{l},\eta}\dot{\xi}^{2}_{\vec{l}\eta} - \frac{1}{2}\sum_{\vec{l}\neq\vec{n};\,\eta,\,\beta} V_{\eta\beta}(\vec{l}-\vec{n})\,\xi_{\vec{l}\eta}\,\xi_{\vec{n}\eta}$$

$$-\sqrt{m\mu}\sum_{\vec{l}\neq\vec{n};\,\eta,\,\beta} \xi_{\eta\beta}\,\tau^{-1}_{\eta\beta}(\vec{l}-\vec{n})\,\dot{\chi}_{\vec{n}\beta} + \frac{\mu}{2}\sum_{\vec{l},\eta}\dot{\chi}^{2}_{\vec{l}\eta} \tag{5.151}$$

where the first term is the kinetic energy of atoms, the second term describes the potential energy of atoms related to their conventional electromagnetic interaction, the third term depicts the interaction of atoms with inerton clouds of other atoms, and the last term is the kinetic energy of inerton clouds that are treated as members of the system studied. $\xi_{\vec{l}\eta}$ (η =1, 2, 3) are three components of the atom displacement from the lattice site whose equilibrium position is determined by the lattice vector $\vec{l}$; $\dot{\xi}_{\vec{l}\eta}$ are three components of the velocity of this atom; $V_{\eta\beta}(\vec{l}-\vec{n})$ are the components of the elasticity tensor of the crystal lattice. Let m and μ be the characteristic mass of an atom and its inerton cloud, respectively. $\chi_{\vec{l}\eta}$ are three components of the position of the inerton cloud for the atom determined by the lattice vector $\vec{l}$, then $\dot{\chi}_{\vec{l}\eta}$ are three components of the velocity of this inerton cloud. $\tau^{-1}_{\eta\beta}(\vec{l}-\vec{n})$ are components (generally they might be tensor quantities) of the rate of collisions between the inerton cloud of the atom determined by the lattice vector $\vec{n}$ and with the atom whose equilibrium position is determined by the vector $\vec{l}$.

In a standard way, we carry out canonical transformations in Eq. (5.151) with respect to collective variables, both for atoms ($\mathcal{A}_{\vec{k}} = (\mathcal{A}_{-\vec{k}})^{*}$) and for inerton clouds ($a_{\vec{k}} = (a_{-\vec{k}})^{*}$):

$$\xi_{\vec{l}\eta} = \frac{1}{\sqrt{Nm}} \sum_{\vec{k}} e_\eta \mathcal{A}_{\vec{k}} \exp\left(i\vec{k}\vec{l}\right) \tag{5.152}$$

$$\chi_{\vec{l}\eta} = \frac{1}{\sqrt{N\mu}} \sum_{\vec{k}} e_\eta a_{\vec{k}} \exp\left(i\vec{k}\vec{l}\right) \tag{5.153}$$

where $e_\eta \equiv e_\eta(\vec{k})$ are components of the polarization vector and N is the number of atoms in the crystal. On rearrangement, the Lagrangian (5.151) takes the form

$$L = \tfrac{1}{2}\sum_{\vec{k},\eta} e_\eta \dot{\mathcal{A}}_{\vec{k}} e_\eta \dot{\mathcal{A}}_{-\vec{k}} - \tfrac{1}{2}\sum_{\vec{k};\eta,\beta} \tilde{V}_{\eta\beta}(\vec{k}) e_\eta \mathcal{A}_{\vec{k}} e_\beta \mathcal{A}_{-\vec{k}}$$
$$- \sum_{\vec{k};\eta,\beta} \tilde{\tau}^{-1}_{\eta\beta}(\vec{k}) e_\eta \mathcal{A}_{\vec{k}} e_\beta a_{-\vec{k}} + \tfrac{1}{2}\sum_{\vec{k},\eta} e_\eta \dot{a}_{\vec{k}} e_\eta \dot{a}_{-\vec{k}} \tag{5.154}$$

where real elements of force matrices are

$$\tilde{V}_{\eta\beta}(\vec{k}) = \frac{1}{m}\sum_{\vec{l}} V_{\eta\beta}(\vec{l}) \exp\left(i\vec{k}\vec{l}\right) \tag{5.155}$$

$$\tilde{\tau}^{-1}_{\eta\beta}(\vec{k}) = \sum_{\vec{l}} \tau^{-1}_{\eta\beta}(\vec{l}) \exp\left(i\vec{k}\vec{l}\right) \tag{5.156}$$

Euler-Lagrange equations for the variables $e_\eta \mathcal{A}_{\vec{k}}$ and $e_\eta a_{\vec{k}}$ are respectively

$$e_\eta \ddot{\mathcal{A}}_{-\vec{k}} + \sum_\beta \left[\tilde{V}_{\eta\beta}(\vec{k}) e_\beta \mathcal{A}_{-\vec{k}} + \tilde{\tau}^{-1}_{\eta\beta}(\vec{k}) e_\beta \dot{a}_{\vec{k}}\right] = 0 \tag{5.157}$$

$$e_\beta \ddot{a}_{-\vec{k}} - \sum_\eta e_\eta \tilde{\tau}^{-1}_{\eta\beta}(\vec{k}) \dot{\mathcal{A}}_{\vec{k}} = 0 \tag{5.158}$$

Differentiating Eq. (5.157) with respect to time and replacing $(-\vec{k})$ for $\vec{k}$ we obtain

$$e_\eta \dddot{\mathcal{A}}_{\vec{k}} + \sum_\beta \left[\tilde{V}_{\eta\beta}(\vec{k}) e_\beta \dot{\mathcal{A}}_{\vec{k}} + \tilde{\tau}^{-1}_{\eta\beta}(\vec{k}) e_\beta \ddot{a}_{-\vec{k}}\right] = 0 \tag{5.159}$$

Substituting $e_\beta \ddot{a}_{-\vec{k}}$ from Eq. (5.158) into Eq. (5.159), we gain the equation for $\mathcal{A}_{\vec{k}}$ which after integration over t changes to

$$e_\eta \ddot{\mathcal{A}}_{\vec{k}} + \sum_\beta W_{\eta\beta}(\vec{k}) e_\beta \mathcal{A}_{\vec{k}} = C \quad (5.160)$$

where C is the integration constant and the force matrix is

$$W_{\eta\beta}(\vec{k}) = \tilde{V}_{\eta\beta}(\vec{k}) + \tilde{\tau}^{-1}_{\eta\beta} \sum_{\eta'} \tilde{\tau}^{-1}_{\eta'\beta}(\vec{k}) \frac{e_{\eta'}}{e_\beta} \quad (5.161)$$

Eq. (5.160) has the form of a standard equation for collective variables of the crystal lattice and it determines three frequencies $\varpi_s(\vec{k})$ (s =1, 2, 3), i.e., three branches of acoustic vibrations: one longitudinal branch (along $\vec{k}$) and two transverse ones (normal to $\vec{k}$). The equation for the frequencies $\varpi_s(\vec{k})$ has the form

$$\| \varpi_s^2(\vec{k}) - W_{\eta\beta}(\vec{k}) \| = 0 \quad (5.162)$$

But in our case, as seen from (5.161), the force matrix $W(\vec{k})$ comprises in addition to the elastic (electromagnetic nature) component $\tilde{V}(\vec{k})$ also the inerton component caused by overlapping the inerton cloud of each atom with the adjacent atoms and proportional to $(\tilde{\tau}^{-1})^2$.

It seems likely that in the crystal lattice under ordinary conditions the correction $(\tilde{\tau}^{-1})^2$ to the elastic matrix $\tilde{V}(\vec{k})$ is small. But given a sufficiently intensive external source of inertons, this correction can substantially increase and then the inerton component can show up explicitly. Indeed, given an external periodic inerton source with a frequency $\Omega_{\vec{k}}$, Eq. (5.158) is replaced by the generalized equation

$$e_\beta \ddot{a}_{-\vec{k}} - \sum_\eta e_\eta \tilde{\tau}^{-1}_{\eta\beta}(\vec{k}) \dot{\mathcal{A}}_{\vec{k}} = f_{\vec{k}\beta} \cos(\Omega_{\vec{k}} t) \quad (5.163)$$

With the permanently acting source, when the external force $f_{\vec{k}} > \tilde{\tau}^{-1} \dot{\mathcal{A}}_{\vec{k}}$, the equation

$$e_\beta \ddot{a}_{-\vec{k}} \approx f_{\vec{k}\beta} \cos(\Omega_{\vec{k}} t) \quad (5.164)$$

follows from (5.163). Integrating (5.164) over t and then substituting $e_\beta \dot{a}_{-\vec{k}}$ from (5.164) into (5.157), we obtain the equation

$$e_\eta \ddot{\mathcal{A}}_{-\vec{k}} + \sum_\beta \tilde{V}_{\eta\beta}(\vec{k}) = \sum_\beta \frac{\tilde{\tau}^{-1}_{\eta\beta}}{\Omega_{\vec{k}}} f_{\vec{k}} \sin(\Omega_{\vec{k}} t) \quad (5.165)$$

At a sufficiently large value of the permanent disturbance, e.g., along the projection e_1, it is easy to find from (5.165) the amplitude $\mathcal{A}_{k_1}^{(0)}$ of collective vibrations of atoms:

$$\mathcal{A}_{k_1}^{(0)} = f_{k_1} \frac{\tilde{\tau}_{11}^{-1} / \Omega_{\vec{k}}}{\varpi^2(\vec{k}) - \Omega_{\vec{k}}^2} \tag{5.166}$$

and, therefore, the amplitude of individual atom vibration

$$\xi_{n_1} = \frac{1}{\sqrt{Nm}} \sum_{k_1} \mathcal{A}_{k_1}^{(0)} \exp(ik_1 n_1) \tag{5.167}$$

(as is generally known, the consideration of friction enables the limited and constant sign of amplitude $A_{k_1}^{(0)}$ in (5.166)). Thus, we can see from expressions (5.167) and (5.166) that given the disturbing force $f_{\vec{k}}$ of the applied inerton field, the amplitude of atom vibrations in the crystal will increase, especially at resonance.

How can we test the obtained theoretical results? It can be done under natural conditions using the Earth's inerton cloud. In fact, since the Earth revolves around the Sun and rotates on its axis, it has to have its own inerton cloud that accompanies the Earth.

Two types of stationary inerton flows can be initiated on the terrestrial globe: (i) the orbital motion around the Sun with the velocity $\upsilon_{01} \approx 30$ km·s^{-1}, and (ii) the proper rotation; with this motion the velocity changes from $\upsilon_{02} = 0$ in the centre of the Earth to $\upsilon_{02} \approx 2\pi R_{\text{Earth}} / 24$ hours ≈ 462 m·s^{-1} at the equatorial surface. Structural bonds, which keep atoms in the globe, lead to the coherence of their motion. If we assume that the mean mass of atoms of the Earth is $\langle m \rangle = 30\, m_p$ (where m_p is the proton mass), then the de Broglie wavelength for the two types of the motion will be $\lambda_1 = h / \langle m \rangle \upsilon_{01} \approx 4 \times 10^{-13}$ m and $\lambda_2 = h / \langle m \rangle \upsilon_{02} \approx 1.5 \times 10^{-11}$ m, respectively. Substituting the values $\lambda_{1(2)}$ and $\upsilon_{01(02)}$ into relation $\Lambda = \lambda c / \upsilon_0$ (2.69), we acquire the amplitudes of inerton clouds of the moving atoms of the terrestrial globe: $\Lambda_1 \approx 8$ nm and $\Lambda_2 \approx 40$ μm. In both cases the overlap of inerton clouds is substantial ($\Lambda_1 / g_0 \sim 5$ and $\Lambda_2 / g_0 \sim 10^4$ where g_0 is the lattice constant), but due to inequality $\Lambda_2 >> \Lambda_1$ the degree of coherence between atoms is greater along the velocity vector $\vec{\upsilon}_{02}$ (i.e., along the East-West line) than along the orbital velocity vector $\vec{\upsilon}_{01}$.

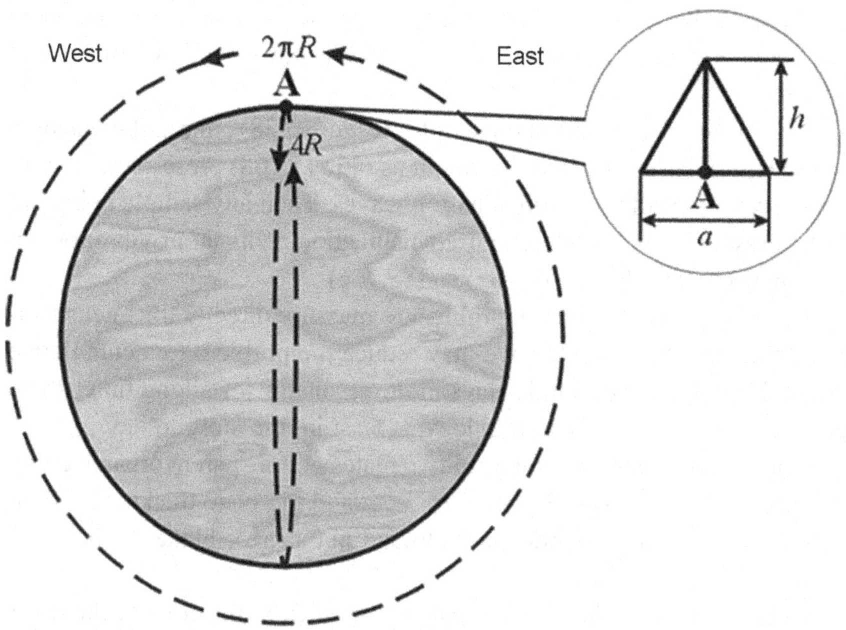

FIGURE 5.28 Generation of inerton waves in a sphere (in particular, the Earth) and the linear resonator of these inerton signals in the point A.

Let us imagine a sphere; if one sharply hits the sphere in a point A on its surface, two types of waves will be produced: longitudinal and transverse which travel across the surface of the sphere (Figure 5.28). Hence two types of inerton waves propagating in the terrestrial globe can be distinguished: (i) radial inerton (which in some moments are parallel/antiparallel with the orbital velocity vector $\vec{v}_{01}$ of the Earth; in the other moments the line of their motion coincides with the radial ray emanating from the Sun), and (ii) tangential inerton waves propagated over the surface zone of the Earth along the equatorial East-West line (i.e., along or against the rotational velocity vector $\vec{v}_{02}$ of the Earth on the equator). In the former case, the inerton wave front travels a distance $l_{\rm rad} = 4R_{\rm Earth}$ in the cyclic period and in the second case $l_{\rm tan} = 2\pi R_{\rm Earth}$, $R_{\rm Earth}$ is the radius of the Earth. From these two expressions we obtain the relation

$$l_{\rm tan} / l_{\rm rad} = \pi / 2 \qquad (5.168)$$

Apparently, relation (5.168) also characterizes the ratio between the wavelengths of the tangential and radial nth harmonics of inerton waves.

Therefore, if we insert in the point A (Figure 5.28) a three-dimensional figure whose base relates to the height as ratio (5.168) prescribes, we will have a resonator of the Earth's inerton waves. Then any sample being put in this resonator will undergo an intensification of its atom vibrations in accordance with relations (5.165) and (5.166).

Of course it is not an easy problem to measure directly the intensification of atom vibrations. But we may reduce the problem to a cumulative effect. Namely, if the sample stays in the resonator for a long time, some changes will occur in its morphology, at least in fine morphology.

Thus the resonator was made of two plates of transparent organic glass; the size of the plate was 20 cm × 16.5 cm and the plate thickness 3 mm. Inside we set a short wooden column and put a razor blade as shown in Figure 5.29.

Then we studied the small fragments taken from the edge of the razor blade by the scanning electron microscope JSM-35 (Japan) operated in secondary electron mode under 25 kV accelerated voltage. A number of blades were tested over a period of about two years. Besides, we examined also the duration of exposition. Below you can see micrographs (Figures 5.30 and 5.31), which show real changes of the fine morphological structure of the edges of the blade that stayed for 30 days in the resonator. These changes show that protuberances on the cutting edge became more extended, as if they underwent an exposure to some force.

Figure 5.32 exhibits the cutting edge of the blade whose edges were orientated to the north and south in the resonator. No changes in the fine morphological structure of the reference specimens are observed.

Thus we can see that the resonator used for the Earth's inerton waves (Figure 5.29) is a model version of an Egyptian pyramid. The Great Pyramid of Giza satisfies exactly the relation (5.168), namely, the ratio base: height = $\pi/2$. When was it built and who were its builders? This is the biggest mystery of our planet linked to ancient civilizations. Modern studies relate the time of construction of the Egyptian pyramids to the age over 10 thousand years ago. Bearing in mind the advanced technology used in their construction, which even to this day is not fully understood, is

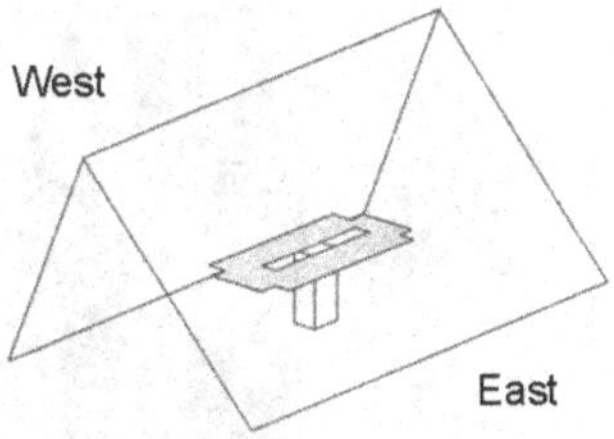

FIGURE 5.29 Resonator of inerton waves of the Earth. The razor blade is put inside the resonator, such that the Earth's inerton waves intensify atoms of the blade at its sharp edges.

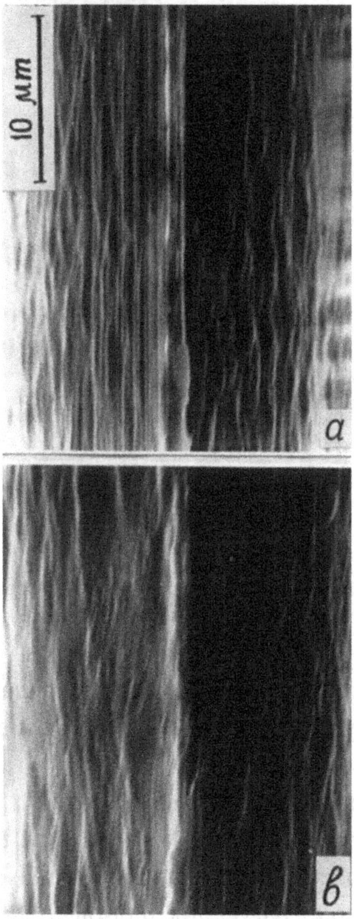

FIGURE 5.30 Micrographs of fragments of the cutting edge of a "Gillette" blade that stayed in the resonator for 30 days: *a* – the reference specimen, *b* – the test specimen.

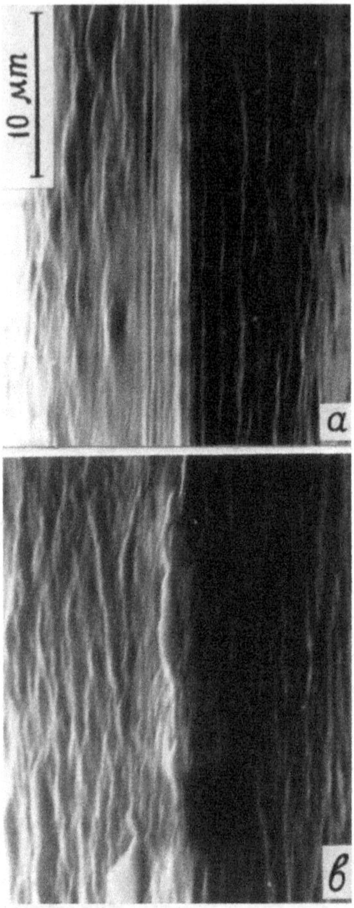

FIGURE 5.31 More micrographs of fragments of the cutting edge of a "Gillette" blade that stayed in the resonator for 30 days: *a* – the reference specimen, *b* – the test specimen.

it possible that the architects had recognized some subtle beneficial effects of inerton resonance within pyramids. It seems the architects knew about the Earth's inerton field in detail.

The rotation of the Earth was also revealed in a precise experiment carried out by Navia et al. [261]. Their goal was to look for signals of anisotropic light propagation as a function of a laser beam alignment to the Earth's motion (solar barycenter motion) obtained by COBE. Two raster search techniques were used. First, in the laboratory frame a laser beam scanned the space due to Earth's rotation. Second, mounted in

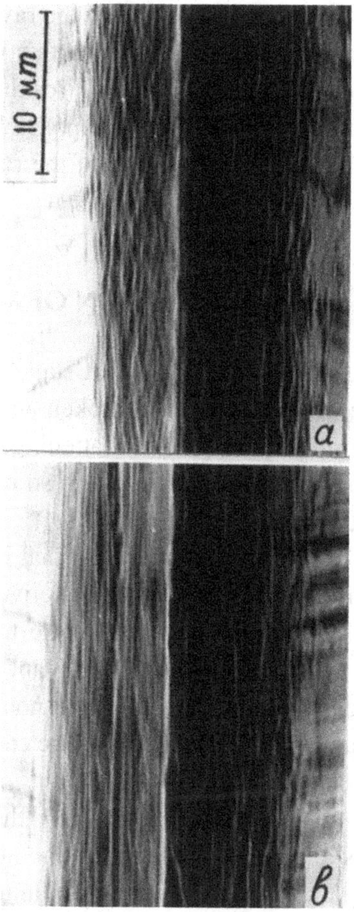

FIGURE 5.32 Micrographs of fragments of the cutting edge of a "Gillette" blade that stayed in the resonator for 30 days. The razor blade was oriented in the resonator to the north and south by its cutting edges. *a* – the reference specimen, *b* – the test specimen.

a turntable system a laser beam scanned actively the space by turning the table. The both methods exhibited that the path of light rays change with the motion of the Earth, and a predominant first order quantity with a $\Delta c / c = -(\upsilon / c) \cdot (1 + 2\bar{a}) \cos\vartheta$ signature with $\bar{a} = 0.393 \pm 0.032$ describes well the experimental results (υ is the velocity of the Earth and c is the speed of light). This result differs by 21% from the result $\bar{a} = -1/2$ predicted by special relativity that strictly prescribes the total isotropy of the ambient space. Thus Navia et al. [261] demonstrated that the motion of the Earth affects the course of the rays (which of course is possible only through the influence of the Earth's inertons on photons). The authors concluded by saying that their results point out that it is impossible to neglect the preferred frame imposed by cosmology, which in substance means that the inerton cloud of the rotating Earth intrudes the mentioned frame.

5.12 CROP CIRCLES: AN ELIMINATION OF MYSTICISM

Studies of crop circles [262–266] show that stalks inside the circles are bent up to ninety degrees without being broken and something had softened the plant tissue at the moment of flattening. Something stretches stalks from the inside; sometimes this effect is so powerful that the node looks as exploded outwards from the inside.

Researchers [262–266] have hypothesized that crop circle formations involve organized ion plasma vortices, which deliver energy components from the lower atmosphere sufficient magnitude to produce bending of stalks, the formation of expulsion cavities in plant stems and significant changes in seedling development. It should be noted that an idea of the origin of crop circles associated with atmospheric energy and/or UFO phenomenon is wildly accepted.

However, we have also to mention paper [267] that criticized the above mentioned hypothesis and results; moreover, the authors concluded that crop circles should be the result of rogues who flattened grass using rope and wood plank. It seems this is rather a typical conclusion of armchair scientists who have never dealt with both crop circles and the real samples of flattened plants.

On the other hand, researchers who study geophysical processes and earthquakes have noted about possible regional hemispherical magnetic fields that might be generated by vortex-like cells of thermal-magmatic energy, rising and falling in the Earth's mantle [268]. Another important factor is magnetostriction of the Earth's crust – the alteration of the direction of magnetization of rocks by directed stress [269, 270]. Moreover, recent study of Yamazaki [271] has suggested a possible mechanism of earthquake triggering due to magnetostriction of rocks in the crust. The phenomenon of magnetostriction in geophysics is stipulated by mechanical deformations of magnetic minerals accompanied by changes of their residual or induced magnetization. These deformations are specified by magnetostriction constants, which are proportional coefficients between magnetization changes and mechanical deformations. A real value of the magnetostriction constant of the crust is estimated as about 10^{-5} ppm/nT, which is a little larger than for pure iron. Yamazaki's [271] calculation shows that effects connected to the magnetostriction of rocks in the crust can produce forces of nearly 100 Pa/year and even these comparatively small stress changes can trigger earthquakes.

Of course, weaker deformities related to magnetostriction of rocks could precede larger deformations. These are the magnetostriction deformations that we [272] put in the foundation of the present study of field circles because the effect of magnetostriction has well been proven as a powerful source of inertons in our laboratory experiments (see Section 5.5).

The thickness of the crust is about 20 km. The mantle extends to an average depth of 3000 km and it is made of a thick solid rocky substance. Due to dynamical processes in the interior of the Earth, magnetostrictive rocks contract with a coefficient of about $C = 10^{-5}$ [271] (C is a proportional coefficient between magnetization changes and mechanical deformations), which is a trigger mechanism for the appearance of a flow of inerton radiation. This flow of inertons shoots up from the depths through channels in the mantle and crust. Such channels are composed of typical terrestrial materials with some non-homogenous inclusions (e.g., limestone in granite) penetrating down tens or hundreds of kilometers from the surface of the terrestrial globe.

A mantle-crust channel can be modeled as a cylindrical tube, which has a cross-section area equal to A. An inerton flow of inertons moves from a deep rocky magnetostrictive source to the surface of the globe. The inner surface of the channel has to reflect inerton radiation, at least partly, so that the inerton flow will continue to follow along the channel to its output, i.e., the surface of the Earth.

The inner surface of a mantle-crust channel can be described by a rotating potential U, holds a flow of inertons spreading along the channel from an underground source. Let μ be the mass of an effective batch of terrestrial inertons from this source, which interact with a grass stalk. The planar motion of such a batch of inertons in the central field is described by the Lagrangian

$$L = \tfrac{1}{2}\mu\left(\dot{r}^2 + r^2\dot{\phi}^2\right) - U(r,\ \dot{\phi}) \tag{5.169}$$

which is written here in polar coordinates r and ϕ; the dot standing for the derivative with respect to time. To model a spreading inerton field, the potential should include a dependence on the angular velocity, $U(r,\ \dot{\phi})$, which means that we involve the proper rotation of the Earth relative to the flow of inertons; besides, a supplement to the angular velocity may be related to the relative motion of colliding rocks in the Earth's crust. The discussed potential can be chosen in the form of the sum of two potentials:

$$U(r,\ \dot{\phi}) = \tfrac{1}{2}\beta_1 r^2 + \tfrac{1}{2}\beta_2 r^2\dot{\phi} \tag{5.170}$$

In the right-hand side of expression (5.170) the first term is a typical central-force harmonic potential, which describes an elastic behavior of the batch of inertons in the channel and the surrounding space; the second term includes a dependence on the azimuthal velocity, which means that it depicts the rotation-field potential. The introduction of this potential allows us to simulate more correctly the reflection of inertons from the walls of the mantle channel, which of course only conditionally can be considered round in cross-section.

The Euler-Lagrange equations of motion are

$$\ddot{r} - r\dot{\phi}^2 + r\beta_1 / \mu + r\dot{\phi}\,\beta_2 / \mu = 0 \tag{5.171}$$

$$r\ddot{\phi} + 2\dot{r}\cdot\left(\dot{\phi} - \beta_2 / (2\mu)\right) = 0 \tag{5.172}$$

These equations can be integrated explicitly or solved numerically at the given initial conditions $r(0)$, $\dot{r}(0)$, $\phi(0)$, $\dot{\phi}(0)$, and the trajectory of motion can be plotted in rectangular coordinates $\{r\cos\varphi,\ r\sin\varphi\}$. The second equation represents the conservation of the angular momentum M:

$$\frac{d}{dt}\left[\mu r^2\cdot\left(\dot{\phi} - \frac{\beta_2}{2\mu}\right)\right] = 0 \ \text{ or } \ M = \mu r^2\cdot\left(\dot{\phi} - \frac{\beta_2}{2\mu}\right) = \text{const} \tag{5.173}$$

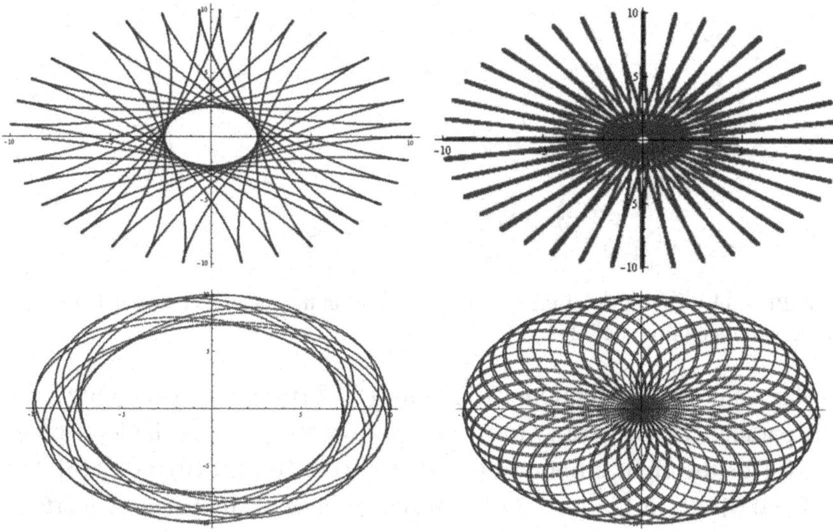

FIGURE 5.33 Trajectories of the motion of inertons in the rotating central field (5.169). (With permission from Ref. 272.)

Figure 5.33 shows four possible trajectories at particular values of the parameters. The parameters used (reading top left to bottom right) are: $\beta_1 / \mu = 1$ s^{-2}; $\beta_2 / \mu = 0.1$, 0.5, and 2 s^{-1}; $r(0) = 10$ m; $\dot{r}(0) = 0$; $\phi(0) = 0$; $\dot{\phi}(0) = 0.01$ and 1 s^{-1}. The velocity $|\dot{\vec{r}}| = \sqrt{\dot{r}^2 + r^2\dot{\phi}^2}$ of the batch of inertons along the flat trajectories at these parameters may reach the maximal value $\upsilon_{\max} \approx 12$ m·s^{-1}. The corresponding acceleration $|\ddot{\vec{r}}| = \sqrt{(\ddot{r} - r\dot{\phi}^2)^2 + (2\dot{r}\dot{\phi} + r\ddot{\phi})^2}$ of the batch of inertons has the maximum value $a_{\max} \approx 10$ m·s^{-2}.

In the case of the Newton-type potential

$$U(r, \dot{\phi}) = -\frac{\gamma}{r} + \frac{\beta_2}{2} r^2 \dot{\phi} \tag{5.174}$$

Then the equations of motion for the Lagrangian (5.169) become

$$\ddot{r} - r\dot{\phi}^2 + \gamma / (\mu r^2) + r\dot{\phi}\, \beta_2 / \mu = 0 \tag{5.175}$$

$$r\ddot{\phi} + 2\dot{r} \cdot \left(\dot{\phi} - \beta_2 / (2\mu)\right) = 0 \tag{5.176}$$

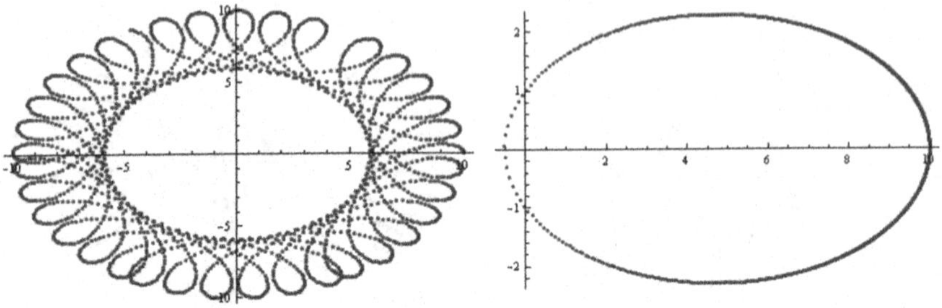

FIGURE 5.34 Trajectory of the motion of inertons in the rotating central field (5.173). (With permission from Ref. 272.)

The solutions to equations (5.175) and (5.176) are shown in Figure 5.34. The parameters used are $\gamma / \mu = 1$ m$^3 \cdot$s^{-2}, $\beta / \mu = 0.1$ s^{-1} (for the left figure) and 0 (for the right elliptic trajectory); $r(0) = 10$ m; $\dot{r}(0) = 0$; $\phi(0) = 0$; $\dot{\phi}(0) = 0.01$ s^{-1}. In Figure 5.33 the acceleration a of the batch of inertons changes from 10 to 15 m$\cdot$s^{-2}. Note the right elliptic trajectory is the solution to the equations of motion of a batch of inertons for the case of simplified potential (5.174), namely, when it is represented only by the Newton-type potential $U(r) = -\gamma / r$.

The conditions under which the stalks of herbaceous plants bend due to mantle/crust inertons can be determined in the following way. A stalk of a plant is modeled as an elastic rod (Figure 5.35), which can be deformed by an external force f distributed uniformly over the rod length l. This external force is caused by a flow of inertons that are emitted from the mutual collisions of subterranean rocks as described above. The rod profile in the projections to the vertical f_y and horizontal f_x axes is described following Ref. [273].

The horizontal force f_y is derived from expressions (Figure 5.35b)

$$x = \sqrt{\frac{IE}{2f_x}} \int_0^{\vartheta} \frac{\sin\vartheta \, d\vartheta}{\sqrt{\sin\vartheta_l - \sin\vartheta}},$$

$$y = \sqrt{\frac{2IE}{f_x}} \left(\sqrt{\sin\vartheta_l} - \sqrt{\sin\vartheta_l - \sin\vartheta} \right) \qquad (5.177)$$

Here $I = \pi R^4 / 4$ is the rod's moment of inertia, R is the rod's radius, and E is the Young's modulus of the rod's material. The length of the rod is explicitly given as

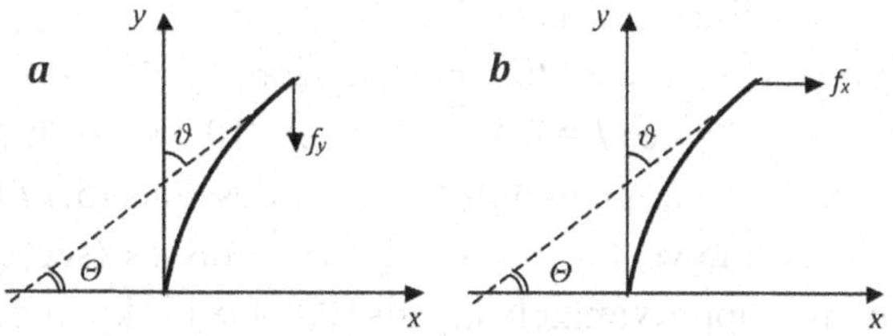

FIGURE 5.35 Elastic rod model (with permission from Ref. [272]).

$$l = \sqrt{\frac{IE}{2f_x}} \int_0^{\vartheta_l} \frac{d\vartheta}{\sqrt{\sin\vartheta_l - \sin\vartheta}} \tag{5.178}$$

In this case the maximum bending angle should be smaller than $\pi / 2$ (no such a force exists that can bend the rod by this angle). So, we select the maximum bending angle at $\vartheta_l = \pi / 3$ and write the corresponding relationship between the rod's length and the acting force:

$$l \approx \sqrt{\frac{IE}{2f_x}}\, 2.61 \text{ and } f_x \approx 3.41\; IE / l^2 \tag{5.179}$$

The vertical force f_y is derived from expressions (Figure 5.35*a*)

$$x = \sqrt{\frac{2IE}{f_y}} \left(\sqrt{1 - \cos\vartheta_l} - \sqrt{\cos\vartheta - \cos\vartheta_l} \right),$$

$$y = \sqrt{\frac{IE}{2f_y}} \int_0^{\vartheta} \frac{\cos\vartheta \, d\vartheta}{\sqrt{\cos\vartheta - \cos\vartheta_l}} \tag{5.180}$$

The length of the rod at the maximum bending $\vartheta_{\max} = \vartheta_l = \pi / 2$ is

$$l = \sqrt{\frac{IE}{2f_y}} \int_0^{\pi/2} \frac{d\vartheta}{\sqrt{\cos\vartheta}} = \sqrt{\frac{IE}{f_y}}\, 1.854 \tag{5.181}$$

Hence, we come to an expression for the force required to bend the rod by a $\pi / 2$ angle:

$$f_y \approx 3.44\; IE / l^2 \tag{5.182}$$

which is practically the same as f_x (5.179).

Let us evaluate the value of the breaking force $f = f_x \cong f_y$. We have to substitute numerical values $l = 0.5$ m, $R = 1.5\times10^{-3}$ m for the rod and the value of elasticity (Young's) modulus E to expressions (5.179) or (5.180). The value of E is known for many different grasses (see quotations in Ref. [272]) and it is approximately equals $(0.8\text{-}1)\times10^9$ kg·m·s^{-2}. In the case of wheat we can take a little bit larger value $E \approx 3\times10^9$ kg·m·s^{-2}, which gives for the horizontal breaking force (5.179)

$$f_{\text{break}} = f_x \approx 3.41 I E_{\text{Young}} / l^2 \approx 0.163\ \text{N} \qquad (5.183)$$

Besides, in addition to the elasticity modulus grass stalks are characterized by a bending stress, yield strength (tensile strength) and shearing stress. These parameters range from 7×10^6 to about 50×10^6 kg·m·s^{-2} and, hence, significantly decrease the real value of f, which is capable of bending stalks. For example, putting for E the value of the maximum tensile stress 50×10^6 kg·m·s^{-2} we obtain the bending non-breaking force

$$f_{\text{bend}} = f_x \approx 3.41 I E_{\text{tens}} / l^2 \approx 0.0027\ \text{N} \qquad (5.184)$$

The gravity force acting on the rod is

$$f_{\text{grav}} = mg = \rho V g = \pi \rho R^2 l g \approx 0.033\ \text{N} \qquad (5.185)$$

where ρ is the rod's material density of about $\rho = 10^3$ kg·m^{-3}, m and V are its mass and volume, and g = 9.8 m·s^{-2} is the acceleration due to gravity.

Thus we may conclude that any extraneous force F applied to a grass stalk will be able to fold the stalk to the ground if the value of the force satisfies inequalities

$$f_{\text{bend}} \le F < f_{\text{break}} \qquad (5.186)$$

Now we have to estimate the intensity of inerton radiation needed to form a crop circle of total area, for instance, $A \approx 100$ m^2. Let m_{rocks} be the mass of the mantle-crust rocks that generate inertons owing to their magnetostriction activity. We have to take into account the magnetostriction coefficient C that describes an extension strain of the rocks. In view of the fact of that low frequencies should accompany geophysical dynamical processes, we can assume that the striction activity of a local group of rocks occurs at a low frequency ν (i.e., rocks collide N times per a unit time Δt producing radiation of inertons). Having these parameters, we can evaluate a flow of mass μ_Σ that is ejected in the form of inerton radiation at the striction of rocks: $\mu_\Sigma \approx NCm_{\text{rocks}}$.

If we put $m_{\text{rocks}} \sim 10^7$ kg, $C \sim 10^{-5}$, and $N = 5$ we obtain $\mu_\Sigma \approx 500$ kg. This mass μ_Σ is distributed along the area of A in the form of a flow of the inerton field. Let each square meter be the ground for the growth of 1000 stalks. Then 10^5 stalks can grow in the area of $A = 100$ m^2. This means that each stalk is able to catch an additional mass $\mu = \mu_\Sigma / 10^5 = 5$ g from the underground inerton flow; this value is of the order of the mass of a stalk itself. Inertons being absorbed by the stalks lead to an increase of the stalks' mass at least for a short time.

Knowing the mass $\mu = 5 \times 10^{-3}$ kg of the batch of inertons, which interacts with a stalk, and the acceleration of this inerton batch $a = 10$ to 15 m·s^{-2}, we can assess the force of inertons, which counteracts the weight of the stalk while the absorbed inertons increase the stalk's mass, and in addition the force flattens the stem: $F = \mu a \approx 0.05$ to 0.075 N. This estimation exceeds not only the threshold bending force f_{bend} (5.184), but also the gravity force f_{grav} (5.185), while still the inerton force F does not physically break the stalk, because the value of F still satisfies the inequalities (5.186).

Once again, the force F is primarily an upward force but with a lateral flattening component. For a while the stalk becomes weightless, which allows the x-component of the force F to deflect the stalk horizontally. Thus, the stalk is twisted to one side by the angular momentum of the rotating inerton field. The stalk also becomes more massive due to absorbed inertons and its tissues have to change (become softer). Therefore, the stalk cannot remain upright and, being forced sideways, it rapidly collapses under its additional weight and weakness of the stem.

A flow of mass, which is going as a pulse of inertons from the interior of the Earth to its surface, partly countervailes the gravitational acceleration at the Earth surface $g = GM_{\text{Earth}} / R^2_{\text{Earth}} = 9.81$ m·s^{-2}. This statement can be verified in places where crop circles appear most frequently.

Basically, crop circles can be treated as inverse inerton astronomy, as they project subterranean inerton flows onto the surface of the Earth.

The model described is a typical kaleidoscope model that gives a static description of inerton structures. A bunch of inertons going from the Earth's interior is reflected from the walls of a particular geometry of mantle/crust channel. Multiple reflections from the walls produce the pattern shown in Figures 5.32 and 5.33. Such a model is an analogy of geometrical optics with light reflecting from mirrors. Combining the rotating central field model and the kaleidoscope model can generate yet more complex patterns.

Sometime at geophysical studies a gravimeter detects gravity anomalies (see, e.g., Ref. [274]), which may point to the presence of sources of an inerton radiation similar to the mentioned above in local places of the Earth's crust and mantle.

KEYWORDS

- **Bose-Einstein condensation**
- **clusters**
- **crop circles**
- **crystal lattice**
- **diffraction**
- **diffractionless light**
- **double-slit experiment**
- **electron droplet**
- **inertons**
- **multiphoton photoelectric effect**
- **pyramid power**
- **sonoluminescence**

CHAPTER 6

QUARKS AND HADRONS IN THE TESSELLATTICE

CONTENTS

6.1 THE DISCOVERY OF QUARKS

In 1961 Gell-Mann [275] and Neeman [276] suggested a "periodic table" of the known hadrons, which also included new predicted particles that were later discovered. Their approach was based on using the $SU(3)$ symmetry as a correct representation of physical reality. The regularities of the $SU(3)$ classification scheme required three new fundamental particles that were recommended by Gell-Mann [277] and Zweig [278] in 1964. Those particles were named 'quarks.' Each of the fundamental quarks was given its own name: "up" or u, "down" or d and "strange" or s; the corresponding antiquarks were also predicted. Thus mesons are built from two quarks (or quark and antiquark), while baryons are composed of three quarks.

The proton was considered as a combination of two up quarks plus a down quark, i.e., uud, while the neutron is represented as udd. In such a manner, charged quarks should be fractional. So Gell-Mann and Zweig assumed that the up quark has the charge $+2e/3$ and each of the quarks d and s has the charge $+e/3$. This allowed Gell-Mann and Zweig to prescribe correct charges to all the known mesons and baryons. In the early 1970s the existence of quarks was proved in experiments on non-elastic scattering of electrons from the interior of neutrons by a powerful electron beam.

In 1964 Greenberg [279] introduced the notion of "color charge" to explain how quarks could coexist inside hadrons in otherwise identical quantum states without violating the Pauli exclusion principle. Greenberg's work gave rise to the foundation of the theory of quantum chromodynamics (QCD). Fermi-Dirac statistics accounts for each (fundamental) flavor of quarks being classified into three colors, which brings to QCD the strong interaction that operates by color charges exchanging color gluons.

Quarks interact through quanta of strong interactions, which are gluons, and they can also interact through $W^{\pm}$ and Z^{0} bosons – carriers of electroweak interactions. All these quanta are called fundamental bosons. The modern Standard Model includes a primary particle called the Higgs boson, which is needed for an abstract formalism to launch a family of massive particles.

A theory of hadrons was developing from earlier partons to modern confinement. The parton model was proposed by Feynman [280] to analyze high-energy hadron collisions. Later on it was found that partons describe quarks and gluons.

There also exist other approaches describing the behavior of quarks in hadrons, which try to bring some physical ideas into the highly abstract formalism of QCD. In work [99] we reviewed approaches based on QCD, the Nambu–Jona-Lasinio model [281], the Skyrme model [282], the MIT bag model [283–286] and the topological soliton model [287], which are the most accepted among physicists.

Nevertheless, in the forefront of particle physics, a significant gap remains in understanding the causes of the stability of baryons, the quark confinement, the nature of spin-½ of baryons and the origin of nuclear forces. All these points will be elucidated in this chapter.

The formalism of QCD was developed based on quantum mechanics and quantum electrodynamics, which themselves were elaborated in abstract phase spaces, not the ordinary physical space. The Standard Model of particle physics combines all fundamental interactions in a unified theory. Nevertheless, this *theory of everything* requires clarification of its basic notions (mass, particle, charge, lepton, quark, Compton wavelength, de Broglie wavelength, wave-particle, matter waves, wave ψ-function, spin, Pauli principle, etc.), which automatically and absolutely demands *the theory of something.*

6.2 DEEPER PRINCIPLES: INTEGER COLORLESS CHARGES

In the Introduction and Section 1.1 we have discussed the development of the concept of space in the second part of 19th century. In particular, we emphasized that the ideas of Riemann and especially Helmholtz that a rigid body is able to move in space without changing its state of rigidity became of paramount importance. The idea of the stability of a moving solid object further allowed researchers to introduce laws of symmetry to the study of the equations of motion. The laws of symmetry make it possible to reduce the number of equations that describe the motion, find conditions for the manifestation of symmetric properties and hence finally admit the analytical solutions.

On the other hand, as has been shown by Bounias and the author, the physical space itself is the origin of matter and a particle is a part of space. Therefore any particle in motion always interacts with space, which results in periodic changes to the particle's shape and properties in its course of motion. Hence particles cannot be considered as completely rigid objects in principle. This automatically means that at least some of the basic postulates of QCD require reconsideration as a matter of urgency. First, the submicroscopic concept cannot support the idea of the fractional charge of quarks, i.e., $\frac{1}{3}$ and $\frac{2}{3}$ of e because the definition of electric charge (Sections 4.1 and 4.2) excludes this possibility; the elementary charge can only be an integer, $\pm e$.

Although the majority of theoretical physicists working in the area of high-energy physics still undeniably believe that the hypothesis of fractal charges ($\pm\frac{1}{3}e$ and $\pm\frac{2}{3}e$) is correct, deep inelastic experiments do not rule out even integer-charge quarks [288–290]: in higher orders of perturbation the integer-charge quark model gives results closer to those of the fractional charge quark model and the properties like factorization of mass singularities, which have been shown for the fractional charge quark model, may also be assumed for the integer-charge quark theory. Rajasekaran and Rindani [288] point out that a clear and unambiguous high-energy test, which distinguishes the one model from the other, has not yet been found: "As long as it leads to almost similar empirical phenomena to that of the fractional-charge quark model, it is going to be very difficult to rule it out experimentally. It may even be the right model! Although exact $SU(3)c$ symmetry appears to be an elegant hypothesis, exact $SU(3)c \times U(1)$ does not look so elegant. Why can't the degenerate gluons and photon mix and break the symmetry?"

Thus in QCD the integer-charge quark is not ruled out, which was shown by conventional field methods [288–291]. Integer-charge quark theories fit experimental data far better than the standard model does [289, 290]. In particular, Ferreira [290] has reviewed the evidence for fractional quark charges and argued that they are not conclusive. On the other hand, since integer-charge quark theories are renormalizable they demonstrate a good comparison with experiment data, which is held for any order of perturbation theory; these theories also predict identical rates for meson radiative decays. In the end Ferreira [290] states: "Regardless of whether one believes in the integer charge quark models or not, it seems clear they do a better job than the Standard Model at describing the two-photon data."

LaChapelle [291] notes that quarks and leptons furnish the same $U_{EW}(2)$ representation, which means that they are manifestations of the same underlying field and "this would seem to require a phase change in certain regions of field phase space. If this characterization turns out to be correct, it would relate to some of the parameters of the Standard Model, and, hopefully, aid in the search for an underlying theory."

Further evidence that quarks possess an integer charge comes from the beta decay (β-decay) that is a type of radioactive decay in which a proton is transformed into a neutron, or vice versa, inside an atomic nucleus as for example in the β^--decay of carbon-14 into nitrogen-14: ${}^{14}_{6}\mathrm{C} \rightarrow {}^{14}_{7}\mathrm{N} + \mathrm{e}^- + \bar{\nu}_e$. The emitted electron has the integer charge "$-e$," which directly indicates that inside the parent nuclide there were only integer charges.

The introduction of integer flexible quarks will require changes to the formulas for rigid mesons and also the formulas for the rigid proton (uud) and neutron (udd). It is really strange, why do researchers adhere to these formulas? Rigid objects are rejected by the quantum mechanical formalism itself that high-energy theorists still use, because by the quantum mechanical rules all particles are fuzzy in a non-determined volume, which the wave ψ-function prescribes. In the submicroscopic approach all these conceptual difficulties are easily removed.

Color charges were introduced in the theory of quarks to account for the quark's coexistence inside hadrons, because the researchers believed that the violation of the Pauli exclusion principle is forbidden at any conditions. Nevertheless, the exclusion principle is only a phenomenological rule that has been clarified in submicroscopic terms in Sections 3.4 and 5.3: The exclusion principle allows the violation at a number of different conditions. Moreover, we will see below that violating the Pauli principle is even necessary in the dynamics of quarks. This means that such complicated part of the theory of quarks as chromodynamics, has to be discarded – QCD should be substituted for the Quark Dynamics (QD) theory (because charges are not fractional or in other words they are colorless).

The structure of quarks in the standard model is described by unitary symmetry $SU(3)$. The earlier $SU(6)$ theory [292, 293] successfully explained many experimental facts, but later was rejected because it was thought that from the fundamental point of view, $SU(6)$ was contradictory. $SU(6)$ theory assumes that quarks obey the Fermi-Dirac statistics, but in reality it looks as if they obey the Bose-Einstein statistics.

This paradox was explained by Nambu [294] on the example of the Ω^- baryon. Its spin is 3/2, and the strangeness –3, so it occupies a state in which the spins of all three S-quarks are parallel. But this state is symmetric under permutation of any pair of particles, in contradiction with the requirement of the Fermi-Dirac statistics. However if for this situation one applies the Bose-Einstein statistics, then for the Ω^- particle (as well as for other baryons) the values derived by using the $SU(6)$ become consistent with experimental data. So, it turns out that in baryons quarks behave as bosons, but the quarks are separated. Thus, it was recognized that the theory of $SU(6)$ connected the properties that were mutually exclusive and therefore it was too unrealistic.

However, in terms of the proposed deterministic submicroscopic theory this imperfection of the *SU*(6) theory becomes its advantage. In fact, in our model quarks obey the Fermi-Dirac statistics. However, in hadrons quarks are ultra-relativistic and hence their clouds of excitations are small and do not overlap.

The absence of overlapping of the quarks' clouds immediately prevents the Pauli exclusion principle. The cloud irradiated by one quark is absorbed by another quark. The situation is similar to the behavior of dilute gases of atoms under laser cooling, when a moving atom emits the atom's inerton cloud and its neighbor completely absorbs the cloud (Section 5.3).

In nuclear physics the proton and neutron are different only in their isospin projection. However, this notion does not seem fundamental but simply is useful in the appropriate algebra.

6.3 THE BEHAVIOR OF QUARKS IN THE TESSELLATTICE

The problems outlined in the previous section clearly demonstrate the requirement for a deeper understanding of the structure of quarks and the nature of their combinations. We will proceed to examine this with reference to their tessellattice interactions.

The static quark potential is presented in the literature by an asymptotic expansion

$$V(r) = \sigma r - \alpha / r + \mu_{\mathrm{r-d}} + O(1/r^2) \tag{6.1}$$

which is proved experimentally. The string tension σ is computed in continuum QCD. The inputs are the standard values of the vacuum condensates. The output is $\sqrt{\sigma} \approx 0.5$ GeV and is very insensitive to quarks. The term α/r is the quantum correction that characterizes the relativistic bosonic string (Figure 6.1) and $\mu_{\mathrm{r-d}}$ is a regularization-dependent mass.

The string tension σ is the subject of intensive theoretical and experimental studies, as was mentioned in Ref. [295]. The value of σ is evaluated as a function of temperature; the string tension is compared with the behavior of parameters of ferromagnets and superconductors relating them to confinement.

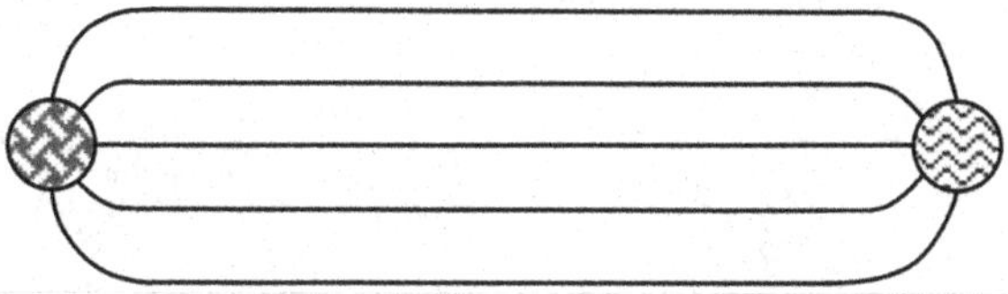

FIGURE 6.1 Bosonic string that unifies two quarks, as it is seen in QCD.

The second term in the quark potential (6.1) includes the coupling constant

$$\alpha_s(Q^2) = \frac{1}{b\ln(Q^2/\Lambda^2_{\text{QCD}})} + \ldots \tag{6.2}$$

where the free parameter in QCD, or the QCD scale parameter Λ_{QCD}, which refers to a particular definition of the effective coupling, may vary from 0.1 to 0.5 GeV, though researchers tend to the value of $\Lambda_{\text{QCD}} = 217 \pm 25$ MeV [295] or rather 220 MeV [296]. Bethke [296] emphasizes that if Q^2 becomes larger, $\alpha_s(Q^2)$ asymptotically decreases to zero, but the constant $\alpha_s(Q^2)$ increases at smaller Q^2; for example, in the case of the Z^0 boson whose energy is 90 GeV, the constant $\alpha_s(Q^2)\big|_{90\text{ GeV}} = 0.12$ and $\alpha_s(Q^2)\big|_{35\text{ GeV}} = 0.14$ [297]. Bethke [296] mentions that $\alpha_s(Q^2)$ can exceed unity for energies in the range 100 MeV to 1 GeV. The energy scale below the order of 1 GeV is called the non-perturbative region where confinement sets in.

The spatial separation between quarks goes as $\lambda\!\!\!^{-} = \hbar / Q$. At a very short distance, which means a high value of Q, the coupling between quarks decreases, vanishing asymptotically. At the limit of a very large value of Q, quarks can be considered to be "free," which is called an asymptotic freedom.

An important provision of the submicroscopic concept is the interaction of a moving particle with the space, i.e., the tessellattice. Therefore a moving quark interacts with the tessellattice. The interaction of quarks with the tessellattice introduces some nonlinearity in the behavior of quarks.

As has been discussed in Sections 1.8.4 and 1.8.5, the quark is in fact a bubble in the tessellattice (Figure 6.2). It is reasonable to assume that by analogy with leptons, the quark's kernel cell constantly exchanges

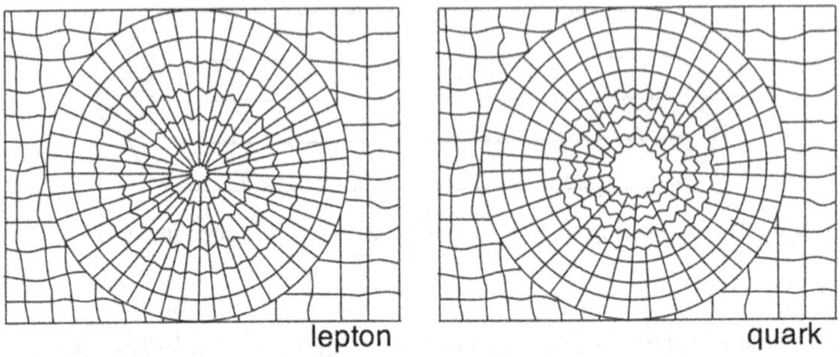

FIGURE 6.2 Lepton and quark in the tessellattice.

with the quark's coat by excitations: the inflated state periodically decomposes and excitations of inflated fragments spread and oscillate in the medium of compressed (massive) cells of the quark's deformation coat. Due to the central symmetry, such oscillations of inflated excitations can be compared to oscillations of a gas in a real bubble, in which the gas oscillations obey the inverse law, $\propto 1/r$. Hence the vibration energy of the quark

$$E_{\mathrm{vib}} \propto 1/r \tag{6.3}$$

These excitations are 'inflated inertons,' or 'inverse inertons'; since they accompany the quark, we may name them *qinertons* (quark-inertons). So the origin of the second term in the quark potential (6.1) is related to the standing wave emitted by the quark itself; the quark emits qinertons that behave as mathematical physics prescribes: the front of the standing spherical wave spreads in line with expression (6.3).

The energy scale below the order of 1 GeV is called the non-perturbative region. This region corresponds to large distances at which the inter-quark coupling increases and in this case, it becomes impossible to detach individual quarks from a hadron. In the quark potential (6.1) the first term, which increases with the distance between quarks, is responsible for the inter-quark coupling, which is called the phenomenon of confinement.

The nature of quark confinement is visualized with the use of an elastic bag (bubble) that allows the quarks to move around freely and this fact was described in a review paper on the MIT bag model [298]: When small bubbles in a vacuum medium are created with a characteristic size of one fermi, the hadron constituent fields may propagate inside the bubbles in normal manner; the bubble (bag) is stabilized against the pressure of the confined hadron constituent fields by vacuum pressure and surface tension.

The naked quark in the tessellattice (Figure 6.2) is unstable and collapses under the pressure of the whole space and in such a case, its final state shall be a stable lepton (the left picture in Figure 6.2). The transformation of a quark into an electron is a direct consequence of its structure, as Figure 6.2 demonstrates but by interacting with each other, two or more quarks jointly form a stable hadron. The stable hadron has to be an analog of a rigid rotator: two bubbles rotating around each other are able to resist the pressure from the entire space. Besides, the merger of bubbles resulting in a united bubble decreases the Young-Laplace pressure, which also stabilizes the system of the two quarks.

6.3.1 THE PHENOMENON OF CONFINEMENT IN THE TESSELLATTICE

Let two bubbles formed in a medium interact (see, e.g., Ref. [299]), which means that their surfaces overlap forming a structure shown in Figure 6.3. At the point of contact the bubbles make a channel with a cross-section $\pi\varepsilon^2$. Disappearance of two borders between the bubbles in a local place means that the energy of the bubbles decreases by a value of $\Delta E = -2\gamma\pi\varepsilon^2$ where γ is the coefficient of the surface tension of the bubble. If we put $2\chi << 2R$, where R is the radius of the bubble and χ is the range of overlapping of the bubbles, this will mean that the total fusion of the bubbles does

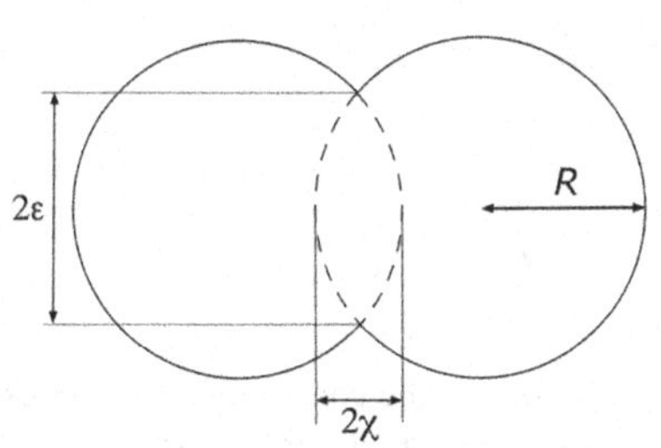

FIGURE 6.3 Agglutination of two bubbles with quarks located at the bubbles' centers.

not occur. Then from the equality $R^2 = \varepsilon^2 + (R-\chi)^2$ we get $\varepsilon^2 \cong 2R\chi$. Hence the energy of attraction of two bubbles becomes

$$\Delta E_{\text{att.}} = -2\pi\gamma\chi \cdot (2R-\chi) \cong -4\pi\gamma\chi R \tag{6.4}$$

and R is the radius of the quark's bubble (which can be associated with the quark's Compton wavelength). In Eq. (6.4) substituting R by r (the distance between the quarks) and putting for the coefficient $4\pi\gamma\chi = \sigma$, we obtain

$$\Delta E_{\text{att.}} \approx -\sigma r \tag{6.5}$$

Combining Eqs. (6.4) and (6.5) we arrive at the static quark potential (6.1).

Thus the confinement, i.e., a linear dependence of the interaction energy of quarks on a distance r, is quite natural and originated from the geometry of the contact of two bubbles, i.e., two quarks solvated with the quark's deformation coat filled with inflated excitations, or qinertons. The interaction proportional $1/r$ is caused by the emission of standing spherical waves of qinertons by a quark in the quark's bubble (Figure 6.4).

Each quark experiences an outside compression pressure

$$\mathcal{P}_{\text{compr}} = \mathcal{P}_0 + 2\gamma / R \tag{6.6}$$

on the side of the tessellattice and the bubble surface.

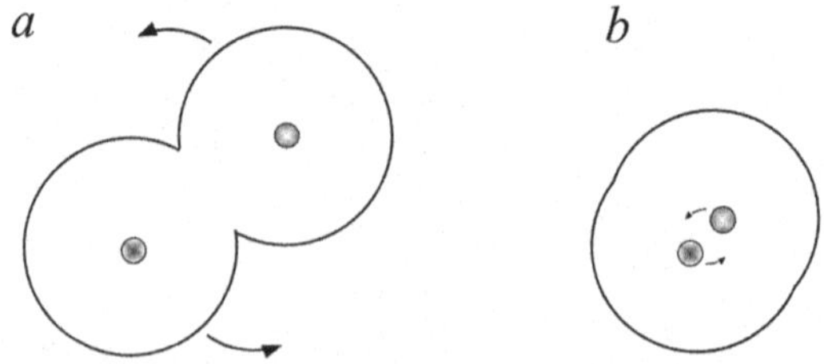

FIGURE 6.4 Interacting quarks: a – the confinement of quarks when the pair of quarks interact with the static potential $V(r) \propto r$; b – free quarks when they interact with a static potential $V(r) \propto 1/r$.

Let us consider merging bubbles. Let N be the number of cells contained in the bubble. Qinertons migrate through these cells and these excitations can be treated as analogous to gaseous molecules in a soap bubble. Therefore we may associate the number N of these excitations in the bubble with its volume $4\pi R^3/3$, the pressure P produced by these excitations and the temperature Θ:

$$P4\pi R^3/3 = Nk_B\Theta \tag{6.7}$$

where k_B is the Boltzmann constant. From Eq. (6.7) we get the pressure in the bubble $P = 3Nk_B\Theta/(4\pi R^3)$. The equality of the compressing $\mathcal{P}_{\text{compr}}$ and the stretching P pressure allows us to derive the number of excitations N via other parameters:

$$N = \frac{4\pi R^3}{3k_B\Theta}\left(\mathcal{P}_0 + \frac{2\gamma}{R}\right) \tag{6.8}$$

The merger of n bubbles is an efficient process. The number of excitations in a resultant bubble with a radius $\mathcal{R}$ is $\aleph = nN$, which allows us to derive an equation

$$\frac{2\gamma}{\mathcal{P}_0} = \frac{nR^3 - \mathcal{R}^3}{\mathcal{R}^2 - nR^2} \tag{6.9}$$

In order to decrease the energy of the bubbles merged together, i.e., reduce pressure on them on the side of the surface and the tessellattice, it is necessary that the numerator and the denominator in the right-hand side of Eq. (6.9) be negative; this is possible when the following inequalities are held (see Ref. [299], p. 37):

$$\mathcal{R}^3 > nR^3, \; \mathcal{R}^2 < nR^2 \tag{6.10}$$

Applying inequalities (6.10) to the problem of quarks, we obtain for the meson and the nucleon, respectively:

$$2^{1/3} < \mathcal{R}/R < 2^{1/2}, \; 3^{1/3} < \mathcal{R}/R < 3^{1/2} \tag{6.11}$$

Therefore, the united bubble is characterized by a lower Young-Laplace pressure on the quark, $2\gamma / \mathcal{R} < 2\gamma / R$.

Besides, the agglutination of the bubbles allows the quarks to gain an angular motion, which seems to become the major stabilizing factor for the existence of bubbles in the agglutinated state. Indeed, a pair of the agglutinated bubbles, which are extremely rapidly rotated form a rigid rotor exhibiting the integral of the moment of momentum $J \neq 0$. Such a characteristic is lacking for a single ball in the tessellattice in which a ball (a cell of the tessellattice) is deprived of the opportunity to rotate. That is why the compressing pressure $\mathcal{P}_{\text{compr}}$ on the side of the tessellattice that attacks the rotating quarks' bubble does not have enough power to collapse the bubble. The equivalent of the sound velocity for the tessellattice is the speed of light c; hence surrounding balls attack the bubbles with this velocity. The energy of hadrons is hundreds of MeV, which means that the velocity of rotating quarks in them $\upsilon_{\text{quark}} \approx 0.997c$. These two velocities are very close to each other and, therefore, only the vortex state of the quarks can keep the bubbles from collapsing.

6.3.2 MESONS AND HADRONS

In studying quark systems, researchers initially consider the interaction between two quarks, then add a third quark, fourth, etc., which is needed to arrange a wave function of the baryon studied. Quarks are treated as points or rather pseudo-points. The main task is the calculation of the eigenvalue and the binding energy of quarks. However, in such an approach the diquark system becomes indistinguishable from an antiquark.

The concept of the diquark and the achievements of submicroscopic mechanics in the realm of leptons allow us to reconsider the approach to the interaction of quarks. Indeed, the submicroscopic consideration of electrodynamics shows that magnetic monopoles are real entities (Figures 4.7 and 4.8), though they are hidden in the inner points of the path of a charged particle, as the charged particle periodically changes its electric state to the monopole state. The surface structure of quarks should be the same as is the case of leptons (Figure 1.9), i.e., the surface spikes are directed inside (the negative charge) or outside (the positive charge). The combed spikes correspond to the magnetic monopole state.

The motion of quarks (Figure 6.4) should also obey submicroscopic mechanics, as described above for the case of leptons, because the quark is surrounded with its own cloud of qinertons and its motion occurs in the tessellattice, which all together is the quark's wave function. This cloud as a whole can be associated with a gluon of QCD. Since quarks are charged particles, their qinertons have to carry electromagnetic properties. So, the electrodynamics of a quark is the same as is the case of the electron and positron (Figures 4.7 and 4.8). This means that the electric charge of quarks is integer: $\pm e$ (neither $\pm\frac{2}{3}e$ nor $\pm\frac{1}{3}e$). Below we put for the quark u the charge $+e$ (i.e., it is u^{+}), for the quark d the charge is $-e$ (i.e., it is d^{-}). The up antiquark has the charge $-e$ (i.e., it is negative, $\bar{u}^{-}$) and the down antiquark has the charge $+e$ (i.e., it is positive, $\bar{d}^{+}$).

Then the structure for the lightest π-mesons can be presented as follows:

$$\pi^{0} = \mathrm{d\,u}\,,\ \pi^{+} = \mathrm{u\,g_{d}}\,,\ \pi^{-} = \bar{\mathrm{u}}\,\mathrm{g}_{\bar{\mathrm{d}}} \tag{6.12}$$

where g_d and $g_{\bar{d}}$ are magnetic monopoles of the quark d and antiquark $\bar{d}$, respectively. Within the $\pi^{\pm}$-meson the magnetic monopoles g_d and $g_{\bar{d}}$ rotate emitting their own qinertons and exchanging with similar qinertons of the quarks u and $\bar{u}$, respectively. In such presentation the π^{-}-meson is the antiparticle to the π^{+}-meson and hence their masses are the same. Formulas (6.12) give automatically known transformations of quarks to leptons: a bubble collapses to a local deformation, from Figure 6.2, right to Figure 6.2, left (also see Figure 1.9, the structure of quarks and leptons). Namely, the transformations look as follows:

$$\pi^{+} \rightarrow \ (\mathrm{u}^{+}) + (\mathrm{g_d}) \rightarrow \begin{cases} e^{+} + \nu_e \\ \mu^{+} + \nu_\mu \end{cases},$$

$$\pi^{-} \rightarrow \ (\bar{\mathrm{u}}^{-}) + (\mathrm{g}_{\bar{\mathrm{d}}}) \rightarrow \begin{cases} e^{-} + \bar{\nu}_e \\ \mu^{-} + \bar{\nu}_\mu \end{cases} \tag{6.13}$$

i.e., two pairs of the quark and antiquark: u and g_d, and $\bar{u}$ and $g_{\bar{d}}$, disintegrate and the quarks and the antiquarks collapse, i.e., they mutate to the appropriate leptons (6.13).

In the Standard Model isospin arguments indicate that the π^0 state is $(\mathrm{u\bar{u}} - \mathrm{d\bar{d}})/\sqrt{2}$. However, what does this mixed state of $\mathrm{u\bar{u}}$ and $\mathrm{d\bar{d}}$ really mean with the addition of the factor $1/\sqrt{2}$? This state shows a mix of wave ψ-functions of the corresponding quarks. What do those wave ψ-functions mean in the formula $(\mathrm{u\bar{u}} - \mathrm{d\bar{d}})/\sqrt{2}$? What are the kinetics of quarks inside the meson? That is, how do the quarks and antiquarks move inside of the π^0-meson that is characterized by such formula? It is not possible to imagine. The formula $(\mathrm{u\bar{u}} - \mathrm{d\bar{d}})/\sqrt{2}$ is simply an abstract 'paper exercise' to satisfy some abstract desire of its promoters.

However, as has been shown above, the ψ-function in the Schrödinger equation (3.59) is the real key component associated with the central particle and its cloud of excitations. In the case of the present approach the quarks inside the π^0-meson are moving by known trajectories (see below). That is why the simple structure of the π^0-meson presented in expression (6.12) is plausible. Indeed, the pion is neutral and can annihilate by the scheme

$$\pi^0 \rightarrow (\mathrm{d}^-) + (\mathrm{u}^+) \rightarrow \begin{cases} \gamma + \gamma \\ e^- + e^+ + \gamma \end{cases} \tag{6.14}$$

Although experimental studies [300] of the decay of the π^0-meson were carried out in detail, they did not disclose an inner structure of this subatomic particle; the major issues that allowed the examination were the conditions at which the pion appeared and the accurate measurement of its lifetime.

In nuclear physics the proton and neutron are different only in their isospin projection, which is non-informative. Instead of this, we may suggest the structure of a nucleon as depicted in Figure 6.4. Namely, instead of the generally accepted view that the structure of the proton and the neutron respectively are p = duu and n = ddu, we may suppose a couple of other versions for the formulas of nucleons. It seems the most plausible are p = duu and n = dug$_{\mathrm{u}}$, or in the explicit form

$$\begin{aligned} p^+ &= (\mathrm{d}^-\mathrm{u}^+,\ \mathrm{u}^+) = (\pi^0,\ \mathrm{u}^+), \\ n^0 &= (\mathrm{d}^-\mathrm{u}^+,\ \mathrm{g_u}) = (\pi^0,\ \mathrm{g_u}) \end{aligned} \tag{6.15}$$

where the structure of π^0 is defined above in expressions (6.12).

Figure 6.5 illustrates how the tessellattice avoids the problem of three bodies, which does not have a steady-state solution: initially two quarks form a stable vortex system; then this system jointly with one more quark/monopole forms another stable vortex system.

As is known, bosons are mediators of neutrino emission and absorption. Their charge manifests itself through the emission or absorption of an electron/positron. The emission of a W^+ or W^- boson by a baryon either raises or lowers its electric charge by one unit, and also changes the spin by one unit. These bosons cause nuclear transmutation. The Z^0 boson is detected as a force-mediator whenever neutrinos scatter elastically from matter and the appearance of Z^0 is not accompanied by the production or absorption of new charged particles. Three bosons $W^{\pm}$, Z^0 and the photon represent the four gauge bosons of the electroweak interaction. Let us see how the bosons of the weak interaction appear in the submicroscopic approach.

The decay of a hadron takes place mostly under an impact of perturbative conditions. In other words, spontaneous pairs of quark-antiquark must stimulate the decay. For instance, in the framework of the submicroscopic approach the decay of the neutron (presented below as (du + g_u), i.e., a combination of quarks d and u, and the magnetic monopole g_u) occurs at the collision with a quark-antiquark pair $u\bar{u}$ by the following formula:

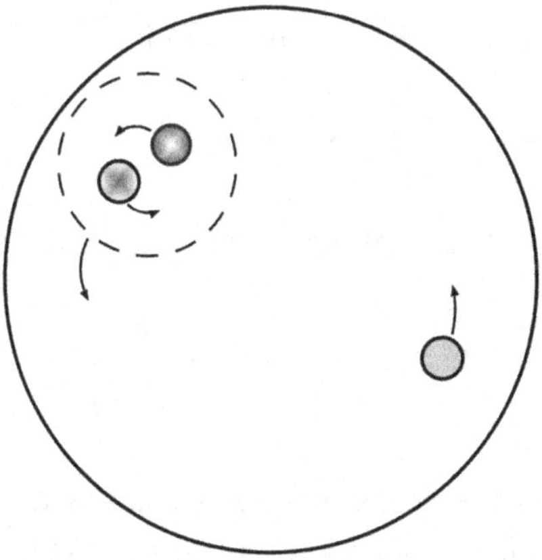

FIGURE 6.5 Nucleon. The pion, i.e., π^0-meson, which can be treated as an inner vortex, is rotated together with quark d (the case of the proton) or with the magnetic monopole g_u (the case of the neutron). The pion and the quark d produce their own vortex in the proton (or the pion and the magnetic monopole g_u produce the vortex in the neutron).

$$\left(\mathrm{d}^- \mathrm{u}^+ + \mathrm{g_u}\right) + \left\{\mathrm{u}^+ \bar{\mathrm{u}}^-\right\} \rightarrow \left(\mathrm{d}^- \mathrm{u}^+ + \mathrm{g_u} + \mathrm{u}^+ + \bar{\mathrm{u}}^-\right)$$
$$\rightarrow \left(\mathrm{du} + \mathrm{u} + \bar{\mathrm{u}}\mathrm{g_u}\right)$$
$$\rightarrow \left(\mathrm{d}^- \mathrm{u}^+ + \mathrm{u}^+\right) + \left(\bar{\mathrm{u}}^- \mathrm{g_u}\right) \quad (6.16)$$

That is, we have obtained

$$n^0 \rightarrow p^+ + W^- \quad (6.17)$$

or in other words, we just revealed the inner structure of the combined particle W^-: $W^- = (\bar{\mathrm{u}}^-\ \mathrm{g_u})$, i.e., it is composed of the antiquark $\bar{\mathrm{u}}$ and the u-quark's magnetic monopole g_u, which rotate around each other. (Note that an idea about the composition of $W^\pm$ and Z^0 bosons has already been expressed [300] in the framework of the Next-to-Minimal Supersymmetric Standard Model.) Under the compressing pressure (6.6) this combined particle collapses in a short time of around 3×10^{-25}s, such that each of the components changes its quark state (Figure 6.2, right) to the appropriate lepton state (Figure 6.2, left). Namely,

$$W^- \equiv \left(\bar{\mathrm{u}}^-\ \mathrm{g_u}\right) \rightarrow \left(e^-,\ \bar{\nu}_\mathrm{e}\right) \rightarrow e^- + \bar{\nu}_\mathrm{e} \quad (6.18)$$

Then the antiparticle to the boson (6.18) is

$$W^+ \equiv \left(\mathrm{u}^+\ \mathrm{g}_{\bar{\mathrm{u}}}\right) \rightarrow \left(e^+,\ \nu_\mathrm{e}\right) \rightarrow e^+ + \nu_\mathrm{e} \quad (6.19)$$

The third boson of the weak force seems to have the structure

$$Z^0 = \left(\mathrm{u}^+ \bar{\mathrm{u}}^-\right) \rightarrow \begin{cases} \gamma + \gamma \\ \mu^+ + \mu^- \end{cases} \quad (6.20)$$

Figure 6.6 depicts a typical structure of a hadron that consists of a quark q and antiquark $\bar{\mathrm{q}}$; such a pattern is typical for π^0-mesons, Z^0-bosons and similar hadrons.

Figure 6.7 pictures a typical structure of a hadron that consists of a quark q and the magnetic monopole $\mathrm{g}_{\bar{\mathrm{q}}}$; such a pattern is typical for $\pi^\pm$-mesons,

$W^{\pm}$-bosons and similar hadrons. Figure 6.7 shows that if a boson is formed by a quark and a monopole, the monopole will remain unchanged in its monopole state throughout the whole lifetime of the boson. This follows from the Maxwell equations because only the electric charge state is able to transform into the monopole state and back again to the charge state, and so on. The monopole state is "frozen" and is unable to transform into the electric state.

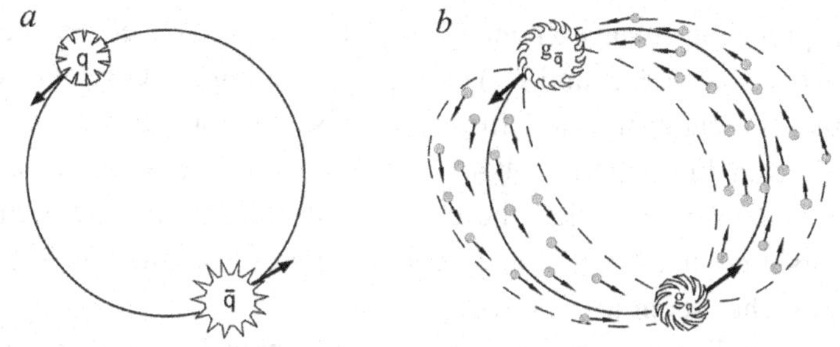

FIGURE 6.6 Hadron formed by the quark q (with the charge $-e$) and the antiquark $\bar{q}$ (with the charge $+e$). *a* – the quark q and the antiquark $\bar{q}$ are found in the initial state; *b* – both quark q and the antiquark $\bar{q}$ have passed the section λ of their paths and their state is transformed: $q \rightarrow g_q$ (i.e., from the quark state to the monopole state) and $\bar{q} \rightarrow g_{\bar{q}}$ (i.e., from the antiquark state to the antimonopole state). Then passing the next section λ these entities interact through spatial inflated excitations, i.e., qinertons, changing the configuration to the initial state (Figure *a*): the quark q and the antiquark $\bar{q}$, respectively; and so on.

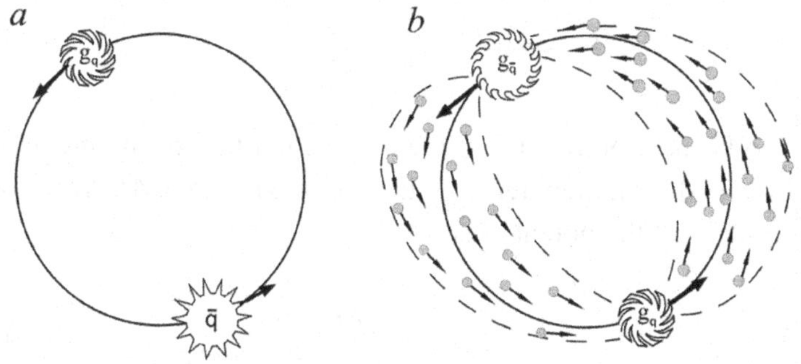

FIGURE 6.7 Hadron formed by the magnetic monopole g_q of the quark q (with the charge $-e$) and the antiquark $\bar{q}$ (with the charge +e). *a* – the magnetic monopole g_q and the antiquark $\bar{q}$ are found in the initial state; *b* – both of the entities, the monopole g_q and the antiquark $\bar{q}$, have passed the section λ of their paths and the state of the monopole continues to be preserved, g_q, and the antiquark is transformed to its monopole state: $\bar{q} \rightarrow g_{\bar{q}}$ (i.e., from the antiquark state to the antimonopole state). Then passing the next section λ these entities interact through qinertons to change their configuration to the initial state (Figure *a*): the magnetic monopole g_q and the antiquark $\bar{q}$, respectively, and so on.

6.4 ON THE STRUCTURE OF A NUCLEON

QCD states that nuclei are bound together by the residual strong force (the nuclear force). On the other hand, because of the well-known difficulty of QCD in the non-perturbative domain, many other effective models reflecting the characteristics of the strong interaction are used to study the behavior of nucleons (see, e.g., the quotations in Ref. [99]).

High precision measurements [302] of the deuteron electromagnetic structure functions (A, B and T_{20}) extracted from high-energy elastic electron-deuteron scattering, and from the cross sections and asymmetries extracted from high-energy photodisintegration $\gamma + d \to d + n$ allowed the authors to conclude that the experiments offered no preference for either a residual quark-gluon interaction or a meson exchange. Moreover, both approaches seem to disagree.

Let us initially unify quarks d and u to the π^0-meson state (6.12), as is shown in Figure 6.5. For this we shall solve the problem of two bounded charged particles, which is also known as the rigid rotor problem (see, e.g., Ref. [160]). In our case, two quarks are additionally bounded through the potential of the strong force. For the Coulomb interaction of the quarks d and u the potential is

$$V_{\rm Coulomb} = -\frac{e^2}{4\pi\varepsilon_0 \,|\, r_{\rm d} - r_{\rm u} \,|} \tag{6.21}$$

where $r_{\rm d}$ and $r_{\rm u}$ are coordinates of the quarks. For the strong interaction of the quarks d and u (in the interior of the bubble, Figure 6.4), which occurs through qinertons, the potential is

$$V_{\rm strong} = -\frac{hc}{|\, r_{\rm d} - r_{\rm u} \,|} \tag{6.22}$$

The ratio $V_{\rm strong} / V_{\rm Coulomb} \approx 2\pi \times 137.1597$ seems to play the role of the inverse fine structure constant for quarks, i.e., the ratio of the radius of the inflated bubble to the radius of the electric polarization that is induced on the cells of the tessellattice involved in the formation of the bubble (compare with expression (1.93)). Due to a small value of the Coulomb potential (6.20) in comparison with the strong interaction (6.22) we may neglect the former. The potential (6.22) is completely non-perturbed and deals with energies less than 10 MeV, therefore, such presentation (6.22) does not contradict data analyzed by Bethke [296, 297]. Then the Lagrangian that describes two bounded quarks reads

$$L = \tfrac{1}{2} m_{\mathrm{d}} \dot{r}_{\mathrm{d}}^2 + \tfrac{1}{2} m_{\mathrm{u}} \dot{r}_{\mathrm{u}}^2 - \frac{hc}{|r_{\mathrm{d}} - r_{\mathrm{u}}|} \tag{6.23}$$

We can pass on to the coordinate of the center of mass $r_{\mathrm{c.m.}} = (m_{\mathrm{d}} r_{\mathrm{d}} + m_{\mathrm{u}} r_{\mathrm{u}}) / (m_{\mathrm{d}} + m_{\mathrm{u}})$ and the relative coordinate $r = r_{\mathrm{d}} - r_{\mathrm{u}}$, which then changes the Lagrangian (6.23) to

$$L = \tfrac{1}{2}(m_{\mathrm{d}} + m_{\mathrm{u}})\, \dot{r}_{\mathrm{c.m.}}^2 + \tfrac{1}{2} \mu\, \dot{r}^2 - \frac{hc}{r} \tag{6.24}$$

where the reduced mass is

$$\mu = m_{\mathrm{d}} m_{\mathrm{u}} / (m_{\mathrm{d}} + m_{\mathrm{u}}) \tag{6.25}$$

Let the center of mass be motionless; then the first term in Lagrangian (6.24) becomes zero. Now we can solve the Schrödinger equation (3.59) having preserved the two last terms in the Lagrangian (6.24) in which we can insert the relativistic masses of the quarks. Thus we can now reduce the problem of being able to specify where the two (quark) masses are in space at a given time by applying the known solution to the same historic problem that related to the particles of the hydrogen atom. In particular, we can write the radius of the orbit for the reduced mass μ_{π^0} (6.25). In the conventional case of the hydrogen atom, this is the Bohr radius

$$r_{0,\,\mathrm{Bohr}} = \frac{\hbar^2}{m_{\mathrm{electron}}\, e^2 / (4\pi\, \varepsilon_0)} \tag{6.26}$$

For the case of the strong potential (6.22) the solution for the radius of the reduced mass μ is

$$r_0 = \frac{\hbar}{2\pi\, \mu\, c} \tag{6.27}$$

Let us first consider a free π^0-meson. Its energy is 135 MeV. We may put for the u and d quarks the rest energy 2.3 and 4.8 MeV, respectively; then their total energies in the π^0-meson correspondingly are $E_{\rm d}$ = 43.732 MeV and $E_{\rm u}$ =91.268 MeV. Hence masses are $m_{\rm d} = E_{\rm d} / c^2$ and $m_{\rm u} = E_{\rm u} / c^2$, and from expression (6.25) we get for the reduced meson mass $\mu_{{\rm free}\ \pi^0} = 7.796 \times 10^{-29}$ kg. Substituting this value into the expression (6.27), we obtain $r_{{\rm free}\ \pi^0} \cong 1$ fm.

Now we can consider a nucleon. In the proton an energetic π^0-meson and a quark u rotate around one another; in the neutron a π^0-meson and a u-quark's monopole $g_{\rm u}$ revolve around each other. We shall emphasize that since two particles with similar parameters are found in the same orbit, the section between them along the orbital path is equal to the de Broglie wavelength for each of the particles.

Given this situation, how is the energy $E_{\rm nucleon}$ =939 MeV distributed among the π^0-meson and its rotating partner u (for proton) or $g_{\rm u}$ (for neutron)? For example, in the case of a proton this energy has to be distributed among the quarks as follows: 939 MeV = $[(2.3\beta + 4.8\beta) + 2.3\beta]$ MeV, where $\beta = (1 - \upsilon^2 / c^2)^{-1/2}$ is the usual factor for the contraction of mass upon its motion. The corresponding masses are $m_{\rm u} = 229.755 / c^2$ =4.096 × 10^{-28} kg and $m_{\rm d} = 479.441 / c^2$= 8.546 × 10^{-28} kg. Then the reduced mass of the π^0-meson (6.25) is $\mu_{\pi^0} = 2.769 \times 10^{-28}$ kg. Hence the radius of the reduced mass (6.27) becomes

$$r_{0,\,\pi^0} \cong 0.202 \text{ fm} \tag{6.28}$$

The reduced mass $\mu_{\rm p}$ of three quarks in the proton is derived from the relation

$$\frac{1}{\mu_{\rm p}} = \frac{2}{m_{\rm u}} + \frac{1}{m_{\rm d}} - \frac{1}{\mu_{\pi^0}} \tag{6.29}$$

where the last term is due to the formation of a subsystem of the π^0-meson by a couple of these quarks. The calculation gives $\mu_{\rm p} \cong 4.096 \times 10^{-28}$ kg. Then from expression (6.27) we obtain the radius of the orbit of this reduced mass

$$r_{0,\,3\text{ quarks}} \cong 0.137 \text{ fm} \tag{6.30}$$

Expressions (6.30) and (6.28) show the radii of orbits of nonperturbed quarks in the proton. Although the repulsive potential (6.21) is very small, it can support the stability of quark orbits in the proton (compare with the problem of cluster formation studied in Chapter 5: a cluster exists only in the presence of attraction and repulsion pair potentials in the system studied).

In a nucleon quarks together with their qinerton clouds form a closed sphere with a radius r_{p}. This radius is formed by the radius (6.28) of the orbit of the reduced mass μ_{π^0} in the center part, which is further extended by the appropriate amplitude Λ of the qinerton cloud of the reduced quark particle (see relation (2.69)):

$$r_{\text{p}} = r_{0,\,\pi^0} + \lambda_{\text{dB}}\, c/\upsilon\big|_{\upsilon \cong c} = r_{0,\,\pi^0} + \pi r_{0,\,\pi^0} = 0.837 \text{ fm} \tag{6.31}$$

where two coupled quarks share the same orbit and hence the quark's wavelength is equal to half of the circumferential length, i.e., $\pi r_{0,\,\pi^0}$.

The radius r_{p} (6.31) does not exceed the radius of the unified bubble created around the three quarks of the nucleon, i.e., the amplitude of the qinerton cloud is less than the Compton wavelength of the nucleon: $r_{\text{p}} < \lambda_{\text{Com, nucl}} = 1.32$ fm. Therefore, qinertons generated by the quarks in motion are located strictly inside the nucleon, namely, in the range limited by the radius $r_{\text{p}} \cong 0.84$ fm.

Outside the range covered by the radius r_{p} the morphology of space is quite different, as the space here is not excited by qinertons. When the proton moves as the whole, its quarks participate in this rectilinear motion and the quasi-single quark u, which is revolving around the π^0-meson (Figure 6.8), generates also inerton-photons. These inerton-photons of the quark u^+ are able to spread outside the proton inducing a conventional Coulomb potential in the ambient space.

Similar quark orbits are also present in the neutron in which one quark u is substituted for its monopole g_{u}. However, due to the absence of the repulsion potential (6.21) in the neutron, this small cluster of three quarks should be unstable especially for free neutrons, which in fact is observed experimentally as beta decay.

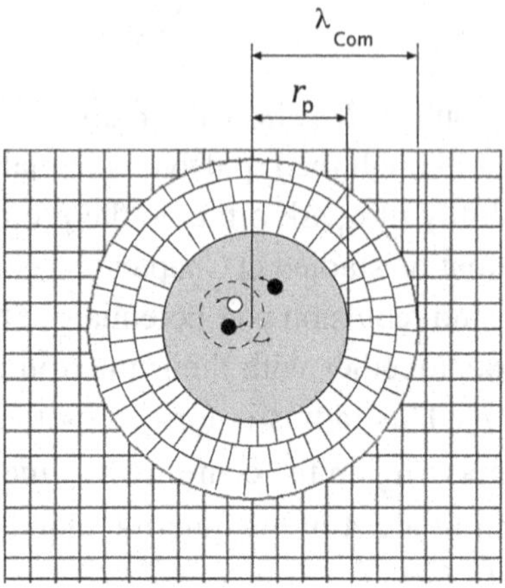

FIGURE 6.8 Structure of a nucleon. The quarks rotate near the center of the nucleon. The quarks' qinerton clouds reach the radius r_p. The membrane that screens the nucleon bubble spreads from $r = r_p$ to $r = \lambda_{Com}$. In the membrane cells of the tessellattice are found in a stretched state and the cohesive forces between these cells are responsible for the surface tension.

Now we can turn to the recent experimental data of Islam et al. [303]: The physical picture is that the proton is comprised of three regions: an outer cloud of $q\bar{q}$ condensed ground state of size $r_{q\bar{q}} \geq 0.86\times10^{-15}$ m, an intermediate shell of baryonic charge of size $r_B = 0.44\times10^{-15}$ m, and a core of size $r_c = 0.2\times10^{-15}$ m, where valence quarks are confined.

The same authors [304, 305] further note that these experimental results allow a description in terms of the topological soliton model of the nucleon. For this purpose they introduce an abstract scalar field ζ of an undetermined nature. Using ζ they arrive at the topological soliton model in which the large mass problem is resolved by tearing the scalar field ζ at the critical size $r_c = 0.2\times10^{-15}$ m, i.e., the pion decay coupling constant f_π (= 93 MeV) [305], drops down sharply to zero at $r < r_c$, which decreases the mass of the soliton by a significant amount; this allows them to associate their model with a chiral bag model. The region $r_c < r < r_B$ was called the shell of topological baryonic charge density. At $r > r_B$, the scalar field ζ decreases smoothly, which makes here the quarks and antiquarks massive and reduces the energy of the Dirac sea. Hence, the region $r_B < r < r_{q\bar{q}}$ should represent a $q\bar{q}$ condensed ground state that forms an outer cloud of the proton.

Such a topological soliton model of the proton was called a 'Condensate Enclosed Chiral Bag' [305]. In the end, the authors asserted: "The consequent discovery of the structure of the proton at LHC at the beginning of the 21st century will be analogous to the discovery of the structure of the atom from high energy α-particle scattering by gold atoms at the beginning of the 20th century."

Nevertheless, from the physical point of view, a non-topological soliton model [306] looks more preferable. Indeed, a bag model exhibits physical characteristics very similar to those of a "gas bubble" immersed in a "medium": the model operates with a constant surface tension and a constant pressure exerted by the medium on the gas in the bubble; besides, the model includes the thermodynamic energy of the gas and the related gas pressure.

The submicroscopic description of the behavior of quarks presented above agrees rather with the physical pattern of a "gas bubble" constructed in Ref. [305].

Let us now compare our theoretical results with the experimental data [305]:

- a core of size $r_{\mathrm{c}} = 0.2 \times 10^{-15}$ m, where valence quarks are confined – this exactly corresponds to the radius (6.28) of the orbit of the reduced particle of the nucleon;
- an outer cloud of $\mathrm{q\bar{q}}$ condensed ground state of size $r_{\mathrm{q\bar{q}}} \geq 0.86 \times 10^{-15}$ m – this conforms to the radius (6.31) that qinertons reach;
- an intermediate shell of baryonic charge of size $r_{\mathrm{B}} = 0.44 \times 10^{-15}$ m – this is the inflection point of the static quark potential (6.1) that in the explicit form, combining expressions (6.5) and (6.22), can be read

$$V(r) = -\,\sigma r - hc / r \tag{6.32}$$

In this potential the two terms are negative and they describe attraction. The first term is negative because it is stipulated by the mechanism of the attraction of bubbles, expressions (6.4) and (6.5). The second term is dictated by a spherical standing wave generated by each of the quarks in the system under consideration: the quark periodically decomposes, i.e., it throws off portions of its inflated state and the standing spherical wave spreads these qinertons along a relief given by the deformation coat of the quark (Figure 6.2, right); it is this peculiar relief that directs two quarks – through their qinertons – to each other. The extremum of the Eq. (6.32) is attained at the solution of the equation, i.e., $dV/dr = 0$, i.e., $\sigma - hc/r^2 = 0$. This equation gives the solution for the inflection point

$$r_{\text{inf}} \cong \sqrt{hc/\sigma} \tag{6.33}$$

that can be identified with the shell of baryonic charge $r_{\text{B}} = 0.44 \times 10^{-15}$ m of Ref. [303]. In the point $r = r_{\text{inf}}$ the potential (6.32) is maximum, as $V''(r_{\text{inf}}) < 0$, which means that at $r = r_{\text{inf}}$ the attraction is minimum.

Thus, the ruling space solves the problem of three bodies in a very original way: Space simply reduces the problem to a system of two bodies, which allows an analytical solution, as has been discussed above.

The solution (6.33) allows an estimate of the value of the surface tension of the quark's bubble. Indeed, for the constant σ we get $\sigma = hc/r_{\text{inf}}^2 \cong 1.027 \times 10^6$ J·m^{-1}. Taking into account expressions (6.4) and (6.5), we obtain the tension of the quark's bubble:

$$\gamma = \sigma/(4\pi\chi) \sim 10^{20} \text{ N·m}^{-1} \tag{6.34}$$

where we set the depth of overlapping $\chi \sim 10^{-16}$ to 10^{-15} m (see Figure 6.3). For example, for some liquid substances typical values of the surface tension at a room temperature are: 0.465 N·m^{-1} (mercury), 0.073 N·m^{-1} (water) and 0.03 N·m^{-1} (soap water).

According to the theory described above a nucleon is a typical bubble and its surface film (interface, or membrane) is specified by the thickness from $r = r_{\text{p}} \approx (0.84 - 0.86)$ fm to $r = \lambda_{\text{Com, nucl}} = h/(m_{\text{nucl}}\, c) = 1.32$ fm. Such a bubble with the membrane is shown in Figure 6.8.

The proton radius has intensively been studied experimentally from the outside, i.e., the proton radius has been examined using electrons, either by electron–proton scattering or by careful measurements of atomic energy levels (see, e.g., a review paper [307]); the result for the root-mean-square (rms) charge radius of the proton was r_{p}^{E} =0.86 to 0.88 fm. Puzzling measurements of the rms charge radius in a muon hydrogen gave the value of the radius

0.84 fm. The proton radius puzzle has become a widely studied challenge (see, e.g., Refs. [308–311]). Lorenz and Meißner [312] summarize the problem as follows: "The results for the proton charge radius r_{p}^{E} are in perfect agreement with the values obtained via a dispersion relation approach and the recent muonic hydrogen measurements. The remaining r_{p}^{E}-discrepancy is the $\sim 4\sigma$ deviation between the average of the spectroscopic measurements in electronic hydrogen and those in muonic hydrogen." Lorenz and Meißner hold the view that the radius r_{p}^{E} =0.84 fm and complete their paper with words about the necessity for new measurements in ordinary hydrogen and the planned muon-proton scattering experiment MUSE. Carlson [307] finishes his review with the words: "Perhaps the discrepancy is due to new physics. Perhaps the explanation is an ordinary physics effect that has been missed. Perhaps, the muonically measured radius will come to be the accepted number."

6.5 NEUTRINO – WHAT IS IT?

The beta decay of a free neutron outside a nucleus has a lifetime of about 15 minutes and is denoted by the radioactive decay $n^0 \rightarrow p^+ + e^- + \bar{\nu}_e$. Where do the electron and the electron antineutrino come from? Nobody knows. Nevertheless, expressions (6.16)–(6.19) clarify the process of transformation in the real space constituted as the tessellattice. Figure 6.9 depicts vividly successive changes in the process of neutron transformation: (*a*) the initial stable state of the neutron; (*b*) the separate orbital monopole g_{u}, which is the axial state of the quark u^+ scattering by the oncoming virtual quark-antiquark pair (u^+, $\bar{\mathrm{u}}^-$). Such pairs are created from the tessellattice only in the electrically charged state and never in the monopole state; (*c*) the quark u^+ substitutes for the monopole g_{u} occupying the orbit of the latter and at the same time the antiquark $\bar{\mathrm{u}}^-$ and the monopole g_{u} leave the nucleon as an unstable pair of $\bar{\mathrm{u}}^-$ and g_{u} known as a virtual particle W^-. The separating particles $\bar{\mathrm{u}}^-$ and g_{u} cannot exist in the tessellattice in the inflated state. That is why the tessellattice immediately squeezes $\bar{\mathrm{u}}^-$ and g_{u} to the state of local stable deformations and they become the electron e^- and the antineutrino $\bar{\nu}_e$, respectively. In this phase transition from one topology to the other, only the particles' volumes change (in Figure 6.2 from right to left); their surface polarizations are preserved.

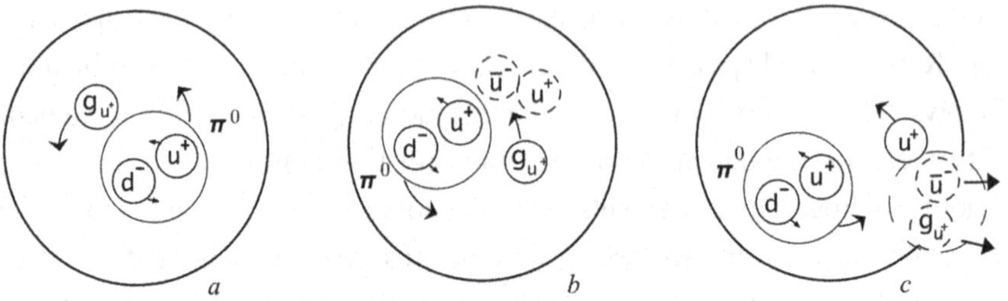

FIGURE 6.9 Beta decay of the neutron to the proton.

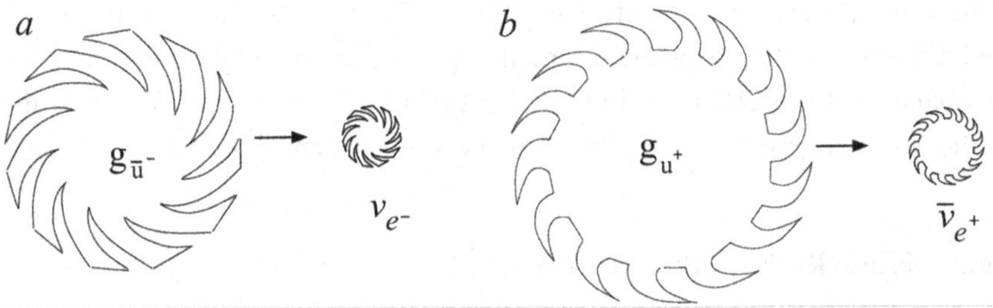

FIGURE 6.10 Transformation of the quark's monopole to the lepton's monopole. *a*: $g_{u} \rightarrow \bar{\nu}_e$ (6.18), i.e., combed spikes are oriented inward; *b*: $g_{\bar{u}} \rightarrow \nu_e$ (6.19), i.e., combed spikes are oriented outward. At the transformation the particle's radius changes from the large size to the small size in comparison with the scale of a degenerate cell of the tessellattice.

At a fast non-adiabatic process (annihilation, explosion, decay) the magnetic monopole is able to escape from the confinement of the hadron breaking the sound barrier in the tessellattice, which is the speed of light *c*. Coming through the barrier the quark state collapses to the lepton state (the transition from Figure 6.2, right, to Figure 6.2, left) and the quark's magnetic monopole becomes the corresponding lepton's magnetic monopole. This lepton's monopole is known as a neutrino that travels with the velocity close to *c*. Figure 6.10 depicts the transformation of the quark's monopole to the lepton's monopole, which are presented in Eqs. (6.18) and (6.19); namely, the monopole $g_{\bar{u}}$ of the quark u^+ is squeezed to the electron antineutrino $\bar{\nu}_e$ and the monopole $g_{\bar{u}}$ of the antiquark $\bar{u}^-$ tightens to the electron neutrino ν_e.

In its motion, the neutrino interacts with cells of the tessellattice and generates a cloud of inertons but the structure of the neutrino's surface does not change – the neutrino still continues to be a lepton's monopole. The motion of a neutrino should obey the de Broglie relationships (2.60). The theory of the neutrino (see, e.g., Refs. [312, 313]) points to the existence of a mass in this particle.

Indeed, any particle in the tessellattice must have a volumetric fractal deformation or inflation (lepton or quark) and/or surface fractal deformation (charge). The surface structure of the neutrino is presented as a monopole state, which is frozen on the surface and cannot be transformed into the initial charge state. Since the surface cannot change, the volume must change. Because during motion a particle interacting with the tessellattice must periodically change its inner state, which occurs in a section equal to the particle's de Broglie wavelength λ_{dB}. Researchers studying reactions of neutrinos point out the value of its mass somewhere circa 0.3 eV, which is around 10^{-9} of the electron rest mass. In terms of the submicroscopic concept this means that a neutrino leaving the hadron obtains a small fragment of a local volumetric deformation from the hadron's deformation coat. Having this mass δm the neutrino starts to move and interacting with the ambient space it periodically changes its mass δm to tension ξ, though its surface state does not change. In other words, the neutrino moves propelled by inertons.

With a neutrino energy from 1 MeV to 1 GeV and the velocity υ equal to about c we may deduce from relationships (2.60) the neutrino's de Broglie wavelength $\lambda_{\text{dB}} \sim 10^{-12}$ to 10^{-16} m and the frequency of the neutrino oscillations $\nu \sim 10^{20}$ to 10^{24} Hz. In transverse directions the neutrino's inertons reach the same distance $\Lambda = \lambda_{\text{dB}}\, c/\upsilon \approx \lambda_{\text{dB}} \sim 10^{-12}$ to 10^{-16} m. The higher the neutrino's energy, the shorter is λ_{dB} and the smaller Λ, though the frequency increases. In each odd section λ_{dB} of the neutrino's path it emits inertons and gradually loses its velocity υ and mass δm; during each even section λ_{dB} the neutrino re-absorbs its inertons and restores its mass, and inertons colliding with the neutrino reset its initial velocity, and so on.

Experimental studies strongly support the three-flavor neutrino oscillation model (see, e.g., Ref. [313–315]). Figures 6.9 and 6.10 show that this is quite possible, because the process of neutrino emission after the appropriate nuclear reaction is the dominating factor. After the decay of a hadron, the emitted neutrino/antineutrino is still in an excited state that lasts for a time Δt – because the particle changes its morphology and such process cannot be immediate. During the relaxation time Δt the neutrino is found in a meso-state, which means that the neutrino's radius may oscillate between three values given by three known lepton flavors: electron, muon and τ-lepton. The phase transition $g \rightarrow \nu$ can be realized into the final stable state only in the time Δt that is unknown. If $\Delta t \sim 1\ \mu s$, then the neutrino turns into its final stable state passing a distance of about 300 km. Transferring to the final state the excited neutrino perhaps is able to emit a low energy inerton. Since the neutrino has a magnetic moment (which is logical, as it is monopole), it is able to scatter by oncoming particles having the magnetic moment. Such a scatter process may additionally excite the neutrino increasing or decreasing its initial mass. Besides, since the neutrino's mass is in the vicinity of 10^{-39} kg, it is able to interact with the mass field induced around massive objects because cells of the tessellattice possess a similar value of mass μ (see Sections 8.4 and 10.2).

If the neutrino starts to move with the speed $\upsilon > c$, then in collisions with oncoming cells of the tessellattice the neutrino will excite them, so that after passing a cell, the cell will quickly relax, generating an inerton and/or photon in transverse directions. Such motion signifies a kind of a real friction (the bremsstrahlung). Tracks of γ-quanta after the passing of a neutrino have never been observed; this means that the speed of neutrinos is less than c (see also Ref. [316]).

6.6 PROTON SPIN

The problem of the proton spin crisis has been discussed in the literature for years (see recent experimental works [317–319]). The quarks inside a proton have their own intrinsic spin. But numerous experiments have confirmed that a directional preference among all these quark spins can only account for only about 25% of the proton's total spin. Modern studies points out that gluons contribute 20–30% to the total spin. So the source of the spin still remains a mystery.

Camay [320] noted that the problem would be solved if one correctly calculates the angular momenta of all the quarks and takes into account the quark spatial motion (which also is in agreement with experiment [318]).

Nevertheless, to resolve the problem of the proton's spin, one first has to have a correct determination of the notion of spin as such.

The determination has been done above (Sections 3.4, 4.3.1 and 4.3.2): basic properties of the spin-1/2 are the presence of the charged state on the surface of a particle and the direction of polarization of the monopole state – left or right. Besides, spin-1/2 is an integral property of a moving particle, which is associated with a libration of the surface fractals, i.e., spikes, which is given by the initial conditions – to the left or right.

It is important to pay attention to the fact that photons that carry fragments of the surface fractals are able to move freely by surfaces of both qinertons and inertons. Qinertons and inertons are volumetric fractals, but photons represent a complete different quality – they are surface fractals. This is why a change of the volume of cells does not affect their relay movement – even hopping from an inflated cell to a contracted cell a photon does not experience a resistance.

Thus when a proton moves as a whole with a velocity υ, a quasi-free quark u^+ inside the proton emits its qinerton-photon cloud that is able to extend far beyond the boundary radius r_p (6.30) to a distance $\Lambda = \lambda_{dB}\, c/\upsilon$ from the proton. The proton's de Broglie wavelength and velocity define the period of libration of the quark u^+: $T = \lambda_{dB}/\upsilon$. For example, at $\upsilon = 10^5\ \mathrm{m{\cdot}s^{-1}}$ the de Broglie wavelength $\lambda_{dB} \approx 4 \cdot 10^{-12}$ m, the distance Λ to which the proton's photons spread is about 10^{-8} m, and in this case $T = 10^{-12}$ s.

Therefore probing the structure of the proton for its spin, one has to use low energy collisions and prolong the duration of the reaction until the proton has taken the full de Broglie wavelength during the period of libration of the quark u^+.

At a higher energy (a few to hundreds of GeV) the length of the proton's de Broglie wavelength λ_{dB} falls within the proton size, i.e., λ_{dB} becomes less than $\lambda_{Com,\,p}$ and even r_p. In this case an experiment will probe only the inner motion of the proton's entities – quarks – and the data will be specified by broad dispersion (a similar situation takes place in femtosecond optics: a probing laser pulse fixes only one "frozen" instantaneous state of a large set of complete vibrations of atoms).

The origin of the neutron spin is exactly the same: a quasi-free monopole g_u creates a vortex jointly with a π^0 meson. The monopole g_u has a constant magnetic charge that does not oscillate to the electric state then to the magnetic state and so on, because such a procedure is possible only in the case when the original state is the electric charge. Thus, the monopole's qinerton cloud carries only qinertons. The monopole state, as the surface effect, is strictly frozen on the surface of the g_u. The interaction with an external magnetic field is possible only owing to the penetration of the magnetic field inside the proton, such that external photons begin to engage with the monopole. This is exactly what the Maxwell equations prescribe (see Section 4.3).

KEYWORDS

- **gluon**
- **meson**
- **monopoles**
- **neutrino**
- **nucleon**
- **proton**
- **qinertons**
- **quarks**
- **tessellattice**

CHAPTER 7

NUCLEONS AND THE NUCLEAR FORCES

CONTENTS

The radius of nuclear forces reaches circa (1 to $2A^{1/3}$)$\times10^{-15}$ m where A is the atomic number, which is compatible with our approach according to which a nucleon has a membrane that separates the nucleon from the ambient space and which spreads from the proton/neutron radius $r_{\rm p} \cong 0.84$ fm to the nucleon's Compton wavelength $\lambda_{\rm Com,\ nucl} = h / (m_{\rm p(n)} c) = 1.32$ fm (Figure 6.8).

This membrane is the edge of the deformation coat of a nucleon. In the deformation coat cells of the tessellattice are in the state of collective vibrations, as has been considered in Section 3.3. In the coat vibrations of all cells co-operate and the total energy of vibrating cells, which is equal to the total energy mc^2 of the nucleon, is quantized (see expressions (3.24) and (3.26)):

$$\hbar\omega_k = mc^2 \tag{7.1}$$

where $k = 2\pi / \lambda_{\text{Com}}$ is the wave number, $\omega_k = ck$ is the cyclic frequency of an oscillator in the k-space and m is the mass of the nucleon.

In motion a nucleon generates its own inerton cloud that accompanies the nucleon. Since in nuclei the energy per nucleon is about 8 MeV, and the speed of a nucleon is about 0.1 of the speed of light c, we may estimate the average nucleon's de Broglie wavelength as about $\lambda_{\text{dB}} \approx 10^{-14}$ m and hence the nucleon's inerton cloud reaches a distance

$$\Lambda = \lambda_{\text{dB}}\, c / \upsilon \approx 10^{-13} \text{ m} \tag{7.2}$$

from the appropriate proton or neutron. A preliminary theory of the inerton interaction of nucleons was developed in Ref. [321]. Below we will discuss the behavior of the deformation coat at the nucleon-nucleon interaction, the behavior of nucleons in a deuteron and a nucleus consisting of a large quantity of nucleons that can be considered as a cluster of protons and neutrons.

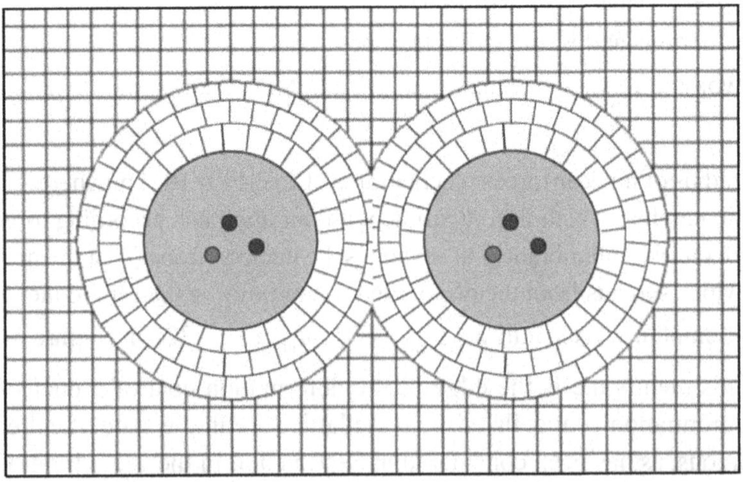

FIGURE 7.1 Attraction of two nucleons is caused by the adhesion of their membranes.

7.1 DEFORMATION COAT OF THE NUCLEON

7.1.1 BEHAVIOR OF THE COAT IN THE PHASE SPACE

Excitations of the deformation coat owing to its spherical form can be boiled down to the well-known task of mathematical physics to describe small vibrations of a gas contained in a sphere (see, e.g., Ref. [322]).

The velocity potential u of the gas fulfills the wave equation

$$\frac{\partial^2 u}{\partial r^2} + \frac{2}{r}\frac{\partial u}{\partial r} = \frac{1}{c^2}\frac{\partial u^2}{\partial t^2}; \tag{7.3}$$

here u for radial vibrations is a function of r and t alone where r is the distance from a vibrating particle of the gas to the center of the sphere and t is the time. Equation (7.3) is solved by substitution

$$u(r,\, t) = w(r)\,\vartheta(t) \tag{7.4}$$

The surface of the sphere is treated as a hard envelope and therefore the normal component of the velocity is equal to zero, which leads to the boundary condition

$$\partial w(r) / \partial r\big|_{r=R_0} = 0 \tag{7.5}$$

where R_0 is the radius of the spherical envelope (the radius of the deformation coat). The solution for $w(r)$ is

$$w(r) = C\,\frac{\sin(\eta\, r + \phi)}{r} \tag{7.6}$$

or in the explicit form

$$w_s(r) = \frac{1}{r}\,\sin\left(\frac{\eta_s + \kappa}{R_0}\, r\right) \tag{7.7}$$

where $\eta_s = (s + \frac{1}{2})\pi - \varepsilon_s$, κ is the positive constant and the function ε_s satisfies equation

$$\varepsilon_s = \frac{2}{(2s+1)\pi} + \frac{4\varepsilon_s^2}{3(2s+1)\pi} \tag{7.8}$$

The partial solution to the problem is

$$u_s(r,t) = U_s \frac{1}{r} \sin\left(\frac{\eta_s + \kappa}{R} r\right) \sin\left(c \frac{\eta_s + \kappa}{R} t + \beta_s\right) \tag{7.9}$$

Nuclear forces appear when deformation coats of separate nucleons overlap (Figure 7.1). Indeed, vibrating cells of one coat begin to interact with those of another. It is a matter of fact that the interaction between two oscillators reduces the total energy of the oscillators. In our situation the overlapping means that boundary condition (7.5) is destroyed: the derivative $\partial w / \partial r$ becomes nonzero at $r = R_0$. The nucleon-nucleon interaction extends the potential $w(r)$ of one participant beyond its spherical envelope to within the potential induced by the other participant. Therefore we can expect that condition (7.5) will now be realized at the other effective distance $r = R_0 + \Delta R$, i.e.,

$$\partial w(r) / \partial r\big|_{r = R_0 + \Delta R} = 0 \tag{7.10}$$

Boundary condition (7.10) alters arguments $(\eta_s + \kappa) r / R_0$ and $(\eta_s + \kappa) ct / R_0$ in expression (7.9) to $(\eta_s + \kappa) r / (R_0 + \Delta R)$ and $(\eta_s + \kappa) ct / (R_0 + \Delta R)$, respectively. This enables us to rewrite the solution (7.9) in the form

$$u_s(r,t) = U_s \frac{1}{r} \sin\left(\frac{\eta_s + \kappa}{R_0 + \Delta R} r\right) \sin\left(c \frac{\eta_s + \kappa}{R_0 + \Delta R} t + \beta_s\right) \tag{7.11}$$

Solution (7.11) shows that the union of nucleons lowers their energy. In fact the multipliers at time t in expressions (7.9) and (7.11) are the respective cyclic frequencies ω_s and $\tilde{\omega}_s$, i.e.

$$\omega_s = c(\eta_s + \kappa) / R_0, \qquad \tilde{\omega}_s = \omega_s \cdot (1 - \Delta R / R_0) \tag{7.12}$$

It is reasonable to assume that proper harmonic vibrations of massive cells in the deformation coat occur at the fundamental tone that is characterized by the frequency $\omega_1 = (\eta_1 + \kappa)c / R_0$ where $\eta_1 \cong 4.494$. If we set $\kappa = 1.785$, we indeed reach the equality

$$\hbar\omega_1 = m_{0n}c^2 \cong m_{0p}c^2 = 938.26 \text{ MeV} \tag{7.13}$$

Since the interaction of nucleons can reduce their energy by $W = 47$ MeV (in agreement with the Fermi gas model, see e.g., Ref. [323]), we can write the equality

$$\hbar\omega_1 \frac{1}{1 + \Delta R / R_0} \cong (938.26 - 47) \text{ MeV} \tag{7.14}$$

that makes it possible to estimate an effective range of the overlapping of deformation coats of two nucleons, $\Delta R \cong 0.053 R_0$. Such an overlapping can virtually draw two nucleons together, but only a little. A deeper penetration into the core size R_c can be achieved only in the case of heavy nuclei when the collective motion of a great number of nucleons is allowed for.

Thus the study conducted in the framework of the phase space has shown that the coupling of nucleons is a beneficial process. Now let us take a look at the behavior of nucleons in the real space.

7.1.2 BEHAVIOR OF THE COAT IN THE REAL SPACE

Vibrations of cells in the deformation coat mean that the values of their masses oscillate in such a way that the local volumetric deformation (the mass) periodically passes into the local tension of the tessellattice. So, we can construct the following peculiar Lagrangian (a function of the square forms of length, mass and time) that describes the behavior of cells in the deformation coat (compare with Sections 3.4 and 3.6)

$$L = \sum_{\mathbf{n}} \left(\tfrac{1}{2}\lambda^2 \dot{\mu}_{\mathbf{n}}^2 + \tfrac{1}{2}\mu_{\mathbf{n}0}^2 \dot{\xi}_{\mathbf{n}}^2 + c\lambda\,\mu_{\mathbf{n}0}\,\dot{\mu}_{\mathbf{n}}\nabla\xi_{\mathbf{n}} \right) \tag{7.15}$$

Here μ_n is the mass of the **n***th* cell that is specified by the radius vector **n** and $\xi_{\mathbf{n}}$ is the tension of that cell; the tension also includes a deviation of the cell from its equilibrium position in the deformation coat. λ is a characteristic diameter of the nucleon, μ_{n0} is the initial mass of the **n***th* cell and c is the speed of light.

The Euler-Lagrange equations of motion written for μ_n and $\xi_{\mathbf{n}}$ are

$$\ddot{\mu}_{\mathbf{n}} + c\mu_{\mathbf{n}0}\lambda^{-1}\nabla\dot{\xi}_{\mathbf{n}} = 0 \tag{7.16}$$

$$\ddot{\xi}_{\mathbf{n}} + c\lambda\mu_{\mathbf{n}0}^{-1}\nabla\dot{\mu}_{\mathbf{n}} = 0 \tag{7.17}$$

These equations are easy to decouple to wave equations. Besides, since this system has radial symmetry, the equations can be rewritten in the spherical coordinates (recall that in the spherical coordinates the Laplace operator $\Delta\mu$ is $\frac{1}{r}\frac{\partial^2}{\partial r^2}(r\mu)$). Omitting the index **n**, we obtain two uncoupled wave equations

$$\frac{\partial^2\mu}{\partial r^2} + \frac{2}{r}\frac{\partial\mu}{\partial r} = \frac{1}{c^2}\frac{\partial^2\mu}{\partial t^2}, \tag{7.18}$$

$$\frac{\partial^2\xi}{\partial r^2} + \frac{2}{r}\frac{\partial\xi}{\partial r} = \frac{1}{c^2}\frac{\partial^2\xi}{\partial t^2} \tag{7.19}$$

In Eqs. (7.18) and (7.19) μ and ξ are functions of the distance r from the central point of the spherical deformation coat and time t. Taking into account that values μ and ξ should oscillate in opposite phases, we arrive at the appropriate initial and boundary conditions

$$\mu(0, 0)|_{r\sim\ell_P} = \mu_0, \ \xi(0, 0)|_{r\sim\ell_P} = 0 \tag{7.20}$$

$$\mu(r, t)|_{r=R_0} = 0; \qquad \xi(r, t)|_{r=R_0} = \xi_0 \tag{7.21}$$

Then the solutions for μ and ξ become

$$\mu(r, t) = \mu_0\frac{r_{01}}{r}\left|\cos\left(\frac{\pi}{2}\frac{r}{R_0}\right)\right|\left|\cos\left(\frac{\pi}{2}\frac{t}{T}\right)\right|; \tag{7.22}$$

$$\xi(r, t) = \xi_0\frac{r_{02}}{r}\left|\sin\left(\frac{\pi}{2}\frac{r}{R_0}\right)\right|\left|\sin\left(\frac{\pi}{2}\frac{t}{T}\right)\right| \tag{7.23}$$

Here, μ_0 is the mass of the deformation coat; ξ_0 is the maximum value of the tension at the boundary $\xi(r, t)|_{r=R_0}$; ω is the cyclic oscillation frequency of cells (i.e., the mass of cells); parameters r_{01} and r_{02} are normalized constants having the dimension of length. Since the value of mass in the solution (7.22) cannot be negative, the solution includes modules of spatial and temporal parts. Time t is a natural parameter, which means that $T = R_0 / c$ and hence $\omega = \pi / T$; this fact is also reflected in the dispersion relationship: $\omega = kc$, where the wave number $k = \pi / R_0$.

A typical frequency for oscillations of the inerton cloud of a nucleon is $\omega \approx 10^{22} - 10^{23}$ s^{-1}. Hence if the time t of the interaction of the nucleon with a partner is longer than the time ω^{-1}, we can consider the solutions (7.22) and (7.23) as averaged in time, namely:

$$\mu(r) = \tfrac{2\sqrt{2}}{\pi}\mu_0 \frac{r_{01}}{r} \left| \cos\left(\frac{\pi}{2}\frac{r}{R_0}\right) \right| \tag{7.24}$$

$$\xi(r) = \tfrac{2\sqrt{2}}{\pi}\xi_0 \frac{r_{02}}{r} \left| \sin\left(\frac{\pi}{2}\frac{r}{R_0}\right) \right| \tag{7.25}$$

These solutions may be further reduced to the mean field approximation:

$$\mu(r) = \mu_0 \bar{r}_{01} / r \tag{7.26}$$

$$\xi(r) \approx \xi_0 \tag{7.27}$$

where we put $\bar{r}_{0i} = 8 r_{0i} / \pi^2$. The parameter $\bar{r}_{01}$ can be related to the Planck length ℓ_{P} or the rms proton charge radius r_p (6.30) and the parameter $\bar{r}_{02}$ is more precisely the boundary of the deformation coat equal to $R_0 \cong \lambda_{\mathrm{Com,\ p(n)}}$.

So the deformation coat with the range of $R_0 \cong \lambda_{\mathrm{Com}} = h / (m_{\mathrm{p(n)}} c)$ represents the actual deformation potential of the nucleon. This is the mass field, or mathematically is a deformation field, which is responsible for the availability of attractive nuclear forces in nuclei. These solutions show that the potential of the nuclear forces is carried by inertons and the potential has the dependence $1/r$. The nucleon-nucleon interaction changes the boundary conditions (7.21); in particular, for the mass the condition is adjusted to $\mu(r, t)|_{r=R_0+\Delta R} = 0$, i.e., the overlapping of two coats extends the range of action of the massive (deformation) potential. This brings about the replacement of R_0 for $R_0 + \Delta R$ in the solution for the mass μ (7.21), (7.22), (7.24) and (7.26).

The amplitude of the inerton cloud Λ much exceeds the object's size, (7.2). In nuclei an average distance between nucleons is about 2 fm, which means that the nucleons' deformation clouds are overlapped, and due to the nucleon's motion, that the nucleons additionally interact through their inerton clouds.

Thus the phenomenon of the attraction, in essence, is caused by the contraction of ambient space that surrounds nucleons. The overlapping of the deformation coats means that one nucleon comes under the influence of the deformation coat of another nucleon. Using experimental data on the minimum value of the potential well (W = 35 to 47 MeV), we can evaluate an increase in the total mass of cells gathered in the deformation coat. Since the total energy $m_{\mathrm{p(n)}}c^2 \to m_{\mathrm{p(n)}}c^2 - \Delta E$, we may write for the total mass of cells in the coat: $m_{\mathrm{p(n)}} \to m_{\mathrm{p(n)}} + \Delta m$ where $\Delta m = m_{\mathrm{p(n)}} \Delta E / (m_{\mathrm{p(n)}} c^2) \cong (0.037 \text{ to } 0.05)\, m_{\mathrm{p(n)}}$. Hence the notion of a "potential well" implies that in the range of space covered by the well the cells of the tessellattice are found in a more contracted state than in the space beyond the potential well.

The tension ξ has practically a constant absolute value ξ_0 in the solution (7.23) (see also expressions (7.25) and (7.27)). It is a linear elastic response on the side of the tessellattice to the appearance of the defect μ in it. The tension plays the role of a restoring force, which guides, or pushes inerton excitations back to the nucleon in the nucleon's deformation coat. Thus, reiterating and reinforcing what has already been said above, the nucleon's deformation coat represents the deformation potential, i.e., the mass field, which is responsible for the attractive nuclear forces between nucleons with the nucleus. To understand this more fully we now need to look at the specific interaction of nucleons in concrete examples.

7.2 NUCLEONS IN THE DEUTERON

The deuteron is the simplest nuclear system; nevertheless, the origin of nuclear forces that bind two nucleons has not been fully understood [302]. The deuteron is characterized by the potential well that has the depth $U = 38.5$ MeV, the binding energy is $\Delta U = 2.22$ MeV and hence the total kinetic energy, as the difference of these two values, is 35.7 MeV.

Let us now look at the problem of the deuteron from the viewpoint of the constitution of the real space and the submicroscopic concept. Each of the two nucleons has the radius $r_{\text{p(n)}} \cong r_{\text{p}} \cong 0.84$ fm (6.30). The availability of the binding energy allows us to assume a possible extension of the radius on a value $r_{\text{p}} \cdot \Delta U / U$, such that a new hard radius of a nucleon in the deuteron becomes $\tilde{r}_{\text{p}} = r_{\text{p}} \cdot (1 + \Delta U / U) \cong 0.89$ nm.

It seems the two nucleons will be involved in the rotation prescribed by the rigid rotor; hence the deuteron is specified with a reduced mass $\mu_{\text{d}} = m_{\text{p}} m_{\text{n}} / (m_{\text{p}} + m_{\text{n}})$. The binding energy ΔE will increase the value of the reduced mass on a correction $\Delta m = m_{\text{p (n)}} \cdot \Delta U / U \cong 0.106 \times 10^{-27}$ kg. Hence, the deuteron is described by a reduced mass $\tilde{\mu} = \mu_{\text{d}} + \Delta m =$ $(0.8368876 + 0.1061) \times 10^{-27}$ kg.

The potential well U can be presented in the form $U = \hbar\omega = h\nu$ and then $\nu = 9.309 \times 10^{21}$ Hz becomes the frequency of a γ-quantum needed for breakup of the deuteron.

From the submicroscopic point of view, the potential energy U is an energy stored in the common deformation coat that two nucleons share. In other words, it is an inner inerton energy of the deformation coat. The inerton potential that holds two nucleons can be presented in the form

$$U(r) = h\nu\, \tilde{r}_{\text{p}} / r \tag{7.28}$$

The Lagrangian of a deuteron can be written as follows

$$L = \tfrac{1}{2}(m_{\text{d}} + m_{\text{u}})\, \dot{r}_{\text{c.m.}}^{2} + \tfrac{1}{2} \tilde{\mu}\, \dot{r}^{2} - h\nu\, \tilde{r}_{\text{p}} / r \tag{7.29}$$

The equilibrium position in the orbit of two nucleons is provided by the equilibrium of the force of the potential $U(r)$ (7.28) and the centrifugal force, namely

$$\frac{h\nu\,\tilde{r}_{\mathrm{p}}}{r^2}=\frac{\tilde{\mu}\upsilon^2}{r} \tag{7.30}$$

The condition of quantization of the orbit, i.e., when only one de Broglie wavelength is placed in the orbit

$$\tilde{\mu}\upsilon r = n\hbar \tag{7.31}$$

allows us to derive a radius of the stationary orbit similar to the case of the Bohr radius for the hydrogen atom:

$$r_{\text{deutron}}=\frac{n^2\hbar}{2\pi\nu\,\tilde{r}_{\mathrm{p}}\,\tilde{\mu}} \tag{7.32}$$

For the first orbit ($n=1$) substituting the numerical values of the parameters above (ν, $\tilde{r}_{\mathrm{p}}$ and $\tilde{\mu}$) we obtain from expression (7.32): $r_{\text{deutron}}=2.148$ fm, which is in an agreement with recent experimental data 2.1413 ± 0.0025 fm (CODATA, 2014).

Thus we have just demonstrated that the deformation coat of a nucleon, which is induced around the nucleon in the ambient tessellattice, and the inerton cloud generated by the nucleon in its motion, cause the nuclear forces that bind the nucleons together. The interaction between nucleons is provided by inertons.

7.3 A NUCLEUS AS A CLUSTER OF PROTONS AND NEUTRONS

Let us consider the stability of the nucleus as reasoned from the statistical description of a system of a large quantity of interacting protons and neutrons.

Nucleons fill shells in a nucleus and move in the field of the same potential. The potential shape is caused by additional deformations of cells in the deformation coats of nucleons, i.e., the attraction occurs owing to (i) the overlapping of deformation coats of nucleons, and (ii) the overlapping of inerton clouds of nucleons, which carry the space volumetric fractal deformation, i.e., mass. It is evident that in the potential field the nucleons move along their proper trajectories and each of the nucleons possesses its own energy, the moment of momentum and spin. If we project such a dynamic behavior of nucleons on a model lattice, we arrive at the pattern in which each nucleon occupies its own knot.

Before applying the aforementioned statistical approach, we shall first clarify the structure of the pair potential that acts between nucleons, and then subdivide the potential into attractive and repulsive components. The Hamiltonian of two kinds of interacting particles has been considered in Section 5.2. The initial Hamiltonian has been boiled down to the formal Hamiltonian (5.24) that in general form includes the paired potentials of attraction V^{att} and repulsion V^{rep}. The potentials take into account the effective paired interaction between nucleons located in states s and s'. The occupation numbers n_s can have only two meanings: 1 (the sth knot is occupied in the model lattice studied) or 0 (the sth knot is not occupied). The signs before positive functions V^{att} and V^{rep} in the Hamiltonian (5.24) directly specify proper signs of attraction (minus) and repulsion (plus).

In considering a nucleus as a cluster of nucleons, we arrive at the appropriate action (5.56). The number of nucleons $\mathcal{N}$ that form the cluster can be calculated when the potentials are given. So let us provide the system under consideration with realistic potentials.

Although the mass distribution in the inerton cloud of a nucleon is governed by law (7.26), the paired interaction at the scale $r < \Lambda$ can be simulated according to harmonic law. Starting from some critical value of nucleons in the system, nucleons having the same energy and momentum will behave like elastic balls. In other words, the nucleons' inerton clouds will elastically interact with each other in the range $r \ll \Lambda$. Therefore, in the model lattice of the system of a large number $\mathcal{N}$ of nucleons the attraction between nucleons, associated with the space deformation of the kinds (i) and (ii) mentioned above, can be presented in the form of two terms, namely

$$V^{\text{att}} = -\hbar\omega_1 \frac{\Delta R}{R_0} + \tfrac{1}{2}\gamma r^2 \qquad (7.33)$$

where $\hbar\omega_1$ is the total energy of a nucleon (7.14); basically, the first term in expression (7.33) should depend on a distance as $1/r$ owing to the radial symmetry, however, since we consider only the effect related not to the interior of nucleons but associated with the interaction of nucleons, such a dependency can be omitted. In the second term in expression (7.33), γ is the elasticity/force constant of the inerton field in the lattice studied. The harmonic potential is often employed in nuclear physics, though so far its origin has remained completely unclear. The potential (7.33) should be identified with the effective attractive potential (5.22), which enters the function $b(\mathcal{N})$ (5.58).

The electromagnetic interaction, i.e., repulsion, which happens between protons, is realized via the protons' deformation coats. Therefore, we shall preserve rule (7.21) for the description of the charge drop in the polarized nucleon's coat with distance r. If we denote the repulsion between protons associated with their electromagnetic interaction as

$$U_{\text{el.-magn.}} = \frac{1}{4\pi\varepsilon_0}\frac{e^2}{r} \tag{7.34}$$

we can rewrite the potential of effective repulsive (5.23) as follow

$$V^{\text{rep}} = \tfrac{1}{2}\left[\left(-\hbar\omega_1\frac{\Delta R}{R_0} + \tfrac{1}{2}\gamma r^2\right) + \left(-\hbar\omega_1\frac{\Delta R}{R_0} + \frac{1}{4\pi\varepsilon_0}\frac{e^2}{r}\right)\right] \tag{7.35}$$

Calculating functions $a(\mathcal{N})$ (5.57) and $b(\mathcal{N})$ (5.58) we get instead of action (5.56)

$$S \cong K \cdot \left\{\left(\frac{3e^2\mathcal{N}^{2/3}}{16\pi\varepsilon_0\, g\, k_B\Theta} - \frac{3\gamma g^2\mathcal{N}^{5/3}}{20k_B\Theta}\right)\mathcal{N}^2 - \ln\mathcal{N}\right\} + (N-1)\ln\xi \tag{7.36}$$

The minimum of action (7.36) is reached at the solution of the equation $\partial S/\partial\mathcal{N} = 0$ (if the inequality $\partial^2 S/\partial\mathcal{N}^2 > 0$ holds). With the approximation $\mathcal{N} \gg 1$ the corresponding solution is

$$\mathcal{N} = \frac{5e^2}{4\pi\varepsilon_0\gamma g^3} \equiv \frac{5}{3}\frac{e^2 n_{\text{nucl}}}{\varepsilon_0\gamma} \tag{7.37}$$

where n_{nucl} is the concentration of the nuclear matter, which can be put $n_{\text{nucl}} = 1.68 \times 10^{44}$ m^{-3} (note the density of nuclear matter is 1.8×10^{18} kg·m^{-3}). The solution (7.37) shows that the number of nucleons $\mathcal{N}$ in a nucleus is set for a given number of charged protons whose repulsion is balanced out by the mean inerton field of all $\mathcal{N}$ the nucleons.

Table 7.1 based on result (7.37) presents some major parameters of a nucleus with $\mathcal{N}$ nucleons. The value $\hbar\omega$ represents the difference between equidistant levels of energy in the solution of the corresponding oscillatory problem. We see from the table below that, as $\mathcal{N}$ increases, it is the elasticity constant γ of the inerton field that mostly changes, which in turn leads to a dramatic decrease in $\hbar\omega$. In fact when $\hbar\omega$ falls to the critical value of 0.75 MeV, which corresponds to the Coulomb repulsion between protons, when the cluster (i.e., the nucleus) begins to decompose.

In the above model, the quantities of protons and neutrons have been considered to be the same. It is obvious that increasing the proportion of protons should soften the elasticity of the inerton field, which, therefore, will decrease to a critical value γ_c leading in further decay of the nucleus. If for example we consider Carbon-12, the *p*:*n* ratio is obviously 6:6, i.e., 1:1. If we increase the ratio by adding one more proton we get an isotope of Nitrogen-13 (the *p*:*n* ratio is 7:6), which is unstable with a half-life of about 10 minutes.

But then, if the quantity of neutrons significantly exceeds that of the protons, the cluster will also become unstable. So it seems there is an upper limit to the weight of such a disproportionate mixture of nucleons that can exist. However, one can say that increasing the proportion of neutrons would

TABLE 7.1 Estimation of Some Parameters (from Ref. [301])

$\mathcal{N}$	γ [N/m]	$\omega = \sqrt{\gamma / m_{\text{nucl}}}$ [s^{-1}]	$\hbar\omega$ [MeV]	λ_{dB} [m]	$R = R_0 \mathcal{N}^{1/3}$ [m]	R / λ_{dB}
50	11.78×10^5	2.65×10^{21}	1.74	2.21×10^{-14}	4.42×10^{-15}	0.2
100	5.89×10^5	1.88×10^{21}	1.24	2.61×10^{-14}	5.57×10^{-15}	0.21
200	2.94×10^5	1.33×10^{21}	0.88	3.05×10^{-14}	7.01×10^{-15}	0.23
250	2.36×10^5	1.19×10^{21}	0.78	4.58×10^{-14}	7.56×10^{-15}	0.17

Here, $m_{\text{nucl}} = m_{\text{p}} \cong m_{\text{n}}$.

mean strengthening (hardening) the elasticity of the inerton field. On the other hand, as follows from Section 5.1, the cluster solution is possible only in the case of two types of particles with opposite interactions – attraction and repulsion. When the system under consideration has a small number of protons, its neutrons can be treated as a practically free Fermi gas, i.e., as quantum particles without any clusterization. Of course there are other factors that may correct the result (7.33). First of all this is the spin interaction between nucleons, which should induce some additional increase in the energy of the system of many nucleons, and the atom's orbital electrons. Nevertheless, a quantitative pattern accounting for the restriction on the mass of nuclei as a function of $\mathcal{N}$ has become evident.

The major reasons for the instability of heavy nuclei can be summarized as follows. The number $\mathcal{N}$ of nucleons that enter a cluster, i.e., a nucleus, is defined by condition (7.37). This condition links the value of $\mathcal{N}$ to the elasticity constant of the inerton field γ in the nucleon, the nuclear density ρ_{nucl}, and the elementary electric charge e of the proton. The value of $\mathcal{N}$ is inversely proportional to the elasticity constant γ of the inerton field in a nucleus: an increase in $\mathcal{N}$ requires a decrease in γ towards the critical value γ_c, such that when $\gamma < \gamma_c$ the inerton field of the nucleus is incapable of holding nucleons in the cluster and hence the nucleus decays.

7.4 NUCLEAR COUPLING OF PROTON AND ELECTRON: SUBATOMS

In the previous sections we have discussed the binding forces between nucleons in the atomic nucleus. Now we will extend our discussion to the interactions between the nucleons and electrons that lead to stable atoms. We begin this thread by exploring the strange nature of subatoms – a combination of nucleons and electrons of mere nuclear size.

In 1993 two papers by Don Borgie et al. [324] and Dufour [325] were published in which the researchers demonstrated having created a new particle – quite stable hydrogen and deuterium that had a nuclear size in the order of femtometers. Then there were further studies of Dufour et al. [327], Mills et al. [328] and Santilli [329]. In those works the researchers also reported details about a new stable subatomic particle with a nuclear size – a combined proton and electron.

In 2002–2014 Taleyarkhan, Nigmatulin and their colleagues [330, 331] presented studies on sonofusion – acoustic cavitation experiments using pure deuterated acetone that generated thermonuclear neutrons, tritium and gamma-quanta.

Putterman's group [332] and Geuther et al. [333] by experimenting with pyroelectric crystals $LiNbO_3$ and $LiTaO_3$ detected 2.5 MeV neutrons. Danon [334] by using a similar device was able to generate X-rays with energy over 200 keV and about 10^5 neutrons.

Cardone et al. [335] reported the emission of neutrons from the mechanical crushing of solid specimens, such as marble and granite, measured by means of Helium-3 neutron detectors. Piezonuclear decay of thorium was measured by Cardone and Mignani [336] during the cavitation of a solution of thorium-228 in water; the transformation occurred at a rate 10^4 times faster than the natural rate of its radioactive decay.

An analysis of these studies was carried out in our work [337] that additionally described discharge experiments of a new kind that were performed to probe the nuclear coupling of protons and electrons by analyzing the resultant emission of neutron flux and gamma quanta. A sophisticated high-voltage set-up had been designed to form discharges in gaseous atmospheres. The apparatus consisted of two units: a converter unit, i.e., generator, and a power supply. The converter unit converted a low voltage signal to a high voltage pulsed signal. The master oscillator generated high-voltage antiphase control pulses with a frequency of 1 kHz and "dead" time of about 3 to 6% of the working cycle. These pulses created a plasma in a gas between the electrodes. Regulation of the output voltage was achieved by changing the voltage V of the power supply between 12 to 35 V. The output voltage V induced an attractant electric field in the gap between the electrodes. Typical values of the current used in our experiments did not exceed 0.1 A.

Discharges were carried out in a special hermetically sealed stainless steel reaction chamber having a volume of about 250 ml. We carried out experiments in a hydrogen atmosphere, at a pressure of 0.6 to 1 bar, and also in atmospheres of air and helium. By our estimate, the discharge chamber filled with hydrogen at normal conditions contained up to 100 times more hydrogen atoms than is the case with air (in the air hydrogen atoms are almost all part of the water molecules).

For the measurement of gamma-radiation, we used a 'Vector' gamma-ray spectrometer designed by our team, which was adjusted to measure an integral activity of gamma radiation over a range up to 3 MeV.

For the measurement of a possible flow of neutrons, we used a neutron radiation detector BDPN-07 (produced by Zakhidprylad Ltd., Lviv, Ukraine). The measurement range of thermal neutron flux density is 10 neutrons/(cm^2·min) to 10^5 neutrons/(cm^2·min). The measurement range of fast neutron flux density is 50 neutrons/(cm^2·min) to 10^5 neutrons/(cm^2·min). We also used a passive-type track-edge fast neutron plastic detector CR-39.

In addition we used a 'Rudra' device for measuring inertons, which measured mass flux, i.e., inertons. Inerton signals were continuously measured when discharges occurred, which were predictable, because free inertons emerge at any fast non-adiabatic physics processes (like collisions of particles).

We observed that the discharge in hydrogen gas and air produced a bright blue light, though in the case of helium gas the light was purple. The light intensity did not change with the change of the voltage level, which we applied to the electrodes.

After launching a discharge in the reactor chamber containing the gas being studied, we gradually increased the output voltage from 12 to 35 V. In the hydrogen atmosphere, between 16.5 and 17.5 V all three detectors arranged near the reaction chamber measured signals. When the voltage level was raised higher than 17.5 V and up to the maximum value of 35 V, no signals were measured by either the 'Vector' gamma-ray measuring device or the BDPN-07 neutron-measuring unit. The 'Rudra' inerton measuring device recorded signals over the entire range of voltage changes.

The excess of gamma-ray radiation above the background was quite noticeable, namely, more than three times in the vicinity of the window.

At a strong discharge, the neutron counter measured 10^3 pulses per second at a distance of 1 m from the window of the reactor chamber. At a distance of about 15 cm from the window the device measured the maximum value that it could measure, 3×10^5 neutrons/(cm^2 min). When the generator was switched off or was turned to a higher or lower voltage level outside the resonance range of 16.5 to 17.5 V, the neutron counter measured only the background neutron radiation.

No tracks were observed in the six used fast neutron plastic detectors CR-39. This points out to the fact that the reaction chamber generated only slow neutrons.

Very similar results were obtained for all three gases: hydrogen, helium and air.

How to explain all those results?

7.4.1 GAMMA RADIATION

Let us initially consider the origin of gamma-quanta. At the collision of a proton with a tungsten atom located on the edge of the cathode, the proton's inerton cloud can be shaken off from the proton whereupon it is seized by an electron from the outer shell of the tungsten atom, such that the proton becomes immobile for a short time until the environment restores its inertons bringing them back to the proton, renovating its mobility.

A moving proton consists of two sub systems: the proton core and the proton's inerton cloud. The mass of each of these two subsystems is the same and equals the proton mass $m_{\rm p}$ (the mass oscillates between these two subsystems). Due to an inelastic scattering of incoming protons by tungsten atoms of the cathode a tungsten atom may absorb the whole of the proton's inerton cloud. Namely, electrons of the tungsten atom absorb the incoming inerton cloud. In the Bohr model of hydrogen-like nuclei with atomic charge Z an average orbit radius of the electron density is

$$r_Z = \frac{4\pi\varepsilon_0 \hbar^2 \bar{n}^2}{e^2 Z m_{\rm e}} \tag{7.38}$$

where for a tungsten atom $Z = 74$; $m_{\rm e}$ is the electron rest mass; e is the elementary electric charge and $\bar{n}$ is the average orbit number. Let us put $r_Z = 10^{-10}$ m.

The tungsten atom's electrons absorbing the whole of the proton's inerton cloud increase their mass to the value of $Zm_{\rm e} + m_{\rm p}$. This instantly transfers the electrons to a deeper orbit with an average radius

$$r_Z^* = \frac{Zm_e}{(Zm_e + m_p)} r_Z \approx 2.58 \times 10^{-13} \text{ m} \tag{7.39}$$

The disturbed orbit (7.39) cannot be short-lived, as the large absorbed mass m_p squeezes the electron shells and induces an additional elastic interaction $m_p r_Z^{*2} \Omega^2 / 2$ in the medium of the electrons' inerton clouds, which stabilize the disturbance. The relationship

$$eV = \tfrac{1}{2} m_p r_Z^{*2} \Omega^2 \tag{7.40}$$

allows us to estimate the frequency of induced inerton oscillations in the electron density of the tungsten atom: $\Omega \approx 1.5 \times 10^{17}$ s^{-1}. Every 10^{-3} seconds discharges brought new inerton clouds of the incoming protons to the disturbed atom and hence the mass of the disturbance gradually increased by portions, $m_p + m_p + m_p + \ldots$ At a threshold number $\aleph$ of incoming protons, which passed their inerton clouds on to the tungsten atom under discussion, the mass becomes $\aleph m_p$. This huge mass is concentrated in the electrons of the tungsten atom, which strongly squeezes the electron system towards the tungsten atom's nucleon. As the result, this disturbance excites the nucleus such that it can only relax again by radiating a gamma quantum from it. Therefore, the kinetic energy of the oscillating mass disturbance is converted to the energy of the emitting γ-quantum

$$\tfrac{1}{2} \aleph m_p \upsilon^2 = h\nu_{\gamma\text{-quant}} \tag{7.41}$$

10^{17}–10^{18} protons impact upon the cathode per second at the current 0.01–0.1 A. The surface area of the cathode edge can be estimated as ~10^{-8} m^2. The number of tungsten atoms in this area is around 10^{11}. Therefore 10^6–10^7 protons impact upon each atom of tungsten per second. A part $\aleph$ of these protons may be inelastically scattered by each tungsten atom. Suppose $\aleph \approx 10^4$. An average velocity typical for electrons of outer shells of an atom may be assumed at 10^5 m·s^{-1}. Substituting these values into expression (7.41) we do get the energy of gamma quantum, $h\nu_{\gamma\text{-quant}}$~ 1 MeV.

7.4.2 NEUTRON RADIATION

Now let us consider the appearance of neutrons. Applying the data of Santilli [329] on his arc plasma, we are able to estimate the energy and the momentum of protons that impacted upon the cathode:

$$E = eV = 4.38 \times 10^{-18}\ \text{J},\ m_{\text{p}}\upsilon_{\text{p}} = 1.2 \times 10^{-22}\ \text{kg·m·s}^{-1} \qquad (7.42)$$

where the voltage was V = 27 V and hence the velocity of protons $\upsilon_{\text{p}} = \sqrt{2eV / m_{\text{p}}} = 7.19 \times 10^{4}$ m s^{-1}. As the cathode and anode were made of tungsten, the kinetic energy and the momentum of a tungsten atom can be estimated as follows

$$E_{\text{W}} = \tfrac{1}{2} m_{\text{W}} \upsilon_{\text{W}}^{2} \cong 2.07 \times 10^{-20}\ \text{J},\ m_{\text{W}}\upsilon_{\text{W}} \cong 1.2 \times 10^{-22}\ \text{kg·m·s}^{-1} \qquad (7.43)$$

where the mass of a tungsten atom is $m_{\text{W}} = 3.05 \times 10^{-25}$ kg; the thermal velocity of a tungsten atom $\upsilon_{\text{W}} = \sqrt{3k_{\text{B}}T / m_{\text{W}}} \cong 3.95857 \times 10^{2}$ m·s^{-1} (here the temperature of heated electrodes T = 1150 K is a fitting parameter).

Comparing expressions (7.42) and (7.43) we can see the momentum resonance between momenta $m_{\text{p}}\upsilon_{\text{p}}$ and $m_{\text{W}}\upsilon_{\text{W}}$. Thus when the proton impacts the tungsten atom that is oscillating in antiphase to its site of the crystal lattice, i.e., when $m_{\text{p}}\vec{\upsilon}_{\text{p}} + m_{\text{W}}\vec{\upsilon}_{\text{W}} = 0$, both particles must stop. Under such conditions the tungsten atom's inerton cloud is stripped from the atom and the latter becomes momentarily immobile until surrounding atoms bring their inertons to the tungsten atom under discussion and restore its mobility. At the same time, a free electron, as the lightest particle in the tungsten-gas interface, captures the tungsten atom's released inerton cloud. This is because the effective cross section of the electron, i.e., the size of the cross section of the electron's inerton cloud ~ $\lambda_{\text{dB}}\Lambda \approx \lambda^{2}_{\text{dB}}\, c / \upsilon_{\text{e}}$, much exceeds that of a proton. Thus the electron absorbing the tungsten atom's inerton cloud becomes heavy. Then merging the proton with this heavy electron will result in the creation of a super heavy hydrogen atom.

In such a super heavy system of the proton and the heavy electron, the electron's inerton cloud is very dense. In this system the reduced mass of the proton and the electron is almost equal to m_{p} (in fact $1/m_{\text{p}} + 1/(m_{\text{e}} + m_{\text{W}}) \cong 1/m_{\text{p}}$). Therefore the proton starts to rotate around the heavy electron; the Bohr radius is

$$r_{\text{p-e}} = \frac{4\pi\varepsilon_0 \hbar^2 n^2}{e^2 m_{\text{p}}} = 2.88 \times 10^{-14} \text{ m} \tag{7.44}$$

where we put $n = 1$. Though the electron orbit (7.44) deeply penetrates into the middle of the proton, the electron still does not reach the critical distance of 2×10^{-14} m that characterizes the quark orbit inside the proton [99]. If the penetration of the electron in the proton is forbidden for some reasons, we may put $n = 2$ or 3 in expression (7.44). Then the radius $r_{\text{p-e}}$ will be larger but is still in the order of femtometers.

In the tungsten-gas interface, the pair of the proton and electron is created with nearly zero kinetic energy.

As has been discussed in Section 7.4.1, in our experiments the gamma-ray radiation has appeared as a result of a peculiar mechanism of scattering of protons by tungsten atoms of the cathode. On the other hand, if a tungsten atom collides with an incoming proton being in the antiphase, i.e., hits on the counter movement, we arrive at the situation that just has been considered above for Santilli's [329] experiment. Namely, if at the collision the equality

$$m_{\text{W}} \upsilon_{\text{W}} = m_{\text{p}} \upsilon_{\text{p}} \tag{7.45}$$

holds, then the proton will itself absorb the tungsten atom's inerton cloud and catching an electron it will form a {proton + heavy electron} pair. Let us consider the left and right hand sides of the relationship (7.45).

The thermal velocity of the tungsten atom at room temperature is $\upsilon_{\text{W}} = \sqrt{3k_{\text{B}}T / m_{\text{W}}} \cong 2.018 \times 10^2$ m·s^{-1} and hence the momentum of the tungsten atom is $m_{\text{W}} \upsilon_{\text{W}} = 6.155 \times 10^{-23}$ kg·m·s^{-1}.

The velocity of the incoming proton is $\upsilon_{\text{p}} = \sqrt{2eV / m_{\text{p}}}$ where the voltage, which corresponds to the resonance interaction, is 17 V. In the experiment we have dealt with a pulsing current, which due to the modulation should generate electric impedance at the edge surface of the cathode. Impedance slows the transfer of charge across the border gas plasma – electrode, which in terms of capacitance reflects the capacity of the electrical double layer on the electrode. Slowing the motion of charges can be expressed as a tangent of the angle between the active and reactive resistance, $\tan\varphi$. Hence the velocity of the proton becomes $\upsilon_{\rm p} = \sqrt{2eV\tan\varphi / m_{\rm p}}$. The resonance condition (7.45) is reached at a fitting value $\tan\varphi = 0.416$. This condition corresponds with the proton velocity $\upsilon_{\rm p} = 3.68\times10^{4}$ m·s^{-1}.

Pairs of {proton + heavy electron} were named *subhydrogen* [337].

In the experiment carried out with a helium gas, we observed practically the same phenomena: gamma quanta and a neutron flux. In this case instead of protons we dealt with α particles. At the momentum resonance the expression (7.45) looks as follows: $m_{\rm W}\upsilon_{\rm W} \cong 4\,m_{\rm p}\upsilon_{\alpha}$, because the mass of α particle is $m_{\alpha} \cong 4m_{\rm p}$. The resonance was reached at V ≈ 15 V, which means that in the expression for velocity $\upsilon_{\alpha} = \sqrt{2\cdot(2e)V\tan\varphi_{\alpha} / (4m_{\rm p})}$ of the α particle we have to put $\tan\varphi_{\alpha} = 0.157$. Hence the resonance conditions have resulted into the creation of {${}^{4}_{2}\mathrm{He}^{2+}$ + two heavy electrons} pairs. This subatomic particle was named a *subhelium* [337].

In Mills' experiments [328] the resonance relationship (7.45) also comes true. In this case we put for the proton $\upsilon_{\rm p} = \sqrt{2eV / m_{\rm p}}$ and at $V = 1.5$ V the proton momentum equals $m_{\rm p}\upsilon_{\rm p} = 1.306\times10^{-23}$ kg·m·s^{-1}. For the tungsten atom, which diffuses in a hydrogen gas, one should include in the expression for the velocity a factor C related to the diffusion coefficient; namely, we have to rewrite the velocity of the tungsten atom as follows: $\upsilon_{\rm W} = \sqrt{3k_{\rm B}TC / m_{\rm W}}$. Putting $C = 0.212$, we calculate for the tungsten atom the same momentum $m_{\rm W}\upsilon_{\rm W} = 1.306\times10^{-23}$ kg·m·s^{-1}.

In the experiments of Taleyarkhan, Nigmatulinet et al. [330, 331] neutrons and tritium atoms were emitted from the collapsed bubble in a quantity of 150–200 for each kind of species per one pulse. This is also in line with the model described above. It seems reasonable to assume that at the collapse, the inerton clouds were stripped from deuteron atoms that were present in the molecules of the boundary layer of the bubble, which separated the liquid from the gas. The boundary layer of the bubble was formed by circa $N \approx 3\times10^{2}$ molecules. Hence each of these molecules was able to release its inerton cloud, such that $N \approx 3\times10^{2}$ heavy inertons transferred the energy $\varepsilon = W / N = 1.5\times10^{-13}$ J to gas entities in the collapsed bubble, where W is the work produced by ultrasound. Some of those entities, namely, molecules of acetone/benzene absorbed the inertons and radiated gamma quanta as has been discussed in Section 7.4.1. The other entities, rather molecules/atoms of deuterium, could emerge in the bubble owing to the destruction of deuterated acetone/benzene molecules by an outside neutron flux or α particles from the uranium salt.

Thus in the gas, a deuterium, which has only one electron, absorbs a heavy inerton cloud and a pair of the {deuteron + heavy electron} is formed, which has been named the *subdeuterium* [337]. The reduced mass of the subdeuterium is equal to almost $2m_{\text{p}}$. Then the size $r_{\text{d-e}}$ of the subdeuterium may be twice the radius (7.44) at the orbit $n = 2$.

Experiments [332–336] with piezocrystals in which ostensibly also neutrons were generated can be accounted for in a similar way [337].

7.4.3 STABLE NUCLEAR PAIRS

The device that detected neutron radiation was operating as a Helium-3 proportional counter that calculated a number of reactions

$$^{3}\text{He} + n^{0}\ (\text{thermal}) \rightarrow {}^{1}\text{H} + {}^{3}\text{H} + \text{Q} \tag{7.46}$$

where Q is a quantum of energy (typically this is a gamma quantum with the energy 764 keV).

In our experiments described above when instead of a neutron we have a subhydrogen, the reaction (7.46) may look like this

$$^{3}\text{He} + \langle n \rangle_{\text{H}}\ (\text{thermal}) \rightarrow {}^{1}\text{H} + \{{}^{2}\text{H}, \langle n \rangle_{\text{H}}\} + \text{Q} \tag{7.47}$$

where $\langle n \rangle_{\text{H}} \equiv \{p^{+}, \text{heavy } e^{-}\}$ is the subhydrogen and $\{{}^{2}\text{H}, \langle n \rangle_{\text{H}}\}$ is the neutral subatomic particle formed by the deuteron and subhydrogen, i.e., the {deuteron + subhydrogen} pair.

When instead of a neutron we have a subhelium, the reaction (7.47) may look as follows

$$^{3}\text{He} + \langle n \rangle_{\text{He}}\ (\text{thermal}) \rightarrow {}^{1}\text{H} + \{{}^{2}\text{H}, \langle n \rangle_{\text{He}}\} + \text{Q} \tag{7.48}$$

where $\langle n \rangle_{\text{He}} \equiv \{{}^{4}_{2}\text{He}^{2+}, \text{two heavy } e^{-}\}$ is the subhelium and $\{{}^{2}\text{H}, \langle n \rangle_{\text{He}}\}$ is the neutral {deuteron + subhelium} pair.

The nuclear pairs $\{^2\mathrm{H}, \langle n \rangle_\mathrm{H}\}$ and $\{^2\mathrm{H}, \langle n \rangle_\mathrm{He}\}$ may imitate a real tritium $^3\mathrm{H}$.

The reduced mass in the present subhelium is about $4m_\mathrm{p}$, which permits one to estimate the size of subhelium by using the expression for the Bohr radius (7.44). If we put the number of orbit $n = 2$, we will get exactly the same radius, as is the case for the subhydrogen (7.44).

The reaction (7.46) in the case of a synthesized subdeuterium, which acts instead of a natural neutron, may look as follows

$$^3\mathrm{He} + \langle n \rangle_\mathrm{D}\ (\text{thermal}) \rightarrow\ ^1\mathrm{H} + \{^2\mathrm{H}, \langle n \rangle_\mathrm{D}\} + \mathrm{Q} \qquad (7.49)$$

where $\langle n \rangle_\mathrm{D} \equiv \{^2\mathrm{H}^+, \text{heavy } e^-\}$ is the subdeuterium and $\{^2\mathrm{H}, \langle n \rangle_\mathrm{D}\}$ is the {deuteron + subdeuterium} pair.

The kinetic energy of the heavy inerton cloud $\varepsilon = 4 \cdot 10^{-13}$ J continues to push the whole pair. Indeed, if the inerton cloud was stripped from a deuteron atom, we may write the kinetic energy of the cloud in the form of $2m_\mathrm{p}\upsilon_\mathrm{inert}^2/2 = 4 \cdot 10^{-13}$ J, which defines the velocity $\upsilon_\mathrm{inert} = 1.55 \times 10^7$ m·s^{-1}. In the subdeuterium the total mass is $2m_\mathrm{p} + (2m_\mathrm{p} + m_e) \cong 4m_\mathrm{p}$ and its center-of-mass can continue to move with the same kinetic energy equal to $4 \cdot 10^{-13}$ J, which is 2.5 MeV (as was revealed in the experiments of Taleyarkhan, Nigmatulin et al. [330, 331]), though with slightly lower velocity 10^7 m·s^{-1}.

The known nuclear reaction

$$p^+ + e^- \rightarrow n^0 + \nu_e \qquad (7.50)$$

cannot really be applied for the subhydrogen, as this reaction occurs only in a large nucleus when the nucleus catches an orbital electron from the K or L shell, which results in the decay of an inner proton to a neutron. In the subhydrogen it is the huge mass of the electron, which protects the proton from the direct Coulomb invasion of the electron [337].

Regarding tritium detected in works [330, 331], we may assume the usual reaction

$$^2_1\mathrm{D} + ^2_1\mathrm{D} \rightarrow\ ^2_1\mathrm{T}\ (1.01\ \mathrm{MeV}) + p^+(3.02\ \mathrm{MeV}) \qquad (7.51)$$

or maybe the nuclear pair $\{^2\text{H}, \langle n \rangle_\text{D}\}$ determined above, which is formed of the deuteron and subdeuterium pair, imitating tritium ${}^2_1\text{T}$.

One more interesting reaction at sonofusion was reported by FitzGerald [338]; experimenting with sonoluminescence in a liquid D_2O, he measured an output of α particles, i.e., the low energy nuclear reaction ${}^2_1\text{D} + {}^2_1\text{D} = {}^4_2\text{He}$ took place. Or maybe that was an imitation of the reaction. FitzGerald's result has not been replicated by other researchers yet, but the reaction is extremely interesting, as it can occur with release of heat above 20 MeV.

Thus each of the considered subatoms (subhydrogen, subdeuterium and subhelium) is formed on the basis of a proton, deuteron, α particle and an electron (or a pair of electrons as is the case for an α particle) in the presence of an inerton cloud released from one of the nearest atoms from the surrounding. Moreover, these particles are able to mate with a proton, neutron and deuteron, forming nuclear pairs whose behavior can be described within the rigid rotor problem.

In the future it will be interesting to examine discharges in different heavier gases in an experiment similar to ours, which would bring new heavier subatoms and new nuclear pairs.

A large mass concentrated in a subatom owing to absorbed inertons is able to affect nuclear reactions overcoming the Coulomb barrier, in particular, the reactions of fission and fusion. Subatoms and nuclear pairs formed with the participation of subatoms are able to imitate nuclear reactions in a way similar to that of usual neutrons. Furthermore, subatoms and nuclear pairs may stimulate nuclear transmutations. In particular, a number of spontaneous transmutations were observed in biological systems [339].

The results obtained point to the fact that nuclear physics and particle physics must more tightly cooperate with condensed matter physics both in physics of nuclear fission and fusion.

It seems that the study of fusion energy in the framework of the pure thermonuclear concept requires a rethinking and reconsideration. Nuclear reactions in rarefied plasma require confinement and control, which is questionable. On the other hand, in dense plasma the conditions for the confinement become more plausible, because in such circumstances an inerton field begins to play a part. Besides, it should be emphasized that in nature we do not observe sustained fusion reactions in a pure plasma state. Stars and in particular the Sun consist rather of a metallic liquid than gaseous plasma [340]; in fact the gravitational attraction that protons experience on the surface of the Sun exceeds by a factor of 10^5 their Coulomb repulsion and the surface temperature of the Sun, which may combine hydrogen in the state of a real liquid.

In the case of low energy nuclear reactions we also meet a number of problems and one of them is associated with non-radiating transformations of particles. The majority of such reactions do not generate heat and they cannot be considered as a prelude for future industrial power plants. The first attempt of Rossi [341] needs a theoretical comprehension, which will allow one to model more successful experiments. Such experiments are quite possible to do involving subatoms, nuclear pairs and the inerton field that is able to control nuclear reactions.

7.5 INERTONS AND RADIOACTIVE ISOTOPES

In our recent studies we revealed that inerton fields produced in rather simple laboratory conditions are able to influence not only chemical, physical and biological processes, but also nuclear reactions [99]. A laboratory apparatus that generates inerton fields is illustrated in Figure 5.13. Using this, we studied samples of radioactively contaminated water.

Gamma spectra of a sample of radioactively contaminated water, 300 ml, initially showed a level of radioactivity of about 10^{-5} $Bq \cdot l^{-1}$. After 30 seconds of processing the water sample with an inerton field having an intensity of a few thousand pulses per second, the level of radioactivity was quenched by 32%, namely, from 4517.9 to 3401.49 pulses per second. Further treatment did not reduce the level of radioactivity of water – we reached a saturation threshold. It seems a further quenching could be possible by increasing the intensity of the applied inerton field.

Cylindrical samples with a length of 15 mm and a diameter of 1 mm made of technical iron were used to generate an inerton field under special conditions, namely, when we provoked their magnetostriction activity. Those iron agents were also themselves affected by an inerton field in the apparatus Lx (Figure 5.13) in the air for 30 seconds. The structure of the control sample, which was studied with the use of a JEOL electron microscope, was (percent by weight): Fe = 87.89%, C=11.72%,

Mn = 0.39%. After exposure to the inerton field generated at the striction by the iron agents, the elemental composition of three samples taken from three points on the surface of an iron agent is shown as below:

- Fe = 68.63%, C = 2.09%, Mn = 0.38%, O = 26.18%, Ni =1.47%, Co = 0.56%, Cr = 0.35% and Ca = 0.34%;
- Fe = 63.04%, C = 3.57%, O = 28.78%, Ni = 1.93%, Hf = 1.33%, Ca = 0.58%, Cr = 0.40% and Cs = 0.36%;
- Fe = 63.16%, C = 3.88%, Mn = 0.20%, O = 29.50%, Ni = 2.44%, Ca = 0.48%, Cr = 0.34%, and Cs = 0.00%.

As we can see new elements have appeared. The transmutation revealed can be related only to the presence of the inerton field.

One more experiment [342] was devoted to testing a new technological method for neutralizing alloys containing heavy contaminants, including radionuclides. The proposed approach seamlessly combined positive properties of both physicochemical methods and nanotechnology using silica-magnetite nanocomposite SiO_2/Fe_3O_4. The said nanocomposite was synthesized directly in an artificially polluted solution (the so-called method of direct sedimentation). The novelty of this method is the absorption of pollutants by the whole volume of synthesized nanoparticles in addition to the conventional absorption of the particles' surface. After several following steps that were needed to remove the water, we were able to collect the pollutant heavy elements and radionuclides in a tiny solid piece of glass.

The effect of the synthesis of nanoparticles with their subsequent volumetric adsorption of pollutants occurred only in the presence of a "soft" inerton field. The container with the liquid sample was surrounded with a set of induction coils connected to a microprocessor unit. An electromagnetic field launched in the setup switched pulses in the coils, which then gradually decayed. During those decay processes an inerton field was emitted irradiating the liquid material in the container. The intensity of the field varied in a wide range, namely, from 5 $mW \cdot cm^{-2}$ to hundreds of $W \cdot cm^{-2}$. The carrier frequency of the electromagnetic pulse modulation varied from 50 Hz to 20 kHz, although the main studies were conducted at a carrier frequency of 4.5 kHz, a pulse duration $\tau = 500 \pm 10$ ms and a repetition rate of 1 ± 0.1 Hz.

The content of Cs, Sr, Co, Cu and Fe in the aqueous solution was measured using the atomic absorption spectrophotometer model AA-8500 (Nippon Jarrell-Ash Co., Ltd., Kyoto, Japan).

Thus, in this last example, an inerton field proved itself as a field catalyst that assisted heavy particles to form clusters that became deposited on the bottom of the vessel.

KEYWORDS

- **deuteron**
- **heavy electron**
- **inertons**
- **nucleons**
- **nucleus**
- **proton**
- **subatoms**
- **subheluim**
- **subhydrogen**
- **subteuterium**

CHAPTER 8

GRAVITY

CONTENTS

The gravitation theory is a theory dealing with attraction of massive bodies. Gravity is a force that makes things move toward each other. In Newtonian physics objects are attracted through the gravitational force. In Einstein-Hilbert's general relativity gravitation is ascribed to space-time curvature instead of a force. General Relativity is a rather formal theory. Instead of a direct force, general relativity operates with the equivalence principle, which equates free fall with inertial motion, such that free-falling inertial objects become accelerated relative to non-inertial observers on the ground.

In practice, general relativity is only required when there is a need for extreme precision. The theory unifies Newton's law of universal gravitation with the theory of special relativity, such that gravity becomes a property of the geometry of space and time, or space-time. In particular, the curvature of space-time is directly related to the four momenta (mass-energy and linear momentum) of whatever matter and radiation are present. The relation is specified by the Einstein-Hilbert field equations, a system of 10 differential equations. In addition the equations need a metric that is a free parameter of the theory. The most popular metrics are the so-called Schwarzschild metric and the Friedmann metric, which are very different and even opposite; nevertheless, the majority of researchers involve them together in the same models for constructing cosmology.

Newton's and Einstein-Hilbert's theories are phenomenological. The scale from which the theories start to work is also unknown, but it is definitely larger than 1 μm, as this size is typical for short-range molecular and Casimir forces. Hence it is impossible to track directly the overlapping of quantum mechanical and gravitational interactions. Since gravitational theories are phenomenological, there is a need to construct a theory of gravity from first principles, a microscopic theory of gravity.

Such microscopic theory must be free of conceptual difficulties of general relativity, which do not have resolutions in the framework of relativity formalism. Seven difficulties were listed in Ref. [70]. Here we mention only three of them:

- a massive object can influence space-time but cannot be derived from it, because the unknown and undetermined parameter mass m is entirely separated from the phenomenological notion of space-time;
- none of the theories offer carriers that are able to transfer the gravitational interaction;
- it is completely unclear how to involve the formalism of general relativity in the study of the gravitational interaction of an ensemble of massive points in which pair potentials must be of paramount importance.

Quantum gravity, as a theoretical discipline, tries to integrate the ideas and formalism of general relativity with the formalism of quantum physics for a description of the microstructure of 'space-time' at the Planck scale, at which all fundamental interactions and constants seem to come together to form units of mass, length, and time [343, 344]. For this purpose – a description of space at the Planck scale – quantum gravity remains outside of the practical realm of standard direct methods of experimental verification. Nevertheless, since this is one of the most modern popular disciplines, we shall review it below. We will examine the main approaches that are being most actively developed by the majority of researchers.

Nevertheless, it is also interesting to review ideas on the notion of mass in gravitation theories. Let us begin by giving a brief outline of these ideas.

8.1 A BRIEF REVIEW OF STUDIES ON THE NOTION OF MASS IN GRAVITATIONAL PHYSICS

An idea that matter could be derived from space was expressed initially by Clifford [33] and then briefly discussed by Einstein, Eddington, Schrödinger and Wheeler [345] though without apparent success. Giulini [346] tried to further develop Clifford's [33] idea in the framework of an abstract approach in which the gravitational mass emerged from topology and the constraints implied by Einstein's equation; Giulini noted that localized configurations of overall non-vanishing mass might be formed from the gravitational field alone. However, such an approach in which both notions – mass and gravitational field are not determined – calls a rhetorical question: which came first – the chicken or the egg? Similar exotic examinations have been conducted by some other researchers, in particular, Asselmeyer-Maluga and Rosé [347].

In 1967, Sakharov [348] suggested that gravitation might not be a fundamental interaction, but a purely quantum phenomenon that arises due to an interaction with vacuum energy. His approach required that gravity, as a fundamental field, was replaced by a sort of "grand unified gauge field" generated by the known elementary particles. Then gravity might emerge as an elasticity of space-time. To continue to receive absolute scale units, he concluded it was necessary to introduce some fundamental mass.

In 1973, Boyer [349] derived an expression for the interaction potential $U(r)$ for the retarded van der Waals forces at all distances r between two particles that interact with the classical background zero-point field, i.e., the electromagnetic quantum oscillations of the vacuum.

Then Boyer's potential $U(r)$ was studied by Puthoff [350]. He initially derived an energy associated with a fluctuating motion of a point charged particle interacting with the zero-point field. This "internal" particle energy was identified with its rest mass energy, which gave an expression for the particle's rest mass: $m = r_0 \hbar\omega^2 / (\pi c^2) = \varepsilon / c^2$ where r_0 is the size of the oscillating dipole studied and w is the frequency of oscillations. Using parameters typical for his approximation, Puthoff obtained the following interesting expression for Boyer's potential:

$$U = -\frac{\chi}{r}\left(\frac{\sin R}{R}\right)^2 \tag{8.1}$$

where $\chi = \hbar r_0^2 \omega^3 / \pi$ and $R = \omega r / c$. The potential (8.1) has the desired $1/r$ dependence required for gravity, which is modulated by the second factor having a fine structure with a spatial periodicity of the Planck length ℓ_{P}.

Thus, following Puthoff, the gravitational interaction is similar to the short-range forces of van der Waals and Casimir and this interaction is stipulated by the interaction of particles with zero-point fluctuations of the vacuum electromagnetic field. That study was the first that seriously tried to shed light on the mechanisms that underline the gravitational force.

Although such approach to the origin of gravity is still interesting, it was criticized [351], as it would violate the second law of thermodynamics (when applied to quantum black holes; but what is a black hole? It is the question in question).

Notwithstanding such criticism, Puthoff's [350] ideas were further developed. Namely, it was proposed to interpret inertia as an electromagnetic resistance arising from the known spectral distortion of the zero-point field in accelerated frames [352]; this theory finally came to a unification of inertial and gravitational mass through the zero-point field substrate, or the electromagnetic quantum vacuum. In work [352] it was shown that the electromagnetic quantum vacuum makes a contribution to the inertial reaction force on an accelerated object. This inertial mass is extended to so-called passive gravitational mass. The inertia and gravitation connection with the vacuum electromagnetic fields was called the quantum vacuum inertia hypothesis, which allowed them to derive the classical Newtonian gravitational force. Thus, according to this theory, the gravitational and inertial mass energy emerges from the vacuum. Then m in Newton's second law of motion $F = ma$ becomes nothing more than a coupling constant between acceleration and an external electromagnetic force.

The energy associated with the fluctuations, Zitterbewegung, is interpreted as the energy equivalent of gravitational rest mass. Since according to the theory [353] the gravitational force is caused by this trembling motion, the authors do not speak any longer of a gravitational mass as the source of gravitation, because the source of gravitation is understood as the motion of a charge, not the attractive power of the thing physicists are used to thinking of as mass. They interpret the equation $E = mc^2$ in terms that mass is not equivalent to energy, but mass is energy.

Nevertheless, the success of those hypotheses was recently strongly criticized. In that article Levin [354] shows that the result [353] is erroneous due to incorrect physical and mathematical assumptions associated with taking a non-relativistic approach. He derives a correlation function in the proper non-relativistic limit and corrects some other important results. He concludes that the interaction of the accelerated oscillator with zero-point field radiation does not produce inertia. Hence, the emergence of the gravitational mass from the electromagnetic vacuum, as claimed by Haisch and Rueda [353], also becomes questionable. The more so that other fundamental forces also then become beyond consideration.

Close to ideas developed by Puthoff, Haisch and Rueda is work of Pietschmann [355] in which mass began to be considered as interaction energy; straightforward calculations of the pion mass difference showed an electromagnetic source in the mass difference: $m_{\pi^+}^2 - m_{\pi^0}^2 = \Delta m^2$, where $\Delta mc^2 \sim 0.5$ MeV.

Further studies with the use of theories of the weak and strong interaction resulted in an idea that mass should be interpreted through the energy of the object studied [356–358]:

$$m^2 = (E^2 / c^4) - (\mathbf{p}^2 / c^2) \tag{8.2}$$

where E is the total energy of the object, $\mathbf{p}$ is the momentum of the object and c is the velocity of light. They [356–358] say that the momentum can also be determined through energy, $\mathbf{p} = \mathbf{v}E / c^2$, where $\mathbf{v}$ is the object's velocity. Okun [356] and other high-energy physicists bring an interesting argument in favor of using the term *energy* instead of the term *mass*: The internal energy of bodies (gas, solid) depends on the temperature, the higher the temperature, the higher the internal energy, but it also changes the amount of mass in the body, which varies in a range 10^{-12} to 10^{-10} kg. Hence, he concludes, that mass is in fact energy.

Thus it seems the mass as such has lost its fundamental properties. Several years ago an idea defining mass through the total energy was seriously considered for the introduction in school programs for pupils in Muscovy. In particular, they wished to teach pupils in terms of Okun [356]: "A body's mass *m* does not depend on a reference system. At the end of the 20th century it is time to finally say farewell to the concept of mass depending on velocity."

A modern definition of mass that agrees with the one given by Okun [356, 357] and Wilczek [358] was recommended for American teachers by Hecht [359]: "The invariant mass of any object – elementary or composite – is a measure of the minimum amount of energy required to create that object, at rest, as it exists at that moment." So, modern fundamental physics dealing with high-energy propose a kind of energy creator, or energy substrate, instead of the previous ether, space-time, vacuum, plenum, etc. Maybe indeed Johann Joachim Becher was right introducing the phlogiston theory in 1667? Such modern activity looks like a version of the fable of the "King's new clothes" (that were invisible but everyone had to believe they could really see them).

8.2 OVERVIEW OF MAIN ROADS OF QUANTUM GRAVITY

Basically quantum gravity is subdivided into two branches of practically independent studies: loop quantum gravity and string theory.

8.2.1 CANONICAL AND LOOP QUANTUM GRAVITY

The canonical quantum gravity program [343, 360–364] treats the space-time metric itself as a kind of field, and attempts to quantize it directly.

General relativity is rewritten in so-called 'canonical' or 'Hamiltonian' form; this is a method typical for quantum field theories. Specifically, the theory starts from the Einstein-Hilbert action for metric tensor fields $g_{\mu\nu}$ of Lorentzian ($s = -1$) or Euclidean ($s = +1$) signature, which propagate on a (D + 1)-dimensional manifold M

$$S = \frac{1}{8\pi G}\int_{\mathrm{M}} d^{\mathrm{D+1}} X \sqrt{|\det(g)|}\, R^{(\mathrm{D+1})} \tag{8.3}$$

where $R(\mathrm{D}+1)$ is the curvature scalar associated with $g_{\mu\nu}$. Expression (8.3) can be put into canonical form for which M has to have the special topology $\mathrm{M} = R \times \sigma$, where σ is a fixed three-dimensional, compact manifold without boundary, and this restriction is lifted in the quantum theory of gravity. Decomposition of the D + 1 (space and time) action (8.3) is reached by special diffeomorphisms.

In a canonical description, one chooses a particular set of configuration variables q_i and canonically conjugated momentum variables P_i, which describe the state of a system at a certain time. Then the action (8.3) is transformed to an action known as the Arnowitt–Deser–Misner action

$$S = \frac{1}{8\pi G}\int_{R} dt \int_{\sigma} d^{\mathrm{D}} \left\{ \dot{q}_{ab} P^{ab} - \left[N^a H_a + |N| H \right] \right\} \tag{8.4}$$

where q_{ab} are coordinates (tensor fields called the first fundamental form on Σ that is an image of the (D + 1)-dimensional manifold M) and P^{ab} are conjugate momenta (complicated functions of q_{ab} and the second fundamental form of Σ), N and N^a are the lapse function and the shift vector field; H_a and H are the spatial diffeomorphism constraint and Hamiltonian constraint, respectively.

The equations of motion of q_{ab} and P^{ab} allow us to interpret the motions that the constraints generate on M geometrically. The reduced Hamiltonian

$$\mathbf{H} = \int_{\sigma} d^{\mathrm{D}} \left[N^a H_a + |N| H \right] \qquad 8.5)$$

is a linear combination of constraints, which allows one to obtain the equations of motion. The Hamiltonian (8.5) allows one to obtain the time-evolution of the variables q_{ab} and P^{ab}.

The quantization proceeds by treating the configuration and momentum variables as operators on a quantum state space (a Hilbert space). The quantum configuration spaces (gauge equivalence classes) are compact Hausdorff spaces, which is important in the context of developing measure theory on them. Then quantum configuration and momentum variables obey certain commutation relations similar to the classical Poisson-bracket relations. The analogy with quantum mechanics is drawn owing to the comparison of the Poisson-bracket with Heisenberg's uncertainty principle, which allows one to decipher complicated combinations of quantum variables (though, as we have argued in Section 3.2, the uncertainty principle is a useful means in the quantum mechanical formalism, which however, does not work in nature).

Loop quantum gravity is a realization of the quantization program established in the 1950s and 1960s by Dirac, Wheeler and De-Witt. They used geometric variables g_{ij}, configuration variables, corresponding to the various components of the 'three-metric' for describing the intrinsic geometry of the given spatial slice of space-time. Their conjugate momenta P_{ij} effectively encode the rate-of-change of the metric's time component, i.e., the extrinsic curvature of the slice.

The configuration variable is a *SU*(2)-connection A_a^i on a 3-manifold Σ representing space. The canonical momenta are given by the densitized triad E_i^a. The latter encode the (fully dynamical) Riemannian geometry of Σ and are the analogy of the 'electric fields' of Yang-Mills theory.

According to Dirac, gauge freedoms result in constraints among the phase space variables and they conversely are the generating functions of infinitesimal gauge transformations. Loop quantum gravity is defined using Dirac quantization. The constraint equations

$$\hat{G}_i \Psi = 0, \qquad \hat{C}_a \Psi = 0, \qquad \hat{S}\Psi = 0 \tag{8.6}$$

have solutions in the so-called physical Hilbert space H_{phys} and they fully govern quantum dynamics. In the case of loop quantum gravity, they represent the quantization of Einstein's equations. States in this auxiliary Hilbert space are represented by wave functions of the connection $\Psi(A)$, which are square integrable with respect to a natural diffeomorphism invariant measure. This space can be decomposed into a direct sum of orthogonal subspaces H_γ. Fundamental excitations are given by the — holonomy $h_\ell(A) \in SU(2)$ along a path ℓ in Σ :

$$h_\ell(A) = P\exp\left(\int_\ell A\right) \tag{8.7}$$

Elements of H_γ are given by functions

$$\Psi_{f,\gamma}(A) = f\left(h_{\ell_1}(A), h_{\ell_2}(A), \ldots\right) \tag{8.8}$$

A special node structure of physical Hilbert space H_{phys}, so-called spin-networks, is defined by a graph γ in Σ, a collection of spins $\{j_\ell\}$—unitary irreducible representations of $SU(2)$—associated with links $\ell \in \gamma$ and a collection of $SU(2)$ intertwines $\{I_n\}$ associated to nodes n $\in \gamma$. The spin-networks (quantum states of geometry) are characterized by the wave functions $\Psi_{\gamma,\{j_\ell\},\{I_n\}}(A)$ describing nodes connected via virtual links.

Quantum loop gravity studies how these states of quantum geometry evolve in time.

In the formalism of general relativity, variations of the metric around the flat space-time metric are non-perturbative, i.e., small (graviton-like) alterations, which also correct for the standard quantum field theory. In the case of loop quantum gravity, basic excitations of the gravitational field are arbitrary and can describe the quantum space-time directly at the Planck scale, where geometry is quantized.

A spin-network is a one-dimensional graph. Spin foams generalize spin networks where instead of a graph one uses a higher-dimensional complex. Such geometry of space-time corresponds to a kind of a lattice. A "Planck-lattice," a space-time cubic lattice of lattice constant equal to the Planck length $\ell_P \cong 10^{-35}$ m, which could model a ground state of quantum gravity of Wheeler's space-time foam, is mentioned in Ref. [363].

Rovelli [364, 365] presented a theory without deriving it from classical general relativity; it is formulated as a generating function for amplitudes associated with a combinatorial structure, as in the definition of quantum electrodynamics in terms of Feynman-graphs. A spin network state is interpreted as a granular space where each node n represents a "grain" of space. The volume of each grain n is υ_n. Two grains n and n_0 are adjacent if there is a link ℓ connecting the two, and in this case the area of the elementary surface separating the two grains is $8\pi\gamma\hbar G\sqrt{j_\ell(j_\ell+1)}$.

Sarfatti [366] outlines a proper progress in the foundations of quantum theory representing his views on Einstein's gravity, with emergent macro-quantum phenomena in the sense of Penrose's "off diagonal long range order" inside the vacuum. This is in line with Anderson's "more is different" and Sakharov's [348] "metric elasticity" approaches. Sarfatti's result is a background-independent non-perturbative model that agrees with loop quantum gravity. He offers a model for the formation of the vacuum condensate "inflation scalar field" filled with bound virtual electron-positron pairs and the system in question is endowed with micro and macro properties.

8.2.2 QUANTUM GRAVITY IN TERMS OF STRING THEORY

String theory (see, e.g., Refs. [343, 344, 367–374]) views the curved space-time of general relativity as an effective modification of a flat background geometry by a massless spin-2 field. Demonstrably an object of the theory is a non-straight line abstract string with more than 10 dimensions. This is the major specialty of conventional quantum field theory in which fundamental objects are mathematical points in space-time. String theory started to develop actively from 1974 when it was found [372, 373] that, among the massless string states, there was a spin-2 particle that interacts like a graviton of general relativity. This means that one string is able to expand momentarily a worldsheet instanton-like linking process to envelop another, producing a phase in the string path integral.

The Euler-Lagrange equations of motions of strings are derived from the action [369, 370]

$$S = -\frac{1}{4\pi\alpha'}\int d\sigma\, d\tau \sqrt{h}\, h^{nm}\, \partial_m X^{\mu}\, \partial_n X_{\mu} \qquad (8.9)$$

where coordinates X_{μ} of the string are also fields X_{μ} in a two-dimensional field defined on the surface where a string sweeps out as it travels in space. The partial derivatives are with respect to the coordinates σ and τ on the worldsheet and h^{mn} is the two-dimensional metric defined on the string worldsheet.

The general solution to the relativistic string equations of motion is

$$X^{i}(\sigma,\tau) = x^{i} + \dot{x}^{i}\tau + i\sqrt{2\alpha'}\sum_{n\neq 0}^{\pm\infty}\frac{\alpha_n^{\,i}}{n}\left(\cos\frac{n\pi c\tau}{L} - i\sin\frac{n\pi c\tau}{L}\right)\cos\frac{n\pi c\tau}{L}; \qquad (8.10)$$

here the parameter α' is called the string parameter and the square root of this number represents the approximate distance scale at which string effects should become observable; L is the string's length.

In the quantum case there are quantum states for strings with negative norm known as ghosts. This is determined by the quantum commutators

$$\left[\alpha_m^{\mu},\ \alpha_n^{\nu}\right] = -m\,\delta_{m+n}\eta^{\mu\nu} \tag{8.11}$$

The quantization (8.11) represents the Poincaré group, through which quantum states of mass and spin are classified in a quantum field theory. Therefore harmonics, or normal modes of the quantized string, express a set of elementary particles. Strings can be opened and closed.

String theory is developed for boson particles. Since fermions should also be included in string theory, the theory is supplemented with supersymmetry, which means that every boson (a transmitter of an interaction) is correlated with an appropriate fermion particle (i.e., the matter particle). The massless graviton is part of bosonic string theory. The vector boson is similar to the photon of electromagnetism or the gauge fields of Yang-Mills theory.

A superstring theory allows an examination in superspace where in addition to the normal commuting coordinates X^m, a set of anticommuting coordinates θ^A are added.

String geometry is very different from that of ordinary point particles; strings can modify classical general relativity only at very short distance scales.

String theory can have higher dimensional objects, which are not strings. To these objects belong so-called p-branes – p-dimensional objects belonging to sources of gauge field strengths, or $(p + 2)$-forms. Superstring theories contain electromagnetism and contain field strengths that are three-forms, four-forms and higher. These field strengths can be described by equations similar to the Maxwell equations.

A special class of p-branes is called D-branes. Basically, a D-brane is a p-brane where the ends of open strings are localized on the brane. In particular, curved space-time can be treated as coupling D-branes to supergravity fields. A D-brane may be characterized as a state of curved space-time, which preserves only half of the original space-time supersymmetries. D-branes play an important role in the string formalism of black holes, especially in counting the quantum states.

In the last several years various duality relations and mathematical string theories were developed: — type I, type IIA, type IIB, (heterotic) $SO(32)$ and (heterotic) $E_8 \times E_8$ — in relation to one another and to 11-dimensional supergravity (a particle theory). All of these approaches received the name 'M-theory.'

The majority of string-theory researchers believe that the developments of string theory are very important and fundamental (see e.g., paper [374] in advocacy of string theory). In particular, they believe in the importance of the Higgs Boson. As is thought, this Higgs particle [375] gives mass to other particles, which in turn form objects and the macroscopic structures in the universe. As recently has been reported by CERN [369], the ATLAS and CMS experiments presented their latest results in the search for the long sought Higgs particle. Both experiments observed a new particle in the mass region around 125–126 GeV. As said CERN Director General, Dr. R. Heuer: "We have reached a milestone in our understanding of nature. The discovery of a particle consistent with the Higgs Boson opens the way to more detailed studies, requiring larger statistics, which will pin down the new particle's properties, and is likely to shed light on other mysteries of our universe."

However, it is still not completely clear, why a highly abstract particle of the Higgs model [375] is associated with the discovered particle [376]? How can this new particle be able to account for the origin of the gravitational interaction? How does it really account for the *phenomenon of mass*? Why and how would it help to explain the origin of the universe? What role does this particle play in the rules of construction of ordinary (physical) space?

8.3 SPACE-TIME GEOMETRY, ETHER, PHYSICAL VACUUM, DYNAMIC SPACE, ...

The origin of space-time from non-commutative geometry and their radical implications were reviewed by Balachandran [377]. Yang [378] starts with the question: What is quantum gravity? He determines quantum gravity as a possibility for quantizing space-time itself. A pin-point is that the gravitational constant G is a natural unit that introduces a symplectic structure of space-time causing a non-commutative space-time at the

Planck scale ℓ_P. The dynamical system is described by a Poisson manifold (M, θ). The Poisson structure defines a Poisson-bracket. Dynamic variables acting on a suitable Hilbert space allow one to replace the Poisson bracket by a quantum bracket and thus the phase space M becomes non-commutative, i.e., $[\hat{x}^A, \hat{x}^B] = i\theta^{AB}$. Then the Planck constant $\hbar$ intentionally is inserted into θ. In such approach space-time and matter fields are both emergent from a universal vacuum of quantum gravity.

Kleinert and Zaanen [379] developed a 4D "world liquid crystal lattice" model. In the model the gravitational interaction between sources of curvature results from the world being a crystal that has undergone a quantum phase transition to a nematic phase by a condensation of dislocations. Finally, the elastic energy presented through the difference of the total distortion from the so-called plastic distortion $u_i^P(\mathbf{x})$ results in the effective gravitational action for the disclination part of the defect tensor. The result represents a simple Euclidean model of pure quantum gravity, which also includes small line deviations $h_{ij} = g_{ij} - \delta_{ij}$ that describe a fluctuating Riemann geometry perforated by a grand-canonical ensemble arbitrarily shaped lines of curvature. So, they have further developed Sakharov's [348] idea of gravity as due to elasticity of space-time.

Verlinde [380] tried to introduce a so-called "entropic force" $F = T\Delta S / \Delta x$. Then manipulating with a few other formulas including a "quant" of entropy $\Delta S = 2\pi k_B$ when $\Delta x = \hbar / (mc)$ he showed how this entropic force generates Newton's gravitational law $U = GMm / R^2$. In this approach there are a couple of errors that require clarification, because the results [380] in turn produced another erroneous work [381] whose authors derived MOND (Milgrom's [382] modified Newtonian dynamics), applied Verlinde's entropic gravity and called their outcome a successful phenomenological consequence of quantum gravity.

Verlinde [380] considers the canonical ensemble in which the statistical integral looks as follows

$$\int\int dE\, dx \exp\{S(E,x)/k_B - (E+Fx)/(k_B T)\} \tag{8.12}$$

He examines the integrand in expression (8.12) assuming that an extreme point for variations E and x is associated with the saddle configuration. The variations change from $-\infty$ to ∞ and hence from the structure of the exponent one can see that it monotonically decreases with increase of x. No points of inflection or saddle points can be seen, which means that the hypothesis of an "entropic force" $F = T\partial S / \partial x$ breaks downs. It is also obvious from thermodynamics, since the conventional canonical ensemble describes particles weakly interacting with the heat bath that is found in an equilibrium state; the equilibrium state cannot generate any force inside of the system studied by definition. Besides, an introduction of an entropy with quantum overtones requires rigorous classical as well as quantum justification.

At the same time Padmanabhan's study [383] shows that a suitably defined accelerated observer possesses an entropy proportional to the appropriate area. The field equations of gravity in a very wide class of theories reduce to the thermodynamic identity $Tds = dE + PdV$ on the horizon, although this result still has no explanation in the conventional approach. In the case where the volume of the system studied is constant, $dE = TdS$. Therefore Padmanabhan [383, 384] suggests for the potential of a micro particle an expression $V = mgx$ that can be presented through the average energy $\langle dE \rangle = mg \langle dx \rangle$; this relation he rewrites via a force $\langle dE \rangle = F \langle dx \rangle$. Then he equates $TdS = F\langle dx \rangle$. This simplest relationship was written only for a very peculiar microscopic case of a small fluctuation.

The problem of thermodynamics of space-time is a somewhat difficult problem in which equilibrium vacuum fluctuations would entail an ill-defined space-time metric [385]. That is why it seems it is too premature to discuss macroscopic quantum phenomena in space-time based on thermodynamics and fluctuations of vacuum.

De Haas [386], studying paradoxes in the early writings on relativity and quantum physics, revealed that a combination of Mie's [387–389] theory of gravity and the de Broglie's [10] harmony of phases of a moving particle results in the principle of equivalence for quantum gravity. De Haas notes that Mie's findings on the description of gravity have been very important and have been used by Hilbert to transform Mie's Hamiltonian variation principle into a general covariant variation principle. De Haas, considering Mie's variational principle, shows that the Mie-de Broglie version of the Hamilton variational principle directly holds in the quantum domain, $\delta \int H d\tau \propto \hbar$, though in the classical limit $\hbar \to 0$ one gets a conventional non-quantum version of the Hamiltonian principle, $\delta \int H d\tau = 0$ Thus the Mie-de Broglie theory of quantum gravity analyzed by de Haas accounts for a principle of equivalence of gravitational m_{grav} and inertial m_{inert} mass. Namely, the equality $m_{grav} = m_{inert}$, which is held in a rest-frame of the particle in question, becomes invalid in a moving reference frame. In the quantum context, this equality should be transformed to the principle of equivalence of the appropriate phases, $\phi_{grav} = \phi_{inert}$. This relation ties up the gravitational and inertial energies of the particle and shows that the gravitational mass is completely allocated in the inertial wave that guides the particle. Mie's original result for the gravitational mass was $m_{grav} = m_0\sqrt{1-\upsilon^2/c^2}$ (see sections 42 to 45 in Refs. [387–389]), though the inertial mass was identified with the total mass, $m_{inert} = m_0/\sqrt{1-\upsilon^2/c^2}$. Thus gravitation as such is in fact a pure dynamic phenomenon, which at the quantum level has to be associated with the matter waves.

Consequently, de Haas [386] revived de Broglie's initial physical interpretation of wave/quantum mechanics, i.e., the pilot wave interpretation of de Broglie, and also introduced a fresh wind into the problem of quantum gravity: a particle moving through space deforms the metrics on a quantum local scale in such a way that the inertial energy flow $E_{inert}\mathbf{v}_{group}$ becomes concentrated in the particle's wave packet, though the gravitational energy flow $E_{grav}\mathbf{v}_{particle}$ becomes dislocated from it. Nevertheless, on a macroscopic scale, the equality between these two kinds of energy is preserved. The result of de Haas sounds in fact very realistic, because it links Mie's source of gravitational energy to the trace of the inertial stress-energy tensor that takes the role of the source of gravitational energy in modern concepts.

The number of approaches to understanding gravity, especially its quantum origin, are huge. But lacking contact with experimental results, these models still remain metaphysical. That is why even all post-mechanical ether models have the same rights as the quantum-gravity models mentioned above.

Winterberg's [390] Planck ether hypothesis considers all of space as filled with an equal number of positive and negative particles, i.e., particles with positive and negative mass. The approach allowed him to derive quantum mechanics, Lorentz invariance, the origin of gravitational and inertial mass and the principle of equivalence, etc. In particular, the gravitational interaction is the result of the attraction of both the attracting and attracted mass, which is composed of Planck mass particles bound in vortex filaments.

Arminjon [391] reviews ether models based on the Lorentz-Poincaré ether theory, considers a scalar theory of gravitation [392] (the theory of gravity is based on a scalar field which is arranged from a change of the local pressure and density of the ether), analyses tests and new predictions of the scalar ether theory and discloses links with classical electromagnetism and quantum theory.

Suntola [393] developed a very original approach to the description of the universe, the so-called Dynamic Universe theory. In his theory locally observed phenomena are derived from the conservation of the zero-energy balance of motion and gravitation in the whole universe. In the theory our universe is the three-dimensional surface of an expanding four-dimensional sphere. The kinetic energy of expansion is balanced by gravitational attraction. Inertial work looks as the reduction of the rest energy due to motion in space, which gives a quantitative explanation to March's principle. By using this approach, Suntola derives correct expressions for the perihelion advance, the bending of light passing near a mass center, the redshift of light and the Shapiro delay. Besides he also describes the Sagnac effect. In his theory the velocity of light is locally variable and can drop at the gravitational interaction. Gravitational phenomena are studied on the basis of Newton's gravitational law.

Sorli [394] suggests that mass has to be generated by a change of density of quantum vacuum, motivated by the view of the non-existence of the Higgs boson (he notes that CERN's reports have so far not been conclusive about the existence of this hypothetical particle).

Cahill [395] constructs an approach called the dynamical 3-space, which he extensively tests against experimental and astronomical observations, considers a possibility to generalize the Maxwell, Schrödinger and Dirac equations, leading to a derivation and explanation of gravity as a refraction effect of the quantum matter waves. He adds to the basic equation one more term associated with the Newton potential of an outside body and considers equations at special conditions, which show the presence of this additional term in the formalisms of Schrödinger, Dirac and others. In a review paper Cahill [396] discusses modern and old experiments dealing with the measurement of anisotropy of electromagnetic waves and the tested change in the speed of light, which depends on the chosen direction.

Múnera [397] briefly reviews some old criticisms of the original 1887 Michelson-Morley experiment, and formulates two new, very specific modern views that expect harmonic variation with a large amplitude of several wavelengths. Close to Múnera's paper is book [398] – a collection of 22 old and modern experimental papers examining a possible influence of the ether wind on received data. Work [399] present results of systematic measurements that can be considered as an actual experimental confirmation of the existence of the ether as a material substrate.

Duffy [14] reviews many approaches pertaining to gravitational physics, which fall under the modern ether concept. The concept is subdivided into three development programs of which the two major ones are: (i) the evolution of relativity, relativistic cosmology and geometrodynamics, which discarded the early 1920 mechanical, passive, rigid ether in favor of geometrized space-time; (ii) the development program associated with quantum mechanics and studies of physical vacuum. Duffy emphasized that the modern ether resembles a "sea of information," which demands new techniques for interpreting it, drawn from information science, computer science and communications theory.

The second of Duffy's [14] mentioned programs overlaps with conventional ideas of quantum and gravitational physicists. In fact, models of quantum gravity have used such terms as Planck lattice, discrete geometry, quantum space-time, network, cellular space-time, geometry, granular space, etc. This means that nobody doubts that at a sub microscopic scale the physical world is constructed as a discrete continuum.

Baurov [400] hypothesizes that instead of the structure of the physical space, time and world of elementary particles, a finite set of discrete objects, byuons with an inner magnetic vector should exist. He demonstrates some experiments that could be explained on the basis of such a hypothesis.

Granular space and the problem of large numbers – how many spatial cells may a canonical particle include – have recently been discussed in a simple way [401]. The notion of granular space is very close of the space as a grid of Wilczek [8] and similar ideas of other researchers.

Following Isham [402], Toh [403] noted that rather than quantizing gravity, one should seek for a quantum theory that would be able to yield general relativity as its classical limit. Then he pointed out the main obstruction – the lack of a starting point to construct such a quantum theory. Isham [402] also noticed that a radically new theory might require the re-examination, from the foundation, the concepts of space, time, and matter.

But it is obvious – if macroscopic objects possess the Newton gravitational potential, there must be a microscopic mechanism responsible for such kind of gravity, i.e., the potential ~1/*r*! So, what quantum/subquantum processes give rise to gravity?

8.4 SUBMICROSCOPIC APPROACH TO GRAVITY AND THE GRAVITATIONAL POTENTIAL ~ 1/*r*

Although the variety of listed theories, views and hypotheses are attractive for theorists, they appear constrained and unresolved because they do not consider mass as a real entity of physical space. A microscopic theory of gravity of course requires a strict determination of the notion of mass, its appearance and interaction with space. That precise definition has been presented in Chapter 1, Section (1.7.2): $m \propto \mathcal{V}^{\text{deg}} / \mathcal{V}^{\text{particle}} > 1$ (1.50), where $\mathcal{V}^{\text{deg}}$ is a volume of a degenerate cell of the tessellattice and $\mathcal{V}^{\text{particle}}$ is the volume of a particle. In other words, mass is a local deformation of space.

Considering the motion of a massive particle, we have seen that it periodically decomposes to inertons, field carriers that carry fragments of the particle's mass. The moving particle is surrounded by its inerton cloud. The overlapping of such clouds results in the quantum mechanical interaction that is described in terms of the formalism of particles' wave ψ-functions. At the level of an elementary particle the gravitation emerges when emitted inertons come back to the particle. However, the particle as such does not attract its inertons; this is the elasticity of the tessellattice, which returns inertons back to the particle.

So, the elasticity of the tessellattice, returning the inertons to the particle, is the gravitation. When the inertons have returned to the particle, the gravitation ceases. When the inertons disperse from the particle, there is an anti-gravitation.

Examining the structure of the ψ-function in Section 3.6, we have derived the solution for the behavior of mass of the system {particle + its inerton cloud}, which exhibits an oscillation of the value of m along the particle path. Then in Section 5.1 we have shown that the mass of entities varies inside the crystal lattice.

The nuclear forces studied in Chapter 7 are caused first of all by the overlapping of deformation coats of nucleons and then by the overlapping of nucleons' wave ψ-functions (i.e., their inerton clouds). The mass of the nucleon's deformation coat experiences oscillations in the form of a standing spherical wave; see Eq. (7.18) and the solution (7.21).

The radial oscillation of the nucleon's deformation coat, which has the solution in the form of a standing spherical wave, is a prototype of gravity that emerges at a macroscopic scale.

In fact a material object that includes N entities (for example, atoms) possesses a set of vibration harmonics. Amplitudes of these oscillations are nothing but the entities' de Broglie wavelengths. The entities' inerton clouds overlap. Due to the overlapping, a total inerton cloud of the object is formed. Inertons, which fill the space between acoustically vibrating entities, accompany these entities and hence produce their own spectrum that has been considered in Section 5.1. Wavelengths of such collective inertonic vibrations can be estimated by expression

$$\Lambda_n = 2gnc/\upsilon_{\text{sound}} \tag{8.13}$$

where $n = 1, 2, 3, \ldots, N/2$, which is the consequence of the relationship $\Lambda = \lambda c/\upsilon$ (2.69).

Acoustic waves of an object are a source of the object's standing waves; standing wave modes arise from the combination of reflection and interference, especially on the border of the object, i.e., its surface, such that the reflected waves interfere constructively with the incident waves. The inerton field, as a filler of gaps between quantum entities in the object, follows the vibrating entities. Hence, inerton waves also form standing waves [404]. The longest standing inerton waves are fundamental waves. If the number N of entities (atoms) is huge in the body studied, then for the body's fundamental standing waves the body shape could be associated with a sphere of a radius R_0.

For instance, the solid sphere with volume 1 cm^3 includes around 10^{22} atoms; putting the velocity of sound $\upsilon_{\text{sound}} \approx 10^3$ $\text{m}\cdot\text{s}^{-1}$ and the distance between atoms 0.5 nm, we obtain for the amplitude (8.13) of the longest inerton wave: $\Lambda_{N/2} \sim 10^{18}$ m – this is a distance to which the sphere's inertons are able to reach.

Oscillations of cells of the tessellattice in the standing inerton waves induced by a material sphere can be described by the Lagrangian that is similar to the Lagrangian that specifies oscillations of the nucleon's deformation coat (7.15), namely:

$$L = \sum_{\vec{l}} \left(\tfrac{1}{2} R_0^2 \dot{\mu}_{\vec{l}}^2 + \tfrac{1}{2} m_0^2 \dot{\xi}_{\vec{l}}^2 + c R_0 m_0 \dot{\mu}_{\vec{l}} \nabla \xi_{\vec{l}} \right) \tag{8.14}$$

Here $\mu_{\vec{l}}$ is the mass of the $\vec{l}$th cell remote at the distance $r_{\vec{l}}$ from the sphere; this mass is carried by inertons that oscillate around the sphere; $\xi_{\vec{l}}$ is the tension of the same cell; m_0 is the total mass of the spherical body with the radius R_0; c is the speed of light.

The Euler-Lagrange equations of motion written for $\mu_{\vec{l}}$ and $\xi_{\vec{l}}$ are

$$\ddot{\mu}_{\vec{l}} + c m_0 R_0^{-1} \nabla \dot{\xi}_{\vec{l}} = 0 \tag{8.15}$$

$$\ddot{\xi}_{\vec{l}} + c m_0^{-1} R_0 \nabla \dot{\mu}_{\vec{l}} = 0 \tag{8.16}$$

These equations are easy to decouple to wave equations in the spherical coordinates (recall that in the spherical coordinates the Laplace operator $\Delta\mu$ is $\frac{1}{r}\frac{\partial^2}{\partial r^2}(r\mu)$):

$$\frac{\partial^2 \mu}{\partial r^2} + \frac{2}{r}\frac{\partial \mu}{\partial r} = \frac{1}{c^2}\frac{\partial^2 \mu}{\partial t^2} \tag{8.17}$$

$$\frac{\partial^2 \xi}{\partial r^2} + \frac{2}{r}\frac{\partial \xi}{\partial r} = \frac{1}{c^2}\frac{\partial^2 \xi}{\partial t^2}; \tag{8.18}$$

here, we omit the index $\vec{l}$ at μ and ξ.

In Eqs. (8.17) and (8.18) μ and ξ are functions of the distance r from the sphere and time t. Since μ and ξ oscillate in opposite phases, the initial and boundary conditions are

$$\mu(0,0)|_{r\sim R_0}=m_0,\ \xi(0,0)|_{r\sim R_0}=0,$$

$$\mu(r,t)|_{r=\Lambda_{N/2}}=0,\ \xi(r,t)|_{r=\Lambda_{N/2}}=\xi_0 \quad (8.19)$$

Then the solutions for μ and ξ are

$$\mu(r,t)=m_0\frac{r_{01}}{r}|\cos(kr)||\cos(\omega t)| \quad (8.20)$$

$$\xi(r,t)=\xi_0\frac{r_{02}}{r}|\sin(kr)||\sin(\omega t)| \quad (8.21)$$

where r_{01} and r_{02} are very small constants with the dimension of length.

In Eqs. (8.20) and (8.21) the parameters μ, k and ω, as well as ξ, k and ω, should be the function of the index n that represents the nth harmonic of the oscillating body. That is, the wave number k is $k_n=\pi/\Lambda_n$, the cyclic frequency ω is $\omega_n=\pi/T_n$ where $n = 1, 2, \ldots N/2$. Since time is considered as a natural parameter, the relation $\Lambda_n=cT_n$ is held, which can also be rewritten as a dispersion equation $\omega_n=k_n c$.

The cyclic frequency of the fundamental harmonic can be determined as a ratio of the radius of the body studied to the speed of light (the velocity of inertons from the total inerton cloud of the body studied): $\omega_0=\pi\cdot c/R$. For a material sphere with a radius $R_1 = 1$ cm the fundamental cyclic frequency is $\omega_{01}\approx 10^{11}$ s^{-1}, for the Earth whose radius is $R_2\approx 6400$ km the fundamental cyclic frequency is $\omega_{02}\approx 10^2$ s^{-1} and for the Sun whose radius is $R_3\approx 6.96\times 10^8$ m the fundamental cyclic frequency is $\omega_{03}\approx 1$ s^{-1}. So if the time t of the interaction of the body studied with a test object, which can be a detector approached to the body, satisfies the inequality $t>>\omega_0^{-1}$, the detector will perceive the body's deformation field as a mean field. This means that in the solution (8.20) $|\cos(\omega t)|$ is a constant, which equals ½ per period T, and this result holds for any n harmonic.

Thus the solution for the mass field (8.20), which mathematically is a deformation field in the tessellattice around the local deformation, is reduced to the form

$$\mu(r,t)=m_0\frac{r_{01}}{r}|\cos(kr)| \quad (8.22)$$

where $r \geq R_0$ and R_0 is the radius of the body.

Since typical distances in gravitation physics $r << \Lambda_{\text{fundam.}}$, where $\Lambda_{\text{fundam.}} = 2R_0\, c / \upsilon$ (see relation (8.13)) and υ is the characteristic velocity of the body and/or its entities, we may also substitute cos $(k\, r)$ in expression (8.22) for unit. Hence we arrive at the mass distribution around the massive body, which is a practically stationary deformation potential

$$\mu(r,\, t) = m_0\, r_{01} / r \tag{8.23}$$

that forms the basis for the Newtonian gravitational physics.

Let us discuss the constant r_{01}; what is its value? Expression (8.23) shows the value of mass of cells of the tessellattice around the massive object with the mass m_0. The mass μ is induced by standing inerton waves of the object; it looks like an inerton aura of the massive object. The minimal length of the quantum motion of a nucleon is its Compton wavelength λ_{Com}. The number of inertons emitted by the object at the passing the nucleon's Compton wavelength is $\lambda_{\text{Com}} / \ell_{\text{P}} \approx 10^{20}$. Then the constant r_{01} can be chosen as $r_{01} = \ell_{\text{P}} / (\lambda_{\text{Com}} / \ell_{\text{P}}) \approx 10^{-55}$ m.

At such value of r_{01} the mass of the inerton aura (8.23) around a massive macroscopic object will look as follows. For example, at $m_0 = 0.01$ kg at a distance $r = 10^{-6}$ m from the object the mass of an appropriate standing inerton becomes $\mu \approx 10^{-51}$ kg. The Earth has the mass $m_0 = 6 \times 10^{24}$ kg and the radius $r = 6.731 \times 10^{6}$ m; hence for the Earth's standing inertons near the globe surface we obtain for the mass of the inerton aura: $\mu \approx 10^{-37}$ kg. The Sun generates the mass of the inerton aura near the Earth equal to $\mu \approx 10^{-36}$ kg, which is 6 orders less than the electron mass.

The tension ξ in Eq. (8.21) does practically not change with r; its role is to guide massive excitations toward the center along the topography of space given by the deformation potential formed by the massive body (see the expression (8.23)). However, the time dependence of ξ (8.21) shows that its value is negligible up to the time compared with the period $T = \Lambda_{\text{fundam.}} / c$.

Let us estimate the value of the tension ξ around a small solid object with the radius $R_0 = 10$ μm. The number of atoms in the object is $N = (R_0 / g)^3 \sim 10^{13}$. Hence the corresponding amplitude of the standing wave (8.13) is $\Lambda_{\text{fundam.}} \sim 10^{9}$ m. Then the appropriate period $T_{\text{fundam.}} = \Lambda_{\text{fundam.}} / c \sim$

a few seconds. Of course changes of ξ with such period $T_{\text{fundam.}}$ could be measured theoretically, but it is impossible to do in practice, as the measurement must be conducted at a distance at least $0.1\Lambda_{\text{fundam.}} \approx 10^8$ m from this tiny object.

The potential energy of the gravitational interaction between two massive objects with masses m_1 and m_2 can be written in the way similar to that presented for the interaction of quarks (6.21) and nucleons (7.28),

$$U = -\frac{m_1}{m_{\text{P}}}\frac{m_2}{m_{\text{P}}}\frac{\hbar c}{r} \tag{8.24}$$

where m_{P} is the Planck mass, i.e., the maximum possible volumetric deformation of a cell of the tessellattice. The expression (8.24) can be written in the standard form via the phenomenological gravitational constant G:

$$U = -G\frac{m_1 m_2}{r} \tag{8.25}$$

which is the known Newtonian potential energy of the gravitational interaction.

If the radius of a massive object $R_0 \geq 2.2$ m, then the number of atoms in it is $N = (R_0 / g)^3 \geq 10^{31}$. Hence from the relation (8.13) we derive that its fundamental standing inerton wave has the wavelength of the order of the observed radius of the universe, 10^{26} m.

On the other hand, the shortest inerton wavelength, which is originated from the shortest acoustic wave in a solid, is $\Lambda_1 = 2gc / \upsilon_{\text{sound}} \sim 1\ \mu\text{m}$.

8.5 THE CASIMIR EFFECT AS AN ACTUAL MANIFESTATION OF QUANTUM GRAVITY

A linear spectrum of acoustic phonons breaks down at the Debye frequency. At higher frequencies the dispersion relation is no longer linear (Figure 8.1). Around the edge of the Brillouin zone, the spectrum of large wave numbers k becomes increasingly discrete, because the distance between the wave

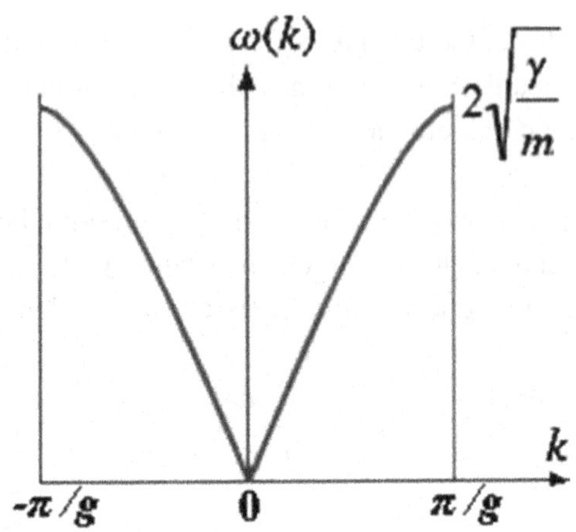

FIGURE 8.1 Spectrum of acoustic phonons that is destroyed near the end of the Brillouin zone.

numbers grows towards the edge of the Brillouin zone. Therefore, the shortest acoustic wave is associated with the edge of the Brillouin zone and it is this fundamental wave that is responsible for the formation of the shortest inerton wave in a solid [405], $\Lambda_1 = 2g\,c / \upsilon_{\text{sound}} \leq 1\ \mu\text{m}$.

Since the spectrum of phonons at the edge of the Brillouin zone is specific, we may anticipate that the edge phonons will also contribute distinctively to the inerton spectrum of the solid object. This can be especially justified in the case of near-surface atoms of the material body. Near-surface atoms may have a more obvious discrete spectrum at the edge of the Brillouin zone than the bulk atoms and the difference of these spectra may give an additional distribution of inertons out of the body's surface.

The phase speed in the simplest 1D case is $\upsilon = (2g/\pi)\sqrt{\gamma/m}$. For example, in the case of gold, its force constant γ may be estimated as varying from 4 to 9 N·m^{-1} [406]. Let us consider γ = 4 N·m^{-1}. The lattice constant of gold is $g = 4.08\times10^{-10}$ m; the mass of the atom of gold is $m_{\text{Au}} = 3.27\times10^{-25}$ kg. Knowing these parameters, we evaluate the phase velocity as $\upsilon \approx 9.08\times10^{2}$ m·s^{-1}. Then the amplitude of vibrations of a pair of atoms at the edge of the Brillouin zone, i.e., the atom's de Broglie wavelength, becomes $\delta r \equiv \lambda = h/(m_{\text{Au}}\,\upsilon) = 2.23\times10^{-12}$ m. The appropriate amplitude of the atom's inerton cloud is $\Lambda = \lambda\, c/\upsilon = 0.74\ \mu$m. This value of Λ should be slightly increased owing to contributions on the side of other discrete modes and the thermal smearing. Nevertheless, the estimated value of Λ accounts for an order of the radius of the boundary effects that can specifically manifest themselves in the vicinity of the body.

Casimir [407] considered two conducting square plates with the size $\mathcal{L} \times \mathcal{L}$ separated by a distance a. One plate is movable and in the first situation the distance a is small and in the second situation it is large. Casimir considered the difference

$$\delta E = (\tfrac{1}{2}\sum_i \hbar\omega_i)_I - (\tfrac{1}{2}\sum_i \hbar\omega_i)_{II} \tag{8.26}$$

between two summations that extend over all possible resonance frequencies in the vacuum cavity confined by these two plates. The geometric size of the cavity $0 \le x \le \mathcal{L}$, $0 \le y \le \mathcal{L}$ and $0 \le z \le a$ determines its possible vibrating modes: $k_x = n_x \pi / \mathcal{L}$, $k_y = n_y \pi / \mathcal{L}$ and $k_z = n_z \pi / a$. To every k_x, k_y, k_z correspond two standing waves. In an explicit form, these standing electromagnetic waves spontaneously excited in vacuum have the form

$$\mathbf{A} = \sum_k c\sqrt{\frac{\pi\hbar}{\omega\mathcal{L}^2}}\,\mathbf{e}(\mathbf{k})\times\left\{A_k e^{-(\omega t-\mathbf{kr})} + A_k^{+} e^{(\omega t-\mathbf{kr})}\right\} \tag{8.27}$$

For large $\mathcal{L}$ wave numbers k_x and k_y can be regarded as continuous variables. Then Casimir presented the difference δE (8.26) in an integral form, which allowed him to obtain the result – an attractive-energy between the two plates:

$$\delta E / \mathcal{L}^2 = -\hbar c \frac{\pi^2}{24\times 30}\cdot\frac{1}{a^3} \tag{8.28}$$

The calculated [407] result (8.28) was verified experimentally by different researchers. Moreover, the Casimir effect also occurs in the case of dielectrics, which points to its universality. Nowadays both theorists and experimentalists continue to intensively study the Casimir effect. Theoreticians have been developing complicated generalized mathematical approaches based on fluctuations of electromagnetic fields, virtual photons and other things that would appear from the zero-point energy and the physical vacuum in general (see Refs. in work [405]).

At the same time, Jaffe and Scardicchi [408] and Jaffe [409] presented a very different viewpoint describing the attraction of two plates. The starting point is the consideration of a scalar field of mass m that satisfies the wave equation $(\nabla^2 + k^2)\,\phi(x) = 0$. The Casimir energy is written as an integral over the difference between the density of states $\delta\rho(k)$ in a domain of conducting planes and the vacuum,

$$\delta E = \tfrac{1}{2}\hbar \int_0^{\infty} dk\,\omega(k)\,\frac{2k}{\pi}\,\mathrm{Im}\int_D d^3x\,\tilde{G}(x,\ x,\ k + i\varepsilon) \tag{8.29}$$

where $\omega = \sqrt{c^2k^2 + m^2c^4/\hbar^2}$ and $\tilde{G} = G - G_0$ is the difference between Green's function in the background of conducting plates and the Green's function in vacuum. To resolve the equation (8.29), they chose Green's function typical for classical geometric optics where G is defined by the sum over optical paths, which includes a combination of abstract factors used in classical ray optics. Then the mass m is approaching zero and they finally acquire Casimir's result (8.28). Thus, the Casimir effect can be considered without reference to zero point energies.

We can further develop Jaffe's view; namely, materializing his fictitious scalar massive field $\phi(x)$ and combining it with the mathematical method suggested by Casimir. Let us come back to the investigation of those boundary inerton effects mentioned at the beginning of this Section. We can consider a rectangular cuboid whose base $\mathcal{L} \times \mathcal{L}$ is located on the surface of a material body and it sticks outward having the height a. Let another cuboid with the same size be disposed inside the material body. What is the difference in the vibrating inerton energy for these two cuboids? We can follow the computational scheme proposed by Casimir [407]:

$$\delta E = \hbar\omega \frac{1}{(\pi/\mathcal{L})^2}\frac{1}{(\pi/a)} \times \left\{ \sum_{n=0}^{\infty}\int_0^{\infty}\int_0^{\infty}\sqrt{k_x^2 + k_y^2 + \frac{\pi^2 n^2}{a^2}}\,dk_x dk_y - \int_0^{\infty}\int_0^{\infty}\int_0^{\infty}\sqrt{k_x^2 + k_y^2 + \frac{\pi^2 n^2}{a^2}}\,dk_x dk_y dn \right\} \tag{8.30}$$

Now turn to polar coordinates ($\sqrt{k_x^2 + k_y^2} = \kappa$)

$$\delta E = \hbar 2\pi v a \frac{\mathcal{L}^2}{\pi^2}\frac{1}{\pi}\left\{\sum_{n=0}^{\infty}\frac{\pi}{2}\int_0^{\infty}\sqrt{\kappa^2 + \frac{\pi^2 n^2}{a^2}}\,\kappa d\kappa - \frac{\pi}{2}\int_0^{\infty}\int_0^{\infty}\sqrt{\kappa^2 + \frac{\pi^2 n^2}{a^2}}\,\kappa d\kappa dn\right\},$$

$$= \hbar c \pi \frac{\mathcal{L}^2}{\pi^2}\left\{\sum_{n=0}^{\infty}\int_0^{\infty}\sqrt{\frac{a^2\kappa^2}{\pi^2} + n^2}\,\frac{\pi}{a}\kappa d\kappa - \int_0^{\infty}\int_0^{\infty}\sqrt{\frac{a^2\kappa^2}{\pi^2} + n^2}\,\frac{\pi}{a}\kappa d\kappa dn\right\} \quad (8.31)$$

Let us introduce a new variable, $u = \kappa^2/(\pi/a)^2$, which changes expression (8.31) to the form

$$\delta E = \hbar c \pi \frac{\mathcal{L}^2}{\pi^2}\cdot\frac{1}{2}\left(\frac{\pi}{a}\right)^3\left\{\sum_{n=0}^{\infty}\int_0^{\infty}\sqrt{u+n^2}\,du - \int_0^{\infty}\int_0^{\infty}\sqrt{u+n^2}\,du\,dn\right\} \quad (8.32)$$

where under the integrands we have only dimensionless variables. Now we need to get rid of divergences in the integrals. Following Casimir [407], we introduce a cutoff function $f(k/k_{\mathrm{m}})$, which equals 1 when $k \ll k_{\mathrm{m}}$, and 0 when $k \gg k_{\mathrm{m}}$. Besides, imputing a new variable, $w = \sqrt{u+n^2}$, we get instead of expression (8.32):

$$\delta E = \frac{\hbar c \pi^2 \mathcal{L}^2}{2a^3}\cdot\left\{\sum_{n=0}^{\infty}\int_n^{\infty}2w^2 f\left(\frac{\pi w/a}{k_{\mathrm{m}}}\right)dw - \int_0^{\infty}\int_n^{\infty}2w^2 f\left(\frac{\pi w/a}{k_{\mathrm{m}}}\right)dw\,dn\right\}$$
$$= \frac{\hbar c \pi^2 \mathcal{L}^2}{a^3}\left\{\sum_{n=0}^{\infty}F(n) - \int_0^{\infty}F(n)\,dn\right\}$$

(8.33)

where

$$F(n) = \int_0^\infty w^2 f[\pi w / (a\, k_{\mathrm{m}})] dw \tag{8.34}$$

If $F(n)$ and all its derivatives tend to 0 as $n \to \infty$, the curly braces in formula (8.33) can be presented as follows [410]

$$\sum_{n=0}^{\infty} F(n) - \int_0^\infty F(n) dn = \frac{1}{2} F(0) - \frac{1}{12} F'(0) + \frac{1}{24 \cdot 30} F'''(0) + \ldots \tag{8.35}$$

Since $F'(n) = -n^2 f[\pi n / (a\, k_{\mathrm{m}})], F''(n) = -2n f[\pi n / (a\, k_{\mathrm{m}})], F'''(n) = -2 f[\pi n / (a\, k_{\mathrm{m}})]$ and the function f is the step function (zero and unity), it is reasonable to put here its mean value, $f = \langle f \rangle = 1/2$. Then we get: $F(0) = 0, F'(0) = 0, F'''(0) = -1$. Substituting these values into the right-hand side of expression (8.35) we obtain the result: $-1 / (24 \quad 30)$. Substituting this value into the expression for δE (8.33), we immediately arrive at the Casimir outcome (8.28).

Thus, the Casimir attraction energy is caused by the deformation of space caused by inertons in the vicinity of the plate's surface. Inertons carry mass and hence they induce a peculiar distribution of the gravitational potential around a body. Shortwave inertons generated by vibrating atoms at the edge of the Brillouin zone are responsible for the gravitational potential energy (8.28), which produces a gravitational force $\mathcal{F} = \hbar c \pi^2 / (240 a^4)$ per square unit of the area. Longwave inertons generated by vibrating atoms (in the acoustic spectrum) induce the conventional Newtonian gravitational potential. Hence, at a short distance of about 1 μm or so the gravitational potential is proportional to $1/r^3$ and at macroscopic distances ≥ 10 μm it is exactly proportional to $1/r$.

8.6 MANIFESTATION OF WAVE EFFECTS IN THE GRAVITATIONAL INTERACTION

In Section 8.4 we have seen that a massive object is characterized by proper oscillations of its mass, which is described by the wave equation (8.17). In the mean field approximation, the behavior of the mass of an object, i.e., the mass together with its inerton cloud, lead to Newton's law of gravitation and the expression (8.25) describes Newton's potential energy of gravity.

The mean field approximation works very well, as it is Newton's well-known gravity, but can a remote observer reveal the wave behavior of a massive object? Let us focus on this problem in some detail. The massive object can be regarded as a sphere. The differential equation that describes the oscillation of the sphere with a radius R_0 is $-\Delta\Psi = k^2\Psi$. Due to the spherical symmetry the solution can be written as

$$\Psi(r, \theta, \phi) = \sum_{l,m} R_l(r)\, Y_{lm}(\theta, \phi) \tag{8.36}$$

where the functions $Y_{lm}(\theta, \phi)$ and $R_l(r)$ satisfy correspondingly the equations (see, e.g., Lee [411])

$$L^2 Y_{lm}(\theta, \phi) = l(l+1) Y_{lm}(\theta, \phi) \tag{8.37}$$

$$-\frac{1}{r^2}\frac{\partial}{\partial r}\left(r^2 \frac{d\Psi_l(r)}{dr}\right) + \frac{l(l+1)}{r^2}\Psi_l(r) = k^2 \Psi_l(r) \tag{8.38}$$

If at the boundary the function $\Psi(R_0, \theta, \phi) = 0$, then Lee [411] gives the total solution in the form

$$\Psi_{lmn}(r, \theta, \phi) = \frac{\varphi_{l+1/2}(k_{nl} r)}{r} A_{ln} Y_{lm}(\theta, \phi) \tag{8.39}$$

where the spherical Bessel functions

$$\varphi_{l+1/2}(z) = \sqrt{\pi z / 2}\; J_{l+1/2}(z) \tag{8.40}$$

However, we wish to study the oscillation of the object's inerton cloud whose mass μ is not zero at the boundary $r = R_0$. Hence we shall look for more comprehensive solutions.

Solutions of these natures were suggested by Kreidik and Shpenkov [412] and Shpenkov [413] who hypothesized that massive objects ranging from an atom of hydrogen to a star induce their gravity by pulsating a spatial "microformation" that is formed in the surrounding space in wave processes. This 'micropulsation' reaches stationary shells around the object. In this hypothesis, the "micropulsation" is present in the gravitational force written as

$$F = \omega_{\text{fund}}^2 \, m_1 \, m_2 \, / \, (C r^2) \tag{8.41}$$

where ω_{fund} is the cyclic frequency of the mass exchange and C is the dimension constant. Then Shpenkov [413, 414] considered the wave equation (8.38) pointing out that positions of the stationary shells should be defined as roots of Bessel functions that therefore define spectral terms in the space around micro- and mega-objects. He noted that such a formation is thereby inseparably associated with the space and that the "microformation" gains mass due to its natural dynamic behavior, pulsations, and the resulting excitations of the pulsating space.

Shpenkov assumed that in the central spherical wave field of an entity, the amplitude of radial oscillations of the entity's spherical shell, which originates from solutions of equation (8.38), has the form

$$A_{\text{sph},\, l}(r) = \frac{A}{kr} \sqrt{\pi k r / 2} \left[J_{l+1/2}(kr) \pm i Y_{l+1/2}(kr) \right] \tag{8.42}$$

Following reference literature [415], Shpenkov [413, 414] has written zeros and extrema of the Bessel cylindrical functions $J_{l+1/2}(kr)$ and $Y_{l+1/2}(kr)$ designating them, correspondingly, as $j_{(l+1/2,\, s)}$, $y_{(l+1/2,\, s)}$, $j'_{(l+1/2,\, s)}$ and $y'_{(l+1/2,\, s)}$. Analogously, zeros and extrema of the Bessel spherical functions have been designated as $a_{l,\, s} = j_{(l+1/2,\, s)}$, $b_{l,\, s} = y_{(l+1/2,\, s)}$, $a'_{l,\, s}$ and $b'_{l,\, s}$. More details concerning the solution in Bessel functions are seen in Refs. [416–418]; in particular, the following discrete values of variables are important: $k\, r_0 = z_{l,\, 1}$ and $k\, r_s = z_{l,\, s}$ where $z_{l,\, 1}$ and $z_{l,\, s}$ are zeros of Bessel functions $J_{l+1/2}(kr)$. This allows one to write the relation between radial shells

$$r_s = r_0 \cdot (z_{l,s} / z_{l,1}) \tag{8.43}$$

where the index l indicates the order of Bessel functions, s denotes the number of the root and they together define the number of the radial shell. Zeros of Bessel functions define possible shells of stationary states.

Introducing rule (8.41), Shpenkov carried out calculations of several parameters, which appear basic in his approach. One of these parameters is the so-called wave gravitational radius $\lambda\!\!\!^{-}_g$, which is linked to the fundamental cyclic frequency also named the gravitational frequency $\omega_{\text{fund}} \equiv \omega_g$, via the dispersion relation $\lambda\!\!\!^{-}_g = c / \omega_g$. The wave gravitational radius was determined by Shpenkov as the value [419]

$$\lambda\!\!\!^{-}_g = 3.274 \times 10^{11} \text{ m} \tag{8.44}$$

Then in accordance with the solution of the radial equation, the roots of Bessel functions will be radii of stable shells as shown below:

$$r = \lambda\!\!\!^{-}_g \, z_{m,n} = 3.274 \times 10^{11} \times z_{m,n} \text{ m} \tag{8.45}$$

Shpenkov [419] (pp. 63, 64) demonstrates that the solution (8.45) is realized in a spectrum of the Keplerian shells-orbits, assuming that the gravitational shells are spherical and, therefore, the orbits around the Sun are circular. For example, for

1) Venus: s =2, $z_{m,n} = j_{1,s}\big|_{s=2}$=3.831706, $r_s\big|_{s=2} = 106.03$ Mkm (the measured semi-major axis of the elliptical orbit is $\bar{r}$ = 108.2 Mkm, the difference is 2%);
2) Earth: s =3, $z_{m,n} = j_{1,s}\big|_{s=3}$=10.17347, $r_s\big|_{s=3} = 153.76$ Mkm (the measured semi-major axis of the elliptical orbit is $\bar{r}$ =149.6 Mkm, the difference is 3%);
3) Jupiter: s =1, $z_{m,n} = j_{0\,s}\big|_{s=1}$=2.4048, $r_s\big|_{s=1} = 783.3$ Mkm (the measured semi-major axis of the elliptical orbit is $\bar{r}$ =778.57 Mkm, the difference is 1%);
4) Saturn: $s = 2$, $z_{m,n} = j_{0\,s}\big|_{s=2}$=5.5201, $r_s\big|_{s=2} = 1807.3$ Mkm (the measured semi-major axis of the elliptical orbit is $\bar{r}$ =1433.45 Mkm, the difference is 26%).

The roots $r_s(\, j_{1,s}\,)$ and $r_s(\, y_{1,s}\,)$ for the orbits of Jupiter, Saturn and Uranus have been obtained from relations $r_s(j_{1,s}) = r_1 \cdot j_{1,s} / j_{1,1}$ and $r_s(y_{1,s}) = r_1 \cdot y_{1,s} / y_{1,1}$.

Similar results have been gained by Shpenkov for asteroids, though different perturbations and deviations from circular orbits were not been taken into account.

Why is the discrepancy for Saturn so large? Maybe Saturn has not always occupied its present orbit. Perhaps other values of roots 's' can also be examined for this planet or maybe the result points to a weak stability of the Saturn orbit and the planet might be able to move to another orbit after a comparatively small disturbance?

Orbits of planetary satellites have also been calculated by using the same approach [419] (pp. 63–65); when computing those orbits, Shpenkov likewise used his wave gravitation radius $\lambda\!\!\!^{-}_g$ (8.44). For instance, in the case of Io, the innermost satellite of Jupiter: $s = 7$, $r_s(j_{1,\,s})\big|_{s=7} = 424.7$ kkm, $r_s(y_{1,\,s})\big|_{s=7} = 395.3$ kkm; the measured semi-major axis of the elliptical orbit is $\bar{r} = 421.8$ kkm.

Thus, Shpenkov considers the parameter $\lambda\!\!\!^{-}_g$ (8.44) as a universal characteristic of both an elementary particle and a celestial body. Nevertheless, he does not explain the origin of the value (8.44), i.e., why has $\lambda\!\!\!^{-}_g$ been taken equal to 3.274×10^{11} m?

On the other hand, we can derive the value $\lambda\!\!\!^{-}_g$ (8.44) proceeding from the first principles laid down in the dynamics of the tessellattice. In fact, we have obtained the wave equation for a body's inerton cloud (8.17), which is based on an acoustic resonance of the body's entities whose inerton clouds form a total inerton cloud of the body with a set of wavelengths given by the relation (8.13). Of course, each body has its own resonance characteristics. However, since celestial bodies have a spherical shape, their resonance characteristics may coincide or they can have common harmonics.

Indeed, with regard to the Sun Robitaille [340] demonstrated forty examples of evidence that the Sun is comprised of condensed matter. Moreover, he noted that the Sun acts as a resonant cavity, which sustains oscillations as sound waves, travelling within its interior. He also emphasized that the Sun has to have relevant radiations from audio to X-ray frequencies, as produced by condensed matter. Ilyanok [420] has defended the model of the liquid Sun though with a gaseous region in the central part.

There is one more proof that the Sun is a condensed matter body. Indeed, if two adjacent hydrogen ions (namely, protons), which are located at the surface, are repulsed by the Coulomb law

$$V^{\text{rep}} = e^2 / (4\pi\varepsilon_0 r) \tag{8.46}$$

and elastically interact (i.e., are practically attracted) in the gravitational field of the Sun by the law

$$V^{\text{att}} = \tfrac{1}{2} m_{\text{p}} \omega^2 r^2 - G M_{\text{Sun}}\, m_{\text{p}} \,/\, R_{0\,\text{Sun}} \tag{8.47}$$

where M_{Sun} is the Sun mass and $R_{0\,\text{Sun}}$ is the Sun radius, then we can use the method described in Sections 5.1 and 5.2 to show that the entities of the Sun are in a condensed state. Namely, we can examine whether ions are able to gather in clusters.

The functions $a(\mathcal{N})$ (5.57) and $b(\mathcal{N})$ (5.58) are respectively

$$a(\mathcal{N}) = \tfrac{3}{2} \frac{e^2}{4\pi\varepsilon_0\, g\, k_{\text{B}} \Theta} \mathcal{N}^{2/3},$$

$$b(\mathcal{N}) = \tfrac{3}{10} \frac{m_{\text{p}}\, \omega^2 g^2}{k_{\text{B}} \Theta} \mathcal{N}^{5/3} - \frac{G M_{\text{Sun}}\, m_{\text{p}}}{R_{0\,\text{Sun}}\, k_{\text{B}} \Theta} \mathcal{N} \tag{8.48}$$

where g is the mean distance between entities (atoms/ions) in the Sun, which could be called the lattice constant of the Sun.

Then the action in the approximation of classical statistics (5.75) is

$$S / K = \tfrac{3}{2} \frac{e^2}{4\pi\varepsilon_0\, r\, k_{\text{B}} \Theta} \mathcal{N}^{5/3} - \tfrac{3}{5} \frac{m_{\text{p}}\, \omega^2 r^2}{k_{\text{B}} \Theta} \mathcal{N}^{8/3} + \tfrac{3}{5} \frac{G M_{\text{Sun}}\, m_{\text{p}}}{R_{0\,\text{S}}\, k_{\text{B}} \Theta} \mathcal{N}^{5/3} \tag{8.49}$$

Considering the equation $\partial S / \partial \mathcal{N} = 0$, in the first approximation we may neglect the term with the lowest degree by $\mathcal{N}$ (namely, the term with $\mathcal{N}^{1/3}$), which results in the simple solution

$$\mathcal{N} \approx \left(\frac{3 G M_{\text{Sun}}}{R_{0\text{Sun}}\, \omega^2 g^2} \right)^{3/2} \tag{8.50}$$

Here, the parameters G, M_{Sun} and $R_{0\,\text{Sun}}$ are known; the lattice constant of the Sun can be in a range $g = 10^{-10}$ to 4×10^{-10} m. The cyclic frequency ω can be estimated from the relation $\tfrac{1}{2} m_{\text{p}}\, \omega^2 \delta r^2 \approx k_{\text{B}} \Theta$, in which we can put the temperature $\Theta = 6000$ K and the amplitude of oscillation of a hydrogen atom near its equilibrium position $\delta r \approx 10^{-10}$ m; this gives $\omega \approx 10^{14}$ s^{-1}.

Substituting all the parameters into the right-hand side of expression (8.50) we obtain for the number of hydrogen atoms gathered in a cluster: $\mathcal{N} = 7.5 \times 10^3$ (at $g = 4 \times 10^{-10}$ m) and $\mathcal{N} = 5 \times 10^5$ (at $g = 10^{-10}$ m). Therefore,

$\mathcal{N} >> 1$, which means that hydrogen atoms in the Sun have a tendency to clustering, which in turn signifies that the Sun is a condensed matter object rather then a gaseous globe.

The radius of the Sun $R_{0\,\text{Sun}} = 6.96\times10^8$ m. The number N of its entities, which are mostly hydrogen atoms (78.5% by modern estimations) that cover the radius, is $N = R_{0\,\text{Sun}} / g$. Substituting this value into the relation (8.13), we get for the fundamental inerton harmonic (the amplitude of the inerton cloud of the Sun)

$$\Lambda_{\text{Sun}} = 2R_{0\,\text{Sun}}\, c / \upsilon_{\text{sound}} \tag{8.51}$$

We shall compare the inerton cloud amplitude (8.51) with Shpenkov's wave gravitational radius $\lambda\!\!\!^{-}_{g}$ (8.44). However, in expression (8.51) the parameter υ_{sound}, which is the sound velocity of entities of the Sun, i.e., atoms or ions, is unknown.

Even though the Sun is a condensed matter ball it has ionized atoms, which means that one could consider its ions and electrons as a solid-state plasma. Plasma is specified by an ion acoustic wave, which is a kind of longitudinal oscillation of the ions and electrons. The ion sound velocity results from the mass of ions and the pressure of the electrons. The expression for such sound velocity in plasma is as below

$$\upsilon = \sqrt{\gamma Z\, k_{\text{B}} T_e / m_i} = 9.79\times10^3 \times \left(\gamma Z T_e m_{\text{p}} / m_i\right)^{1/2} \tag{8.52}$$

In this expression we may put the ion mass equal to the proton mass and then $m_p / m_i = 1$ (note in the first approximation we may consider the Sun's helium ions that account for 19.7% of the Sun's mass as four protons); the charge number $Z = 1$. The value of the electron temperature T_e we may relate to a temperature of the slow solar wind, i.e., a stream of charged particles emitted from the Sun, which is in a range $(1.3 - 1.6)\times10^6$ K; let us put $T_e = 1.35\times10^6$ K. The adiabatic constant $\gamma = 1+2 / n_{\text{free}}$ depends on the degrees of freedom. When a vibrational degree is involved, γ becomes less than 4/3. Hence putting $n_{\text{free}} = 8$, we get $\gamma = 5/4$. Substituting the named values of the parameters into expression (8.52), we obtain the sound velocity for hydrogen ions in the Sun,

$$\upsilon = 1.272\times10^6 \ \text{m·s}^{-1} \tag{8.53}$$

Since the Sun is a condensed matter liquid, the adiabatic constant γ may approach a unit, which however does not change significantly the value of sound velocity (8.53).

Now substituting the sound velocity (8.53) into the expression for the amplitude of the Sun's inerton cloud (8.51), we obtain the same value for the fundamental harmonic (8.44) that has been used by Shpenkov [419]. Thus all the calculations of stationary orbits around the Sun, which were conducted by Shpenkov basing on the Bessel solutions, are right.

For orbits of planetary satellites we can used an expression similar to the Sun's fundamental inerton harmonic (8.51). For example, in the case of the Earth, its fundamental inerton harmonic (the amplitude of the inerton cloud of the Earth) is

$$\Lambda_{\text{Earth}} = 2R_{0\ \text{Earth}}\, c / \upsilon_{\text{sound}} \tag{8.54}$$

The sound velocity of the Earth measured from seismic activity of the globe varies in a range $(8 - 12)\times10^3$ m·s^{-1}; since iron is the most common element in the globe, the average speed deviates more towards high values. If we put $\upsilon = 1.16\times10^4$ m·s^{-1} and insert it and the radius of the Earth $R_{0\ \text{Earth}} = 6.371\times10^6$ m into the relation (8.54), we again arrive at Shpenkov's value (8.44).

Thus, the calculations of Shpenkov carried out on the basis of Bessel solutions for the wave equation is valid. The wave equation (8.17) for the astronomical body's inerton cloud, which possesses spherical symmetry, has the solutions presented by Bessel functions that correctly determine orbital positions of the body's satellites.

We have to mention about other important studies dealing with planetary orbits, which additionally support the results presented above. Some researchers [421–426] found that the distribution of the semi-major axes of planets rotated around stars, or mean distances from the star to the planets, exhibit the same quantum character; besides, Chechelnitsky [422] defends an idea of the wave universe concept.

Christianto [425] describes this quantum character in quite an elegant way, which practically leads to the same results obtained by previous researchers. The force balance relation of Newton's equation of motion for a planet with a mass m on the orbit around the Sun is

$$GM_{\text{Sun}}\, m / r^2 = m\upsilon^2 / r \tag{8.55}$$

Using Bohr-Sommerfeld's hypothesis of quantization of angular momentum of an electron in the hydrogen atom, Christianto introduces a new constant for a macroscopic rotational motion:

$$m\upsilon 2\pi r = n\hat{J} \tag{8.56}$$

namely, the momentum of inertia becomes a quantum value, $\hat{J}$, where n = 1, 2, 3, … Then following the elementary Bohr theory, Christianto obtains the known simple solution for the orbit radius (similar relations were derived also by other researchers, in particular, Nottale et al. [423])

$$r_n = n^2\hat{J}^2 / (4\pi^2 GMm^2) \tag{8.57}$$

This relation can be rewritten in terms of a velocity υ_{fund}, i.e. $r_n = n^2 GM / \upsilon_{\text{fund}}^2$. Nottale et al. [423] determine the value of this "fundamental" velocity as $\upsilon_{\text{fund}} = 2\pi\, GMm / \hat{J} = 144$ km·s^{-1}, though other researchers [421] talk about the fundamental velocity equal to 24 km·s^{-1}, which is 6 times less then Nottale's υ_{fund}.

Christianto [425] and Smarandache and Christianto [426] compared their own predictions based on the relation (8.57) and predictions of five other research teams with direct observations. They presented tables showing that the predicted orbital radii for all 9 planets of the solar system practically coincide with the observed radii.

Thus, the length of a planet's orbit is the planet's de Broglie wavelength; the planet moving through the tessellattice generates an inerton cloud (a total cloud of all the planet's entities) and the planet constantly travels together with its inerton cloud. However, the orbital position of the nth planet is given not by the new "fundamental" parameter $\hat{J}$ or the "fundamental" velocity υ_{fund} in expressions (8.56) and (8.57), but the left hand side of these expressions because it is the values of r_n, as roots of the radial component of the wave equation written in the spherical coordinates, that determines the proper positions of planets, which has been discussed above (in relation to Shpenkov's expression (8.45)).

Over the last few years astronomers have been finding hundreds of exoplanets around other stars that have some most peculiar orbits – Jupiter-sized planets orbiting so close to their star that their orbital period ("year") is measured in (Earth's) days. Since the astronomers have to be fairly confident about the mass of those stars, and from their observations of the exoplanets, calculated masses of those exoplanets should be available. It will be interesting to apply the equations considered above also to those exoplanets. Such studies would give further supplementary confidence in the evaluated orbits.

KEYWORDS

- **Casimir effect**
- **inertons**
- **mass**
- **Newton's gravitation law**
- **origin of gravity**
- **quantum gravity**
- **wave effects in gravitation**

CHAPTER 9

AN IMPORTANT CORRELATION TO NEWTON'S GRAVITATIONAL LAW

CONTENTS

In Chapter 8 we have clarified the origin of gravity – standing inerton waves generated by a massive object, or more exactly, created by vibrating entities inside the object. These standing inerton waves form a stationary relief of the space around each object, i.e., a deformation (mass) potential that is written in physics as Newton's gravitational potential (8.25).

On the other hand general relativity suggests that gravitation is a purely geometric effect of curved space-time, not a force of nature that propagates in space (see, e.g., Van Flandern [427]). The stationary topography around the object formed by the object's standing inerton wave is exactly the purely geometric effect of curved space-time. It is the motion of the object's standing inerton waves, which being averaged to the state of the mean field, induces the geometric effect of curved space-time.

Poincaré [428] noted that the force acting on attraction of a body at time t depends on the position and velocity of the body. Later on Poincaré [9] wrote that the mass as a coefficient of attraction depends on the velocity. Poincaré referred to Laplace: if the body has a velocity υ, then the resultant force of attraction between two bodies is not directed along the straight line that unites the two bodies; the resultant force should have a small angle of deviation from the direct line. Recently this has been accented again; Guy [429] has argued that the law of gravitation has to take into account the relative speeds of moving masses.

Practically all alternative approaches to the description of macroscopic phenomena predicted by the formalism of general relativity have involved the mobility of a test object in the presence of a large central mass (see e.g., quotations in papers [430, 431]). The total expression for gravitation should include the velocity of the attractive object. The submicroscopic approach points to the fact that the gravitational interaction between objects must consist of two terms [430, 70]: (i) the radial inerton interaction between masses M and m, which in the mean field approximation results in the classical Newtonian gravitational force

$$F_{\text{Newton}} = G\frac{Mm}{r^2} \tag{9.1}$$

and (ii) the tangential inerton interaction between the central attracting mass M and the orbiting attracted mass m, which is specified by the tangential component of the motion of the test mass m.

Indeed, components of the inerton cloud's velocity in the vicinity of the particle moving with the velocity $\vec{\upsilon}$ are: $\vec{\upsilon}$ along the particle's path and $\vec{c}$ in the transverse directions (Figure 9.1). The same should be valid for a moving macroscopic object, because individual inerton clouds of vibrating entities in the object overlap forming a total inerton cloud. In the total inerton cloud, inertons migrate by the same rule, as is the case for inertons of a separate particle: they move far away from the object and then return back to it.

Let a satellite with a mass m enveloped in its total inerton cloud rotate around an attracting mass M (Figure 9.2). The inerton cloud of the orbital mass m touching the central mass M is partly absorbed by it, which results in the reciprocal interaction between masses M and m. Components of the velocity of the satellite's inerton cloud are $\vec{c}$ along the radial line and $\vec{\upsilon}_{\perp}$

in the tangential direction. Hence the total inerton velocity in the satellite's inerton cloud is

$$\hat{c} = \sqrt{c^2 + \upsilon_{\perp}^2} \tag{9.2}$$

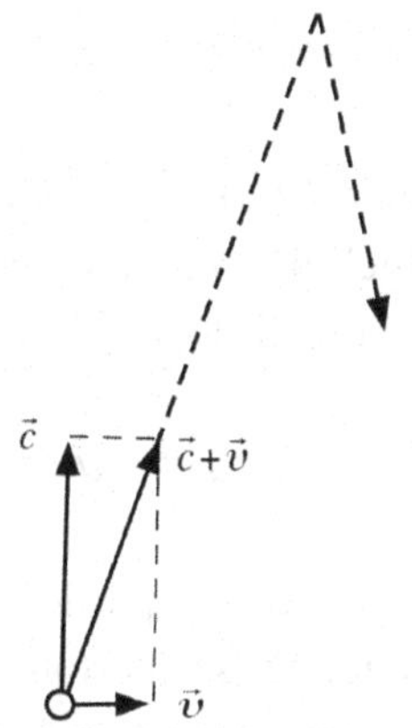

FIGURE 9.1 Components of the particle's inerton velocity: $\vec{\upsilon}$ is along the particle path, $\vec{c}$ is in the transverse direction to the particle's path.

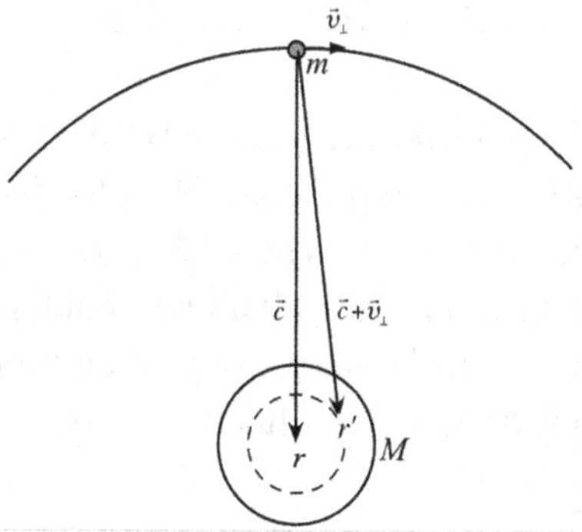

FIGURE 9.2 Orbital motion of the mass m around the central massive object with the mass M. Orthogonal components of the velocity of the inerton cloud of the orbital mass are equal to $\vec{c}$ and $\vec{\upsilon}_{\perp}$. Due to the tangential component $\vec{\upsilon}_{\perp}$ the line of attraction between the masses M and m shifts from the center of the mass M and the effective radius r' of the attraction becomes shorter than r.

Then the kinetic energy of these inertons is $mc^2 \cdot (1 + \upsilon_\perp^2 / c^2)$. Note in Section 3.5 we have already discussed the speed c_{in} of free inertons, i.e., inertons that have been completely released from the inerton cloud; their speed was preliminarily estimated [156] as the value of two orders of magnitude greater than the speed of light *c*.

If the speed of bound inertons is *c*, then the square of a linear element should be written in line with the Minkowski metric,

$$ds^2 = c^2 dt^2 - d\vec{r}^2 \tag{9.3}$$

However, since the speed $\hat{c}$ (9.2) is present in the motion of the particle's cloud of inertons, which exceeds the speed of light, the square of a linear element should appear as follows

$$ds'^2 = c^2 dt'^2 - d\vec{r}'^2 \tag{9.4}$$

The time t is the proper time of the particle and t' is the proper time of the particle's inertons. Since these inertons in the cloud travel faster than *c*, the relationship between these two times has to be as follows: $\sqrt{1 + \upsilon_\perp^2 / c^2}\, dt' = dt$. Then equating the elements ds^2 and ds'^2 to zero we derive an equation

$$dr'^2 = dr^2 / (1 + \upsilon_\perp^2 / c^2) \tag{9.5}$$

Now we can solve the problem of how to introduce the tangential velocity into the Newtonian expression (9.1) for the radial attraction of two gravitating masses *m* and *M*. The real trajectories of inertons of the orbital mass *m* do not go along the radial line *r* but have a small shift ahead following the velocity $\vec{c} + \vec{\upsilon}$; in other words, their trajectory is going along the line r'. Integrating right and left hand sides of Eq. (9.5) we get

$$r'^2 = r^2 / (1 + \upsilon_\perp^2 / c^2) \tag{9.6}$$

Substituting in expression (9.1) r^2 for r'^2 from expression (9.6) we obtain

$$F = G\frac{Mm}{r^2}(1 + \upsilon_\perp^2 / c^2) \tag{9.7}$$

Then for the gravitational interaction of the orbital mass m with the central mass M we have

$$U = -G\frac{Mm}{r}\cdot\left(1 + \frac{r^2\dot{\varphi}^2}{c^2}\right) \tag{9.8}$$

where we substitute for the tangential velocity $\upsilon_\perp = r\dot{\varphi}$.

In contrast to orthodox quantum theory and general relativity, the submicroscopic concept allows us to derive the Newton's gravitational potential (8.25) and introduce the corrected version of Newton's law of gravitation (9.8). Expression (9.8) immediately opens the gateway to the solutions in the framework of the tessellattice for such problems as the motion of Mercury's perihelion, the deflection of starlight by the Sun, the gravitational redshift of spectral lines [430], and the Shapiro time delay effect [431].

9.1 MOTION OF MERCURY'S PERIHELION

Classical mechanics yields the following equations describing the motion of a body with a mass m in the gravitational field induced by a large central mass M

$$I = mr^2\dot{\varphi}; \tag{9.9}$$

$$E_{\text{cl.}} = \tfrac{1}{2}m\dot{r}^2 + \tfrac{1}{2}mr^2\dot{\varphi}^2 - G\frac{Mm}{r} \tag{9.10}$$

Eqs. (9.9) and (9.10) are the classical integrals of the motion – the angular momentum and the energy, respectively. However, as follows from the above consideration, in Eq. (9.10) we have to change the potential gravitation energy (8.25), i.e., the last term in the right-hand side of expression (9.10), to the corrected expression (9.8). Then the two Eqs. (9.9) and (9.10) are transformed to

$$I = mr^2\dot{\varphi} \tag{9.11}$$

$$E = \tfrac{1}{2}m\dot{r}^2 + \tfrac{1}{2}mr^2\dot{\varphi}^2 - G\frac{Mm}{r}\cdot\left(1+\frac{r^2\dot{\varphi}^2}{c^2}\right) \tag{9.12}$$

Note that here the dot over r and φ means the differentiation by the proper time t of the body, i.e., t is the natural parameter that is proportional to the body's path. The system of equations (9.11) and (9.12) are identical to the equations of motion of a body in the Schwarzschild field obtained in the framework of the general theory of relativity. Equations (9.11) and (9.12) can be rewritten in the form

$$r^2 d\varphi / dt = \hbar \tag{9.13}$$

$$(dr/dt)^2 + r^2(d\varphi/dt)^2 - 2GM/r = Q^2 + \frac{2GMr}{c^2}(d\varphi/dt)^2 \tag{9.14}$$

If we multiply Eq. (9.14) by $(dt/d\varphi)^2$ and substitute this factor itself from Eq. (9.13), we obtain the differential equation

$$(dr/d\varphi)^2 = Q^2r^4/\hbar^2 + 2GMr^3/\hbar^2 - r^2 + 2GMr/c^2 \tag{9.15}$$

where $Q^2 = 2E/m$.

The solution to Eq. (9.15) is known (see, e.g., Bergmann [432], pp. 214–217) and it shows that it is the last term in expression (9.12), which displaces the perihelion of the planetary orbit by the amount

$$\Delta\varphi = 6\pi\, G^2M^2/\hbar \tag{9.16}$$

as is directly observed in Mercury's orbit around the Sun.

9.2 THE DEFLECTION OF LIGHT

As has been discussed in Chapter 4, a photon is not a canonical particle, but a quasi-particle, a local excitation of the tessellattice, which migrates in space by hopping from cell to cell. The photon carries the polarization state of the cell and it does not possess an inerton cloud at all. Therefore, since a photon does not disturb the ambient space with a cloud of inertons, it cannot experience the radial component of the gravitational field of a massive object (because there is no overlapping with the inerton cloud of the massive object). Hence, the radial component $-GMm/r$ is absent in the interaction between the massive object and an incident photon. Nevertheless, the tangential component $-GMmr\dot{\phi}^2/c^2$ associated with the true motion of the photon must still be preserved.

Although a free photon does not possess a rest mass, the photon, as a local perturbation of the tessellattice, is still characterized by a momentum and hence it has a kinematic effective mass m (it is known that the photon's momentum is $\hbar\vec{k}$, however, the notion also follows directly from the "moment of junction" considered in Section 1.5.1). That is why the behavior of the photon in the gravitational field of mass M has to be defined by the following pair of equations

$$I = mr^2\dot{\varphi} \tag{9.17}$$

$$E = \tfrac{1}{2}m\dot{r}^2 + \tfrac{1}{2}mr^2\dot{\varphi}^2 - G\frac{Mmr\dot{\varphi}^2}{c^2} \tag{9.18}$$

where the time t is treated as the natural parameter proportional to the photon path, which is very important for the invariance of the theory.

Again Eqs. (9.17) and (9.18) can be reduced exactly to the same input equation for the study of the bending of a light ray in the Schwarzschild field, which is obtained in the framework of the formalism of general relativity, namely,

$$(dr/d\varphi)^2 = Q^2r^4/\hbar^2 - r^2 + 2GMr/c^2 \tag{9.19}$$

Equation (9.19) is different from Eq. (9.15) only in the absence of the term $2GMr^3/\hbar^2$. The solution to Eq. (9.19) is well known (see, e.g., Bergmann [432], pp. 218–221); it yields the following angle deviation of the ray from the direct line

$$\Delta\varphi \approx 4\frac{GM}{c^2 r} \tag{9.20}$$

9.3 RED SHIFT OF SPECTRAL LINES

Let us consider a simple task. Let l and m be, respectively, length and mass of a mathematical pendulum and let φ be the angle of the deviation of the pendulum from the equilibrium. The pendulum is located on the surface of a planet with the radius r. In this case the kinetic energy of the massive point is

$$K = \tfrac{1}{2} m\, l^2 \dot{\varphi}^2 \tag{9.21}$$

and the potential energy is

$$U = -G\frac{Mm}{r + l\cdot(1-\cos\varphi)}\cdot\left(1 + \frac{l^2\dot{\varphi}^2}{c^2}\right) \tag{9.22}$$

(to write the expression, we have used Newton's corrected law (9.8)). Because of the small variable φ one can write the energy $E = K + U$ of the massive point as follows

$$E \cong \tfrac{1}{2} m l^2 \dot{\varphi}^2 - G\frac{Mm}{r} - G\frac{Mm}{r}\cdot\left(-\frac{l\varphi^2}{2r} + \frac{l^2\dot{\varphi}^2}{c^2}\right) \tag{9.23}$$

In the case of the potential depending on the velocity the equation of motion is determined by the Euler-Lagrange equation [114]

$$\frac{d}{dt}\frac{\partial K}{\partial \dot{q}} - \frac{d}{dt}\frac{\partial U}{\partial \dot{q}} - \frac{\partial K}{\partial q} + \frac{\partial U}{\partial q} = 0$$

where in our case $q \equiv \varphi$ and t is the proper time of the oscillating massive point. In the explicit form it yields φ

$$\left(l^2 + 2G\frac{M}{r}\frac{l^2}{c^2}\right)\ddot{\varphi} + G\frac{M}{r} l\varphi = 0 \tag{9.24}$$

If we designate $(2\pi\nu_0)^2 = 2GM/(rl)$, we can write instead of Eq. (9.24) the equation

$$\ddot{\varphi} + \frac{(2\pi\nu_0)^2}{1+2GM/(c^2 r)}\varphi = 0 \qquad (9.25)$$

In Eq. (9.25) assuming the inequality $r_0 = 2GM/c^2 << r$ we acquire the renormalized frequency of the pendulum

$$\nu \approx \left(1 - \frac{GM}{c^2 r}\right)\nu_0 \qquad (9.26)$$

The scheme described above may easily be applied to vibrating atoms (ions) located on the surface of a star. This means that expression (9.26) determines the so-called gravitational red shift of spectral lines

$$\delta\nu \cong -\frac{GM}{c^2 r}\nu_0 \qquad (9.27)$$

The result (9.27) is in complete agreement with that derived in the framework of general relativity (see, e.g., Bergmann [432], p. 222).

9.4 THE GRAVITATIONAL TIME DELAY EFFECT (THE SHAPIRO EFFECT)

Shapiro [433, 434] studied the return of signals transmitted from the Earth to Venus and revealed the time delay effect, or gravitational time delay effect, which is considered now as one of the four available classic solar system tests of general relativity. In 1964, Shapiro had calculated that there would be a time delay for radar signals reflected off Mercury and Venus back to Earth when their path was too close to the Sun because their signal path would become longer than the straight-line path between the planets due to deflection by the gravitational field of the Sun (see Figure 9.3). The first tests, performed in 1966–1967 successfully matched Shapiro's predicted time delay of 200 μs, as was also

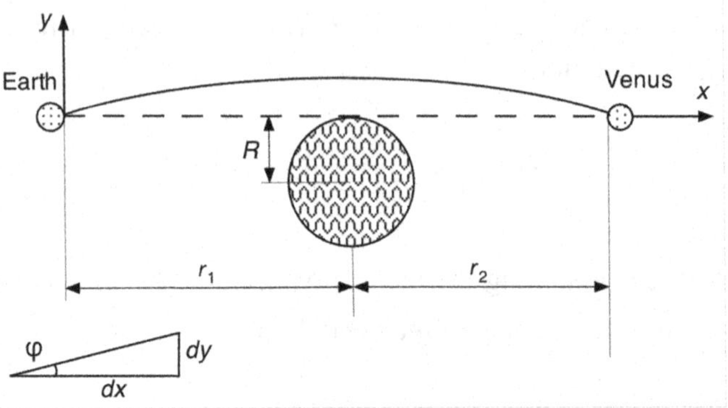

FIGURE 9.3 Photon path from one planet to the other, which passes near the Sun. The X-axis line (dotted line) is the assumed path of photons. The real path (solid line) is distorted by the Sun's gravitational field. Below in the text we use the following designations: the distance from Earth to the Sun is r_1, the distance from the Sun to Venus is r_2 and R is the radius of the Sun.

calculated [433, 434] on the basis of the Schwarzschild-Hilbert metric of the Sun (see details of the calculations, e.g., in Weinberg [435]). Subsequently, the time delay effect was observed for the binary pulsar PSR 1913+16 [436] and the phenomenon has been applied to direct measurements made from probes on the surface of Mars and orbiting Jupiter and Saturn. Such a delay is explained by a general-relativistic delay in the time it takes a radar signal to travel to Mercury (or Venus) and back. But why does it really happen?

Let us consider a path of photons that travel from one planet to the other one and come back passing near the Sun [431] (Figure 9.3). A conventional consideration is based on a variational technique. Time has to be treated as a natural parameter,

$$t = \int \frac{ds}{c} \tag{9.28}$$

where ds is the interval length of the path of photons and c is the constant, the velocity of light. A ray of light, which passes close to a gravitating body with the mass M and the radius R, has to be deflected in compliance with expression (9.20).

9.4.1 CLASSICAL MATHEMATICAL VARIATIONAL PROCEDURE

Following classical mathematics, namely, the variational procedure, we can write

$$t = \int \frac{ds}{c} = \int \frac{\sqrt{dx^2 + dy^2}}{c} = \int \frac{\sqrt{1 + [\varphi'(x)]^2}}{c} dx \tag{9.29}$$

where the function $y(x)$ is defined in expression (9.20) and Figure 9.3:

$$y(x) \equiv \varphi(x) = \frac{4GM}{c^2 \sqrt{R^2 + x^2}} \tag{9.30}$$

and, therefore, its derivative becomes

$$\varphi'(x) = -\frac{4GMx}{c^2 (R^2 + x^2)^{3/2}} \tag{9.31}$$

This approach does not require any of the uncertainties encountered at extremely small distances, as prescribed by the formalism of quantum mechanics because in submicroscopic considerations the minimal distances of the tessellattice require that nature is deterministic.

Substituting the function (9.31) into expression (9.29) we obtain

$$t = \frac{R}{c} \int d\chi \sqrt{1 + \left(\frac{4GM}{c^2 R}\right)^2 \frac{1}{(1+\chi^2)^3}} \cong \frac{R}{c} \int d\chi + \frac{R}{2c} \int \left(\frac{4GM}{c^2 R}\right)^2 \frac{d\chi}{(1+\chi^2)^3}$$

$$= \frac{R}{c} \left(2 \int_{\text{Earth}}^{\text{Sun}} d\chi + 2 \int_{\text{Venus}}^{\text{Sun}} d\chi \right) + \frac{R}{2c} \left(\frac{4GM}{c^2 R}\right)^2 \left(\begin{array}{l} 2 \int_{\text{Earth}}^{\text{Sun}} \frac{d\chi}{(1+\chi^2)^3} \\ +2 \int_{\text{Venus}}^{\text{Sun}} \frac{d\chi}{(1+\chi^2)^3} \end{array} \right) \tag{9.32}$$

where $\chi = x / R$ is the dimensionless variable.

From expression (9.32) we get for the time delay

$$\delta t \approx \frac{R}{2c}\left(\frac{4GM}{c^2 R}\right)^2 2\left(\frac{\chi}{\sqrt{1+\chi^2}}\Bigg|_{\chi=0}^{\chi=r_1/R} + \frac{\chi}{\sqrt{1+\chi^2}}\Bigg|_{\chi=0}^{\chi=(r_1+r_2)/R} \right) \tag{9.33}$$

where r_1 and r_2 are distances from the Earth to the Sun and from Venus to the Sun, respectively. Physical constants are: $G = 6.673\times10^{-11}$ m³·kg⁻¹·s⁻², $M = 1.99\times10^{30}$ kg, $r_1 = 149.6\times10^9$ m, $r_2 = 108.2\times10^9$ m, $R = 0.695\times10^9$ m and $c = 3\times10^8$ m·s⁻¹. Substituting these constants into expression (9.33) we get an estimation of the time delay:

$$\delta t \approx 3.3\times10^{-11}\ \text{s} \tag{9.34}$$

which is 5 orders less than the experimental result and the value obtained by Shapiro [433, 434] on the basis of Schwarzschild metric's components.

9.4.2 FRACTAL CHANGES IN THE PHOTON PATH

A submicroscopic consideration allows a deeper alternative examination into the proper time of migrating photons. Photons have to interact with the mass body via the second term of expression (9.8). This interaction with the body's total inerton cloud changes the path of photons near the body, expression (9.20), which one can perceive as a local curvature of space. Since real space is organized as the tessellattice of topological balls, curvature of space can easily be illustrated by changes in the geometry of cells of the tessellattice around a massive object. Then in this case the proper time (9.28) of photons becomes

$$t = \int \frac{ds}{c} = \int \frac{dx}{c} + \int \varphi \frac{dx}{c} \tag{9.35}$$

and thus the time delay, i.e., the second term in the right hand side of expression (9.35), appears as follows

$$\Delta t = \int \varphi(x) \frac{dx}{c} \tag{9.36}$$

where the angle of deflection $\varphi(x)$ is defined in expression (9.30). Calculating the integral in (9.36) we obtain

$$\begin{aligned}\Delta t &\approx \frac{4GM}{c^3}\left(2\int_0^{r_1/R}\frac{d\chi}{\sqrt{1+\chi^2}}+2\int_{r_1/R}^{(r_1+r_2)/R}\frac{d\chi}{\sqrt{1+\chi^2}}\right)\\ &= \frac{4GM}{c^3}2\left[\ln\left(\chi+\sqrt{1+\chi^2}\right)\Big|_0^{r_1/R}+\ln\left(\chi+\sqrt{1+\chi^2}\right)\Big|_{r_1/R}^{(r_1+r_2)/R}\right]\\ &\approx \frac{4GM}{c^3}2\ln\left(2\frac{r_1+r_2}{R}\right)\approx 2.6\times10^{-4}\ \text{s}\end{aligned} \tag{9.37}$$

Note the numerical value of the result (9.37) coincides with Shapiro's outcome [433].

In the tessellattice, which represents an inner structure of real physical space, volumetric fractal changes of cells are associated with the physical notion of mass. However, what is the gravitating body's inerton cloud doing around the body? It distributes a mass potential around the body, which results in the induction of Newton's gravitational law (8.25) and its corrected form near an orbital satellite (9.8). In turn the induction of mass in the space around the body means the appearance of volumetric fractal changes in appropriate cells of the tessellattice.

This means that the tessellattice is really contracted around the body. Therefore, in the Earth-Venus situation displayed in Figure 9.3, the photon path encounters additional cells in comparison with the case of a degenerate space (when the massive Sun is absent between the Earth and Venus). Note such investigation is in agreement with general rules of fractal geometry (see, e.g., Ref. [61]), which makes it possible to measure a curve by means of the number of balls that cover it.

Now putting the size of a topological ball of degenerate space (a cell of the undisturbed tessellattice) equal to the Planck's size $\ell_{\mathrm{P}}=\sqrt{\hbar G/c^3}$ $\cong 1.616\times10^{-35}$ m, we may estimate the number of cells that create the time delay (9.38). The number of cells, which form a path for photons that hop from cell to cell with the constant velocity c, is

$$N = \frac{1}{\ell_{\mathrm{P}}} \int_{0}^{r_1+r_2} dx = \frac{r_1 + r_2}{\ell_{\mathrm{P}}} \sim 10^{46} \tag{9.38}$$

An additional number of cells involved in the path due to the cells' fractal volumetric shrinking caused by the solar mass M and the interaction of photon with the gravitational field of this mass via the second term in expression (9.8), is

$$\Delta N = \frac{1}{\ell_{\mathrm{P}}} \int_{0}^{r_1+r_2} \varphi(x)\, dx \approx \frac{4GM}{c^2 \ell_{\mathrm{P}}} 2\ln\left(2\frac{r_1 + r_2}{R}\right) \sim 10^{39} \tag{9.39}$$

Thus, the Sun's gravitational field shrinks the tessellattice, such that the number N of cells in a rectilinear path between the Earth and Venus increases by the value of ΔN. Photons hop with the same speed c from cell-to-cell migrating in the tessellattice. But in the presence of the Sun the number of cells that photons have to pass travelling between the Earth and Venus grows to the value of $N + \Delta N$. This is the reason of the gravitational time delay effect, as it is seen from the submicroscopic viewpoint, which takes into account the peculiarities of fractal geometry that rules the tessellattice.

9.5 SUMMARY ON CLASSICAL GRAVITY

Predictions of general relativity, which were tested experimentally, are based on the Schwarzschild-Hilbert metric with the components

$$g_{11} = -\frac{1}{1 - 2GM/(c^2 r)}, \quad g_{22} = -r^2, \quad g_{33} = -r^2 \cos^2 \vartheta,$$

$$g_{44} = 1 - 2GM/(c^2 r) \tag{9.40}$$

All other metrics do not allow a comparison with the experiments discussed above, namely: the anomalous precession of Mercury's perihelion, the bending of light near the Sun, the red shift of spectral lines in the presence of gravity, and the radar echo delay. General relativity explains these phenomena via changes of the flat metric to the Schwarzschild-Hilbert metric in which two new nonlinear terms g_{11} and g_{44} appear: $-g_{11}^{-1} = g_{44} = 1 - 2GM/(c^2 r)$. This metric was developed on the concept of general covariance.

General relativity's basic principle is the expression of a physical law in a generally covariant formalism, such that it takes the same form (preferably in tensor terms) in all coordinate systems. General covariance implies the absence of coordinates in nature, which are treated as artificial, and that is why they should not play any role in the construction of fundamental laws.

On the other hand, in microscopic physics coordinates do play an important role. In particular, in the mean field approximation of gravitational interaction, the derivation of expression (9.8) from the first sub-microscopic principles is fundamental, as it shows that the gravitational attraction of two objects depends on both the distance r between the objects and their relative velocity υ.

The covariant formalism of general relativity does not experience the subtle details of the system studied. It transfers an external perturbation introduced by a secondary mass m into the metric of the central object with mass M. In other words, the covariant formalism embraces the whole system with all its inner interactions (e.g., two gravitationally interacting bodies) as one object and presents internal interactions of the system as a result of only the central object. Thus, the covariant formalism of general relativity brings impression to the phenomenon; namely, it shows that this or that effect is related to gravity but cannot account for the actual mechanism of the phenomenon.

Such a weakness of general covariance stems from its basic principles: an object is quite rigid in space and nothing happens to the object when it moves; this has been discussed and analyzed in the beginning of the book. Therefore the introduced symmetries are largely not justified. This means that conclusions drawn from some predictions of general covariance require a detailed reexamination on the basis of microscopic views. In particular, in literature on general relativity researchers derive the so-called gravitational (or Schwarzschild's) radius $r_{\text{grav}} = 2GM/c^2$ assuming the correctness of the equation $g_{11} = 0$ (see expressions (9.40)), which leads to the existence of so-called black holes.

However, as can be seen from expression (9.8), r is the distance between two interacting masses, a central mass M and a secondary mass m. Hence in the metric (9.40) the distance r has to be interpreted as a distance measured from the central object to where the secondary body perturbs the flat metric in the ambient space of the central object.

Thus, if an object is spherically symmetric, its metric also has to be symmetric and must be flat. Changes to the metric emerge as a result of a secondary body that perturbs the metric of the object studied. Therefore, the value r in expressions (9.40) cannot be smaller than the distance between two interacting bodies. This remark works against advocates of black holes since black holes are not a direct consequence of Einstein's theory. Loinger [437–441] and Crothers [442–450] (see also Rabounski [451] in support of Crothers) came to the conclusion that, even based on the formalism of general relativity, black holes are not an outcome of Einstein's general relativity. Crothers [450] writes: "Although cosmologists have devised mathematical-like methods ... to produce their black hole, all their methods violate the rules of pure mathematics and so they are inadmissible... Moreover, since material sources cannot be both present in and absent from Einstein's field equations by the very same mathematical constraint, the whole theory of black holes is fallacious."

Moreover, Loinger [437, 440] and Crothers [446] convincingly argue that hypothetical gravitational waves, as a solution of the nonlinear equation of motion, which do not have mass, cannot exist in real space-time because such a solution is theoretically constructed on the basis of a gravitational pseudo-tensor, but a pseudo-tensor is not a real tensor at all and, hence, cannot be manifested in real space.

Nevertheless, the astrophysicists of LIGO Scientific Collaboration and Virgo Collaboration [452] have claimed that they observed a gravitational wave from a binary black hole merger. They note that the source lies at a luminosity distance of 410 Mpc corresponding to a redshift $z = 0.09$. According to them the source emitted a gravitational wave only once and exactly in the direction of their antenna. They probably followed a supposition of Baumgarte and Shapiro [453] who had hypothesized a merger of two black holes aiming to present a theoretical basis for the emission of a gravitational wave from the coalescence of a binary black hole. But that work [453] is just a set of symbols and a collection of phrases without any rigorous meaning and definitions; this is like a science fiction that cannot possibly be correlated with academic science.

Thus, starting from two non-existing things, the astrophysicists [452] were able to observe a real splash in their detector but are unable to explain the events that had occurred in the said distant place in the universe. It is their fantasy to relate inexplicable cosmological processes to the generation of a delusive 'gravitational wave.' All the activity of the astrophysicists [452] is reminiscent of a certain mystique. Crothers [454] strongly criticized the declared results [452]. Drawing on rigorous theoretical principles Crothers proved that the LIGO-Virgo Collaborations neither detected gravitational waves, nor proved the collision and merger of black holes. Loinger and Marsico [455] pointed out that "the undulatory solutions of the Einstein homogeneous field equations do not possess a true, generally covariant, energy-tensor, which is different from zero, i.e., they do not posses a physical reality" and the researchers also argued that "the black hole cannot 'swallow' anything."

Regarding the situation with the measured signal [452], it is reasonable to assume that the researchers had instead measured a weak inerton signal that came to the detector from one of the many possible places, such as the nearest building, the interior of the Earth, the atmosphere, the Sun, a distant star, etc. Besides, the researchers [452] did not use an advanced method of signal measurement, which would have enabled them to identify the location of the source. An appropriate method is outlined below in the section devoted to inerton astronomy. Unfortunately, a tunnel vision, which prevails among the majority of researchers studying these complex problems, does not allow them a broader viewpoint to assess the challenge.

The submicroscopic consideration of gravitational phenomena can neither find nor support the existence of black holes and gravitational waves. In particular, spin 2, which is ascribed to gravitons, means a real rotation of such illusory wave(s) having an angular momentum with the quantum number 2. What material can rotate with such angular momentum if the gravitational wave does not have mass?

Only inertons have the possibility to manifest themselves in gravitational phenomena. In the next Chapter we will analyze cosmological effects related to interference and absorption of inertons.

KEYWORDS

- **correlation to gravitational law**
- **deflection of light**
- **gravitational time delay effect**
- **Mercury's perihelion**
- **red shift of spectral lines**

CHAPTER 10

COSMOLOGY

CONTENTS

Physical cosmology analyzes the whole universe through both mathematics and observation, and particularly studies the origin of the Universe. Currently the majority of researchers support the idea that the universe began with the Big Bang after which the emerged matter began to disperse; the moment of the explosion is estimated to have occurred about 13.8 billion years ago. What is the form of the Universe in such understanding? It is only matter, i.e., elementary particles unified in nuclei, atoms and then in macroscopic matter with further clustering in nebulas, galaxies and so on, and all this occurs in a completely empty space.

However, particles can be created in any point of such empty space, which is convincingly demonstrated in particle accelerator experiments. This absolute fact completely throws into question the idea that all matter can only have originated in a primary burst of something in a point source billions of years ago.

In the present book we argue that the ordinary physical space itself is a substrate that has the structure of a mathematical lattice – the tessellattice – and we have argued in Section 1.8.10 that "big bangs" can be created in any part of the tessellattice. In particular, the creation of a particle-antiparticle pair is also a big bang, as it involves a huge number of cells of the tessellattice as also would have happened in the primary Big Bang. Those cells would have necessarily become partly deformed, changed equilibrium states, acquired tension and repolarized in the manners described in Chapters 1 and 3.

The presence of the tessellattice, as an arena in which all events take place, requires a reconsideration of some basic principles of the cosmological model.

10.1 COSMIC MICROWAVE BACKGROUND RADIATION

The standard model of cosmology requires an infinite flat universe that is expanding under the pressure of dark energy. The phenomenon of Cosmic Microwave Background radiation (CMB radiation) discovered in 1964 [456] is considered to be the most important argument in support of the Big Bang model of the Universe. Since then the hot Big Bang model gradually gained popularity and displaced both the early Steady State models [457, 458] and the ether theories [459].

The Big Bang appeared as a development of the Friedmann exact solution of Einstein-Hilbert's field equations of general relativity. Later on the Friedmann–Lemaître–Robertson–Walker (FLRW) metric presented a homogeneous, isotropic expanding (or contracting) universe. However, the Friedmann model looks very artificial [70], as a local point theory is applied to the whole universe and the theory is based on a non-physical metric – time $t \equiv x_4$ that is introduced into the spatial components of the metric

$$ds^2 = R^2(x_4)\,(dx_1^2 + \sin^2 x_1\, dx_2^2 + \sin^2 x_1 \sin^2 x_2\, dx_3^2) + M^2(x_1,\ x_2,\ x_3,\ x_4)\, dx_4^2 \tag{10.1}$$

where $M^2(x_1,\ x_2,\ x_3,\ x_4)$ is a function (not a mass); $R(x_4)$ is proportional to the radius of curvature of space. Hence the radius of curvature R can change over time. Therefore a change of the radius with time is already incorporated in the metric (10.1). Nowadays instead of Friedmann's designation $R(x_4)$ the curvature of space is written as $a(t)$ and called a scale factor.

However, in Riemannian geometry, the scalar curvature (or the Ricci scalar) R is the simplest curvature invariant (scalar) of a Riemannian manifold. Its geometrical meaning is reduced to a single real number determined by the intrinsic geometry of the manifold near the point in which R is calculated. The scalar curvature R represents the difference in volumes of a geodesic ball written in a curved Riemannian manifold and a ball in Euclidean space. Thus R is a number, for example, it might be the number $\pi = 3.14159...$ Even so according to Friedmann, this number can depend on time t! In the metric (10.1), substituting the scalar factor $R \equiv \text{const}$ with a factor depending of the time variable, $a(t) \equiv 3.14159(t)$, one may of course try to solve the Einstein-Hilbert field equations (in which π gradually increases, e.g., initially from 1 to 2 then to 3.14, and then in the future to 4, then 5 and so on...). Finally, the FLRW metric brings about the expansion/contraction of a. Notwithstanding this outcome, we have the right to ask: How can such approach, and especially its result, be related to physics?

The question is particularly appropriate since the FLRW metric is not able to describe any of the crucial tests of general relativity, which have been performed using the Schwarzschild-Hilbert metric, namely, the precession of Mercury, gravitational lensing, gravitational redshift, and the gravitational time delay effect.

Analyses of measurements of CMB are considered to be of paramount importance for cosmology. NASA's COBE (1989–1994) [460] and WMAP (2001–2010) satellites [461–463] and the Planck (2009–2013) satellite of the European Space Agency [464] showed that the CMB spectrum is nearly that of a perfect blackbody with a temperature of 2.725 K. The researchers claim this observation matches the predictions of the hot Big Bang theory.

However, those satellite results were strongly criticized by Robitaille [465–468] who pointed out that the experiments, or rather their processing, "are fraud" and the satellites' infrared antennas recorded signals of the Earth's oceans. He [465] wrote: "the obscured resonances at ~4.63 ppm in the water spectrum would still have a signal to noise (ratio) of ~5:1, if the water line had not contaminated this region… For WMAP, the signal to noise is less than 2:1, and the signal of interest is located at the same frequency of the contamination… Signal suppression, by a factor of 100, or more, while still viewing the underlying signal, depends on the ability to control the source. This has been verified in numerous laboratories where the sample is known and where the correct answer can be readily ascertained. As such, it is impossible for the WMAP team to remove the galactic foreground given the dynamic range situation between the contaminant and the signal of interest… Relative to the signal to noise ratio, the WMAP team is unable to confirm that the anisotropic "signal" observed at any given point is not noise… The requirements for image stability in cosmology are well beyond the reach of both COBE and WMAP." "…with the aid of the Planck satellite, the electromagnetics laboratories of the world should be able to confirm or refute the existence of a ~3 K cosmic signal. The key to this puzzle rests in the understanding of the LFI (low frequency instrument) and reference targets… Enough evidence is already beginning to build … indicating that physics, astrophysics, and geophysics stand on the verge of a significant reformulation. In any event, the definitive proof that the monopole of microwave background belongs to the Earth has now been provided [468]." Robitaille also emphasized that the satellites' detector for which the signal to noise ratio is extremely low, only 2:1, must be aimed at the source of radiation for no less than 10 minutes but that was unrealistic in the conditions of an outer space when the aiming time of the detector was only a few seconds. Such conditions do not allow one to process the data obtained by the satellites with reasonable accuracy and, therefore, the results presented are not only insufficiently accurate but surprisingly close to those which were obtained in terrestrial laboratories.

The criticism of Robitaille [465–468] is quite serious and means that the existence of CMB is very debatable. In any case such a radiation, if any, can hardly have a spectrum that is typical for a classical blackbody. Indeed, measurements of bodies whose photon spectral range is close to the expected by the Planck formula occur within a typical wavelength range of 0.1–15 microns, though the wavelength of the disputed very cold CMB is about 5000 microns. Can we apply the Planck distribution under such conditions? Moreover, the blackbody spectrum itself has recently been revised by Robitaille and Crosthers [469] who argue that each body has its own 'blackbody spectrum,' as reflectivity also must be taken into account (which was neglected by Max Planck).

A list of the top 30 problems with the Big Bang was published in 2002 [470]. In that paper the microwave "background" was treated in terms of "the temperature of interstellar space," which was introduced by Eddington [471] who calculated the minimum temperature that any body in space would cool to, given that it is immersed in the radiation of distant starlight. With no adjustable parameters, he obtained 3.2 K (later refined to 2.8 K [464], which is the same as the observed, the so-called temperature of CMB). In such a way the intergalactic matter is considered as a "fog," which naturally accounts for the microwave radiation, including its blackbody-shaped spectrum.

This intergalactic fog explains the observed ratio of infrared to radio intensities of radio galaxies, which Lerner [472] presented in his book on plasma cosmology. Namely, the radiation emitted by distant galaxies diminishes with increasing wavelengths, since the longer wavelengths are scattered by the intergalactic medium. Lerner explained that the radiation energy has to be thermalized and isotropized by magnetically confined plasma filaments that pervade the intergalactic medium. Since the filaments efficiently scatter radiation with wavelengths longer than about 100 microns, such radiation from distant sources will be absorbed or scattered, and thus will decrease more rapidly with distance than the radiation with wavelengths shorter than 100 microns. Such absorption has been demonstrated by comparing radio and far-infrared radiation from galaxies at various distances – the more distant the galaxy, the greater the absorption effect. Therefore, microwave radiation could not be coming directly to us from a distance beyond all the galaxies.

Lerner [472] (see also Ref. [473]) notes that the alignment of the CMB radiation anisotropy is mainly associated with the direction along the major axis of the (roughly cylindrical) Local Supercluster and to a lesser extent at right angles to this axis, where less high-density matter is encountered. The quadrupole and octupole radiations are concentrated on a ring around the sky and are essentially zero along the preferred axis. The direction of this axis is identical with the direction toward the Virgo cluster and lies exactly along the axis of the Local Supercluster filament of which our Galaxy is a part. So, any anisotropy of measured signals is mainly aligned with the Local Group, i.e., is a local contamination that distorts the signal.

Plasma cosmology denies the original Big Bang and operates within an eternal space that allows a large-scale structure of the universe. In the plasma model, superclusters, clusters and galaxies are formed from magnetically confined plasma vortex filaments.

One more approach to understanding the origin of the measured background temperature was proposed by Aspden [474, 475]. He suggested that particles of the vacuum/ether lattice have a small mass (about 1/25 of the mass of the electron). But in having mass these particles must interact with the gravitational potential of the Sun and the Earth. Due to this interaction the ether lattice particles acquire the kinetic energy, which generates a temperature. Aspden's hypothesis is in fact correct. However, our estimates of the mass of the inerton aura around the Earth (see Section 8.4) show that such temperature ($k_{\rm B}T = GMm/r$ where M is the mass of the Sun/Milky Way and m is the appropriate induced mass in a cell of the ambient tessellattice) is a few orders less than 2.725 K. Of course one can say that if we put a larger value for the constant r_{01} (8.23), we will achieve the induced mass m in a cell of the tessellattice, which is close to the mass calculated by Aspden [474, 475], i.e., $m \approx 3.6\times10^{-32}$ kg. However, cells with such large mass will impede the motion of particles in the universe, which is not observed (recall the mass of electron, 9.1×10^{-31} kg, is close to Aspden's). Therefore, the induced mass in the inerton aura of the Earth is much smaller. Such mass can be in the order of 10^{-35}–10^{-33} kg, which is caused by the Sun, as is shown in Section 8.4. Then the effects due to the interaction $GM_{\rm Sun}m/r$ may be responsible for subsequent peaks in the measured thermal spectrum around 2.7 K assuming that the observed spectrum is correct.

Thus, the measured temperature of about 2.7 K may only be related to the infrared radiation of the Earth's oceans, which was pointed out by Robitaille [465–468], or be related to the temperature of interstellar space, which was considered by Eddington [471, 470] and Lerner [472, 473]. Some small corrections may be stipulated by the gravitational interaction of the Sun with standing inertons, which are present in the interstellar space of the Milky Way, and could be studied with an appropriate equipment.

10.2 DARK MATTER

Researchers have published so much about their studies, claims and discoveries of dark matter that the public has become increasingly confused by the continual absence of any clear or precise description of this phenomenon that satisfies their common sense. Unsurprisingly, therefore, the general community has responded with their own attempts to resolve this mystery. For example, a deep-space mystery movie series, titled Dark Matter, was broadcast by a Canadian TV company in the summer of 2015: a fictional group of six people with no memory of who they are and where they came from wake up on a starship … and begin trying to figure out what had happened to them.

Thus, the time for an accurate and intelligible description of dark matter is urgently required.

Against this background to solving the problem of dark matter, scientists are continuing to present new hypotheses. For example, Livio and Silk [476] point out: "Dark matter is living up to its name. In spite of decades of compelling evidence from astronomical observations showing the existence of matter that neither emits nor absorbs electromagnetic radiation, all attempts to detect dark matter's constituents have failed." Then they continue: "The presence of dark matter is inferred from its gravitational effects. Stars and gas clouds in galaxies, and galaxies in clusters, move faster than can be explained by the pull of visible matter alone. Light from distant objects may be distorted by the gravity of intervening dark material. The pattern of large-scale structures across the Universe is largely dictated by dark matter."

The majority of researchers associate hypothetical so-called weakly interacting massive particles (WIMPs) with the source of dark matter. They hypothesize that these particles have masses that are a few tens to thousands of times that of the proton. Accordingly, physicists think that candidates for WIMPs are some of the supersymmetric particles, supplements for the electron, proton and quarks. There are also other hypothetic supersymmetric particles – axions, warm dark matter particles, and hot dark matter particles, etc. All this is needed for the Big Bang concept with its early universe and its continual expansion. However, so far no supersymmetric particles have been found even in the Large Hadron Collider

(LHC). In connection with this, Schifman [3] recommended that physicists stop developing the supersymmetry theory, which increasingly looks like a mistaken concept. This automatically denies the Big Bang and the idea of the early universe becomes erroneous.

Luminet et al. [477] presented the universe as a simple geometrical model of a finite, positively curved Poincaré dodecahedral space, rather than an infinite, simply connected, flat space. The Poincaré dodecahedral space offers no free parameters in its construction; it is rigid, meaning that geometrical considerations require a completely regular dodecahedron. Roukema et al. [478] emphasize that in the universe with the form of Poincaré's dodecahedral space, the spatial correlation functions take precedence over angular correlation functions. Nevertheless, recent analysis of cosmological parameters based on studies of the Planck satellite [479] does not solely support this Poincaré dodecahedral space but allows several alternative cosmological topologies.

Now let us come to our consideration of space that is filled with mass generated by the standing inerton waves of cosmic bodies. The first link to this subject is work by Delort [480]: if mass is an available component of space (in his terms the ether has mass), then it provides a direct way to understanding dark matter and, moreover, it makes it possible to derive the baryonic Tully-Fisher's law.

So, what form does this dark matter take? Let us consider a model in which the inner entities of stars vibrate near their equilibrium positions [70, 481]. Due to the interaction with cells of the tessellattice the entities emit inertons, which collectively form the total inerton cloud of the star, inducing the star's Newton potential

$$U = -Gm / r \tag{10.2}$$

The star's inerton cloud is distributed around the star as a standing inerton wave, which spreads infinitely, i.e., as far as the boundary of the universe. In an ensemble of stars the inerton clouds of each star have to overlap. This means that standing inerton waves of opposing sources should undergo an elastic interaction. For example, we can observe a similar effect on the surface of water when water waves of two sources interfere, generating wavelets.

Therefore, the standing inerton waves that induce the gravitational potential (10.2) of stars additionally initiate an elasticity in the interstellar space owing to the mutual interference of counter propagating waves emanating from neighboring stars.

The Hamiltonian of interacting stars can be written in the form (5.24) prescribed by the lattice model

$$H(n) = \sum_{s} E_s n_s - \tfrac{1}{2}\sum_{s,s'} V_{ss'}^{\text{att}}\, n_s n_{s'} + \tfrac{1}{2}\sum_{s,s'} V_{ss'}^{\text{rep}}\, n_s n_{s'} \tag{10.3}$$

Here, s is the position of the appropriate knot in the lattice, which can be either occupied (the filling number $n_s = 1$) or not occupied (the filling number $n_s = 0$) by a star. E_s is the additive part of a star's energy (the kinetic energy) in the sth state. The potential $V_{ss'}^{\text{att}}$ represents the paired energy of attraction (due to the Newtonian gravitational energy $-Gm_s m_{s'} / r_{ss'}$) and the potential $V_{ss'}^{\text{rep}}$ is the paired energy of elastic repulsion (i.e., $\frac{1}{2} m_s \omega_s^2 r_{ss'}^2$). For the purpose of this model, all stars are considered to be of identical mass $m_s = m_{s'} = m$. The signs before positive functions $V_{ss'}^{\text{att}}$ and $V_{ss'}^{\text{rep}}$ in the Hamiltonian (10.3) directly specify proper signs of attraction (minus) and repulsion (plus).

The statistical sum of the system of interacting stars

$$Z = \sum_{\{n\}} \exp\left(-H(n) / k_{\text{B}}\Theta\right) \tag{10.4}$$

can be rewritten via the action S, which depends on three functions (see Section 5.2.1 for details),

$$Z = \text{Re}\frac{1}{2\pi i}\int D\varphi \int D\psi \oint dz\, \exp\left[S(\varphi,\ \psi,\ z)\right] \tag{10.5}$$

The action S, which allows a stellar cluster to be treated as an object obeying the Bose statistics, is written as below (see expression (5.59))

$$S \approx K \cdot \{a(\mathcal{N}) - 2b(\mathcal{N})\} \cdot \mathcal{N}^2 \tag{10.6}$$

where the functions a and b are defined as follows (5.57) and (5.58), respectively:

$$a(\mathcal{N}) = 3 \int_1^{\mathcal{N}^{1/3}} \tilde{V}^{\text{rep}}(gx) x^2 \, dx \tag{10.7}$$

$$b(\mathcal{N}) = 3 \int_1^{\mathcal{N}^{1/3}} \tilde{V}^{\text{att}}(gx) x^2 \, dx \tag{10.8}$$

In the case of the Boltzmann statistics the action for a cluster of $\mathcal{N}$ stars has the form (5.75)

$$S = K \cdot \{[a(\mathcal{N}) - 2b(\mathcal{N})]\mathcal{N} + \mathcal{N} \ln \xi_0\} \tag{10.9}$$

where ξ_0 is the fugacity.

Knowing the explicit form of the action (10.6) or (10.9), one can derive the equation for the number of stars gathered in a cluster, i.e., a galaxy: $\partial S / \partial \mathcal{N} = 0$, which in addition requires holding of the inequality $\partial^2 S / \partial \mathcal{N}^2 |_{\mathcal{N} = \mathcal{N}_{\text{in cluster}}} > 0$ (the minimum of the action S).

The absolute value of pair potentials that enter equations (10.6)–(10.9) appears as follows

$$V^{\text{att}}(gx) = \frac{Gm^2}{gx} + \frac{GMm}{R} \tag{10.10}$$

$$V^{\text{rep}}(gx) = \tfrac{1}{2} m \omega^2 (gx)^2 \tag{10.11}$$

where m is the mass of a star; g is the lattice constant, i.e., the distance between neighboring stars in the model lattice; x is the dimensionless distance defined through the relation $r = gx$; ω is the cyclic frequency of oscillations of the mass m near its equilibrium position. Besides, in expression (10.10) we add the second term, which is the background/mean gravitational field in the presence of which the galaxy was formed, where M is the total mass of the whole system of stars and R is the effective radius of the system.

It should be noted that a quadratic potential has already been used in the conformal theory of gravity to fit the observed behavior of galactic rotation curves [482]. However, the origin of the potential $V(r) = \gamma_0 c^2 r^2 / 2$

used by the researchers [482] was modeled as an effect of cosmology on individual galaxies though all parameters had a completely different meaning originating from an abstract metric of a theory of long-range action despite its quantization.

The solutions for clusters of stars can be obtained for all the dimensions [481]: 3D, 2D and 1D. All three cases were observed: bulk galaxies [483], disc galaxies [484] and rings/arcs galaxies [485, 486], respectively.

Perhaps the most interesting solutions are those for disc galaxies (2D) and bulk galaxies (3D). Starting from the action (10.6), we obtain for a disc galaxy the following solution

$$\mathcal{N} = \frac{16}{3}\frac{GM}{Rg^2\omega^2} \tag{10.12}$$

which means that all the stars of the disc galaxy are distributed as a plane cluster with the appropriate number $\mathcal{N}$ of stars. Let us consider some typical values of the mass, radius and distance between stars in the galaxy: $M = 10^8 M_{\text{Sun}}$, $R = 10$ kpc, $g = 1$ pc. Note these values of the parameters permit us to neglect the contribution of the term *Gm/g* to the solution (10.12). The cyclic frequency ω of oscillations of a star near its equilibrium position in the cluster can be estimated from the equality of the centripetal acceleration that attracts a star to the center of the galaxy and the holding force that acts between stars in the cluster:

$$GM / R^2 - \omega^2 g = 0 \tag{10.13}$$

From Eq. (10.13) we get

$$\omega = \left(\frac{GM}{R^2 g}\right)^{1/2} \approx 2.1\times10^{-14}\ \text{s}^{-1} \tag{10.14}$$

Let us now evaluate the acceleration that each star experiences in the plane cluster:

$$a = GM / R^2 \approx 1.4\times10^{-11}\ \text{m}\cdot\text{s}^{-2} \tag{10.15}$$

This value of the acceleration satisfies the conditions prescribed by Milgrom [382] in his MOND (Modified Newtonian Dynamics) theory (see also, e.g., Refs. [487, 488]): $a << a_0 \cong 1.21\times10^{-10}$ m·s^{-2}. Thus we do not need to assume an incomprehensible modernization of Newton's law, i.e., the substitution of the force **F** = ***ma*** by a significantly smaller force of undetermined nature $\mathbf{F} = m\cdot(\mathbf{a}-\Delta\mathbf{a})$ at $a << a_0$. Stars are distributed as clusters and each star is strongly bonded with the other $\mathcal{N}-1$ stars in the cluster. This bonding countervails the centripetal acceleration. Expression (10.13) demonstrates this balance of two competing forces. That is why a Keplerian law, $V \sim 1/\sqrt{r}$, is substituted for Milgrom's constant orbital velocity $V^4 = GMa_0$, which satisfies Tully-Fisher's law: the luminosity L of a spiral galaxy is proportional to the 4th power of the velocity v of stars in this galaxy.

Let us now discuss the 3D cluster solution

$$\mathcal{N} = \left(\frac{5}{2}\frac{GM}{Rg^2\omega^2}\right)^{3/2} \tag{10.16}$$

The parameters can be chosen as follows: the total mass $M \sim 10^{14} M_{\text{Sun}}$, the radius $R \approx 250$ kpc and the central mass density $\rho_0 = 3.85\times 6M_{\text{Sun}}\text{kpc}^{-3}$. Assigning for the mass of a star $m = M_{\text{Sun}}$, we obtain the mean distance between stars: $g = (M_{\text{Sun}}/\rho_0)^{1/3} = 4.25\times10^{16}$ m. Then the stability of the cluster in respect to its gravitational collapse is determined by the relation (10.13): the gravitational attraction of the stars to the cluster's centroid is retained by the elasticity of inerton waves in the cluster. The corresponding cyclic frequency of oscillations of stars at their equilibrium positions owing to the overlapping of their inerton waves is

$$\omega = \left(\frac{GM}{R^2 g}\right)^{1/2} \approx 3.82\times10^{-13}\ \text{s}^{-1} \tag{10.17}$$

The acceleration to the centroid, which each star experiences in the spherical cluster, is

$$a = GM/R^2 \approx 2.24\times10^{-8}\ \text{m·s}^{-2} \tag{10.18}$$

The acceleration (10.18) is opposite to the inequality $a << a_0$ needed for the use of MOND. Besides, the acceleration (10.18) is not equilibrated by the acceleration caused by the elastic interaction in the cluster of $\mathcal{N}$ stars: $\omega^2 g \approx 6.2 \times 10^{-9}$ m·s^{-2}.

By this means, the origin of so-called dark matter is perceived in the mutual interaction of stars through their standing inerton waves. Stars are enveloped in the elastic tessellattice of space. This elastic interaction via inertons is also responsible for the non-homogeneous distribution of stars, i.e., they assemble in clusters. Hence in a galaxy there must exist a non-equilibrium density distribution of mass. It is this regularity that has to be responsible for the correct rotation curve [489, 490]. By introducing a complicated distribution of mass in a disc galaxy, Feng and Gallo [489, 490] managed to obtain the correct rotational curve profile even without an introduction of dark matter. But where is the origin for a non-homogeneous distribution of mass? The answer is obvious: the elastic interaction affects the density distribution and influences the rotational curve profile bringing it to the law $\upsilon^4 = GMa_0$.

The basic concepts of gravitational lensing must also be modified – perhaps a point mass approach with a correction based on MOND or another model will require a substitution by an approach resting on the involvement of elastically interacting masses. In particular, it seems the deflection angle $\varphi = 4Gm/(c^2 r)$ of a point mass m, which includes the absolute value of the gravitational potential Gm/r, can be modified as follows

$$\tilde{\varphi} = \frac{4}{c^2}\left(\frac{Gm}{r} + \tfrac{1}{2}\omega^2 r^2\right) \tag{10.19}$$

In a cluster the second term in expression (10.19) tends to align the space deformed by the first term. This has to be typical for 2D clusters (i.e., clusters in disc galaxies). In the case of 3D clusters (rich clusters in galaxies) the second term may even prevail over the first one, namely, the second term prolongs the deflection angle φ for larger distances at which the first Newtonian term becomes negligible.

The solutions (10.12) and (10.16) include the mean-field gravitational potential induced by all the stars in the cluster (i.e., a galaxy) under consideration, *Gm/r*. What does that look like in space and can it be related to a mysterious black hole situated in the center of an appropriate galaxy?

10.2.1 BLACK HOLES

The presence of a black hole is determined through the measure of mass of the small region studied. If one finds a large mass concentrated in a small volume, and if the mass is dark, i.e., cannot be seen, then there is a real suspicion that there is a black hole there. Dopita [491] describes black holes as follows: Massive black holes lurk in the centers of elliptical and disc galaxies. Relativistic jets ejected by these black holes interact with the interstellar gas in their host galaxies, generating strong radiative shocks, and induce powerful bursts of star formation. If continued, such bursts would convert all the gas contained in the protogalaxy into stars in a roughly dynamical (orbital or collapse) timescale (~ 10^8 years). Astronomers have found that the central black hole and the stellar bulge of its host galaxy are intimately associated. In other words, black holes influence the galaxy's environment; there is a remarkable correlation between stellar dispersion velocity and black hole mass.

Astronomers [492] have revealed at the center of the Milky Way, between the constellations of Sagittarius and Scorpius, a bright and very compact astronomical radio source named Sagittarius A* (Sgr A*) with a mass of 4.3×10^6 solar masses and a size 44 million km. This source is associated with a real black hole.

Doeleman et al. [493] have also reported the discovery of a supermassive black hole with a mass of 6×10^9 solar masses in the galaxy M87 with a radius of $R_{\text{b.h.}} = 2\times10^{13}$ m, which generates a powerful jet of electromagnetic radiation in the range of sub millimeter wavelengths.

How does this compare with our inerton concept? A disc or elliptic galaxy can be described as a cluster of stars that possess a most distinctive property. Due to these stars extreme proximity to each other they occupy a small region in which their inerton standing waves interfere. Their inerton waves deliver mass to the central part of the cluster, a centroid, but in doing so the waves cannot spread further and, therefore, the mass μ is not transferred to the tension ξ, as Eqs. (8.17) and (8.18) prescribe. This centroid region appears as an endpoint of the standing inerton waves of each star of the galaxy; inertons have to reflect from this dense region. Such a region is of course accessible for observation and could be interpreted as a black hole. Nevertheless, inertons do not disappear in this point but establish a special stationary relief of mass from stars to the centroid endpoint supported by the equilibrium of the centripetal acceleration (generated by orbital rotation of the stars within the cluster) and the centrifugal acceleration (due to the linear response to the inerton wave pressure produced by the elastic tessellattice).

In Figure 10.1 the distribution of mass around stars is shown. If a star is motionless or moves in a straight line, the distributed mass potential is symmetric (Figure 10.1*a*). If stars rotate, i.e., move in orbits, they induce strictly asymmetric mass potentials (Figure 10.1*b*). For example, let us examine the Milky Way, whose disc radius is $R = 15$ kpc $= 4.629\times10^{20}$ m and the total mass is around $M = 2.6\times10^{12} M_{\text{Sun}}$. In this case using the relation (8.23) we get for the inerton mass in the center of the galaxy $\mu_{\text{inside}} \approx 10^{-33}$ kg (which is one order greater than the Sun's inerton mass at the Sun's surface), while outside the galaxy each star at the same distance

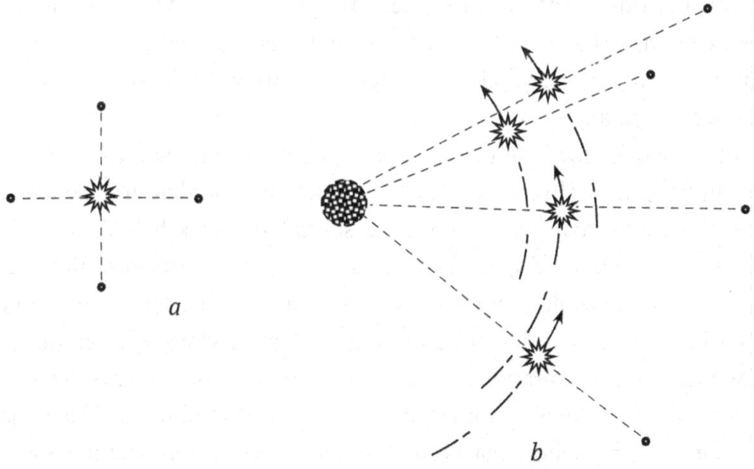

FIGURE 10.1 Distribution of mass around stars caused by standing inertons (see expression (8.23)). *a* – the symmetric mass aura around the motionless star or the star that moves rectilinearly; *b* – strictly asymmetric distribution of mass around rotating stars. The dense inerton spot is located in the center part of the galaxy.

R induces in the space only the mass $\mu_{\text{outside}} \approx M_{\text{Sun}} r_{01} / R \sim 10^{-46}$ kg, which is 13 orders less than μ_{inside} (here, we assume for simplicity that the mass m of each star is equal to the solar mass M_{Sun}).

Geometric and physical conditions impose some requirements to the dense inerton region at the centroid of the Milky Way, namely, a high density ρ and a small radius $R_{\text{b. h.}}$ (this 'black hole' radius is $R_{\text{b. h.}} = 2.2 \times 10^9$ m [492]). Putting, for instance $\rho \approx 2 \times 10^8$ kg·m^{-3}, we obtain the mass of the 'black hole' $M_{\text{b. h.}} \approx \frac{4\pi}{3} R^3_{\text{b. h.}}\, \rho \approx 8.6 \times 10^{36}$ kg, which is 4.3×10^6 solar masses as has been mentioned above for the black hole of the Sgr A* source [492] in the Milky Way. The chosen density of inertons means that their concentration in the 'black hole' is $n = \rho / \mu_{\text{inside}} \sim 10^{41}$ m^{-3} and the total number of inertons that form this 'black hole' is $M_{\text{b.h.}} / \mu_{\text{inside}} \sim 10^{67}$. In a solid the inerton concentration is around 10^{30} m^{-3} (the concentration of cells in the tessellattice is $\ell_{\text{P}}^{-3} \sim 10^{104}$ m^{-3}).

Since the Milky Way has around 10^{12} stars and the inerton concentration in each star is no less than that in a solid, we indeed may anticipate that at the centroid of the galaxy the inerton concentration reaches $10^{41} - 10^{42}$ m^{-3}, which is enough to accumulate there the mass equal to $4.3 \times 10^6 M_{\text{Sun}}$. This mass of inertons in the centroid, i.e., $M_{\text{b. h.}} = 4.3 \times 10^6 M_{\text{Sun}}$, which is used in expressions (10.10) to (10.18), is directly equivalent to the proportion of invisible dark matter that is required to explain the rotational dynamics of the whole galaxy.

Thus, the described region of the dense mass in space at the center of each disc/elliptic galaxy (Figure 10.1, *b*) is an actual candidate for the observed unusual astrophysical peculiarities, the so-called 'black holes'. In fact, this dense inerton region of space possesses a huge mass. Besides, this mass is twisted in a vortex following the stars. Therefore, in the presence of a gas in this region, one can anticipate collisions of fast inertons with atoms of the gas, which will lead to short-lived inerton excitations of atoms with the resultant emissions of intense photons across a wide spectral range. The situation is similar to such phenomena as sonoluminescence (ultraviolet radiation, see Section 5.10), sonofusion (gamma radiation, see Section 7.4) and the generation of X-ray (at the formation of electron droplets, see Section 5.6).

The magnitude of the mass of an inerton in the center of our galaxy $\mu_{\text{inside}} \approx 10^{-33}$ kg indicates the existence of a natural threshold value of the number of stars, their mass and the size of the galaxy. In fact approaching the value of μ_{inside} to the mass of the electron m_e will mean the effective saturation of the galaxy by stars, because at $\mu_{\text{inside}} \to m_e$ the space in the central part of the galaxy becomes so rigid that any free motion of material particles will be impossible. For example, μ_{inside} becomes equal to $0.1m_e$ at the total mass $M = 10^{16} M_{\text{Sun}}$ of a galaxy with the same radius $R = 15$ kpc. In such a viscous space, the motion of a free electron is hardly possible.

Of course in the present Chapter we have only outlined the phenomenon of the dense inerton region. A tremendous amount of work is needed to formulate and study the problem(s) in detail.

10.2.2 VISCOSITY OF SPACE

Space having inerton mass generates a number of other interesting phenomena. One of them is the long-standing problem about the additional enigmatic acceleration of artificial satellites in the solar system, which has been debated for years. In particular, researchers have discussed some Doppler shift anomalies for the Pioneers 10 and 11, Voyager 1 and 2, Galileo, NEAR, Cassini, Stardust, Rosetta, Hayabusa, and MESSENGER, and many other flyby anomalies [494].

Based on data for the Pioneers 10 and 11 trajectory determination and more detailed studies of all the systematics, a total error budget, i.e., an acceleration pulling the satellite Pioneer 10 to the Sun was determined as $a_{\text{Pioneer}} = 8.74 \times 10^{-10}$ m·s^{-2} [495]. However, later on the same researchers changed their opinion [496]: they mention that since the spacecraft was facing the Sun, the solar energy was absorbed primarily by the high-gain antenna that largely shadowed the rest of the spacecraft from direct solar irradiation; the absorbed energy increased the high-gain antenna temperature and was then emitted as infrared radiation. The contribution of this thermic radiation emanating from this side of the antenna was estimated to be of the same magnitude as its Doppler shift.

However, the hypothesis of the thermal origin of the Pioneer 10 anomaly [496] has a significant flaw, because with increasing distance of the satellite from the Sun, the heating by the Sun would decrease and hence the effect of the thermic radiation of the antenna should have become weaker. Unfortunately, the authors of the hypothesis did not pay attention to this fact.

Lämmerzahl et al. [494] have discussed other anomalies, such as the increase of the astronomical unit (the distance between the Earth and the Sun), the quadrupole and octupole anomaly, and dark energy and dark matter.

It seems likely that all the anomalies mentioned above are directly associated with the inerton elasticity of space, which is generated by the gravitational interaction of cosmic objects. In fact, a spacecraft moves through a viscous network formed by the standing inerton waves of interacting stars, planets and other cosmic objects. Of course the viscosity of space has to slow the movement of satellites (through the switching of terms $\kappa\dot{r}$, $\frac{1}{2}\gamma r^2$ and the Navier-Stokes equation), which so far has never been taken into account by the teams that directed spacecraft. In particular, expressions (10.15) and (10.18) allow for the introduction of the viscosity of space in our galaxy related to the Milky Way's "dark matter."

So any planning of spacecraft trajectories (and the necessity for mid-course corrections) requires consideration of the viscosity of space, which is caused by the presence of standing inerton waves. In other words, the vacuum in which spacecraft move does have a significant component of mass.

10.3 DARK ENERGY

Observational cosmology has established a list of cosmological parameters, which were formed on the basis of the Friedmann model, determined to one or two significant figure accuracy [497, 498]. One of these is the cosmological constant, represented by the letter Λ (lambda) that is currently associated with a vacuum energy or dark energy in empty space that is used to explain the contemporary accelerating expansion of space against the attractive effects of gravity. The cosmological constant Λ is characterized in terms of negative pressure

$$P = -\rho c^2 \qquad (10.20)$$

This pressure is added to the stress-energy tensor of general relativity, which causes accelerating expansion in the point of space studied.

The fraction of the total energy density in the universe related to dark energy, is currently estimated to be $\Omega_\Lambda = 69.2 \pm 1.2\%$ [499].

As we mentioned above, the Friedmann model and the related Big Bang theory, which are not bound to the structure of space, meet severe difficulties (see also Ref. [68]), namely: the problem of singularity, the problem of horizon, the flatness problem, the monopole problem and so on.

In Chapter 1 we discussed our universe as a huge cluster of topological balls, which are arranged as a spheroid with the tessellattice structure. In the tessellattice the size of a ball has gradually to increase from the center to the periphery, as for instance, expression (1.94) prescribes. In the continuous approximation at a distance r from the center the size of the appropriate cell of the tessellattice can be presented in the form $\ell(r) \to \ell \cdot (1 + r / R)$ where ℓ is the size of the cell in the center of the universe, which can be equal to the Planck length ℓ_P. Let us introduce the following designation for the dimensionless parameter: $x = r/R$.

Why do the size of balls increase from the center to the periphery? It is obvious that any substance has a boundary and, hence, that must include our universe too. The universe is similar to a fluidic matter ball, a huge droplet. That is why it is characterized by the Young-Laplace pressure, $2\gamma/R$, where R is the radius of the universe and γ is the surface tension of the universe. For this reason the Young-Laplace pressure dominates in the universe and hence the total pressure in the universe varies from the center to the periphery.

By definition, the pressure is the force $\boldsymbol{F}$ that affects the area A. The surface of a ball located in the center of the universe has the area $\pi\ell^2$. A ball located at a distance r from the center of the universe has the area $\pi\ell^2 \cdot (1 + x)^2$. Therefore the ratio of the pressure P_0, which is experienced by a cell in the center of the universe, to the pressure P, which is experienced by a cell situated at the distance r from the center, results in the relation

$$P = P_0 / (1 + x)^2 \tag{10.21}$$

Previously [70] it has been considered how the density of a massive object may be changed in its movement from one point to another by using the expression (8.22) that describes the distribution of the mass field around the object. Nevertheless it seems it is even more relevant to consider the density of a particulate cell that is responsible for the appearance of mass.

By definition, the mass of a particulate cell relates to the ratio of the degenerate volume and the volume of the particle, $m_0 \propto \mathcal{V}^{\text{deg}} / \mathcal{V}^{\text{particle}}$. That is why the density of the massive particulate cell is $m_0 / \mathcal{V}^{\text{deg}} \propto 1 / \mathcal{V}^{\text{particle}}$, which means that the density of a particulate cell is rather constant, ρ_0; it is not changed in its motion from point to point.

The invariable of the density of a particulate ball, ρ_0 (the same for a ball in which an inerton excitation is located), and the relation (10.21) allow us to reformulate the cosmological equation (10.20) as below

$$\frac{P_0}{(1+x)^2} = \rho_0 c^2 \tag{10.22}$$

from which we obtain

$$c(x) = \frac{\sqrt{P_0 / \rho_0}}{(1+x)} \tag{10.23}$$

Here, putting $\sqrt{P_0 / \rho_0} = c$ for the speed of light near the center of the universe, the relation (10.23) immediately shows that with a distance r approaching the boundary of the universe R the speed of light drops down reaching the minimum value at the boundary equal to $\frac{1}{2}c$ (recall $x = r / R$).

Let us now consider Hubble's law, which is expressed by the equation $\upsilon = H_0 D$ [500], where H_0 is the Hubble constant and D is a distance; υ is the recessional velocity in km·s^{-1}. The parameters H_0 and D are not directly measured. They are found through a supernova brightness, which provides data on its distance, and the redshift $z = \Delta\lambda / \lambda$ of its spectrum of radiation (the fainter and smaller a galaxy appears, the higher is its redshift, which typically is due to the Doppler effect). Hubble tied the brightness and redshift parameter z suggesting a surprising correlation

$$H_0 D = zc = \frac{\lambda_{\text{measured}} - \lambda_{\text{source}}}{\lambda_{\text{source}}} c \tag{10.24}$$

which suggests that more distant galaxies are moving faster away from us.

Thus, by the relationship (10.24), the more distant a galaxy is, the faster it is moving away from us, which means that the recessional velocity of a galaxy is proportional to its distance from us, $\upsilon = H_0 D$. The relationship (10.24) has been proved experimentally and is considered now as the major argument supporting the expansion of our universe.

However, let us rewrite the relationship (10.24) in forms

$$\lambda_{\text{source}} = \frac{\lambda_{\text{measured}}}{1 + H_0 D / c}, \; c / \nu_{\text{source}} = \frac{c / \nu_{\text{measured}}}{1 + H_0 D / c} \qquad (10.25)$$

From the second version of expressions (10.25), we can see another explanation of the phenomena of the recessional velocity is also possible. Namely, we may shift the change in the frequency, $\nu_{\text{source}} \rightarrow \nu_{\text{measured}}$, to a change in the speed of light: at a distance *D* from a terrestrial observatory the velocity of light can be different from the value of *c* measured in the central part of the universe. The appropriate relationship becomes

$$c(D) = c \cdot \frac{1}{1 + H_0 D / c} \qquad (10.26)$$

It is easy to see that the relationship (10.26) can be related to expression (10.23). In fact, let us check the equality $1 / R = H_0 / c$. Applying the Hubble constant H_0 = 68 km·s^{-1}×Mpc $= 2.2 \times 10^{-18}$ s^{-1} and the speed of light $c = 2.9979 \times 10^8$ m·s^{-1}, we obtain for the radius of the universe $R_{\text{univ}} = 1.36 \times 10^{26}$ m that is very close to the 46 billion light years declared by astrophysicists for the radius of the universe. Hence the tessellattice model of cosmology naturally interprets Hubble's law as a gradual decrease of the speed of light towards the periphery of the universe. Therefore, the observed redshift is a direct consequence of the speed of light having decreased owing to the progressive reduction of the concentration of cells of the tessellattice in the peripheral framework of the universe (compare with the behavior of the sound speed in the sea in which there is a tendency towards increasing sound speed at increasing depth, due to the increasing pressure in the deep sea). This also means that there is no dark energy in the universe, which allegedly accelerates galaxies after the hypothetical Big Bang.

The problem of the so-called accelerating expansion of the universe with high-redshift supernovae $z \approx 0.5$ [501, 502] has a natural elucidation in the framework of the present submicroscopic concept. The velocity of light is a function of the distance r from the center of the universe and the velocity $c(r)$ gradually decreases and especially dramatically drops near the boundary of the universe reaching the minimum value $\frac{1}{2}c$ at the boundary. This means that the observed "accelerating expansion of the universe" can be accounted for by an abrupt dropping of the speed of photons in the vicinity of the boundary of the universe where the 42 Type Ia supernovae were detected. In fact, the derivative of the speed $c(r)$ (10.23) with respect to distance r is negative,

$$c(r) = c/(1 + r/R) \Rightarrow c'(r) = -\frac{c/R}{(1 + r/R)^2} < 0 \tag{10.27}$$

which means a deceleration of photons at that point. However, if one does not take into account an abrupt reduction of the speed of light near the periphery, the supernovae could be perceived by an observer on the Earth to be accelerated in the direction from the center of the universe.

Summarizing the section, we shall state that the phenomenon of dark energy relates to the shape of the universe. The spherical universe being a substrate – a huge cluster with the tessellattice structure – experiences the surface tension described by the Young-Laplace pressure $\sim 1/R$ that induces a relevant inhomogeneity in the universe, which in turn influences the motion of different deformations (i.e., matter). As a result, the speed of light c becomes a function of the radius R of the universe and twice decreases near the universe's surface.

10.4 INERTON ASTRONOMY

At a time when some scientists have been exploring space with the help of high-precision equipment installed on spacecrafts to identify the correct cosmological parameters, other scientists working in terrestrial laboratories have been investigating instabilities of cosmological physical factors that affect different physical, chemical and biological processes.

Particularly productive researches on the detection of unstable cosmological factors were conducted by Shnoll et al. [503–506]. For example, their analysis of a six-year data-set of alpha-activity of ^{239}Pu, with a total of about 60,000 measurements, showed regular patterns in histograms (recording counts per minute), namely, the highest probability of similar nearest neighbors (the effect of near-field) and a clear circadian periodicity. They also observed a change in timing between measurements of α-activity of ^{239}Pu on a ship in the Indian Ocean and in the laboratory near Moscow in 1988. Furthermore, there was a direct correlation with the position of the Moon in the sky (especially at the rising and setting of the Moon) and the form of histogram data of alpha decay of ^{239}Pu.

A similar coincidence of histograms was observed in recorded measurements of γ-activity of ^{137}Cs at the Columbus Polarity Therapy Institute in Columbus (USA) by M. Sue Benford and J. Talnagi and in α-activity of ^{239}Pu by S. Shnoll in Pushchino (Muscovy) during the period of 18–20 February 2001, separated by a distance of 12,000 km.

In the study of histograms (constructed for measurements of alpha activity of ^{210}Po) periodic signal repetitions were detected, as "macroscopic fluctuations," which were associated rather with a sidereal day (1436 minutes) but not the solar day (1440 minutes). Therefore, many months of continuous studies demonstrated periodic cosmic fluctuations in the intensity of other fundamental physical processes that manifested themselves every 546 hours (22.75 days), 576 hours (24 days), 648 hours (27.0 days) and 672 hours (28 days).

Later on Shnoll and co-workers carried out experiments with collimators directed towards the Polar Star and the Sun, the West and the East, as well as with clockwise and counterclockwise rotations, then with GCP-generators, electronic noise generators, etc.

Shnoll [504] concluded that factors determining the shapes of modeled histograms may help to reveal determining factors of physical processes, which in turn can be attributed to the motion of the Earth through cosmic space.

Similar results have been obtained also by other research teams [507–510], which associate the observed phenomenon of variation of the decay rate of radioactive nuclei with the annual variation of the distance to the Sun and solar tides.

Our R&D team detected fluctuations of the decay constant of tritium caused by non-stationarity cosmic factors of unknown origin; we found that the constant varies from 0.24 to 0.25 decays per second with periodicity around 24 hours [70].

The described changes in nuclear decay related to non-stationary and/or periodic cosmic processes can be caused only by the presence of inertons.

Evgen Andreev (Inst. of Physics, Kyiv) reported in private communication his results on the use of liquid flocculating agents for cleaning water. He conducted studies every day throughout the whole of January in 2013 and revealed the following picture. Initially flocculants precipitated impurities available in the water to the bottom of the test-tube, which was observed until January 15. However, after January 15, the picture changed: from day to day a cloud of impurities gradually rose to the surface, though the cloud became very blurred. From January 19 the cloud of impurities gathered at the surface as a thin layer. During the next four days the cloud of impurities floated at the water surface, but it was not so thin as during January 19.

There is an interesting coincidence here with January 19 also being the Eastern Orthodox Church day of Epiphany (St. John the Baptist Day), a day of purification by water ('water immersion or christening') that can be traced back thousands of years in the territory of modern Ukraine.

The above-described physical phenomenon can be related to the Earth passing through an inerton anomaly possibly caused by some cosmic bodies in the solar system or even from the regular emitters within our galaxy because such an observed behavior of water is possible only if the water sample absorbs inertons, which we have discussed in Chapter 5. It will be interesting to repeat Andreev's experiment at another laboratory. If there is an annual pattern to this result, due to solar inertons, then it seems in half a year the Earth has again to cross this global inerton anomaly (or could it be annual?) and the sample of water studied will restore its initial properties that were observed by Evgen Andreev in the beginning of January. When does it happen? In July, after Midsummer, which might be associated with Ukrainin's ancient holiday of Kupala (many rites of the holiday of Ivan Kupala who originated from Yuvan Gopal, young Krishna, are connected with water, fertility and purification).

One more related study of cosmic factors is the pioneering work of Kozyrev [511, 512] who demonstrated the presence of a remote influence of a star (Procyon), which affected a signaling sensor (based on a resistor incorporated in an electrical bridge circuit) put in the focal volume of the telescope, much earlier than the star became visible in the telescope, i.e., it responded to a superluminar signal. Those results of the 1970s then were confirmed by other researchers [513, 514] who noted that both Kozyrev's and their own observations of planets, stars and galaxies showed that in fact there exists a kind of interaction, which modern physics does not consider. In their opinion the investigation of this kind of the interaction has an important bearing on the development of ideas of physics about the real constitution of space-time.

So, what was the reason for Kozyrev's enigmatic superluminal rays? It seems plausible that they are related to free inertons irradiated by stars, i.e., inertons escaping from the appropriate inerton clouds of stars and planets at some inner fast non-adiabatic processes.

Free inertons liberated from the inerton clouds of particles can migrate through space until other massive objects absorb them. Since inertons are carriers of the property of mass of matter, it will be interesting to launch astrophysical observatories incorporating appropriate inerton antennas, receivers, electronics and software. Inerton observatories will be able to yield pictures of the universe derived from inerton radiation (instead of focusing attention on the existing laboratories constructed for the detection of hypothetical gravitational waves, incomprehensible dark matter particles and other exotic particles).

Figure 10.2 demonstrates the 'Rudra' device designed by our research team, which measures inerton signals. In this device we use two kinds of the antenna: a ferrite rod-type core and a disc-type piezoceramic. The antenna is direction sensitive. The antenna absorbs a signal and the density of the antenna's material increases locally for a short time. As we have shown in Chapter 5 (in particular see Sect. 5.4.3), the absorption of inertons decreases the amplitude of oscillations of the appropriate atoms; this has to result in the separation of the appropriate local mode. In a magnetic material, due to the magnetoelastic coupling between the magnetic moments and the lattice the spin waves are modified in two different ways, i.e., the static deformations of the crystal, and the dynamic time-dependent

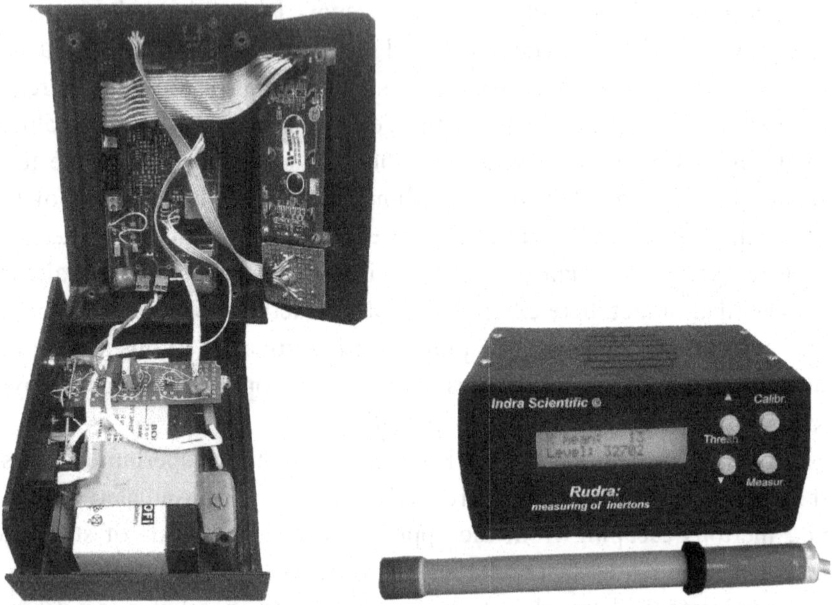

FIGURE 10.2 The 'Rudra' inerton measuring device that counts inerton signals received by the antenna per second (i.e., measures the intensity) and shows the spectrum of absorbed inerton signals from a few Hertz to 100 kHz.

modulations of the magnetic moments that interfere with the lattice vibrations (a similar situation takes place in a piezoelectric material, which is also characterized by the piezoelectric-elastic coupling). Besides, the exchange integral J, which characterizes the overlapping of spin wave functions of magnetic atoms, in the submicroscopic consideration appears as the overlapping of inerton clouds of the magnetic atoms studied, i.e., J depends on the number of inertons and their mass.

The sensitivity of a radio wave antenna is restricted by the intensity $10^{-6}\,\mathrm{W{\cdot}m^{-2}}$, though the threshold of hearing sound intensity is $5\cdot10^{-13}\,\mathrm{W{\cdot}m^{-2}}$. The intensity of an inerton signal, which falls on an antenna, can be evaluated as follows. The intensity $\mathrm{W{\cdot}m^{-2}}$ = energy/(time × volume). In our case the energy is μc^2; the speed of inertons can be equal to the speed of light (for bound inertons) or can be larger (for free inertons) and for the inerton mass we can put $\mu = 10^{-50}$ kg. The time interval chosen for the measurement is 1 s. The volume is a volume occupied by the crystal lattice cell, $g^3 \approx (3\times10^{-10})^3$ m^3. These values allow us to estimate roughly the intensity of inerton signals that fall on an antenna: $I = 10^{-5}$ to 10^{-1} W·m^{-2}, which is an intensity that can easily be measured by a conventional ferrite antenna with a coil or piezoceramic chip.

In the 'Rudra' device, an amplifier increases changes of the current induced by inertons to a level sufficient to record the signal. The frequency divider, interchanging switch and timer perform the adaptation between the intensity of the signal and the discharge grid of the recorder. The device is designed on the principle of a counter: it counts arriving inertons that strike the antenna per second (or several seconds). The signal is then converted from the time domain to the frequency domain through a Fourier conversion. Digitizing the analog signal makes it possible to determine the frequencies of inertons present in the measured signal. The frequency is related to the inerton mass through a conventional relationship: $h\nu = \mu c^2$.

When the antenna is put in a metal box that screens it of electromagnetic waves, the device continues to operate counting inerton signals. However, if a standard radio antenna used for conventional radio reception is placed in the same metal box, the receiver stops to function altogether.

We have also designed a device that can be called an inerton gravity-gradient meter. Its antenna consists of two quartz crystals functioning at the frequency of 10 MHz with an accuracy of 10^{-12} Hz. Each crystal is set in its proper metal casing that completely screens the detector from the electromagnetic environment. A change in the orientation and/or position of one of the detectors immediately shows a disharmony between the resonance frequencies of 10 MHz. Since the device measures only inertons, this disharmony correlates with the density of an inerton pulse or inerton flow received by the detector.

In Figures 10.3–10.5, we present results of our measurements as a function of the antenna orientation and the time of measurement. The measurements were performed with a predecessor of the 'Rudra' device and were on purpose conducted under difficult conditions in a windowless room in a concrete building. Obvious measurements of the Earth's inerton field exhibit a pronounced orientation effect: signals picked up along the East-West line are more intense than signals received along the North-South line. The gradual decrease in the intensity of inertons along the East-West line and the increase of the intensity along the North-South line with time is associated with the proper rotation of the Earth when the

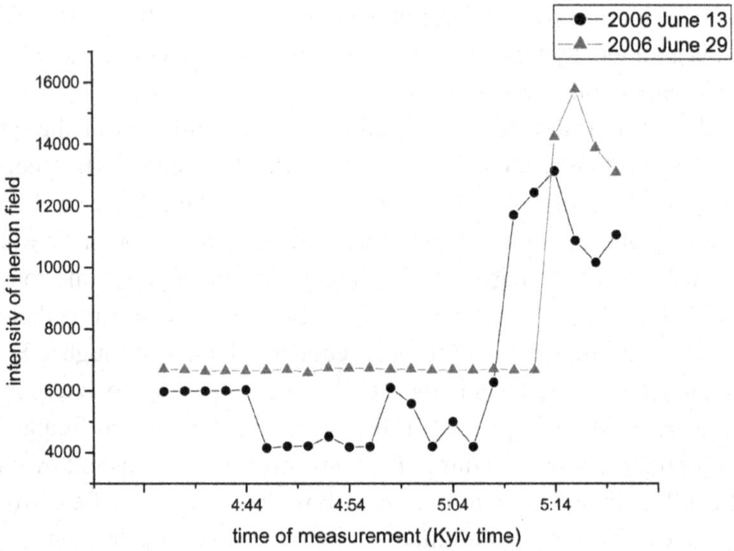

FIGURE 10.3 Intensity of inerton signals versus time of measurement. The ferromagnetic antenna was tuned along the East-West line. Sharp increases correspond to the time of sunrise (on both occasions the nights were clear and days were sunny).

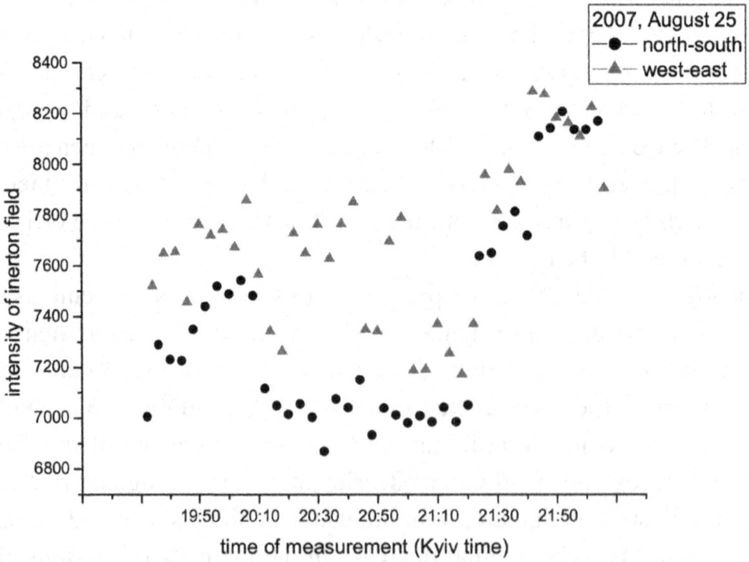

FIGURE 10.4 Intensity of inerton signals versus time of measurement. The ferromagnetic antenna was tuned along the East-West line (triangles) and along the North-South line (circles). The inerton atmosphere was relatively noisy.

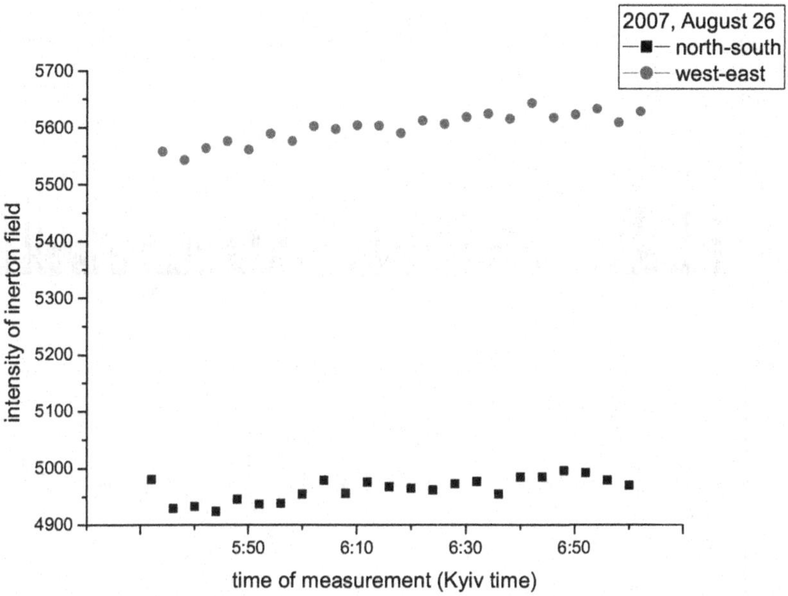

FIGURE 10.5 Intensity of inerton signals versus time of measurement. The ferromagnetic antenna was tuned along the East-West line (circles) and along the North-South line (squares). A clear morning.

antenna orientation with respect to the Sun gradually changes. In the case of the East-West orientation of the antenna, the dip angle of inertons coming from the Sun begins to depart from normal, which results in a decrease in absorption of inertons. In case of the North-South orientation of the antenna, the dip angle of inertons coming from the Sun tends to the normal giving rise to an additional absorption of inertons.

With the measurement of terrestrial inertons we can trace periodic modulations caused by the influence of solar inerton rays, Figures 10.6–10.8. Fluctuations may be associated with the influence of solar inerton activity. Solar inertons introduce appropriate changes to the intensity of terrestrial inertons. These changes vary from 5 to 20% probably owing to the rotation of the Earth. The most significant rise in inerton intensity is observed during the daylight period of the astronomical day. Meteorological factors also severely affect the measurements, which means that clouds saturated with water, the motion of clouds strengthened by wind and the rain are additional sources of inerton radiation on the surface of the Earth.

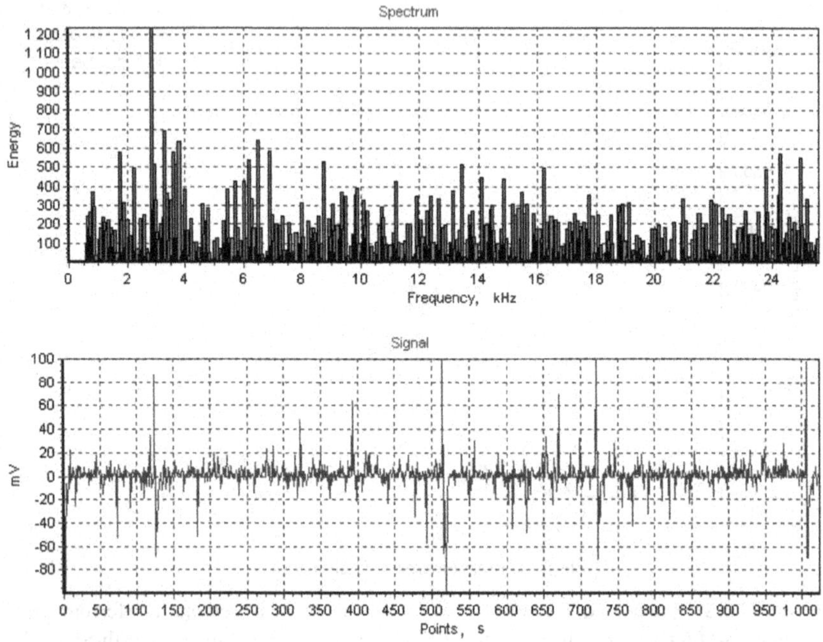

FIGURE 10.6 Record of inerton signals (bottom) and its frequency spectrum decoding (top) in the dark morning time at 5:57 a.m. In the upper histogram the grey bars show the spectrum of inerton impulses while the black bars depict gradients of these impulses. The antenna is the piezoceramic disc.

Figures 10.6–10.8 exhibit the results of measurements obtained with our modern version of the 'Rudra' device with the orientation of the antenna along the East-West line. The measurements were carried out in a cabin in a village south of Kyiv, Ukraine (latitude: 50 North, longitude: 30.5 East). There were no electromagnetic power sources in a radius of 25 km. The date of the experiment: 26 November 2011.

On orientation of the antenna's surface along the East-West line the device recorded 82 ± 10 impulses per second with a ferrite antenna with a coil and 90 ± 10 impulses per second for the piezoceramic sensor. At the orientation of the two antennas along the North-South line, there were recorded 76 ± 10 and 69 ± 10 impulses per second, respectively.

One hour later in the dawn time we could see that the intensity of inertons significantly increased, as the energy of inerton impulses reached 5000 a.u. (arbitrary units). When the antennas were aligned along the East-West line

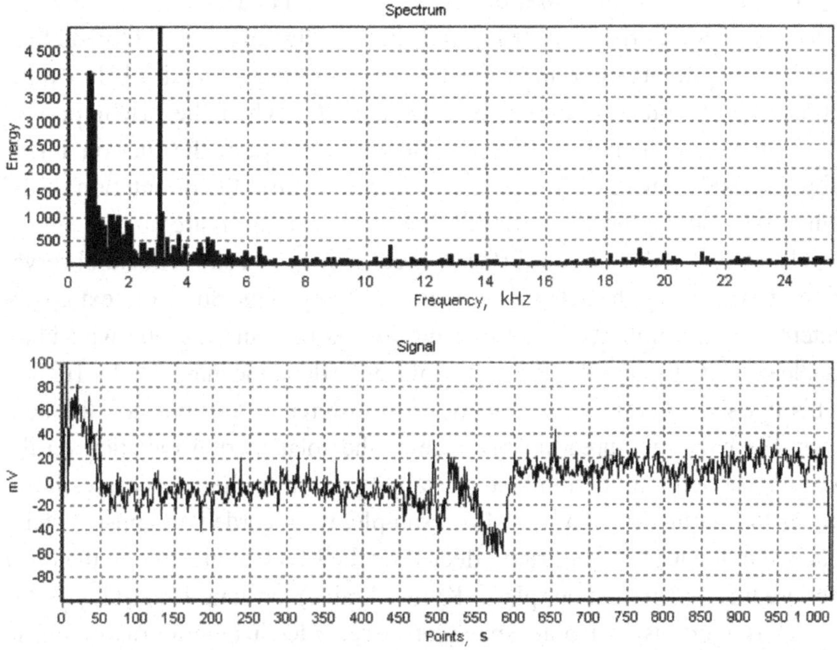

FIGURE 10.7 Record of inerton signals (bottom) and its frequency spectrum decoding (top) in the grey morning time at 6:56 a.m. The antenna is the ferrite rod inserted in a coil.

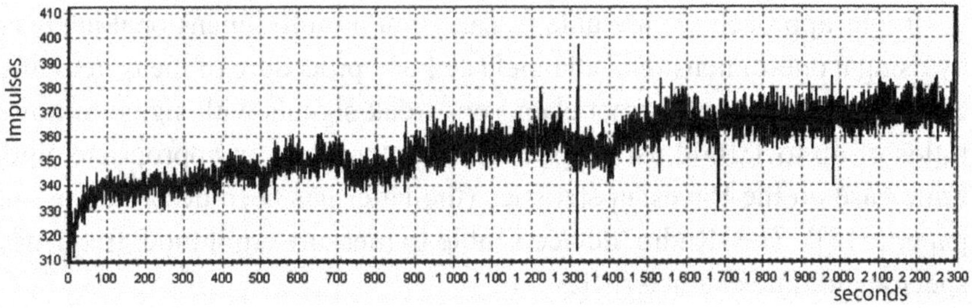

FIGURE 10.8 Experiment started in 6:55 a.m. (just after the sunrise) and lasted 2300 seconds. The trend depicts the increase of the intensity of inerton impulses coming from the Sun. The antenna is the ferrite rod inserted in a coil.

the intensity of the inerton field increased by 400% (340 ± 30 impulses per second), while with the orientation along the North-South line it showed an increase 82% (92 ± 10 impulses per second).

Based on our measurements, we may conclude that the spectral decomposition of the recorded signals in the range of a few Hz to 100 kHz exhibits the greatest inerton intensity between about 2 and 4 kHz (influenced by the Sun, as shown in Figure 10.7). When the orientation of the antennas coincides with the East-West direction, the inerton intensity across the range of frequencies is greater than if the orientation of the antennas is along the North-South line (as shown in Figure 10.6).

The 'Rudra' device can further be redesigned for astronomical observations. In particular, the behavior of inerton activity of the Sun looks extremely interesting and important; because the Sun is not only a globe with inner nuclear reactions, it is also the source of a huge dynamic mass, 1.99×10^{30} kg. This mass irradiates flows of inertons of different intensities and frequencies. We have measured an influence of the solar inerton radiation on the Earth's background inerton field. Instability of atmospheric processes influences the Earth's inerton aura. For example, we recorded an unusual sharp change in the inerton spectrum affected by the Moon at the time of the lunar eclipse that occurred at our place (Kyiv, Ukraine), on June 15, 2011.

Thus, inertons from outer space can be recorded in laboratories on Earth. We may study the intensity of inerton signals, the direction from which they arrive, their frequency, the degree of homogeneity of signals and we may also measure the velocity of free inertons, which by preliminary estimation, is about two orders higher than the speed of light c.

In our approach we are able to carry out a multidimensional analysis by using a one-dimensional channel (i.e., one procedure of measurement). Subsequently we can extract even very weak signals with signal to noise ratios of up to 1:1000 using special technology and an appropriate algebraic base on the Bayesian statistics (the latter has been described in our paper [515]). The 'Rudra' device is able to measure amplitude, frequency spectra and time characteristics.

These findings allow one to claim the possibility of designing the first inerton observatory, which will be able to give amplitude, spectral and time characteristics of non-stationary inerton signals in real time. Such a system will record inerton signals from the Earth's atmosphere, from the Earth, and the Sun, as well as from distant objects in the celestial sphere by means of different methods, namely:

- measurement of anomalies by using detectors on the basis of a piezo-ceramic disc or and/or a ferrite rod with a copper coil in a range of Hz up to hundreds of MHz;
- measurement using a high precision inerton gravity-gradient meter made on the basis of quartz crystals; the vertical gravity gradient data can be gauged with an accuracy better than 0.1 Eötvös;
- measurement of cosmic anomalies by means of the point-by-point gauging of background radiation (when a single measurement can be transmitted to a multidimensional analysis of signals with a sensitivity of 10^{-10} impulses);
- measurement using of Terra-hertz receivers.

In conditions of fixed or variable geometry an inerton observatory will distinguish an anomaly that by its characteristics is significantly weaker than the general background (i.e., a very weak signal will be separated from the noise of the Earth, a star, nebula, etc.). All this can be done with single measurements that will also include directional-sensors. Although the noise can be significantly higher than the signal that we will be searching for, our method [515] will allow one to extract a real signal(s) from the recorded spectrum. Thus, with such a system consisting of a tracking facility and instruments to investigate the four experiments mentioned above, an era of inerton astronomy may be launched that will more meaningfully resolve the mysteries of dark energy and dark matter, and fully establish the submicroscopic mechanics that define the material and energetic components of our universe.

KEYWORDS

- **black holes**
- **cosmic microwave background radiation**
- **dark energy**
- **dark matter**
- **inerton astronomy**

CHAPTER 11

PROSPECTS FOR FURTHER STUDIES AND APPLICATIONS

Physical cosmology is one of the most attractive and enigmatic disciplines, which studies the largest-scale structures and dynamics of the universe. It is concerned with fundamental questions about the origin, structure, evolution, and ultimate fate of the universe. In the past, it was a branch of metaphysics and religion.

Modern physical cosmology began with Einstein-Hilbert's theory of general relativity a hundred years ago. Although being highly abstract, the theory had several predictions that were proved by astronomical observations. These advances made it possible to speculate about the origin of the universe. Among the many approaches, nowadays the most popular is the concept in which the universe, having started from the Big Bang, is expanding and accelerating. Dark matter and dark energy have been added as some new elements, which manifest themselves but are not components of general relativity. This so-called Standard Model of Cosmology is considered as a tremendous success of modern astrophysical and physical science.

Scientific American [516] published a map of Einstein's influence on the study of fundamental physics related to gravitation and cosmology on the basis of articles put on the arXiv.org website in 2014. Statistics show the number of papers devoted to specific topics: black holes was about 500; physical cosmology was about 400; string theory and gravitation in general was above 300 each; dark matter and spacetime was about 250 each; special relativity, time, and quantum gravity was around 200 each; gravitational waves, Big Bang, Standard Model, Hawking radiation, dark energy, and symmetry varied from 100 to 150 each.

Nevertheless, as has been demonstrated in the previous chapters, these kinds of researches are rather naive, as they completely ignore the fact of the existence of a substrate in which all the nebulas, stars, planets, and gases drift. This primary substrate that represents our universe as such, which has a spherical or spheroidal shape, is the tessellattice with a founding element, a topological ball.

Indeed, astrophysicists do not observe massive objects in the universe that are older than about 13.8 billion years. Does this mean that 13.8 billion years is the age of the universe? Probably not, because any physical (chemical or biological) system has its own lifetime. For example, the life expectancy of a queen bee is about 7 years. Does it mean that bees have lived only for 7 years? No, it does not, because modern studies show that species of bees exist already for more than 40 million years. So the same is true for stars. After stars have burned out, and appropriate galaxies cease to exist, their residuals scatter in all directions and interact with other groups of astronomical objects and new galaxies are created.

The author's study would be supplemented by some constructions developed for the vacuum considered as an ether with a lattice structure, which has been discussed by many researchers, particularly, in some aspects by Aspden [517].

An interesting approach to the description and understanding of the parameters of the universe was proposed by Rubio [518]. His model presents the universe as an elastic sphere filled with electrons (the smallest stable particles) that have their size determined by Thompson's classical electron radius. He does not consider the possibility of dropping the speed of light at the boundary of the universe and in his model there are no cells of space. But if the model includes cells and permits their contraction, the suggested radius of the universe would be decreased by two orders reaching the observable value of the order of 10^{26} m. Rubio's approach is important, as it gives a simple explanation of the origin of major fundamental constants. There are of course other approaches dealing with physical parameters and of considering the universe as a substrate/ether, which require a study and deep analysis that takes into account the presented tessellated universe.

Since the end of the 1960s scientists began to scan the sky "listening" for non-random patterns of electromagnetic emissions (like radio and television waves) in order to detect signals of other civilizations elsewhere in the universe. The search for extraterrestrial intelligence (SETI), as a NASA program, began in 1977. Radio telescopes with diameters over 20 meters have been used and new projects increased the number of channels of extraterrestrial assay to approach billions. Frequencies have been varying from kilohertz to terahertz with a resolution of 0.5 hertz per channel. Besides, as from 2016 Australia and the USA play a crucial role in the "Breakthrough Listen" project's hunt for extraterrestrial life, with the Parkes Observatory in Australia and the Green Bank Observatory in West Virginia, USA. The history of SETI is presented in a recent book by Linde [519] together with new research and methods being used for detection and communication.

All these programs are aimed at searching in the realm of electromagnetic waves, i.e., photons. However, developed extraterrestrial civilizations may hardly use the photon channel for their own search for aliens. Why? Because we are talking about advanced civilizations. This automatically means that they would preferentially use an inerton channel for listening to cosmic signals.

Of course, the majority of the readers have heard about telepathy, i.e., transmission of thoughts at a distance – distances which can reach thousands of kilometers. For example, Murphy [520] describes a number of physical and psychical experiments in which the force, or energy of consciousness acts on the surrounding world (see also, experiments described by Atkinson [521]). In particular, Murphy emphasizes that a mysterious "fifth force," i.e., a "carrier wave" of consciousness that is virtually impossible to shield and possibly can travel even faster than the speed of light over very long distances. Murphy portrays these thought waves as scalar waves in the vacuum/ether/zero point field/time-space/implicate order.

Anyone may discuss with extrasensory people their perception of thoughts from a person or persons located in a diametrically opposite point of the Earth. For example, sitting in a room in Kyiv, Ukraine, one can estimate the reaction to this or that article, news, idea, etc., of some given person in Australia. Everybody knows that is impossible to do through the electromagnetic channel in principle. But it is quite realistic to do via the inerton channel, because inerton waves have a much better penetration ability. Besides, inertons carry not only scalar mass; they are also carriers of spatial fractals, i.e., information.

The inerton field is broadly comparable with the electromagnetic field – they are both real fields that obey wave equations – and therefore we may anticipate that in the future we will have actual inerton observatories exploring the universe.

In particular, it will be interesting to observe the sky using inerton receivers in the millimeter range of the so-called CMB (Cosmic Microwave Background) radiation. Furthermore, measuring equipment operating on spacecraft which measure signals at a ratio of signal to noise at the level 2:1, can be significantly improved by our measuring method [515] of precision measurement, such that one will be able to correctly measure signals at the ratio of signal to noise equaling 1:1000.

Before applying inertons to the search for extraterrestrial beings, one of the first tests has to be the measurement of inerton signals above model pyramids and then over Egyptian pyramids. Dunn [522] studying the Great Pyramid of Giza came to the conclusion that one of its major purposes was to serve as a beacon for spacecraft (perhaps of extraterrestrial intelligence), as the Great Pyramid was built as a hydrogen maser functioning on the bandwidth frequency of hydrogen gas (i.e., the 1420 MHz radiation that originates from the transition between the two levels of the hydrogen 1s ground state, which is slightly split by the interaction between the electron spin and the nuclear spin; this is the so-called hyperfine structure). Dunn's [522, 523] research convincingly shows that while absent from the archaeological record, highly refined tools, techniques, and even mega-machines must have been used in ancient Khemit, later (since 1700–1600 BCE) known as Egypt.

At that time the Great Pyramid was built as a resonator of the Earth's inerton field, which has been discussed in Section 5.11. Our preliminary examinations of a model pyramid with the use of the 'Rudra' device (the pyramid was oriented by the sides of the world and had the base $a = 5$ m and height $h = 3.2$ m) showed that the intensity of the pyramid's inerton field was 2–4 times greater than the background inerton field outside the pyramid. The inerton field was especially intense at the pyramid's corners with the maximum being at the top (both inside and outside the pyramid). So in the past the Great Pyramid of Giza could also have been used as a beacon of inerton signals emitted in a narrow spectral range into Outer Space.

Thus, spaceships of extraterrestrial beings could receive both kinds of artificial signals from the Earth, which were emitted from the Great Pyramid: an electromagnetic radio signal at the 21-cm wavelength of Hydrogen and an inerton signal in a specific range of frequencies (although so far that frequency range is unknown).

It would be interesting to consider an inerton field as a source contributing to possible suppression of tornadoes and hurricanes on land and the sea. In any case an inerton field is able to bring some new prospects into the theory of waves, namely, nonlinear nonlocal equations describing the mentioned phenomena.

Let us briefly outline a few new directions in applied science and new technologies, which researchers will be able to develop based upon the submicroscopic deterministic concept that involves inertons. First of all, using inertons is a very new method of communication. The first device for receiving inerton signals in the kHz range has already been designed (Figure 10.2) together with an inerton field generator (Figure 5.13). However, the generator was designed to amplify chemical and physical reactions in the interior of a reaction chamber, not to generate inerton communication signals. Nevertheless, there is a real possibility to develop wireless communication via the inerton field. The final apparatus, which will send and receive inerton signals, will look like a typical mobile phone or at least resemble a pocket radio. Inerton signals easily pass through water and, in principle, their penetration through matter is much better in comparison with typical electromagnetic signals. This means that the power consumption for generating inerton signals will also be less. Besides, inerton signals cannot be measured and deciphered by equipment that is applied for such purposes in conventional (electromagnetic) electronics. Inertons signals seem to propagate faster than the speed of light.

Another direction of study is the phenomenon of antigravity. Some independent enthusiastic researchers have been trying to attain success in this line of study, although the majority of academic scientists ignore their attempts principally because they are dominated by the theory of general relativity that somehow is seen as something absolute, perfect and immutable. Nevertheless, as has been shown in this book, gravitation is based on dynamic processes associated with inertons, i.e., it is inertons that induce the gravitational potential of a massive object.

Inertons are emitted in all directions from oscillating atoms or the mechanical motion of massive bodies. If the body is located on the surface of the Earth, we may distinguish inertons emitted in both directions – to the center of the globe and to the sky. However, if we managed to interrupt the mechanical vibration of an atom in the direction towards the center (for example, by using an electric or magnetic trigger) leaving only the mechanical vibration in the direction to the sky, we will obtain a propulsion force, an inerton propulsion, which will be directed away from the surface of the Earth.

One more promising approach for study may be based on the employment of magnetostriction and piezostriction materials. Indeed, at the moment of striction those materials emit a batch of inertons. If we set the direction of actuation for the striction of the material, for example, only along the X-axis, the inerton propulsion will provide a push in the opposite direction. The larger the striction coefficient C is, the more powerful is the inerton propulsion. A typical value of C for usual magnetostriction materials is 10^{-5} (iron) and 10^{-4} (nickel). However, for some ceramic compounds C reaches the value of 10^{-3}. If the material having such a large value of C features a fast recovery time (maybe in a range of microseconds), one may anticipate a macroscopic propulsion force with an acceleration of at least a few g, where $g = 9.8\ \mathrm{m \cdot s^{-2}}$ is the free fall acceleration on Earth.

In addition, a combination of electromagnetic and inerton properties of the system being studied may lead to new effects, such as the saving of energy and/or increasing the functional efficiency of some machines or engines.

Finally, we have to mention an important direction for future research dealing with low energy nuclear reactions. In this area the submicroscopic deterministic concept plays an important role since an inerton field may play the role of a control factor. Subatomic particles and nuclear pairs of a tiny nuclear size, which have been discussed in Chapter 7, are easily able to overcome the Coulomb barrier in atoms and nuclei because they will fall onto the Coulomb barrier as practically neutral points. That allows us to go ahead planning experiments with predetermined parameters. In the case of radioactive nuclei, generated subatomic particles may destroy the nuclei, quenching their radioactivity. In the case of fusion, a tenfold increase in the energy output might be planned.

We also have to mention the necessity of a purely mathematical study of the structure of space. For example, Shpenkov [524] discusses the nature of physical units and the law of decimal base. He argues that the numerical wave field of affirmation-negation, with some fundamental basis B and period $2\pi \log_B e$ is one of the elementary levels of the informational field of the ideal facet of the universe. For human hands the choice for B is 10, so he talks about quantum periods of fields of decimal base $\Delta = 2\pi \lg e$. He also discussed how Nature choses the left and right for which sixteen equal intervals are required: the left tenth metameasure (the subdominant) and the right one (the dominant). In art, a similar selection of measures was called the golden ratio, which is formally (conventionally) regarded as an irrational ratio. Shpenkov discusses that the law of decimal base $\Delta = 2\pi \lg e$ causes the golden ratio.

On the other hand, Bounias argued that the golden ratio is the result of self-organization of a set (the information is presented in an implicit form in Ref. [525]). In fact, if a set consists of N elements, then the minimum interval between the elements is $1/N$. The minimum interval in the set equates to its greatest stability. If the set possesses asymmetry, then the question arises, which asymmetry is still able to support this smallest interval?

We may remove the normalization, i.e., division by N, and then the interval will be equal to unit, 1. The difference between two intervals formed by nearest members x and $(x - 1)$ or $(x + 1)$ and x may preserve this distance, i.e., unit. Explicitly, the difference between the intervals is: $1/(x-1) - 1/x = 1$ or $1/x - 1/(x+1) = 1$, respectively. The solutions to the equations are $x_1 = \pm 1.61803...$ and $x_2 = \mp 0.61803...$. Hence, the solution $\phi = 1.61803\ldots$ postulates the golden ratio, which is a peculiar asymmetric division of the whole interval while maintaining its length.

The golden ratio ϕ has a number of different presentations and particularly via the limit of the Fibonacci number sequence (see, e.g., Wolfram MathWorld [526])

$$\phi = \lim_{n \to \infty} \frac{F_n}{F_{n-1}}$$

where the Fibonacci numbers are defined by the linear recurrence equation $F_n = F_{n-1} + F_{n-2}$, which are 1, 2, 3, 5, 8, 13, 21, 34, 55, 89, … Mathematical examination of the Fibonacci function still continues [527].

Since the golden ratio ϕ provides the greatest stability of the set in question, it must be realized in all animate and inanimate matter. The ratio ϕ occurs ubiquitously throughout Nature forming logarithmic spirals and other shapes that underlie the process of growth. Stakhov [528, 529] explains that Nature seems to solve the problem of harmony from two sides and adds the results. Once the amount of 1 is reached, space makes a jump and begins to build again. In Nature, the golden ratio ϕ is constructed by a particular rule. In many cases space does not use the golden ratio immediately. The ratio is approached by successive iterations. Space utilizes the Fibonacci series for the generation of the golden ratio. Fibonacci sequences, as a series of peculiar fractals, govern the optimal organization of phyllotaxis, i.e., the arrangement of leaves on an axis or stem [530, 531]; the Fibonacci relation is also used in the crystal lattice [529]; the golden ratio ϕ manifests itself in the Bohr radius, the bond length of the hydrogen molecule, the ozone molecule, and the bond length in alkali metal hydrides [532, 533].

An examination of space at the elementary level is very important because it makes it possible to apply the obtained knowledge to information medicine, physiology and education. We also learn many interesting things from reading religious literature.

For example, Christianto [534] studying the connection of religion with physical reality notes that the Word of God who made everything came into being through the Voice of God, because it was spoken in the beginning, just as Genesis told us. If the Voice was spoken, then He was sound. If He is sound, then He is also vibration and frequency. In other words, everything came into being from sacred frequency or sacred sound.

In Hinduism, the source of everything in the universe is believed to have come into being from cosmic sound [535]. Probably the majority of knowledge came to eastern religions from the Vedas that seem to keep a deep knowledge about space and the universe encoded, even including information on fundamental physics and astrophysics [536, 537].

In Chapters 8 to 10 we indeed have seen that the universe, as a body, is characterized by a fundamental velocity – the speed of light – that is described by the wave equation. Inerton waves generated by any celestial body pervade each galaxy and are able to travel across the whole universe. These waves are a source of information.

A very new historical discovery [538] has shown that the first civilization, which started circa 8000 years ago, represented the culture of communal primary Aratta – a developed country of pre-slavery, pre-capitalist and pre-totalitarian state structure. This type of culture is still partly preserved in folk traditions. Their commune culture rested on a *figurative-intuitive perception of the world*, which through the world archetypes subconsciously brought culture to an information field (i.e., the all-powerful, all-seeing god of the Slavic, Indian and Iranian Vedas). Aratta's people followed natural civilized behavior in the mode of an "autopilot." The community's efforts strived for connection to this "autopilot," learning the natural harmonic behavior of space. The laws of space automatically lead a person through life.

Modern culture is characterized by secondary, totalitarian civilizations (slave-holding, feudal, capitalist and present day socialist society); this culture is based on a *logical-analytic world outlook*, which through the means of rationalism leads culture towards material manifestations of the real world. This type of culture is changing over to "manual control," which substituted the culture of "autopilot." The result of such cultural transfer is dramatic: in the 21th century mankind is still in the throes of routine religion, international conflicts and terrible wars. The system of material control over people is constantly increasing: passports, identity cards, fingerprints, iris photos, video observation, and in the future probably will include indication of blood group, urine analysis, semen analysis, and perhaps other things too. Governments of the developed countries try to maximally preserve the modern materialistic doctrine, which means the conservation of *Homo sapiens* in its present very imperfect state.

However, humans are composed not only of matter but also of a field component. Everyone has their own aura with a specific inerton spectrum and set of fractal properties. And mankind, whether they want it or not, begins once more to revert to the mode of "autopilot," but this time at the level of up-to-date technologies and a modern outlook. Governments must loosen the reins of materialism and seriously take up the upbringing and development of the field component of humanity. Spiritual and yogic practices developed under the supervision of scientists and sages should be supported and put into everyday activity (and a manifesto of scientists [539] must also be perceived seriously).

KEYWORDS

- **antigravity**
- **culture of "autopilot"**
- **extraterrestrial intelligence**
- **field component of humanity**
- **inerton communication**
- **mathematical study of space**
- **nuclear fusion**
- **parameters of the universe**

BIBLIOGRAPHY

1. Wilczek, F. (2013). Triumph, window, clue, and inspiration: The Higgs particle in context, *MIT Phys. Ann.*, 39–46.
2. Wilczek, F. (2000). QCD made simple, *Phys. Today 53*(8), 22–28.
3. Shifman, M., Frontiers beyond the standard model, arXiv:1211.0004 [physics.pop-ph].
4. Compton, A. H. (1923). A quantum theory of the scattering of X-rays by light elements, *Phys. Rev. 21*(5), 483–502.
5. Feynman, R. P. (1985). *QED: The Strange Theory of Light and Matter*. A. G. Mautner Memorial Lectures, Princeton University Press, Princeton, New Jersey.
6. Mecklenburg, M., & Regan, B. C. (2011). Spin and the honeycomb lattice: Lessons from grapheme, *Phys. Rev. Lett. 106*, 116803. Erratum 229901.
7. Merali, Z. (2013). The origins of space and time, *Nature 500*, 516–519.
8. Wilczek, F., What is space? Physics @MIT 30. (2009). http://web.mit.edu/physics/people/faculty/docs/wilczek_space06.pdf.
9. Poincaré, H. (1908). La Dynamique de l'électron, *Revue générale des Sciences pures et appliquées 19*, 386–402. (in Muscovian translation: *Henri Poincaré. Selected Transactions in Three Volumes*, Vol. 3. Ed.: Bogolubov, N. N., Nauka, Moscow (1974). pp. 487–515).
10. de Broglie, L. (1925). Recherches sur la théorie des Quanta, *Ann. de Phys.* 10 série, *t. III*, 22–128, Janvie-Février English translation by A. F. Kracklauer: *On the theory of quanta*, Lulu.com, Morrisville, NC (2007).
11. de Broglie, L. Interpretation of quantum mechanics by the double solution theory, *Ann. Fond. L. de Broglie 12*, 399–421. (1987).
12. Bohm, D. (1952). A suggested interpretation of the quantum theory in terms of "hidden" variables. Part I, *Phys. Rev. 85*, 166–179. Part II, *ibid.*, 180–193. (1952).
13. Krasnoholovets, V. (2004). On the origin of conceptual difficulties of quantum mechanics, in *Developments in Quantum Physics*, Eds.: F. Columbus & Krasnoholovets, V., pp. 85–109, Nova Science Publishers, New York arXiv.org: physics/0412152.
14. Duffy, M. C. (2008). Ether as a disclosing model. Ether theory of gravitation: why and how? in *Ether Space-Time and Cosmology, Vol. 1. Modern Ether Concepts, Relativity and Geometry*, Eds.: Duffy, M. C., Levy, J., & Krasnoholovets, V., pp. 13–45, PD Publications, Liverpool.
15. Heine, V. (1970). *Group Theory in Quantum Mechanics. An Introduction to its Present Usage*, Pergamon Press, New York.
16. Joshi, A. W. (2005). *Elements of Group Theory for Physics*, New Age International (P) Ltd., New Delhi.
17. Atiyah, M. (1989). Topological quantum field theories, *Publications Mathématiques de l'IHÉS 68* (68), 175–186.

18. Nakahara, M. (2003). *Geometry, Topology and Physics*, 2nd edition, Brewer, D. F., Series Editor, *Graduate Student Series in Physics*, IOP Publishing Ltd., Bristol.
19. Eschrig, H. (2011). *Topology and Geometry for Physics. Lecture Notes in Physics*, Vol. 822. Springer, Heidelberg.
20. Mandelbrot, B. B. (2012). *The Fractal Geometry of Nature*, Macmillan Publishers Ltd.
21. Nottale, L. (1997). Scale-relativity and quantization of the universe. I. Theoretical framework, *Astron. Astrophys. 327*, 867–889.
22. Gouyet, J.-F. (1996). *Physics and Fractal Structures*, Masson, Springer, Paris, New York.
23. Barnsley, M. F. (2012). *Fractals Everywhere: New Edition*, Dover Publication, Inc., Mineola, New York.
24. Riemann, B. (1854). On the hypotheses which lie at the bases of geometry, *Nature* Vol. VIII, Nos. 183, 184, pp. 14–17, 36, 37. (translated by Clifford, W. K., *Mathem. Papers*, pp. 55–71. (1867)); also http://www.maths.tcd.ie/pub/HistMath/People/Riemann/Geom/WKCGeom.html).
25. Scholz, E. (1999). The concept of manifold, 1850–1950, in *History of Topology*, Ed.: James, I. M., Elsevier Science B. V., Amsterdam pp. 25–64.
26. Patton, L., & Helmholtz, H. (2014). *The Stanford Encyclopedia of Philosophy* (Fall 2014 Edition), Ed.: Zalta, E. N., http://plato.stanford.edu/archives/fall2014/entries/hermann- helmholtz/.
27. Helmholtz, H. (1868). Über die tatsächlichen Grundlagen der Geometrie, *Nachrichten königlichen Gesellschaft der Wissenschaften zu Göttigen 15*, 193–221, June. Reprinted in *Wissenschaftlische Abhadlungen 2*, 618–638. (1883). tr. On the facts underlying geometry, in *Epistemological Writings*, Eds.: Hertz, P., & Schlick, M., *Boston Studies in the Philosophy of Science 37*, pp. 39–71, Reidel, Dordrech and Boston (1977).
28. Helmholtz, H. (1870). Über den Ursprung und die Bedeutung der geometrischen Axiome, in *Populäre wissenschaftliche Vorträge, Bruanschwieg, 2.* tr. On the origin and significance of geometrical axioms, *Mind 1*(3), pp. 301–321 (Jul. 1876); also http://www.ub.uni-heidelberg.de/helios/fachinfo/www/math/txt/Helmholtz/geo2e.pdf.
29. Lie, S. (1888). *Theorie der Transformationsgruppen*. 1st Part, Leipzig/Berlin, Reprint: Leipzig/Berlin, 1930.
30. Poincaré, H. (1913). *Science and Hypothesis*, The University of Adelaide G. B. Halsted; Ch. 4: Space and Geometry, https://ebooks.adelaide.edu.au/p/poincare/henri/science-and-hypothesis/.
31. Kuyk, W. (1977). *Complementarity in Mathematics*, Springer Science + Business Media Dordrecht, D. Reidel Publishing Company, Dordrecht pp. 113.
32. Bunge, M., & Mayenez, A. G. (1976). A relational theory of physical space, *Int. J. Theor. Phys. 15*(2), 961–972.
33. Clifford, W. K. (1882). On the space-theory of matter, in *Mathematical Papers by William Kingdom Clifford*, Ed.: Robert Trucker, Macmillan and Co., London pp. 21–22.
34. Giulini, D., Matter from space, arXiv:0910.2574 [physics.hist-ph].

35. Asselmeyer-Maluga, T., & Rosé, H. (2012). On the geometrization of matter by exotic smoothness, Gen. Relat. Grav. 44(11), 2825–2856.
36. Aleksandrov, A. D., Space, (2010). *The Great Soviet Encyclopedia*, 3rd edition (1970–1979). The Gale Group, Inc.
37. Vernadsky, V. (2007–2008). On the states of physical space, *21th Century Science & Technology*, Winter Issue, pp. 10–22; http://www.21stcenturysciencetech.com/Articles%202008/States_of_Space.pdf.
38. Planck, M., A speech in Florence, Italy in 1944, entitled "Das Wesen der Materie" (The Essence of Matter). From: Archiv zur Geschichte der Max-Planck-Gesellschaft, Abt. Va, Rep. 11 Planck, Nr. 1797.
39. Bounias, M. (1990). *La Création de la vie: De la matière à l'esprit*, Editions du Rocher, Paris.
40. Bonaly, A., & Bounias, M. (1993). A topological model for fundamental structures, *Ultra Scientist Phys. Sciences 5*(2), 175–184.
41. Bonaly, A., & Bounias, M. (1994). On mathematical links between physical existence, observability and information: towards a "theorem of something," *Ultra Scientist Phys. Sciences 6*(2), 252–259.
42. Bonaly, A., & Bounias, M. (1995). The trace of time in Poincaré sections of a topological space, *Phys. Essays 8*(2), 236–244.
43. Bonaly, A., & Bounias, M. (1996). Timeless space is provided by the empty set, *Ultra Scientist Phys. Sciences 8*(1), 66–71.
44. Bounias, M., & Bonaly, A. (1996). On metrics and scaling: physical coordinates in topological spaces, *Ind. J. Theor. Phys. 44*(4), 303–321.
45. Bounias, M. (1997). Definition of some properties of set-differences, instans and their momentum, in the search for probationary spaces, *Ultra Scientist Phys. Sciences 9(2),* 139– 145.
46. Bounias, M., & Bonaly, A. (1997). The topology of perceptive functions as a corollary of the theorem of existence in closed spaces, *BioSystems 42*, 191–205.
47. Bounias, M., & Bonaly, A. (1997). Some theorems on the empty set as necessary and sufficient for the primary topological axioms of physical existence, *Phys. Essays 10*(4), 633– 643.
48. Bounias, M. (2000). The theory of something: a theorem supporting the conditions for existence of a physical universe, from the empty set to the biological self, Ed.: Dubois, D. M., *Int. J. Comput. Anticipatory Systems 5–6*, 1–14.
49. Bounias, M. (2000). A theorem proving the irreversibility of the biological arrow of time, based on fixed points in the brain as a compact, D-complete topological space, *Amer. Inst. Physics' Con. Proc., Symposium 8*, 20–24.
50. Bounias, M. (2001). Indecidability and incompleteness in formal axiomatics as questioned by anticipatory processes, Ed.: Dubois, D., *Int. J. Comput. Anticipat. Systems 8*, 251–274.
51. Bounias, M. (2002). On spacetime differential elements and the distribution of biohamiltonian components, *Spacetime & Subsrtance 3*(1), 15–19. arXiv:physics/0205087.
52. Krasnoholovets, V., & Ivanovsky, D. (1993). Motion of a particle and the vacuum, *Phys. Essays 6*(4), 554–563. arXiv:quant-ph/9910023.
53. Krasnoholovets, V. (1997). Motion of a relativistic particle and the vacuum, *Phys. Essays 10*(3), 407–416. arXiv:quant-ph/9903077.

54. Bounias, M., & Krasnoholovets, V. (2003). Scanning the structure of ill-known spaces: Part 1. Founding principles about mathematical constitution of space, Eds.: Feng, L., B. P. Gibson and Yi Lin, *Kybernetes: The Int. J. Systems & Cybernetics 32*, No. 7/8, 945–975. arXiv:physics/0211096.
55. Bounias, M., & Krasnoholovets, V., Scanning the structure of ill-known spaces: Part 2. Principles of construction of physical space, Eds.: Feng, L., Gibson, B. P. & Yi Lin, *ibid. 32*, No. 7/8, 976–1004. (2003). arXiv:physics/0212004.
56. Bounias, M., & Krasnoholovets, V. (2003). Scanning the structure of ill-known spaces: Part 3. Distribution of topological structures at elementary and cosmic scales, Eds.: Feng, L., Gibson, B. P. & Yi Lin, *ibid. 32*, No. 7/8, 1005–1020. arXiv:physics/0301049.
57. Bounias, M., & Krasnoholovets, V. (2004). The universe from nothing: A mathematical lattice of empty sets, Ed.: Dubois, D. M., *Int. J. Comput. Anticipat. Systems 16*, 3–24. arXiv:physics/0309102.
58. James, G., & James, R. C. (1992). *Mathematics Dictionary*, Van Nostrand Reinhold, New York, pp. 267–268.
59. Tricot, C. (1999). *Courbes et dimension fractale*, Springer-Verlag, Berlin, Heidelberg, pp. 240–260.
60. Weisstein, E. W. (1999). *CRC Concise Encyclopedia of Mathematics*, Springer-Verlag, New York, pp. 473–474.
61. Feder, J. (1988). *Fractals*, Plenum Press, New York, p. 16.
62. Schwartz, L. (1991). *Analyse I: théoriedes ensembles et topologie*, Hermann, Paris pp. 30– 35.
63. Aczel, P. (1997). *Lectures on Non-Well-Founded Sets. Lecture-Notes 9*, CSLI Publications, Stanford.
64. Barwise, J., & Moss, J. (1991). Hypersets, *Math. Intelligencer 13*(4), 31–41.
65. Rubin, J. E. (1967). *Set Theory for the Mathematician*, Holden-Day, New York, p. 81.
66. Ciesielski, K. (1997). *Set Theory for the Working Mathematician*, Cambridge University Press, Cambridge, England, p. 37.
67. Arunasalam, V. (1997). Einstein and Minkowski versus Dirac and Wigner: covariance versus invariance, *Phys. Essays 10*(3), pp. 528–532.
68. Lerner, E. J., Falomo, R., & Scarpa, R. (2014). UV surface brightness of galaxies from the local Universe to z ~ 5, *Int. J. Mod. Phys. D 23*(6), 1450058.
69. Anders, M. et al. (2014). (LUNA Collaboration), First direct measurement of the 2H(α, γ)6Li cross section at Big Bang energies and the primordial lithium problem, *Phys. Rev. Lett. 113*, 042501.
70. Krasnoholovets, V. (2014). Submicroscopic viewpoint on gravitation, cosmology, dark energy and dark matter, and the first data of inerton astronomy, in *Recent Developments in Dark Matter Research*, Eds.: Kinjo, N., & Nakajima, A., Nova Science Publishers, New York, pp. 1–61.
71. Wu, X.-B., Wang, F., Fan, X., Yi, W., Zuo, W., Bian, F., Jiang, L., McGreer, I. D., Wang, R., Yang, J., Yang, Q., Thompson, D., & Beletsky, Yu. (2015). An ultraluminous quasar with a twelve-billion-solar-mass black hole at redshift 6.30", *Nature 518*, 512–515.

72. Watson, D., Christensen, L., Knudsen, K. K., Richard, J., Gallazzi, A., & Michałowski, M. J. (2015). A dusty, normal galaxy in the epoch of reionization, *Nature 519*, 327–330.
73. Kubel, H. (1997). The Lorentz transformation derived from an absolute aether, *Phys. Essays 10*(3), 510–523.
74. Rothwarf, A. (1998). An aether model of the universe, *Phys. Essays 11*(3), 444–66.
75. Blokhintsev, D. I. (1976). *Principles of Quantum Mechanics*, 5th edition, Nauka, Moscow, §9, p. 47; in Muscovian.
76. Hales, T. C. (2007). Jordan's proof of the Jordan curve theorem, *Studies in Logic, Grammar and Rhetoric 10*(23), 45–60.
77. Haddad, W. M. (2012). Temporal asymmetry, entropic irreversibility, and finite-time thermodynamics: from Parmenides–Einstein time-reversal symmetry to the Heraclites entropic arrow of time, *Entropy 14*, 407–455.
78. Bolivar-Toledo, O., Candela Sola, S., & Munoz Blanco, J. A. (1985). Nonlinear data transforms in perceptual systems, *Lecture Notes in Computer Sciences 410*, 1–9.
79. Schwartz, L. (1966). *Théorie des distributions*, Hermann, Paris.
80. Bourbaki, N. (1990). *Théorie des ensembles*, Masson, Paris, Cps. 1–4, pp. 352, 376.
81. Pensrose, R. (2010). *Cycles of Time: An Extraordinary New View of the Universe*, The Bodley Head, London, p. 202.
82. Lobo, F. S. N., Martinez-Asenciob, J., Olmo, G. J., & Rubiera-Garcia, D. (2015). Planck scale physics and topology change through an exactly solvable model, *Phys. Lett. B 731*, 163–167.
83. Bounias, M., & Krasnoholovets, V. (2002). Science addendum to: I. Harezi, *The Resonance in Residence. An Inner and Outer Quantum Journey*, Ilonka Harezi, USA.
84. Krasnoholovets, V. (2006). The tessellattice of mother-space as a source and generator of matter and physical laws, in *Einstein and Poincaré: The Physical Vacuum*, Ed.: Dvoeglazov, V. V., Apeiron, Montreal, pp. 143–153.
85. 't Hooft, G. (1999). Quantum gravity as a dissipative deterministic system, *Class. Quant. Grav. 16*, 3263–3279.
86. Rüegg, A. (1985). *Probabilities and Statistics*, Presses Polytechniques Romandes, Mausanne, Switzerland, pp. 52–87.
87. Barnsley, M. F. (2013). *Fractals Everywhere: New Edition*, Courier Corporation, USA p. 442.
88. McClure, M. (2006). Newton's method for complex polynomials. A preprint version of a "Mathematical graphics" column from *Mathematica in Education and Research*, pp. 1–15.
89. *The NIST Reference on Constants, Units, and Uncertainty*. US National Institute of Standards and Technology, 2014. http://physics.nist.gov/cgi-bin/cuu/Value?research_for=atomnuc!
90. Born, M. (1969). *Atomic Physics*, Eighth Edition, Blackie and Son Limited, London-Glasgow, Appendices. X. The Compton effect, pp. 380–382.
91. Thomson, J. J. (1906). *Conduction of Electricity Through Gases*, Cambridge University Press.
92. Sommerfeld, A. (1916). Zur Quantentheorie der Spektrallinien, *Ann. Physik 356*(51), 1–94.

93. Kragh, H. (2003). Magic number: A partial history of the fine-structure constant, *Arch. Hist. Exact Sci. 57*, 395–431.
94. Lunn, A. C. (1922). Atomic constants and dimensional invariants, *Phys. Rev. 20*, 1–14.
95. Oldershaw, R. L., The meaning of the fine structure constant, arXiv:0708.3501 [physics.gen- ph].
96. Oldershaw, R. L., The meaning of the fine structure constant, http://vixra.org/abs/1112.0016.
97. Michaud, A. (2007). Field equations for localized individual photons and relativistic field equations for localized moving massive particles, *Int. IFNA-ANS J.* (Kazan State University, Kazan, Muscovy) *13*, No. 2 (28), 123–140. Also viXra:0907.0013.
98. Shpenkov, G. P. (2015). *Dialectical View of the World. The Wave Model (Selected Lectures)*, Vol. 6. *Topical Issues,* http://shpenkov.janmax.com/Vol.6.TopicalIssues.pdf.
99. Krasnoholovets, V. (2016). Quarks and hadrons in the real space, *J. Adv. Phys*. *5*(2), 145–167.
100. Dimopoulos, S., Raby, A. A., & Wilczek, F. (1991). Unification of couplings, *Phys. Today 44*(10), 25–33.
101. Barger, V., Jiang, J., Langacker, P., & Li, T. (2005). Gauge coupling unification in the Standard Model, *Phys. Lett. B 624*, 233–238.
102. Singh, K. S., & Singh, N. N. (2015). Effects of the variation of SUSY breaking scale on Yukawa and gauge couplings unification, *Adv. High Energy Phys*. Article ID 652029, 8 pp.
103. Klepp, J., Sponar, S., & Hasegawa, Y. (2014).Fundamental phenomena of quantum mechanics explored with neutron interferometers, *Prog. Theor. Exp. Phys*. 082A01 (61 pp.).
104. Shadbolt, P., Mathews, J. C. F., Laing, A., & J. L. O'Brien, (2014). Testing foundations of quantum mechanics with photons, *Nature Physics 10*, 278–286. arXiv:1501.0371 [quant-ph].
105. Landau, L. D., & Lifshitz, E. M. (1974). *Quantum Mechanics*, Nauka, Moscow, in Muscovian.
106. Özkan, Ö., Didiş, N., & Taşar, M. F. (2009). Students' conceptual difficulties in quantum mechanics: potential well problems, *Hacettepe University J. Education 36*, 169–180.
107. Krasnoholovets, V. (2001). On the way to submicroscopic description of nature, *Ind. J. Theor. Phys*. *49*(2), 81–95. arXiv:quant-ph/9908042.
108. Krasnoholovets, V., Can quantum mechanics be cleared from conceptual difficulties?; arXiv:quant-ph/0210050.
109. Gao, J. (2015). Future large colliders in Asia – a personal perspective, *LC Newsline* 28 May 2015, http://newsline.linearcollider.org/2015/05/28/future-large-colliders-in-asia-a-personal- perspective/.
110. de Broglie, L. (1956). *Une tentative d'Interprétation causale et non linéaire de la Mécanique Ondulatoire: la Théorie de la Double Solution*, Gauthier-Villars, Paris, (English translation: *Non-linear Wave Mechanics: a Casual Interpretation*, Elsevier, Amsterdam, (1960)).

111. de Broglie, L. (1982). *Les Incertitudes d'Heisenberg et l'Interprétation Probabiliste de la Mécanique Ondulatoire*, Gauthier-Villars, Paris, English translation: *Heisenberg's Uncertainty Relations and the Probabilistic Interpretation of Wave Mechanics with Critical Notes of the Author*, Kluwer Academic Publishers, Dordrecht (1990). Muscovian translation: Mir, Mowsow, (1986). pp. 42–43.
112. Valentini, A. (2009). Beyond the quantum, *Phys. World 22*(11), 32–37. arXiv:1001.2758 [quant-ph].
113. Bohm, D., & Vigier, J. P. (1954). Model of the causal interpretation of quantum theory in terms of a fluid with irregular fluctuations, *Phys. Rev. 96*, 208–216.
114. Haar, Ter D. (1974). *Elements of Hamiltonian Mechanics*, Nauka, Moscow, p. 60; in Muscovian (translated from *Elements of Hamiltonian Mechanics*, Second edition, Pergamon Press, 1971).
115. Dubrovin, B. A., Novikov, S. P., & Fomenko, A. T. (1986). *Modern Geometry: Methods and Applications*, Nauka, Moscow, p. 48; in Muscovian.
116. Landau, L. D., & Lifshits, E. M. (1986). *Hydrodynamics*, Nauka, Moscow, p. 15; in Muscovian.
117. Mandelstam, L., & Leontowitsch, M., (1928). Zur Theorie der Schrödingerschen Gleichung, *Z. Physik 47*, Nrn. 1–2, 131– 136.
118. Gamow, G. (1928). Zur Quantentheorie des Atomkernes, *Z. Physik 51*, Nrn. 3–4, 204–212.
119. Gurney, R. W., & Condon, E. U. (1928). Quantum mechanics and radioactive disintegration, *Nature 122*, 439.
120. Gurney, R. W., & Condon, E. U. (1929). Quantum mechanics and radioactive disintegration, *Phys. Rev. 33*(2), 127–140.
121. Brillouin, L. (1926). La mécanique ondulatoire de Schrödinger: une méthode générale de resolution par approximations successives, *Comptes Rendus de l'Academie des Sciences 183*, 24–26.
122. Kramers, H. A. (1926). Wellenmechanik und halbzählige Quantisierung, *Z. Physik 39*, Nrn. 10–11, 828–840.
123. Wentzel, G. (1926). Eine Verallgemeinerung der Quantenbedingungen für die Zwecke der Wellenmechanik, *Z. Physik 38*, Nrn. 6–7, 518–529.
124. Razavy, M. (2003). *Quantum Theory of Tunneling*, World Scientific, Singapore.
125. Turok, N. (2014). On quantum tunneling in real time, *New J. Phys. 16*, 063006 (13 p.).
126. Landauer, R., & Martin, Th. (1994). Barrier interaction time in tunneling, *Rev. Mod. Phys. 66*(1), 217–228.
127. Heisenberg, W. (1927). Über den anschaulichen Inhalt der quantentheoretischen Kinematik und Mechanik, *Z. Physik **43***, Nrn. 3–4, 172–198.
128. Kennard, E. H. (1927). Zur Quantenmechanik einfacher Bewegungstypen, *Z. Physik 44*, Nrn. 4–5, 326.
129. Robertson, H. P. (1929). The uncertainty principle, *Phys. Rev. 34*, 163–164.
130. Fermi, E. (1965). *Notes on Quantum Mechanics*, Mir, Mowsow, p. 357; Muscovian translation.
131. Dehmelt, H., & Ekstrom, P. (1973). Proposed g-2/dvz experiment on stored single electron or positron, *Bull. Am. Phys. Soc. 18*, 727.

132. Van Dyck, R. S. Jr., Ekstrom, P., & Dehmelt, H. G. (1976). Axial, magnetron, cyclotron and spincyclotron beat frequencies measured on single electron almost at rest in free space (Geonium), *Nature 262*, 776–777.
133. Van Dyck, R. S. Jr., P.Schwinberg, B., & Dehmelt, H. G. (1986). Electron magnetic moment from geonium spectra: Early experiments and background concepts, *Phys. Rev. D 34*, 722– 736.
134. Hofer, W. A. (2012). Heisenberg, uncertainty, and the scanning tunneling microscope, *Frontiers of Physics 7*, 218–222. arXiv:1105.3914.
135. Hofer, W. A. (2012). An extended model of electrons: experimental evidence from high- resolution scanning tunneling microscopy, *J. Physics: Conference Series 361*, 012023, 9 pp.
136. Hofer, W. A. (2014). Elements of physics for the 21st century, *J. Phys.: Conference Series 504*, 012014.
137. Boyd, R. N., Refutation of Heisenberg uncertainty regarding photons, http://worlds-within- worlds.org/refutationofheisenberg.php.
138. Dumitru, S. (2010). Do the uncertainty relations really have crucial significances for physics? *Progress in Phys. 4*, 25–29.
139. Dirac, P. A. M. (1963). The evolution of physicist's picture of nature, *Scient. American 208*(5), 45–53.
140. Krasnoholovets, V. (2000). On the nature of spin, inertia and gravity of a moving canonical particle, *Ind. J. Theor. Phys. 48*(2), 97–132. arXiv:quant-ph/0103110.
141. Krasnoholovets, V. (2002). Submicroscopic deterministic quantum mechanics, Ed.: Dubois, D., *Int. J. Comput. Anticipat. Systems 11*, 164–179. arXiv:quant-ph/0109012.
142. Hofer, W. A. (2011). Unconventional approach to orbital-free density functional theory derived from a model of extended electrons, *Found. Phys. 41*, 754–791.
143. Oudet, X. (1992). L'aspect corpusculaire des électrons et la notion de valence dans les oxydes métalliques, *Ann. Fond. L. de Broglie 17*, 315–345.
144. Oudet, X. (1995). L'état quantique et les notions de spin, de fonction d'onde et d'action, *Ann. Fond. L. de Broglie 20*, 473–490.
145. Oudet, X. (1996). Atomic magnetic moments and spin notion, *J. App. Phys. 79*, 5416–5418.
146. Lévy-Leblond, J.-M. (1967). Nonrelativistic particles and wave equations, *Comm. Math. Phys. 6*, 286–311.
147. Greiner, W. (2000). *Quantum Mechanics: An Introduction*. 4th edition, Springer, Berlin- Heidelberg-New York-Barcelona-Hong Kong-London-Milan-Singapore-Tokyo, pp. 355–366.
148. Denkmayr, T., Geppert, H., Sponar, S., Lemmel, H., Matzkin, A., Tollaksen, J., & Hasegawa, Y. (2014). Observation of a quantum Cheshire Cat in a matter-wave interferometer experiment, *Nature Commun. 5*, 4492 doi: 10.1038/ncomms5492.
149. Stuckey, W. M., Silberstein, M., & McDevitt, T., Quadratic interaction in the quantum Cheshire Cat experiment, arXiv:1410.1522 [quant-ph].
150. Corrêa, R., Santos, M. F., Monken, C. H., & Saldanha, P. L. (2015). 'Quantum Cheshire Cat' as simple quantum interference, *New J. Phys. 17*, 053042. arXiv:1409.0808 [quant-ph].
151. Baeßler, S. (2009). Gravitationally bound quantum states of ultracold neutrons and their applications, *J. Phys. G: Nucl. Part. Phys. 36*(10), 104005, 17 pp.

152. Born, M. (1926). Das Adiabatenprinzip in der Quantenmechanik, *Z. Physik 40*, 167–192.
153. Born, M. (1954). Nobel Lecture, December 11. in *Nobel Lectures, Physics* 1942–1962, Elsevier Publishing Company, Amsterdam, 1964, pp. 256–267.
154. de Broglie, L. (1967). Sur la Dynamique du corps à masse propre variable et la formule de transformation relativiste de la chaleaur, *Compt. Rend. 264 B*(16), 1173–1175.
155. Chang, D. C., & Yi-Lee, K. (2015). Study on the physical basis of wave-particle duality: Modelling the vacuum as a continuous mechanical medium, *J. Mod. Phys. 6*, 1058–1070.
156. Krasnoholovets, V., & Tane, J.-L. (2006). An extended interpretation of the thermodynamic theory including an additional energy associated with a decrease in mass, *Int. J. Simulation and Process Modelling 2*, Nos. 1/2, 67–79. arXiv:physics/0605094.
157. Briner, G., Ph. Hofmann, Doering, M., Rust, H. P., Bradshaw, A. M., Petersen, L., Ph. Sprunger, Laegsgaard, E., Besenbacher, F., & Plummer, E. W. (1997). Looking at electronic wave functions on metal surfaces, *Europhys. News 28*, 148.
158. Mirhosseini, M., Magaña-Loaiza, O. S., Rafsanjani, S. M. H., & Boyd, R. W. (2014). Compressive direct measurement of the quantum wave function, *Phys. Rev. Lett. 113*, 090402.
159. Wei, W., Xie, Z., Cooper, L. N., Seidel, G. M., & Maris, H. J. (2014). Study of exotic ions in superfluid helium and the possible fission of the electron wave function, *J. Low Temp. Phys. 178*, 78–117.
160. Sokolov, A. A., Loskutov, Yu. M., & Ternov, I. M. (1965). *Quantum Mechanics*, Prosveshchenie, Moscow, p. 222; in Muscovian.
161. Buckholtz, T. J. (2015). *Theory of Particles Plus the Cosmos Small Things and Vast Effects*, CreateSpace Independent Publishing Platform, North Charlstone, SC.
162. Roychoudhuri, C., A.Kracklauer, F., & Creath, K. (Eds.). (2008). *The Nature of Light: What is a Photon?* CRC Press, Boca Raton, Florida.
163. Keller, O. (2014). *Light – The Physics of the Photon*, CRC Press, Boca Raton, Florida.
164. Bussey, P. J. (1993). The phonon as a model for elementary particles, *Phys. Lett. A 176*, No. 3–4, 159–164.
165. Feynman, R. P., & Hibbs, A. R. (1965). *Quantum Mechanics and Path Integrals*, McGraw- Hill, New York, Ch.8.
166. Krasnoholovets, V. (2002). On the notion of the photon, *Ann. Fond. L. de Broglie 27*(1), 93–100. arXiv:quant-ph/0202170.
167. Krasnoholovets, V. (2003). On the nature of the electric charge, *Hadronic J. Suppl. 18*(4), pp. 425–456. arXiv:physics/0501132.
168. Bronstein, I. N., & Semendyaev, K. A. (1981). *Mathematical Handbook*, Eds.: Grosche, G., & Ziegler, W., Tojbner, Leipzig and Nauka, Moscow, p. 288; in Muscovian.
169. Fikhtengolts, G. M. (1947). *The Course of Calculus and Differential Calculation.* Vol. 1, OGIZ – Gosudarstvennoe Izdatel'stvo Tekhniko-Teoreticheskoy Literatury, Moscow, Leningrad, p. 478; in Muscovian.
170. Griffiths, D. J. (1999). *Introduction to Electrodynamics*, 3rd Edition, Prentice-Hall, Inc., Upper Saddle River, New Jersey, p. 416.
171. Tamm, I. E. (1976). *Fundamentals of The Theory of Electricity*, Ninth Edition, Nauka, Moscow, § 96, p. 431; in Muscovian.

172. Jackson, J. D. (1998). *Classical Electrodynamics*, Third Edition, John Wiley & Sons, USA Sect. 6.11, p. 273.
173. Zeeman, P. (1897). On the influence of magnetism on the nature of the light emitted by a substance, *Phil. Mag*. Series 5, *43*(262), 226–239.
174. Gerlach, W., & Stern, O. (1922). Der experimentelle Nachweis der Richtungsquantelung im Magnetfeld, *Z. Phys*. *9*(6), 349–352.
175. Krasnoholovets, V., & Lev, B. (2002). Systems of particles with interaction and the cluster formation in condensed matter, *Cond. Matter Phys*. *6*(1), 93–100. arXiv:cond-mat/0210131.
176. Khachaturian, A. G. (1974). *The Theory of Phase Transitions and the Structure of Solid Solutions,* Nauka, Moscow, p. 100 (in Muscovian).
177. Pethick, C. J., & Smith, H. (2008). *Bose-Einstein Condensation in Dilute Gases*, Second Edition, Cambridge University Press, Cambridge, New York, Melbourne, Madrid, Cape Town, Singapore, São Paulo.
178. Krasnoholovets, V. (2010). Variation in mass of entities in condensed media, *Appl. Phys. Research* *2*(1), 46–59.
179. Ketterle, W. (2002). When atoms behave as waves: Bose-Einstein condensation and the atom laser, 2001 Nobel Lecture in Physics. *Rev. Modern Phys*. *74*(4), 1131–1151.
180. Kittel, C. (1996). *Introduction to Solid State Physics*, 7th Edition, John Wiley & Sons, Inc., New York, Chichester, Brisbane, Toronrto, Singapore, p. 71.
181. Strobbe, H., & Peschel, G. (1997). Experimental determination of the static permittivity of extremely thin liquid layers of water dependent on their thickness, *Colloid and Polymer Sci. 275*, 162–69.
182. Kondrachuk, A. V., Krasnoholovets, V. V., Ovcharenko, A. I., & Chesnokov, E. D. (1994). Determination of the water structuring by the pulsed NMR method, *Khim. Fiz. 12,* 1006– 1010. (1993). English translation: *Sov. Jnl. Chem. Phys*. *12*, 1485–1492.
183. Krasnoholovets, V., Tomchuk, P., & Lukyanets, S. (2003). Proton transfer and coherent phenomena in molecular structures with hydrogen bonds, *Advan. Chem. Phys.*, Ed.: Prigogine, I., & Rice, S. A., John Wiley & Sons, Inc., *125*, 351–548.
184. Tane, J.-L. (2014). An important scientific question linked to the interpretation of a very simple process, *The General Sci. J.* http://gsjournal.net/Science- Journals/Research%20Papers-Relativity%20Theory/Download/5726.
185. Tane, J.-L. (2014). An extended interpretation of the concept of entropy opening a link between thermodynamics and relativity, *Natural Science 6*, 503–513.
186. Krasnoholovets, V., Skliarenko, S., & Strokach, O. (2006). The study of the influence of a scalar physical field on aqueous solutions in a critical range, *J. Mol. Liquids 127*, 50–52.
187. Krasnoholovets, V., Skliarenko, S., & Strokach, O., On the behavior of physical parameters of aqueous solutions affected by the inerton field of Teslar technology, *Int. J. Mod. Phys. B 20*(1), 111–124. (2006). arXiv.org:0810.2005 [physics.gen-ph].
188. Andreev, E., Dovbeshko, G., & Krasnoholovets, V. (2007). The study of influence of the Teslar technology on aqueous solution of some biomolecules, *Research Lett. in Phys. Chem.*, Article ID 94286, 5 pp.
189. Dovbeshko, G., Gridina, N., Krasnoholovets, V. (2013). Holographic interferometry as a reliable tool visualizing an influence of low-energy fields on biological systems, *Research & Reviews in BioSciences* *7*(10), 383–393.

190. Emoto, M. (2006). *The Secret Life of Water*, Simon & Schuster UK Ltd., London.
191. Dekhtiaruk, V., Krasnoholovets, V., & Heighway, J., Biodiesel manufacture, International patent No. PCT/GB2007/001957 (Intern. filing date 25 May 2007; London, UK).
192. Shoulders, K. R., Energy conversion using high charge density, U.S. patent 5,018,180, May, 1991.
193. Shoulders, K., & Shoulders, S. (1996). Observations of the role of charge clusters in nuclear cluster reactions, *J. New Energy 1*(3), Fall.
194. Beckmann, P. (1990). Electron clusters, *Galilean Electrodynamics*, Sept./Oct. *1*(5), 55–58.
195. Ziolkowski, R. W., & Tippett, M. K. (1991). Collective effect in an electron plasma system catalyzed by a localized electromagnetic wave, *Phys. Rev. A43*(6), 3066–3072.
196. Mesyats, G. A. (1996). Ecton processes at the cathode in a vacuum discharge, *Proc. XVIIth Int. Symposium on Discharges and Electrical Insulation in Vacuum*, Berkeley, CA, July 21– 26. pp. 721–731.
197. Lisitsyn, I., Akiyama, H., & Mesyats, G. A. (1998). Role of electron clusters – ectons – in the breakdown of solids dielectrics, *Physics of Plasma 5*(12), 4484–4487.
198. Kukhtarev, N., Kukhtareva, T., Edwards, M., Penn, B., Frazier, D., Abdeldayem, H., Banerjee, P. P., Hudson, T., & Friday, W. A. (2002). Photo-induced optical and electrical high-voltage pulsations and pattern formation in photorefractive crystals, *J. Nonlin. Opt. Phys. Materials 11*(4), 445–453.
199. Bayssie, M., Brownridge, J. D., Kukhtarev, N., Kukhtareva, T., & Wang, J. C. (2005). Generation of focused electron beam and X-rays by the doped LiNbO3 crystals, *Nuclear Instruments and Methods in Physics Research, B 241*, 913–918.
200. Krasnoholovets, V., Kukharev, N., & Kukhtareva, T. (2006). Heavy electrons: Electron droplets generated by photogalvanic and pyroelectric effects *Int. J. Mod. Phys. B 20*(16), 2323–2337. arXiv:0911.2361 [quant-ph].
201. Wigner, E. (1932/4). On the quantum correction for thermodynamic equilibrium, *Phys. Rev. 40*, 749–759. On the interaction of electrons in metals, *Phys. Rev. 46*, 1002–1011.
202. Andrei, E. Y., Ed. (1997). *Two-Dimensional Electron Systems: On Helium and other Cryogenic Substrates*, Kluwer Academic Publishers, Dordrecht.
203. Ehrler, O. T., & Neumark, D. M. (2009). Dynamics of electron solvation in molecular clusters, *Acc. Chem. Res. 42*(6), 769–777.
204. Eggers, J. (1997). Nonlinear dynamics and breakup of free-surface flows, *Rev. Mod. Phys. 69*(3), 865–930.
205. Guido, S., & Greco, F. (2004). Dynamics of a liquid drop in a flowing immiscible liquid, *Rheology Reviews*, pp. 99–142.
206. Yarin, A. L. (2006). Drop impact dynamics: splashing, spreading, receding, bouncing… *Annual Review of Fluid Mechanics 38*, 159–192.
207. Putz, M. V. (2010). The bondons: The quantum particles of the chemical bond, *Int. J. Mol. Sci. 11*, 4227–4256.
208. Putz, M. V., & Ottorino, O. (2012). Bondonic characterization of extended nanosystems: Application to graphene's nanoribbons, *Chem. Phys. Lett. 548*, 95–100.

209. Putz, A.-M., & Putz, M. V. (2012). Spectral inverse quantum (spectral-IQ) method for modeling mesoporous systems: application on silica films by FTIR, *Int. J. Mol. Sci. 13*, 15925–15941.
210. Putz, M. V., Duda-Seiman, C., Duda-Seiman, D. M., & Bolcu, C. (2015). Bondonic chemistry: Consecrating silanes as metallic precursors for silicenes materials, in *Exotic Properties of Carbon Nanomatter: Advances in Physics and Chemistry*, Eds.: Putz, M. V., & O. Ori. Volume 8, the series Carbon Materials: Chemistry and Physics, Springer, pp. 323–345.
211. Epstein, P. S., & Ehrenfest, P. (1924). The quantum theory of the Fraunhofer diffraction, *Proc. Natl. Acad. Sci. USA. 10*(4), 133–139.
212. Ehrenfest, P., & Epstein, P. S. (1927). Remarks on the quantum theory of diffraction, *Proc. Natl. Acad. Sci. USA. 13*(6), 400–408.
213. X.-Wu, Y., Zhang, B.-J., Yang, J.-H., Chi, L.-X., Liu, X.-J., Wu, Y.-H., Wang, Q.-C., Wang, Y., Li, J.-W., & Guo, Y.-Q. (2010). Quantum theory of light diffraction, *J. Mod. Optics 57*(20), 2082–2091. arXiv:1011.3593 [quant-ph].
214. Panarella, E., Nonlinear behavior of light at very low intensities: The "photon clump" model, in *Quantum Uncertainties: Recent and Future Experiments and Interpretations*. Eds.: Honig, W. M., Kraft, D. W., & Panarella, E. (1987). *Proceedings of NATO, NATO ASI Series, Series B: Physcis 162*, Plenum, New York, London, pp. 105–167.
215. Dontsov, Yu. P., & Baz, A. I. (1967). Interference experiments with statistically independent photons, *JETF* (*J. Exper. Theor. Phys.*) *52*(1), 3–11. in Muscovian.
216. Krasnoholovets, V. (2010). Sub microscopic description of the diffraction phenomenon, *Nonlin. Opt. Quant. Opt. 41*, 273–286. arXiv:1407.3224 [physics.gen-ph].
217. Wolfe, J. P. (2005). *Imaging Phonons: Acoustic Wave Propagation in Solids*, Cambridge University Press, New York, p. 80. Eds.: Truell, R., Elbaum, C., & Chick, B. B., *Ultrasonic Methods in Solid State Physics*, Academy Press, New York, London, 1969, Sections 35–38.
218. Trucker, J. W., & Rampton, V. W. (1972). *Microwave Ultrasonics in Solid State* Physics, North-Holland Publ. Co., Amsterdam.
219. Bron, W. E., Patel, J. L., & Schairch, W. L. (1979). Transport of phonons into diffusive media, *Phys. Rev. B 20*(12), 5394–5397.
220. Tamura, Sh. (1985). Spontaneous decay rates of LA phonons in quasi-isotropic solids, *Phys. Rev. B 31*(4), 2574–2577.
221. Eisenmeger, W., Lasman, K., & Dottinger, S. Eds.: *Phonon Scattering in Condensed Matter*, (Springer, Berlin, 1984).
222. Born, M., & Wolf, E., *Principle of Optics*, 3rd ed. (Pegamon Press, Oxford, 1965), p. 397.
223. Jeffers, S., Wadlinger, R., & Hunter, G. (1991). Low-light-level diffraction experiments: No evidence for anomalous effects, *Canad. J. Phys. 6*, 91471–1475.
224. Jeffers, S., & Sloan, J. (1992). A low light level diffraction experiment for anomalies research, *J. Scientific Exploration 6*(4), 333–352.
225. Sinton, A. M., P.Gardenier, R., & Bates, R. H. T. (1986). Reinvestigation of optical interference at low light levels, *Speculations in Science and Technology 9*(4), November, 269–278.

226. Jeffers, S. (1987). Discussion III. Complied and interpreted by Kraft, D. W., & Panarella, E., in *Quantum Uncertainties: Recent and Future Experiments and Interpretations*. Eds.: Honig, W. M., Kraft, D. W., & Panarella, E. *Proceedings of NATO, NATO ASI Series, Series B: Physcis 162*, Plenum, New York, London, pp. 197–202.
227. Young, T. (1807). *A Course of Lectures on Natural Philosophy and the Mechanical Arts*. Vol. 1. William Savage. Lecture 39, pp. 463–464.
228. Pereira, S. F., Vegter, N., & Wendrich, T., Making discrete photon effects visible in an optical interference experiment, http://www.tnw.tudelft.nl/en/about-faculty/departments/imaging-physics/research/researchgroups/optics-research-group/education/experimental-projects/photons-in-an-optical-interference-experiment/.
229. Jin, F., Yuan, S., De Raedt, H., Michielsen, K., & Miyashita, S. (2010). Corpuscular model of two-beam interference and double-slit experiments with single photons, *J. Phys. Soc. Jpn. 79*, 074401 (30 pp.).
230. Kolenderski, P., Scarcella, C., Johnsen, K. D., Hamel, D. R., Holloway, C., Shalm, L. K., Tisa, S., Tosi, A., Resch, K. J., & Jennewein, T. (2014). Time-resolved double-slit interference pattern measurement with entangled photons, *Scientific Report 4*, Article number: 4685.
231. Shafiee, A., Massoudi, A., & Bahrami, M. (2008). On a new formulation of micro-phenomena: The double-slit experiment, arXiv:0810.1034 [quant-ph].
232. Möllenstedt, G., & Jönsson, C. (1959). Elektronen-Mehrfachinterferenzen an regelmä*ß*ig hergestellten Feinspalten, *Z. Phys. 155*, 472–474.
233. Jönsson, C. (1961). Elektroneninterferenzen an mehreren künstlich hergestellten Feinspalten, *Z. Phys. 161*, 454–474.
234. Bach, R., Pope, D., Liou, S.-H., & Batelaan, H. (2013). Controlled double-slit electron diffraction, *New J. Phys. 15*, 033018.
235. Zeilinger, A., Gähler, R., Shull, C. G., Treimer, W., & Mampe, W. (1988). Single- and double-slit diffraction of neutrons, *Rev. Mod. Phys. 60*(4), 1067–1073.
236. Arndt, M., Nairz, O., Vos-Andreae, J., Keller, C., van der Zouw, G., & Zeilinger, A. (1999). Wave–particle duality of C60 molecules, *Nature 401*, 680–682.
237. Juffmann, T., Milic, A., Müllneritsch, M., Asenbaum, P., Tsukernik, A., Tüxen, J., Mayor, M., Cheshnovsky, O., & Arndt, M. (2012). Real-time single-molecule imaging of quantum interference, *Nature Nanotechnology 7*, 297–300.
238. Zhao, B. S., & Schöllkopf, W. (2012). Fundamental physics: Molecules star in quantum movie, *Nature Nanotechnology 7*, 277–278.
239. Hornberger, K., Gerlich, S., Haslinger, P., Nimmrichter, S., & Arndt, M. (2012). Colloquium: Quantum interference of clusters and molecules, *Rev. Mod. Phys. 84*, 157–173.
240. Eibenberger, S., Gerlich, S., Arndt, M., Mayor, M., & Tüxenb, J. (2013). Matter–wave interference of particles selected from a molecular library with masses exceeding 10,000 amu, *Phys. Chem. Chem. Phys. 15*, 14696–14700.
241. Agostini, P., & Petite, G. (1988). Photoelectric effect under strong irradiation, *Contemp. Phys. 29*, 57–77.
242. Keldysh, L. V. (1964). Ionization in the field of a strong electromagnetic wave, *JETP 47*, 1945–1957. in Muscovian.
243. Reiss, H. R. (1970). Semiclassical electrodynamics of bound systems in intense fields, *Phys. Rev. A 1*, 803–818.

244. Lompre, L. A., Mainfray, G., Manus, C., & Thebault, J. (1977). Multiphoton ionization of rare gases by a tunable-wavelength 30-psec laser pulse as 1.06 μm, *Phys. Rev. A 15*, 1604–1612.
245. Panarella, E. (1974). Theory of laser-induced gas ionization, *Found. Phys. 4*, 227–259.
246. Panarella, E. (1987). Effective photon hypothesis vs. quantum potential theory: theoretical predictions and experimental verification, in *Quantum Uncertainties: Recent and Future Experiments and Interpretations*. Eds.: Honig, W. M., Kraft, D. W., & Panarella, E. *Proceedings of NATO, NATO ASI Series, Series B: Physcis 162*, Plenum, New York, London, pp. 237–269.
247. Kupersztych, J., & Raynaud, M. (2005). Anomalous multiphoton photoelectric effect in ultrashort time scales, *Phys. Rev. Lett. 95*, 147401.
248. Krasnoholovets, V. (2001). On the theory of the anomalous photoelectric effect stemming from a substructure of matter waves, *Ind. J. Theor. Phys. 49*(1), 1–32. arXiv:quant- ph/9906091.
249. Berestetskii, V. B., Lifshitz, E. M., & Pitaevskii, L. P. (1980). *Quantum Electrodynamics*, Nauka, Moscow, p. 28; in Muscovian.
250. Putterman, S. J., & Weninger, K. R. (2000). Sonoluminescence: How bubbles turn sound into light, *Annu. Rev. Fluid Mech. 32*, 445–476.
251. Brenner, M. P., Hilgenfeldt, S., & Lohse, D. (2002). Single-bubble sonoluminescence, *Rev. Mod. Phys. 74*, 425–484.
252. Cheeke, J. D. N. (2012). *Fundamentals and Applications of Ultrasonic Waves*, Second edition, CRC Press, Taylor & Francis Group, Ch. 12, Cavitation and sonoluminescence, pp. 187–213.
253. Keller, J. B., & Miksis, M. (1980). Bubble oscillations of large amplitude, *J. Acoust. Soc. Am. 68*, 628–633.
254. Vazquez, G., Camara, C., Putterman, S., & Weninger, K. (2001). Sonoluminescence: nature's smallest blackbody, *Opt. Lett. 26*, 575–577.
255. Kappus, B., Khalid, S., Chakravarty, A., & Putterman, S. (2011). Phase transition to an opaque plasma in a sonoluminescing bubble, *Phys. Rev. Lett. 106*, 234302.
256. Kappus, B., Bataller, A., & Putterman, S. J. (2013). Energy balance for a sonoluminescence bubble yields a measure of ionization potential lowering, *Phys. Rev. Lett. 111*, 234301.
257. Zhang, W.-J., & An, Y., Abnormal ionization in sonoluminescence, *Chin. Phys. B 24*, No. 4. (2015). 047802 (7 pp.).
258. Krasnoholovets, V. (2016). Sound into light: on the mechanism of the phenomenon of sonoluminescence, *J. Adv. Phys. 5*(2), 168–175.
259. Brennan, T. E., & Fralick, G. C. (2015). The timing of sonoluminescence, arXiv:1111.5229 [physics.gen].
260. Krasnoholovets, V., & Byckov, V. (2000). Real inertons against hypothetical gravitons. Experimental proof of the existence of inertons, *Ind. J. Theor. Phys. 48*(1), 1–23. arXiv:quant-ph/0007027.
261. Navia, C. E., Augusto, C. R. A., Franceschini, D. F., Robba, M. B., & Tsui, K. H. (2007). Search for anisotropic light propagation as a function of laser beam alignment relative to the Earth's velocity vector, *Progress in Physics, 1*, 53–60.

262. Levengood, W. C. (1994). Anatomical anomalies in crop formation plants, *Physiologia Plantarum 92*, 356–363.
263. Levengood, W. C., & Bruke, J. (1995). Semi-molten meteoric iron associated with a crop formation, *J. Scient. Exploration 9(2),* 191–199.
264. Bruke, J. A. (1998). The physics of crop formations, *MUFON Journal*, October, pp. 3–7.
265. Levengood, W. C., & Talbott, N. P. (1999). Dispersion of energies in worldwide crop formations, *Physiologia Plantarum 105*, 615–624.
266. Haselhoff, E. H. (1999). Opinions and comments on Levengood WC, Talbott NP Dispersion of energies in worldwide crop formations, *Physiologia Plantarum J. 105*, 615–624; *111*, 123–125. (2001).
267. Grassi, F., Cocheo, C., & Russo, P. (2005). Balls of light: The questionable science of crop circles, J. Scientific Exploration 19(2), pp. 159–170.
268. Pratsch, J.-C. (1990). Relative motion in geology: some philosophical differences, *J. Petroleum Geology 13*(2), 229–234.
269. Munk, W. H., & Macdonald, G. J. F., *The Rotation of the Earth. A Geophysical Discussion* (Cambridge University Press, London, 1975).
270. Jeffreys, H., *The Earth: Its Origin, History and Physical Constitution* (Cambridge University Press, London, 1976).
271. Yamazaki, K. (2007). Possible mechanism of earthquake triggering due to magnetostriction of rocks in the crust, *American Geophysical Union, Fall Meeting 2007*, abstract #S33B-1307.
272. Krasnoholovets, V., & Gandzha, I. (2012). A submicroscopic description of the formation of crop circles, *Chaotic Modeling and Simulation 2*, 323–335.
273. Landau, L. D., & Lifshits, E. M. (1987). *The Theory of Elasticity*, Nauka, Moscow, pp. 106– 107; in Muscovian.
274. Monroe, J. S., & Wicander, R. (2015). *The Changing Earth: Exploring Geology and Evolution* Seventh edition, Cengage Learning, Stamford, p. 252.
275. Gell-Mann, M. (1961). The eightfold way: A theory of strong interaction symmetry, *Caltech Synchrotron Laboratory Report* No. CTSL-20.
276. Ne'eman, Y. (1961). Derivation of strong interactions from a gauge invariance, *Nucl. Phys. 26*(2), 222–229.
277. Gell-Mann, M. (1964). A schematic model of baryons and mesons, *Phys. Lett. 8*, 214–215.
278. Zweig, G. (1964). An SU3 model for strong interaction symmetry and its – breaking, *CERN Report* No. 8419/TH.412, Geneva, Fractionally charged particles and SU6, in *Symmetries In Elementary Particle Physics*, Ed.: Zichichi, A., Academic Press, New York, 1965, p. 192.
279. Greenberg, O. W. (1964). Spin and unitary spin independence in a paraquark model of baryons and mesons. *Phys. Rev. Lett. 13*, 598–602.
280. Feynman, R. P., The behavior of hadron collisions at extreme energies, in *High Energy Collisions: Third International Conference at Stony Brook, New* York (Gordon & Breach, New York, 1969), pp. 237–249.
281. Nambu, Y., & Jona-Lasinio, G. (1961). Dynamical model of elementary particles based on an analogy with superconductivity, *Phys. Rev. 122*(1), 345–358. *Phys. Rev. 124*(1), 246–254.

282. Skyrme, T. H. R. (1961). A Non-linear field theory, *Proc. R. Soc. Lond. A 260*, No. 1300, 127–138. A unified field theory of mesons and baryons, *Nucl. Phys. 31*, 556–569. (1962).
283. Chodos, A., Jaffe, R. L., Johnson, K., Thorn, C. B., & Weisskopf, V. F. (1974). A new extended model of hadrons. *Phys. Rev D 9*, 3471–3495.
284. Chodos, A., Jaffe, R. L., Johnson, K., & Thorn, C. B. (1974). Baryon structure in the bag theory, *Phys. Rev. D 10*(8), 2559–2604.
285. Johnson, K. (1975). The M.I.T. bag model, *Acta Phys. Polonica B 6*(6), 865–892.
286. T. DeGrand, Jaffe, R. L., K. Johnson, & J. Kiskis, (1975). Masses and other parameters of the light hadrons, *Phys. Rev. D 12*(7), 2060–2076.
287. Rho, M., Goldhaber, A. S., & Brown, G. E. (1983). Topological soliton bag model for baryons, *Phys. Rev. Lett. 51*, 747–750.
288. Rajasekaran, G., & Rindani, S. D. (1982). Integer-charged quark model and electron-positron annihilation into three jets, *Progr. Theor. Phys. 67*(5), 1505–1531.
289. Ferreira, P. M. (2002). Can we build a sensible theory with broken charge and colour symmetries?, arXiv:hep-ph/0210024.
290. Ferreira, P. M. (2013). Do LEP results suggest that quarks have integer electric charges?, arXiv:hep-ph/0209156.
291. LaChapelle, J. (2004). Quarks with integer electric charge, arXiv:hep-ph/0408305.
292. Van Hieu, N. (1967). *Lectures on the Theory of Unitary Symmetry in Elementary Particle Physics*, Atomizdat, Moscow, 344 p.; in Muscovian.
293. Rumer, Yu. B., & Fet, A. I. (1970). *The Theory of Unitary Symmetry*, Nauka, Moscow, in Muscovian.
294. Nambu, Y. (1985). *Quarks: Frontiers in Elementary Particle Physics*, World Scientific, Singapore.
295. Fritzsch, H. (2009). The fundamental constants in physics, *Phys. Usp. 52*, 359–367. arXiv:0902.2989 [hep-ph].
296. Bethke, S. (2007). Experimental tests of asymptotic freedom, *Prog. Part. Nucl.Phys. 58*, 351–386. arXiv:hep-ex/0606035.
297. Bethke, S. (2010). Data preservation in high energy physics – why, how and when?, *Nucl. Phys. Proc. Suppl. 207–208*, 156–159. arXiv:1009.3763 [hep-ex].
298. Hasenfratz, P., & Kuti, J. (1978). The quark bag model, *Phys. Reports 40*(2), 75–179.
299. Geguzin, Ya. E. (1985). *Bubbles*, Nauka, Moscow, p. 41; in Muscovian.
300. Miskimen, R. (2011). Neutral pion decay, *Ann. Rev. Nucl. Particle Sci. 61*, 1–21.
301. Csaki, C., Shirman, Yu., & Terning, J. (2011). A Seiberg dual for the MSSM: Partially composite W and Z, *Phys. Rev. D 84*, 095011.
302. Gilman, R., & Gross, F. (2002). Electromagnetic structure of the deuteron, *J. Phys. G: Nucl. Part. Phys. 28,* R37–R116.
303. Islam, M., Luddy, R., Kašpar, J., & Prokudin, A. (2009). Picturing the proton by elastic scattering, *CERN Courier 49*(10), 35.
304. Islam, M. M., Kašpar, J., Luddy, R. J., & Prokudin, A. V. (2009). Proton-proton elastic scattering at LHC and proton structure, in *13th International Conference on Elastic and Diffractive Scattering: Moving Forward into the LHC Era*, CERN, Geneva, Switzerland, 29 Jun–03 Jul, pp. 48–54 *CERN Document Server*, http://cdsweb.cern.ch/record/1247028.

305. Islam, M. M., Kašpar, J., & Luddy, R. J. (2011). Proton Structure and Prediction of Elastic Scattering at LHC at Center-of-Mass Energy 7 TeV, *Proc. Eleventh Workshop Non- Perturbat. Quant. Chromodynamics*, l'Institut Astrophysique de Paris June 6–10, Eds.: Mueller, B., M. A. Rotondo and Ch.-I Tan, http://www.slac.stanford.edu/econf/C1106064/.
306. Friedberg, R., & Lee, T. D. (1977). Fermion-field nontopological solitons. II. Models for hadrons, *Phys. Rev. D 16*, 1096–1118.
307. Carlson, C. E. (2015). The proton radius puzzle, *Progr. Particle and Nucl. Phys. 82*, 59–77.
308. Lorenz, I. T., H.-Hammer, W., & U.-Meißner, G. (2012). The size of the proton: Closing in on the radius puzzle, *Eur. Phys. J. A: Hadrons and Nuclei 48*(11), 1–5 (#151).
309. Pohl, R., Gilman, R., Miller, G. A., & Pachucki, K. (2013). Muonic hydrogen and the proton radius puzzle, arXiv:1301.0905 [physics.atom-ph].
310. Bernauer, J. C. (2014). Proton charge radius and precision tests of QED arXiv:1411.3743 [nucl-ex].
311. Horbatsch, M., & Hessels, E. A. (2015). Evaluation of the strength of electron-proton scattering data for determining the proton charge radius, arXiv:1509.05644.
312. Lorenz, T., & Meißner, U.-G. (2014). Reduction of the proton radius discrepancy by 3 sigma, *Phys. Lett. B 737*, 57–59. arXiv:1406.2962 [hep-ph].
313. Mohapatra, R. N., Antusch, S., Babu, K. S., Barenboim, G., Chen, M.-C., de Gouvêa, A., de Holanda, P., Dutta, B., Grossman, Y., Joshipura, A., Kayser, B., Kersten, J., Keum, Y. Y., King, S. F., Langacker, P., Lindner, M., Loinaz, W., Masina, I., Mocioiu, I., Mohanty, S., Murayama, H., Pascoli, S., Petcov, S. T., Pilaftsis, A., Ramond, P., Ratz, M., Rodejohann, W., Shrock, R., Takeuchi, T., Underwood, T., & Wolfenstein, L. (2007).Theory of neutrinos: a white paper, *Rep. Prog. Phys. 70*(11), 1757.
314. Altarelli, G., & Feruglio, F. (2004). Models of neutrino masses and mixings, *New J. Phys. 6*, 106.
315. Dwyer, D. A. (2014). The neutrino mixing angle θ_{13}: Reactor and accelerator experiments, *Phys. Dark Universe 4*, 31–35.
316. Cohen, A. G., & Glashow, S. L. (2011). New constraints on neutrino velocities, arXiv:1109.6562 [hep-th].
317. Kuhn, S. E., Chen, J.-P., & Leader, E., (2009). Spin structure of the nucleon – status and recent results, *Prog. Part. Nucl. Phys. 63*, 1–50. arXiv:0812.3535 [hep-ph].
318. Adare, A. et al. (2011). (PHENIX Collaboration). Cross section and parity-violating spin asymmetries of $W^{\pm}$ boson production in polarized $p+p$ collisions at s =500 GeV, *Phys. Rev. Lett. 106*(6), 062001, 6 pp.
319. Aggarwal, M. M. et al. (2011). (STAR Collaboration). Measurement of the parity-violating longitudinal single-spin asymmetry for $W^{\pm}$ boson production in polarized proton-proton collisions at s=500 GeV, *Phys. Rev. Lett. 106*(6), 062002, 6 pp.
320. Comay, E. (2011). Spin, isospin and strong interaction dynamic, arXiv:1107.4688 [physics.gen-ph].
321. Krasnoholovets, V. (2006). Reasons for nuclear forces in light of the constitution of the real space, *Scientific Inquiry 7*(1), 25–50. arXiv:1104.2484 [physics.gen-ph].
322. Koshlyakov, N. S., Gliner, E. B., & Smirnov, M. M. (1970). *Equations in Partial Derivatives of Mathematical Physics*, Vysshaya Shkola, Moscow, pp. 184–1987; in Muscovian.

323. Sytenko, O. H., & Tartakovsky, V. K. (2000). *The Theory of Nucleus*, Lybid, Kyiv, p. 185; in Ukrainian.
324. Borghi, Giori, C., & Dall'Ollio, A. A. (1993). Experimental evidence of emission of neutrons from cold hydrogen plasma, *American Institute of Physics* (*Phys. At. Nucl.*) *56*, No. 7.
325. Dufour, C. J. (1993). Cold fusion by sparking in hydrogen isotopes, *Fusion Technol. 24*, 205–228.
326. Dufour, J., Murat, D., Dufour, X., & Foos, J. (2004). Exothermic reaction induced by high density current in metals – Possible nuclear origin, *Ann. Fond. L. de Broglie 29*, 1081–1093.
327. Dufour, J. J., Dufour, X. J. C., & Vinko, J. D. (2013). Pico-chemistry: The possibility of new phasein some hydrogen/metal systems, *Int. J. Mod. Phys. B 27*, 1362038, 8 pp.
328. Mills, R., Nansteel, M., & Ray, P. (2002). Argon-hydrogen-strontium discharge light source, *IEEE Transact. Plasma Sci. 30*(2), 639–653.
329. Santilli, R. M. (2006). Confirmation of Don Borghi's experiment on the synthesis of neutrons from protons and electrons, *Preprint IBR-EP*-39 of 12–25–06; arXiv.org: physics/0608229.
330. Taleyarkhan, R. P., West, C. D., Cho, J. S., Lahey, R. T. (Jr.), Nigmatulin, R. I., & Block, R. C. (2002). Evidence for nuclear emissions during acoustic cavitation, *Science 295*, 1868– 1873.
331. Nigmatulin, R. I., Lahey, R. T. (Jr.), Block, R. C., Taleyarkhan, R. P., & West, C. D. (2014). On thermonuclear processes in cavitation bubbles, *Uspekhi Fizich. Nauk 184*, 947–960. in Muscovian.
332. Naranjo, B., Gimzewski, J. K., & Putterman, S. (2005). Observation of nuclear fusion driven by a pyroelectric crystal, *Nature 434*, 1115–1117.
333. Geuther, J., Danon, Y., & Saglime, F. (2006). Nuclear reactions induced by a pyroelectric accelerator. *Phys. Rev. Lett. 96*, 054803.
334. Danon, Y., (2012). Pyroelectric crystal D-D and D-T neutron generators. *J. Instrumentation* (*JINST*) *7*, C04002, 4 pp.
335. Cardone, F., Carpinteric, A., & Lacidogna, G. (2009). Piezonuclear neutrons from fracturing of inert solids, *Phys. Lett. A 373*, 4158–4163.
336. Cardone, F., Mignani, R., & Petrucci, A. (2009). Piezonuclear decay of thorium, *Phys. Lett. A 373*, 1956–1958.
337. Krasnoholovets, V., Zabulonov, Yu., & Zolkin, I. (2016). On the nuclear coupling of proton and electron, *Univer. J. Phys. Appl. 10*(3), 90–103.
338. FitzGerald, F. B. (2009). Sonoluminescence and sonofusion – how and why, *The General Science J.*, 16 Jan 2009; http://www.gsjournal.net/Science-Journals/Essays/View/1922.
339. Kervran, C. L. (1998). *Biological Transmutations*, Ed.: J. de Langre, second printing, Happiness Press, Magalia, California.
340. Robitaille, P.-M. (2013). Forty lines of evidence for condensed matter—the Sun on trial: liquid metallic hydrogen as a solar building block, *Progress in Physics 4*, 90–142.
341. Rossi, A. (2009). International Patent Application WO 2009125444, Method and Apparatus for carrying out nickel and hydrogen exothermal reactions.

342. Kadoshnikov, V., Lytvynenko, Yu., Zabulonov, Yu., & Krasnoholovets, V. (2016). Nanocomposites for decontamination of multicomponent technogenic dilutions, *J. Nucl. Phys., Material Sciences, Radiation and Applications* *3*(2), 279–292.
343. Woodard, R. P. (2009). How far are we from the quantum theory of gravity?, arXiv:0907.4238[gr-qc].
344. Weinstein, S., & Rickles, D. (2015). *Quantum gravity*, Stanford Encyclopedia of Philosophy, http://plato.stanford.edu/entries/quantum-gravity/.
345. Misner, C., Thorne, K., & Wheeler, J. (1973). *Gravitation*, Freeman, San Francisco.
346. Giulini, D. (2009). Matter from space, arXiv:0910.2574v2 [physics.hist-ph] (37 p.).
347. Asselmeyer-Maluga, T., & Rosé, H. (2012). On the geometrization of matter by exotic smoothness, *Gen. Relat. Gravit.* *44*(11), 2825–2856. arXiv:1006.2230v6 [gr-qc].
348. Sakharov, A. D. (1967). Vacuum quantum fluctuations in curved space and the theory of gravitation, *Doklady Akad. Nauk SSSR* *177*(1), 70–71; in Muscovian.
349. Boyer, T. H. (1973). Retarded van der Waals forces at all distances derived from classical electrodynamics with classical electromagnetic zero-point radiation, *Phys. Rev. A* *7*(6), 1832– 1840.
350. Puthoff, H. E. (1988). Zero-Point Fluctuations of the Vacuum as the Source of Atomic Stability and the Gravitational Interaction, *Proc. of the British Soc. for the Philosophy of Science, Intern. Conf.* "Physical Interpretations of Relativity Theory," Imperial College, London, Ed.: M. C. Duffy (Sunderland Polytechnic).
351. Davies, P. C. W. (1982). Spontaneously generated gravity and the second law of thermodynamics, *Phys. Lett. B* *110*(2), 111–113.
352. Haisch, B., Rueda, A., & Puthoff, H. E. (1994). Inertia as a zero-point field Lorentz Force, *Phys. Rev. A 49*, 678–694.
353. Haisch, B., & Rueda, A. (2005). Gravity and the quantum vacuum inertia hypothesis, *Ann. Phys.* [Leipzig] *14*(8), 479–498.
354. Levin, Y. S. (2009). Inertia as a zero-point-field force: Critical analysis of the Haisch-Rueda- Puthoff inertia theory, *Phys. Rev. A 79*, 012114. [14 pp.].
355. Pietschmann, H. (1988). Progress in our notion of mass, *Acta Phys. Hungar. 64*, No. 1–3, 7– 13.
356. Okun, L. B. (1989). The concept of mass. Mass, energy, relativity, *Usp. Fiz. Nauk* *158*(3), 511; in Muscovian.
357. Okun, L. B. (2000). On the letter of R. I. Khrapko "What is mass?" *Usp. Fiz. Nauk* *170*(12), 1366–1371; in Muscovian.
358. Wilczek, F. (2006). The origin of mass, *Mod. Phys. Lett.* A *21*, 9, 701–12.
359. Hecht, E. (2011). On defying mass, *The Physics Teacher* – January, *49*(1), 40–44.
360. Thieman, T. (2001). Introduction to modern canonical quantum general relativity, arXiv:gr- qc/0110034.
361. Perez, A. (2003). Spin foam models for quantum gravity, *Classical Quant. Grav. 20*, R43- R104. arXiv:gr-qc/0301113.
362. Livine, E. (2011). Spinfoam framework for quantum gravity, arXiv:1101.5061 [gr-qc].
363. Preparata, G. (1995). Quantum gravity, the Planck lattice and the standard model, arXiv:hep- th/9503102.

364. Rovelli, C. (2011). A new look at loop quantum gravity, *Class. Quant. Grav. 28*, 114005.
365. Rovelli, C. (2011). Zakopane lectures on loop gravity, arXiv:1102.3660 [gr-qc].
366. Sarfatti, J. (2004). Wheeler's world: It from bit? in *Progress in Quantum Physics Research*, Eds.: Columbus, F., & Krasnoholovets, V., Nova Science Publishers, New York, pp. 41–84.
367. Subir, S. (2013). Strange and stringy, *Scient. Amer. 308*(44), 44–51.
368. Szabo, R. J. (2002). BUSSTEPP lectures on string theory. An introduction to string theory and D-brane dynamics, arXiv:hep-th/0207142.
369. The Official String Theory Web Site: http://superstringtheory.com/.
370. Mohaupt, T. (2003). Introduction to string theory, *Lect. Notes Phys. 631*, 173–251.
371. Marolf, D. (2004). Resource Letter NSST-1: The nature and status of string theory, *Am. J. Phys. 72*, 730–741.
372. Yamaguchi, M. (2011). Supergravity based inflation models: A review, arXiv:1101.2488 [astro-ph.CO].
373. De Klerk, D., Murugan, J., & Uzan, J. P. (2011). The catenary revisited: from Newtonian strings to superstrings, arXiv:1103.0788 [physics.class-ph].
374. Zapata Marín, O. (1995). On facts in superstring theory, arXiv:0905.1439 [physics.hist-ph].
375. Higgs, P. W. (1964). Broken symmetries and the masses of gage bosons, *Phys. Rev. Lett. 13*(16), 508–509.
376. CERN experiments observe particle consistent with long-sought Higgs Boson, http://press.web.cern.ch/press-releases/2012/07/cern-experiments-observe-particle-consistent-long-sought-higgs-boson.
377. Balachandran, A. P. (2002). Quantum spacetimes in the year 2002, *Pramana 59*(2), 359–368.
378. Yang, H. S. (2010). Emergent geometry and quantum gravity, *Mod. Phys. Lett. A 25*, 2381– 2397.
379. Kleinert, H., & Zaanen, J. (2004). Nematic world crystal model of gravity explaining absence of torsion in spacetime, *Phys. Lett. A 324*, 361–365.
380. Verlinde, E. (2011). On the origin of gravity and the laws of Newton, *HEJP*, *1104* 029. also arXiv:1001.0785 [hep-th].
381. Ho, C. M., Minic, D., & Ng, Y. J. (2011). Quantum gravity and dark matter, *Gen. Relativ. Gravit. 43*, 2567–2573.
382. Milgrom, M. (1983). A modification of the Newtonian dynamics as a possible alternative to the hidden mass hypothesis, *Astrophys. J. 270*, 365–370.
383. Padmanabhan, T. (2010). Thermodynamical aspects of gravity: new insights, *Rep. Prog. Phys. 73*, 046901.
384. Padmanabhan, T. (2010). Equipartition of energy in the horizon degrees of freedom and the emergence of gravity. *Mod. Phys. Lett. A 25*, 1129–1136.
385. Jacobson, T. (1995). Thermodynamics of spacetime: The Einstein equation of state, *Phys. Rev. Lett. 75*, 1260–1263.
386. de Haas, E. P. J. (2004). The combination of de Broglie's harmony of the phases and Mie's theory of gravity results in a principle of equivalence for quantum gravity, *Ann. Fond. L. de Broglie 29*, No. 4., 707–726.

387. Mie, G. (1912). Grundlagen einer Theorie der Materie. Part 1, *Ann. Phys. 37*, 511–534.
388. Mie, G. (1912). Grundlagen einer Theorie der Materie. Part 1, *Ann. Phys. 39*, 1–40.
389. Mie, G. (1913). Grundlagen einer Theorie der Materie. Part 3, *Ann. Phys. 40*, 1–66.
390. Winterberg, F. (2002). *The Planck Aether Hyposethis*, Carl Friedrich Gauss Academy of Science Press, Reno, Nevada.
391. Arminjon, M. (2008). Ether theory of gravitation: why and how? In *Ether Spacetime and Cosmology, Vol. 1. Modern Ether Concepts, Relativity and Geometry.* Eds.: Duffy, M. C., Levy, J., & Krasnoholovets, V., PD Publications, Liverpool, pp. 139–201.
392. Arminjon, M. (2002). The scalar ether-theory of gravitation and its first test in celestial mechanics, *Int. J. Mod. Phys. A 17*, 4203–4208.
393. Suntola, T. (2012). *The Dynamic Universe. Toward a Unified Picture of Physical Reality*, Third edition, Physics Foundations Society, Finland.
394. Sorli, A. (2011). Change of density of quantum vacuum might generate mass, http://vixra.org/abs/1111.0015.
395. Cahill, R. T. (2009). Dynamical 3-space: a review, in *Ether Spacetime and Cosmology, Vol. 2. New Insights into a Key Physical Medium*. Eds.: Duffy, M. C., & J. Levy, Apeiron, Montreal, pp. 135–200.
396. Cahill, R. T. (2006). A new light-speed anisotropy experiment: absolute motion and gravitational waves detected, *Progress in Physics 4*, 73–92.
397. Múnera, H. A. (2005). The evidence for length contraction at the turn of the 20th century: nonexistent, in *Einstein and Poincaré: The Physical Vacuum,* Ed.: Dvoeglazov, V. V., Apeiron, Montreal, pp. 77–92.
398. Atsyukovskiy, V. A., Ed. (1993). Ether wind, Energoatombzdat, Moscow, in Muscovian.
399. Galaev, Yu. M. (2002). The measuring of ether-drift velocity and kinematic ether viscosity within optical waves band, *Spacetime & Substance 3*(5), 207–224.
400. Baurov, Yu. A. (2002). Structure of physical space and nature of de Broglie waves (theory and experiment), *Ann. Fond. L. de Broglie*, *27*(3), 443–461.
401. Konushko, V. I. (2011). Granular space and the problem of large numbers, *J. Mod. Phys. 2*, 289–300.
402. Isham, C. J. (1993). Prima facie questions in quantum gravity, ArXiv:gr-qc/9310031.
403. Toh, T.-C. (1996). Quantum gravity: A brief review, in *Proc. First Australian Conference on General Relativity and Gravitation*, Ed.: D. L. Wiltshire (Inst. for Theor. Phys., Univ. Adelaide, South Australia, 12–17 Feb., 1996), pp. 196–209.
404. Krasnoholovets, V. (2008). Reasons for gravitational mass and the problem of quantum gravity, in *Ether, Space-time and Cosmology, Vol. 1. Modern Ether Concepts, Relativity and Geometry*. Eds.: Duffy, M. C., Lévy, J., & Krasnoholovets, V., PD Publications, Liverpool, pp. 419–450; also arXiv:1104.5270 [physics.gen-ph].
405. Krasnoholovets, V. (2014).Inerton field effects in nanosystems, in *Quantum Nanosystems: Structure, Properties and Interactions*, Ed.: Putz, M. V., Apple Academic Press, Toronto, pp. 59–102.
406. Valkering, A. M. C., Mares, A. I., Untiedt, C., K. Babaei Gavan, Oosterkamp, T. H., & J. M. van Ruitenbeek, (2005). A force sensor for atomic point contacts, *Rev. Scientific Instruments 76*, 103903, 5 pp.

407. Casimir, H. B. G. (1948). On the attraction between two perfectly conducting plates, *Proc. K. Ned. Akad. Wet. 51*, 793–795.
408. Jaffe, R. L., & Scardicchi, A. (2004). The Casimir effect and geometric optics, *Phys. Rev. Lett. 92*, 070402. arXiv:quant-ph/0310104.
409. Jaffe, R. L. (2005). The Casimir effect and the quantum vacuum, *Phys. Rev. D 72*(2), 021301. arXiv:hep-th/0503158.
410. Kac, V., & Cheung, P. (2001). *Quantum Calculus*, Springer, p. 87 and 93.
411. Lee, T. D. (1962). *Mathematical Methods of Physics. A Course of Lectures Given at Columbia University*. Columbia University Press, New York, Ch. 7, Sect. 3.
412. Kreidik, L., & Shpenkov, G. (2000). Roots of Bessel functions define spectral terms of micro- and megaobjects, *Book of Abstracts, XIII International Congress on Mathematical Physics* (17–22 July 2000, Imperial College, London), London, p. 118.
413. Shpenkov, G. P. (2008). On the superluminal speed in view of the dialectical model of the universe, *Infinite Energy 13*(77), 29–34.
414. Shpenkov, G. P. (2014). *Dialectical View of the World: The Wave Model (Selected Lectures)*. Vol. 3. *Dynamic Model of Elementary Particles,* Part 2. *Fundamentals,* Lecture 1, pp. 8–21, http://shpenkov.janmax.com/Vol.3.DynamicModel-2.pdf.
415. Olver, F. W. J., Ed. (1960). *Royal Society Mathematical Tables*, Vol. 7, *Bessel Functions*, Part. III, Zeros and Associated Values, Cambridge University Press, Cambridge, England.
416. Kreidik, L., & Shpenkov, G. (2001). *Atomic Structure of Matter-Space*, Bydgoszcz.
417. Shpenkov, G. P. (2006). An elucidation of the nature of the periodic law, in *The Mathematics of the Periodic Table*, Eds.: Rouvray, D. H., & King, R. B., Nova Science Publishers, New York, pp. 119–160.
418. Shpenkov, G. P. (2011). Physics and chemistry of carbon in the light of shell-nodal atomic model, in *Quantum Frontiers of Atoms and Molecules*, Ed.: Putz, M. V., Nova Science Publishers, New York, pp. 277–324.
419. Shpenkov, G. P. (2013). *Dialectical View of the World: The Wave Model (Selected Lectures)*. Vol. 3. *Dynamic Model of Elementary Particles,* Part 1. *Fundamentals,* Lecture 6, http://shpenkov.janmax.com/Vol.2.DynamicModel-1.pdf.
420. Ilyanok, A. M. (1999). Quantum astronomy. Part 1. Energy of stars and the Hollow Sun (nonthermonuclear approach), arXiv:astro-ph/9912537.
421. Rubčić, A., & Rubčić, J. (1998). The quantization of the solar-like gravitational systems, *Fizika B 7*(1), 1–14.
422. Chechelnitsky, A. (2001). Hot points of the wave universe concept, arXiv:physics/0102036.
423. Nottale, L., Schumacher, G., & Lefèvre, E. T. (2000). Scale-relativity and quantization of exoplanet orbital semi-major axes, *Astronom. Astrophys. 361*, 379–387.
424. Ilyanok, A., & Timoshenko, I. A. (1999). Quantization of masses in the solar system, physics/9912537. arXiv:physics/0201057/2002.
425. Christianto, V. (2004). Comparison of predictions of planetary quantization and implications of the Sedna finding, *Apeiron 11*(3), 82–98.
426. Smarandache, F., & Christianto, V. (2006). Schrödinger equation and the quantization of celestial systems, *Progress in Phys. 2*, April, 63–67.
427. Van Flandern, T. (1998). The speed of gravity – what the experiments say, *Meta Research Bulletin 6*, 49. (1997) [*Physics Letters A 250*, Nos. 1–3, 1–11.].

428. Poincaré, H. (1906). Sur la dynamique de l'électron, *Rendiconti del Circolo matematico di Palermo 21*, 129–176. also: *Oeuvres*, t. *IX*, 494–550. (in Muscovian translation: *Henri Poincaré. Selected Transactions in Three Volumes*, Vol. 3. Ed.: Bogolubov, N. N., Nauka, Moscow, 1974, pp. 429–486).
429. Guy, B. (2010). A modified law of gravitation taking account of the relative speeds of moving masses. A preliminary study, http://hal.archives-ouvertes.fr/hal-00472210.
430. Krasnoholovets, V. (2009). On microscopic interpretation of phenomena predicted by the formalism of general relativity, in *Ether Space-Time and Cosmology, Vol. 2: New Insights Into a Key Physical Medium*. Eds.: M. C., J. Lévy (Apeiron, 2009), pp. 417–431; also in *Apeiron 16(3)*, 418–438.
431. Krasnoholovets, V. (2014). On the gravitational time delay effect and the curvature of space, Ed.: Dubois, D. M., *Int. J. Comput. Anticipat. Systems 27*, 137–147.
432. Bergmann, P. G. (1976). *Introduction to the Theory of Relativity*, 2nd edition, Dover Publication, Inc., New York.
433. Shapiro, I. I. (1964). Fourth test of general relativity, *Phys. Rev. Lett. 13*(26), 789–791.
434. Shapiro, I. I., Pettengill, G. H., Ash, M. E., Stone, M. L., Smith, W. B., Ingalls, R. P., & Brockelman, R. A. (1968). Fourth test of general relativity: Preliminary results, *Phys. Rev. Lett. 20*(22), 1265–1269.
435. Weinberg, S. (1972). *Gravitation and Cosmology: Principles and Application of General Theory of Relativity*, John Wiley & Sons, Inc., New York, London, Sydney, Toronto, pp. 201–207.
436. Taylor, J. H., Weinberg, J. M. (1989). Further experimental tests of relativistic gravity using the binary pulsar PSR 1913+16, *Astrophys. J. 345*, 434–450.
437. Loinger, A. (2002). *On Black Holes and Gravitational Waves*, La Goliardica Pavese.
438. Loinger, A. (2002). Relativistic motion, *Spacetime & Substance 3*(3), 129–129.
439. Loinger, A. (2003). On Einsteinian orbits of celestial bodies, *Spacetime & Substance 4*(2), 74–75.
440. Loinger, A. (2007). *More on BH's and GW's*. III, La Goliardica Pavese.
441. Loinger, A., & Marsico, T. (2003). The supermassive center of our galaxy et cetera, *Spacetime & Substance 4*(2), 80–81.
442. Crothers, S. J. (2005). On the ramifications of the Schwarzschild space-time metric, *Progress in Physics 1*, 74–80.
443. Crothers, S. J. (2010). The black hole, the Big Bang: A cosmology in crisis, *Proc. NPA*, vixra:1103.0047.
444. Crothers, S. J. (2011). The black hole catastrophe: A short reply to Sharples, J. J., *Hadronic J. 34*, 197–224. also vixra:1206.0080.
445. Crothers, S. J. (2012). Proof of no "black hole" binary in Nova Scorpii, *Global J. Sci. Frontier Research Phys. and Space Science 12*, No. 4. also vixra:1206.0080.
446. Crothers, S. J. (2012). General relativity – a theory in crisis, *Global J. Sci. Frontier Research Phys. and Space Science 12*, No. 4. also vixra:1207.0018.
447. Crothers, S. J. (2012). Proof of the invalidity of the black hole and Einstein's field equations vixra:1212.0060.
448. Crothers, S. J. (2015). Black hole escape velocity – a case study in the decay of physics and astronomy, vixra:1508.0066.

449. Crothers, S. J. (2015). A few things you need to know to tell if a mathematical physicist is talking nonsense: the black hole – a case study, vixra:1508.0007.
450. Crothers, S. J. (2016). The Painlevé-Gullstrand 'extension' – a black hole fallacy, *Am. J. Modern Phys. 5*(1), 33–39.
451. Rabounski, D. (2008). On the current situation concerning the black hole problem, *Progress in Phys. 1*, 101–103.
452. Abbott, B. P. et al. (2016). (LIGO Scientific Collaboration and Virgo Collaboration), Observation of gravitational waves from a binary black hole merger, *Phys. Rev. Lett. 116*, 061102.
453. Baumgarte, T. W., & Shapiro, S. L. (2011). Binary black hole mergers, *Physics Today 64*(10), 32–37.
454. Crothers, S. J. (2016). A critical analysis of LIOGO's recent detection of gravitational waves caused by merging black holes, *Hadronic J. 39*, 1–32. also http://vixra.org/abs/1603.0127.
455. Loinger, A., & Marsico, T. (2016). A detailed confutation of LIGO's statements on the 150914-signal, https://www.researchgate.net/publication/301542915, April 20.
456. Penzias, A. A., & Wilson, R. W. (1965). A measurement of excess antenna temperature at 4080 Mc/s, *Astrophys. J. 142*, 419–421.
457. Hoyle, F., Burbidge, G., & Narlikar, J. V. (1993). A quasi-steady state cosmological model with creation of matter, *Astrophys. J. 410*, 437–457.
458. Hoyle, F., Burbidge, G., & Narlikar, J. V. (1994). Further astrophysical quantities expected in a quasi-steady state Universe, *Astronomy Astrophys. 289*, no. 3, 729–739.
459. Thornhill, C. K. (2001). A non-singular ethereal cosmology, *Hadronic J. Suppl. 16*, 203–262.
460. COBE satellite, (2009). http://lambda.gsfc.nasa.gov/product/cobe/.
461. Hinshaw, G., Weiland, J. L., Hill, R. S., Odegard, N., Larson, D., Bennett, C. L., Dunkley, J., Gold, B., Greason, M. R., Jarosik, N., Komatsu, E., Nolta, M. R., Page, L., Spergel, D. N., Wollack, E., Halpern, M., Kogut, A., Limon, M., Meyer, S. S., Tucker, G. S., & Wright, E. L. (2009). Five-year Wilkinson microwave anisotropy probe (WMAP1) observations: data processing, sky maps, & basic results, *Astrophys. J. Suppl. 180*, 225–245. also arXiv:0803.0732 [astro-ph].
462. Hinshaw, G., Larson, D., Komatsu, E., Spergel, D. N., Bennett, C. L., Dunkley, J., Nolta, M. R., Halpern, M., Hill, R. S., Odegard, N., Page, L., Smith, K. M., Weiland, J. L., Gold, B., Jarosik, N., Kogut, A., Limon, M., Meyer, S. S., Tucker, G. S., Wollack, E., & Wright, E. L. (2013). Nine-year Wilkinson microwave anisotropy probe (WMAP) observations: cosmological parameter results, *Astrophys. J. Suppl. 208*(2), article id. 19, 25 pp. arXiv:1212.5226 [astro-ph.CO].
463. WMAP satellite. (2011). Tests of Big Bang: The CMB, http://map.gsfc.nasa.gov/universe/bb_tests_cmb.html.
464. Rocha, G., Contaldi, C. R., Colombo, L. P. L., Bond, J. R., Gorski, K. M., & Lawrence, C. R. (2010). Application of XFASTER power spectrum and likelihood estimator to Planck, *Mon. Not. R. Astron. Soc. 414*, 823–846. arXiv:1008.4948 [astro-ph.CO].
465. Robitaille, P.-M. (2007). WMAP: A radiological analysis, *Progress in Physics 1*(1), 3–18.

466. Robitaille, P.-M. (2009). COBE: A radiological analysis, *Progress in Physics 4*(1), 17–42.
467. Robitaille, P.-M. (2010). Calibration of microwave reference blackbodies and targets for use in satellite observations: An analysis of errors in theoretical outlooks and testing procedures *Progress in Physics 3*(1), 3–10.
468. Robitaille, P.-M. (2010). The Planck satellite LFI and the microwave background: importance of the 4K reference targets, *Progress in Physics 3*(1), 11–18.
469. Robitaille, P.-M., & Crothers, S. J. (2015). "The theory of heat radiation" revisited: a commentary on the validity of Kirchhoff's law of thermal emission and Max Planck's claim of universality, *Progress in Physics 11*(2), 120–132.
470. The top 30 problems with the Big Bang. (2002). *Meta Research Bulletin 11*, 6–13. *Apeiron 9*(2), 72–90.
471. Eddington, A. S. (1988). *The Internal Constitution of the Stars*, Cambridge University Press, Cambridge; reprint of 1926. Chapter 13, p. 371.
472. Marmet, P. (1995). The Origin of the 3 K radiation. *Apeiron 2*/1, 1-4.
473. Lerner, E. J. (1991). *The Big Bang Never Happened*, Times Books, New York.
474. Lerner, E. J., & Almeida, J. B., Eds. (2006). *1st Crisis in Cosmology Conference: CCC-1*, American Inst. of Physics Conference Proc., Vol. *822*, Melville, New York.
475. Aspden, H. (1980). *Physics Unified*, Sabberton Publications, P.O. Box 35, Sothhampton, England, p. 177.
476. Aspden, H. (2006). *Creation: The Physical Truth*, Book Guild Publishing, Brighton, England, p. 74.
477. Livio, M., & Silk, J. (2014). Physics: Broaden the search for dark matter, *Nature 507*, 29–31.
478. Luminet, J.-P., Weeks, J., Riazuelo, A., Roland, L., & Uzan, J.-P. (2003). Dodecahedral space topology as an explanation for weak wide-angle temperature correlations in the cosmic microwave background, *Nature 425*(6958), 593–595. arXiv:astro-ph/0310253.
479. Roukema, B. F., Buliński, Z., Szaniewska, A., & Gaudin, N. E. (2008). A test of the Poincare dodecahedral space topology hypothesis with the WMAP CMB data, *Astronomy and Astrophysics 482*(3), 747–753. arXiv:0801.0006 [astro-ph].
480. Planck Collaboration. (2014). Planck 2013 results. XXVI. Background geometry and topology of the Universe, *Astronomy and Astrophysics 571*, A26, 23.
481. Delort, T. (2011). Derivation of the Tully-Fisher's law from the theory of ether, arXiv:1108.5929 [physics.gen-ph].
482. Krasnoholovets, V. (2011). Dark matter as seen from the physical point of view, *Astrophys. Space Sci. 335*, no. 2, 619–627.
483. Mannheim, P. D., & J. G. O'Brien, (2011). Impact of a global quadratic potential on galactic rotation curves, *Phys. Rev. Lett. 106*, no. 12, 121101.
484. Clowe, D., Gonzalez, A., & Markevich, M. (2004). Weak-lensing mass reconstruction of the interacting cluster 1E 0657–558: Direct evidence for the existence of dark matter, *Astrophys. J. 604*, 596–603.
485. Famaey, B., & McGaugh, S. (2012). Modified Newtonian dynamics (MOND): Observational phenomenology and relativistic extensions, arXiv:1112.3960 [astro-ph.CO].

486. Fort, B., & Mellier, Y. (1994). Arc(let)s in clusters of galaxies, *Astronomy and Astrophys. Rev. 5*, 239–292.
487. Jee M. J. M., Ford, H. C., Illingworth, G. D., White, R. L., Broadhurst, T. J., Coe, D. A., Meurer, G. R., van der Wela, A., Benítez, N., Blakeslee, J. P., Bouwens, R. J., Bradley, L. D., Demarco, R., Homeier, N. L., Martel, A. R., & Mei, S. (2007). Discovery of a ring-like dark matter structure in the core of the galaxy cluster Cl 0024+17, *Astrophys. J. 661*, 728–749.
488. Famaey, B., & McGaugh, S. (2012). Modified Newtonian dynamics (MOND): Observational phenomenology and relativistic extensions, arXiv:1112.3960 [astro-ph.CO].
489. McGaugh, S. S. (2015). A tale of two paradigms: the mutual incommensurability of CDM and MOND, *Can. J. Phys. 93*(2), 250–259.
490. Feng, J. Q., & Gallo, C. F. (2008). Galactic rotation described with various thin-disk gravitational models, arXiv:0804.0217 [astro-ph].
491. Feng, J. Q., & Gallo, C. F. (2011). Modeling the Newtonian dynamics for rotation curve analysis of thin-disk galaxies, *Res. Astronomy and Astrophysics 11*(12), 1429–1448.
492. Dopita, M. A. (2005). The physics of galaxy formation, in *The New Cosmology*, Ed.: Colless, M., World Scientific, Singapore, pp. 117–128.
493. Reynolds, C. (2008). Astrophysics: Bringing black holes into focus, *Nature 455*, No. 7209, 39–40.
494. Doeleman, S. S., Fish, V. L., Schenck, D. E., Beaudoin, C., Blundell, R., Bower, G. C., Broderick, A. E., Chamberlin, R., Freund, R., Friberg, P., Gurwell, M. A., Ho, P. T. P., Honma, M., Inoue, M., Krichbaum, T. P., Lamb, J., Loeb, A., Lonsdale, C., Marrone, D. P., Moran, J. M., Oyama, T., Plambeck, R., Primiani, R. A., Rogers, A. E. E., Smythe, D. L., SooHoo, J., Strittmatter, P., Tilanus, R. P. J., Titus, M., Weintroub, J., Wright, M., Young, K. H., & Ziurys, L. M. (2012). Jet launching structure resolved near the supermassive black hole in M87, *Science 338*, No. 6105, 355–358. arXiv:1210.6132 [astro-ph.HE].
495. Lämmerzahl, C., Preuss, O., & Dittus, H. (2008). Is the physics within the Solar system really understood? arXiv:gr-qc/0604052.
496. Anderson, J. D., Laing, P. A., Lau, E. L., Liu, A. S., Nieto, M. M., & Turyshev, S. G. (2002). Study of the anomalous acceleration of Pioneer 10 and 11, *Phys. Rev. D 65*, 082004 (55 p.) arXiv:gr-qc/0104064.
497. Turyshev, S. G., Toth, V. T., Kinsella, G., Lee, S.-C., Lok, S. M., & Ellis, J. (2012). Support for the thermal origin of the Pioneer anomaly, *Phys. Rev. Lett. 108*, 241101 (5 pp.) arXiv:1204.2507 [gr-qc].
498. Freedman, W. L. (2000). Determination of cosmological parameters, *Phys. Scripta 85*, 37–46. arXiv:astro-ph/9905222.
499. Lahav, O., & Liddle, A. R. (2014). The cosmological parameters, arXiv:1401.1389 [astro- ph.CO].
500. BICEP2/Keck, Planck Collaborations, A Joint Analysis of BICEP2/Keck Array and Planck Data, *Phys. Rev. Lett. 114*, 101301. (2015). arXiv:1502.00612 [astro-ph.CO].
501. Hubble, E. (1929). A relation between distance and radial velocity among extra-galactic nebula, *Proc. Nat. Acad. Scien. 15,* no. 3, 168–173.

502. Schmidt, B. P., Suntzeff, N. B., Phillips, M. M., Schommer, R. A., Clocchiatti, A., Kirshner, R. P., Garnavich, P., Challis, P., Leibundgut, B., Spyromilio, J., Riess, A. G., Filippenko, A. V., Hamuy, M., Smith, R. C., Hogan, C., Stubbs, C., Diercks, A., Reiss, D., Gilliland, R., Tonry, J., Maza, J., Dressler, A., Walsh, J., & Ciardullo, R. (1998). The high-z supernova search: Measuring cosmic deceleration and global curvature of the universe using type IA supernovae, *Astrophys. J. 507*(1), 46–63. arXiv:astro-ph/9805200.
503. Perlmutter, S., Aldering, G., Goldhaber, G., Knop, R. A., Nugent, P., Castro, P. G., Deustua, S., Fabbro, S., Goobar, A., Groom, D. E., Hook, I. M., Kim, A. G., Kim, M. Y., Lee, J. C., Nunes, N. J., Pain, R., Pennypacker, C. R., Quimby, R., Lidman, C., Ellis, R. S., Irwin, M., McMahon, R. G., Ruiz-Lapuente, P., Walton, N., Schaefer, B., Boyle, B. J., Filippenko, A. V., Matheson, T., Fruchter, A. S., Panagia, N., Newberg, H. J. M., & Couch, W. J. (1999). (The Supernova Cosmology Project), Measurements of Ω and Λ from 42 high-redshift supernovae, *Astrophys. J. 517*(2), 565–586. arXiv:astro-ph/9812133.
504. Shnol, S. E. (2001). Macroscopic fluctuations are possible consequences of fluctuations of space-time. Arithmetical and cosmological physical aspects, *Ross. Khimich. Zh. 65*(1), 12–15.
505. Shnoll, S. E. (2009). *Cosmophysical Factors in Random Processes*, Svenska fysikarkivet, Täby, in Muscovian; Second edition, supplemented: American Research Press, Rehoboth, New Mexico, 2012.
506. Shnoll, S. E., Rubinshtejn, I. A., Zenchenko, K. I., Shlekhtarev, V. A., Kaminsky, A. V., Konradov, A. A., & Udaltsova, N. V. (2005). Experiments with rotating collimators cutting out pencil of α-particles at radioactive decay of ^{239}Pu evidence sharp anisotropy of space, *Progr. Phys. 1*, 81–84. arXiv:physics/0501004 [physics.space-ph].
507. Shnol, S. E., Zenchenko, T. A., Zenchenko, K. I., Pozharski, E. V., Kolombet, V. A., & Konradov, A. A., Regular changes of fine structure of statistical distributions as a consequence of cosmological physical factors, *Uspekhi Fiz. Nauk 170*(2), 214–218. (2000).
508. Thomas, B. (2011). The sun alters radioactive decay rates. *ICR News. Posted on http://icr.org* September 3, 2010, accessed July 25.
509. Sturrock, P. A., Buncher, J. B., Fischbach, E., Gruenwald, J. T., D. Javorsek II, Jenkins, J. H., Lee, R. H., Mattes, J. J., & Newport, J. R. (2010). Power spectrum analysis of BNL decay rate data, *Purdue University News Service*. http://www.purdue.edu/newsroom/research/2010/100830FischbachJenkinsDec.html.
510. Chaffin, E. (2011). Nuclear half-lives and changes in the strength of the nuclear force. *Presented at the Third Annual Creation Research Society Conference at Trinity Baptist College*, Jacksonville, FL, July 23.
511. Steinitz, G., Piatibratova, O., & Kotlarsky, P. (2011). Possible effect of solar tides on radon signals, *J. Environm. Radioactivity 102*, No. 749–765.
512. Kozyrev, N. A. (1977). Astronomical observations by means of the physical properties of time, in: *Flaring Stars: Proceedings of the Symposium Dedicated to the Opening 2.6-m Telescope of the Byurakan Astrophysical Observatory*, Byurakan, October 5–8, 1976. Yerevan, pp. 209–227; in Muscovian.

513. Kozyrev, N. A., & Nasonov, V. V. (1978). A new method of determination of trigonometric parallaxes on the basis of measurements of difference between apparent and visible position of stars, in: *Asronometry and Celestial Mechanics*, Akademiya Nauk SSSR, Moscow, Leningrad, pp. 168–179; in Muscovian.
514. Lavrentiev, M. M., Yeganova, I. E., Lutset, M. K., & Fominykh, S. F. (1990). About distant influence of stars on a resistor. *Proceed. Acad. Scien. USSR 314*(2), 352–355. in Muscovian.
515. Lavrentiev, M. M., Gusev, V. A., Yeganova, I. A., Lutset, M. K., & Fominykh, S. F. (1990). About the registration of an actual position of the Sun, *Proc. Acad. Sci. USSR 315*(2), 368–370. in Muscovian.
516. Zabulonov, Yu., Burtniak, V., & Krasnoholovets, V. (2016). A method of rapid testing of radioactivity of different materials, *J. Radiation Research and Appl. Sciences 9.* http://dx.doi.org/10.1016/j.jrras.2016.03.001.
517. The Editors (Mariette Di Christina, editor in chief and the board) (2015). Relativity's reach, *Scient. Amer. 313*(3), 46–49.
518. Aspden, H., *Aether Science Papers*, Sabberton Publications, P.O. Box 35, Southhampton SO16 7RB, England (1996).
519. Rubio, L. (2012). The whole universe in three numbers, http://vixra.org/abs/1205.0058 [Relativity and Cosmology].
520. Linde, P. (2016). *The Hunt for Alien Life. A Wider Perspective*, Springer.
521. Murphy, B. D. (2012). *The Grand Illusion: A Synthesis of Science and Spirituality.* Vol. 1, Balboa Press, USA.
522. Atkinson, W. W. (2010). *Telepathy. Its Theory, Facts and Proof,* Publisher: YOGeBooks by Roger L. Cole.
523. Dunn, C. (1998). *The Giza Power Plant: Technologies of Ancient Egypt*, Bear & Company, Rochester, Vermont.
524. Dunn, C. (2010). *Lost Technologies of Ancient Egypt: Advanced Engineering in the Temples of the Pharaohs*, Bear & Company, Rochester, Vermont, Toronto.
525. Shpenkov, G. P. (2014). *Dialectical View of the World. The Wave Model. Vol. 4. Units of Measurement*, p. 13; http://shpenkov.janmax.com/Vol.4.PhysicalUnits.pdf.
526. Bounias, M., & Bonaly, A. (1996). On metric and scaling: physical coordinates in topological spaces, *Ind. J. Theor. Phys. 44*(4), 303–321.
527. Golden Ratio, http://mathworld.wolfram.com/GoldenRatio.html.
528. Gandhi, K. R. R. (2012). Exploration of Fibonacci function, *Bull. Math. Sciences & Applications 1*(1), 77–84.
529. Stakhov, A., & Rozin, B. (2005). On a new class of hyperbolic functions, *Chaos, Solitons and Fractals 23*, 379–389.
530. Stakhov, A., & Aranson, S. (2011). Hyperbolic Fibonacci and Lucas functions, "golden" Fibonacci goniometry, Bodnar's geometry, and Hilbert's fourth problem, *Applied Mathematics 2*, 181–188.
531. Pennybacker, M., & Newell, A. C. (2013). Phyllotaxis, pushed pattern-forming fronts, and optimal packing, *Phys. Rev. Lett. 110*, 248104.
532. Heyrovska, R. (2015). Sorry Bohr, Ground state energy of hydrogen atom is not negative, vixra:1506.0064 [Chemistry].

533. Heyrovska, R. (2015). New interpretation of the structure and formation of ozone based on the atomic and Golden ratio based ionic radii of oxygen, http://vixra.org/abs/1503.0269 [Nuclear and Atomic Physics].
534. Otto, H. H. (2015). Pyroelectric $Bi_{5-x}(Bi_2S_3)_{39}I_{12}S$: Fibonacci superstructure, synthesis options and solar cell potential, *World J. Conden. Mat. Phys. 5*, 66–77.
535. Christianto, V. (2015). *A Biblical Theory of Everything. Inspired by the Johannine Prologue*, Lambert Academic Publishing, pp. 7–8.
536. Ludji, I. (2014). God as sound-consciousness, *Indones. J. Theology 2*(1), 65–77.
537. Kak, S. (1994). *The Astronomical Code of the Rigveda*, Aditya Prakashan, New Delhi.
538. Roy, R. R. M. (1999). *Vedic Physics. Scientific Origin of Hinduism*, Golden Egg Publishing, Toronto.
539. Shilov, Yu. (2015). *Ancient History of Aratta-Ukraine: 20,000 BCE 1,000 CE*, Create Space Independent Publishing Platform, UK.
540. Bounias, M., Wolff, K. E., Tsirigotis, G., Krasnoholovets, V., Kljajic, M., Chandler, J., Burdyuzha, V., & Bonaly, A. (2006). Scientists' responsibility and scientific concern for evolution of planet Earth: a manifesto on action for the world's peace and harmony, Ed.: Dubois, D. M., *Int. J. Comput. Anticipatory Systems 19*, 3–16.

INDEX

C

F

H

I

N

O

P

Q

R

Printed in the United States
by Baker & Taylor Publisher Services